Data Communications and Network Security

Data Communications and Network Security

Houston H. Carr
Auburn University

Charles A. Snyder
Auburn University

McGraw-Hill Irwin

Boston   Burr Ridge, IL   Dubuque, IA   Madison, WI   New York   San Francisco   St. Louis
Bangkok   Bogotá   Caracas   Kuala Lumpur   Lisbon   London   Madrid   Mexico City
Milan   Montreal   New Delhi   Santiago   Seoul   Singapore   Sydney   Taipei   Toronto

McGraw-Hill
Irwin

DATA COMMUNICATIONS AND NETWORK SECURITY

Published by McGraw-Hill/Irwin, a business unit of The McGraw-Hill Companies, Inc., 1221 Avenue of the Americas, New York, NY, 10020. Copyright © 2007 by The McGraw-Hill Companies, Inc. All rights reserved. No part of this publication may be reproduced or distributed in any form or by any means, or stored in a database or retrieval system, without the prior written consent of The McGraw-Hill Companies, Inc., including, but not limited to, in any network or other electronic storage or transmission, or broadcast for distance learning.

Some ancillaries, including electronic and print components, may not be available to customers outside the United States.

This book is printed on acid-free paper.

1 2 3 4 5 6 7 8 9 0 QPD/QPD 0 9 8 7 6

ISBN-13: 978-0-07-297604-5
ISBN-10: 0-07-297604-7

Editorial director: *Brent Gordon*
Executive editor: *Paul Ducham*
Executive marketing manager: *Rhonda Seelinger*
Project manager: *Bruce Gin*
Production supervisor: *Gina Hangos*
Designer: *Jillian Lindner*
Photo research coordinator: *Ira C. Roberts*
Cover design: *Jillian Lindner*
Typeface: *10/12 Times New Roman*
Compositor: *International Typesetting & Composition*
Printer: *Quebecor World Dubuque Inc.*

Library of Congress Cataloging-in-Publication Data
Carr, Houston H., 1937–
    Data communications and network security/ Houston H. Carr, Charles A. Snyder.—1st ed.
        p. cm.
    ISBN-13: 978-0-07-297604-5 (alk. paper)
    ISBN-10: 0-07-297604-7 (alk. paper)
    1. Computer networks—Security measures. 2. Wireless communication
system—Security measures. I. Synder, Charles A. II. Title.
TK5105.59.C365 2007
005.8—dc22   2006043316

www.mhhe.com

About the Authors

Houston H. Carr

Houston H. Carr, Ph.D., BSEE, MMS, MBA, PE, CPE, has taught in colleges of business at the graduate and undergraduate levels for 21 years, teaching telecommunications 18 of these years. He has published over 40 research articles in the area of telecommunications and end user computing. Prior to joining academe, Dr. Carr worked in engineering and business management positions at General Dynamics on the F-111 and F-16 aircraft programs.

Charles A. Snyder

Charles A. Snyder, Ph.D., BFA, MBA, MSEcon, has taught in colleges of business at the graduate and undergraduate levels for 27 years, teaching telecommunications 5 of these years. He has published over 65 research articles. He retired from 20 years' service in the USAF, with positions in operations, command, and staff. For eight years, he was in command, control, communications, and computing assignments.

Preface

The connectivity provided by voice and data communications is absolutely vital to the continuance of any organization.

Telecommunications, the transfer of information (or data) over a distance, is a broad concept including voice and other analog technologies along with digital formats. This book focuses on *data communications*. We concentrate primarily on the digital environment of machine-to-machine communications and the management necessary to control it. We have included more detailed information on analog/voice subjects in the appendix.

This book builds from the fundamentals of digital data communications technology, through topology and protocols using *wired and wireless* capabilities, to the *management concerns for security, privacy, and risk*. We provide both the depth of technological issues and the breadth of management control and network administration. The reader is provided knowledge about the technology and its management, with insights about the risks involved.

The premise of the book, and data communications in general, is that the connectivity provided by data communications is necessary for modern organizations to communicate with suppliers, partners, and customers; it is vital for the conduct of business in a network-centric global economy. The communications range from locally clustered computers to a worldwide network of components. Data communications is not a luxury; secure data communications is vital to the functioning and existence of the organization.

A driving force behind much data communications today is the focus on the Internet and the WWW. Students must cover the Internet in detail as well as the media and technologies that make the Internet and other networks possible. We understand, however, that data communications is much more than the Internet and the WWW. Some organizations do not connect their internal networks to the Internet; some do so only peripherally; while some make this network-of-networks the central communications flow for their organization.

Clarification

Organizations cannot survive without reliable data communications.

The book's chapters are arranged by topic area, often related to layers of the OSI seven-layer model and/or the layered TCP/IP model. We separate wired and wireless technologies in order to give a better understanding of each. We have structured the book (see inside front cover) to provide a logically flowing investigation of data communications but *provide an ability to select the order of coverage desired*. Therefore, the structure allows the coverage to move from *basics to wired network* material, *then* to wireless *or* broadband technologies *or* security. The more detailed management material is separated so that it can be used as desired or omitted. We provide voice communications and emerging technologies as *appendices* so they may be included with flexibility.

CHAPTER SUMMARIES

The book is composed of 14 chapters divided into six topic areas, plus an Introduction chapter and two optional topic chapters. A brief description of each follows.

Introduction Chapter

This (optional) chapter sets the stage for why students and managers study data communications and network security. It lays the business and technical foundation for further enquiry.

Part One. The Basics of Communications

This opening section discusses basic precepts of data communications: digital signals, digital representation of analog signals, circuit sharing, the various media employed, and the utility of architecture, nodes, and standards.

Chapter 1. Basics of Communications Technology

This chapter explores the basis and underlying technology that make machine-to-machine communication and networking possible, without requiring a technical background. It discusses the basic models that provide an understanding of how voice and data communication work while differentiating between the analog and digital worlds.

Chapter 2. Media and Their Applications

Channels are paths over which signals travel; media are the physical circuits on which channels reside. This chapter reviews wired and wireless media. Although wired media, in general, have greater bandwidth, the wireless domain is becoming a dominant form because it frees us to be more productive and competitive. Wireless covers the spectrum of radio, microwave, and satellite; GPS and infrared (IR); IEEE 802.11 and Bluetooth; fixed, movable, and moving.

Chapter 3. Architecture, Models, and Standards

This chapter provides an important foundation for the provision of the systems that make both voice and data communications possible and coherent within an organization. Models show the separate systems' functional parts so that the resulting layers enable standards to be applied within the particular layer without impacting the other layers. Understanding these models and their interrelationship allows the reader to follow processes that occur within and across them.

Part Two. Network Basics

Networks are classified by topology, protocol, forms of connectivity, reach, and the equipment components that make them work. While most of the network basics refer to wired networks, wireless technology is introduced.

Chapter 4. Building a Network: Topology and Protocols

This chapter describes how networks are physically constructed, physically joined, and electronically connected. There are several equipment technologies that need to be understood while providing a vital resource for the organization.

Chapter 5. Network Form and Function

This chapter describes the reach of networks, for example, how they extend from the local organization to the world. If these networks are to achieve the desired attributes of reliability, accessibility, performance, and security, they must have adequate error detection and electrical power. Synchronous and asynchronous communications are covered.

Part Three. Wide Area Networks: The Internet

Wide area networks extend the reach of the organization from near to far. The Internet is the ultimate network and often connects and extends the organizational private networks.

Chapter 6. From LANs to WANs: Broadband Technology

This chapter discusses higher bandwidth networks more in depth, moving from LANs to WANs, and the related issues, such as packet switching and other techniques for improving bandwidth use. It examines larger, geographically dispersed, and more complex networks, networks of greater and greater bandwidth.

Chapter 7. The Internet, Intranets, and Extranets

The Internet has evolved to be a major factor in the way people, businesses, and governments communicate. This chapter discusses Internet components and WWW technology so that future evolution may be comprehended adequately. Internal Internet, that is, intranets, and their extension, that is, extranets, complete the global reach of organizations and their partners. This chapter focuses on the technology of the Internet, the next chapter on its applications.

Chapter 8. Internet Applications

The examination of the Internet, intranets, and extranets continues, focusing on how this technology is used in applications. The concepts of e-business, e-commerce, and common applications are explored, along with some issues of Internet usage.

Part Four. Wireless Networks

This section addresses the basic technologies of the wireless (radiated) domain along with management and usage issues. The concerns of security are addressed in the next section.

Chapter 9. Wireless Networks: The Basics

Wireless capabilities provide channels without need for right-of-way permissions and often give the user untethered mobility. This chapter discusses technologies of fixed, portable, and nomadic forms of wireless networks. Wireless communications is rapidly becoming the technology of choice.

Chapter 10. Wireless Networks: Issues and Management

Wireless network management issues of interference and range, regulation, points of failure, total costs of ownership, maintenance, and support are discussed in this chapter. There should be an implementation plan that seeks logical integration with the organization's wired architecture.

Part Five. Security

Security is the denial of unauthorized access (intrusion) and the protection of assets. Wired networks are inherently more secure than wireless as physical attachment is required for direct access. However, people and organizations must be able to provide a reasonable level of security for their facilities, systems, networks, and data no matter the connection. This section focuses on the idea that security is a combination of prevention, detection, and correction for any network system.

Chapter 11. Network Security

In this chapter, overall security is the focus; concerns for wireless security are explored in the next chapter. Diverse threats and responses as well as prevention measures required to protect networks and their data are discussed. Security is the front line in the environment of information warfare.

Chapter 12. Wireless Network Security

The special security issues and threats to wireless networks are the focus in this chapter. Wireless systems provide enhanced services to organizations but mean different media and very different modes of attack must be addressed.

Part Six. Network Management and Control

The control and administration of data communications include the management concerns as well as the technology itself. Failing to consider the areas of monitoring performance, security, and new projects places the organization in jeopardy.

Chapter 13. Monitoring and Control of Network Activity

This chapter addresses the management and technical aspects of control of the networks. Two primary objectives of network management are to (1) satisfy systems users and (2) provide cost-effective solutions to an organization's telecommunications requirements. Active management of all the network resources is required if the telecommunications infrastructure is to support the organization's purposes.

Chapter 14. Network and Project Management

This chapter provides an outline of telecommunications systems analysis and design methodology to assist in developing, implementing, and operating data networks. The systems development life cycle (SDLC) approach is implemented.

Appendix A. Analog Voice Capabilities

The plain old telephone systems (POTS) may seem like ancient history, but it connects most homes and offices in the industrial world, providing a channel for low-speed data communications as well as vital voice communications. This chapter is provided as a review of its capabilities.

Appendix B. Epilogue: Emerging Technologies, Innovation, and Risks

This chapter presents three ideas, as a way to end the book: (1) What are the forces that support or constrain the adoption of technology, in general, and data communications, in particular? (2) What are some of the application trends in this field? (3) What should we learn from *Hurricane Katrina*?

DISTINGUISHING FEATURES AND BENEFITS

Our book provides students and managers the following features and benefits:

- Data communications is a business resource and is at the heart of any organization's future.
  - Most business functions depend on data communications.
  - All organizational activities depend on security.
  - The management of telecommunications is a part of business strategy.
  - Telecommunications, properly utilized, can give a competitive advantage.
  - Telecommunications, like MIS, is vital to the conduct of business.
  - Telecommunications allows decentralization of the decision process.
  - Coverage is provided on management of telecommunications projects.
- Data communications means wired and wireless communications.
  - Wired and wireless communications offer different advantages.
  - Wired and wireless communications have different security concerns.
- Network security is a portion of risk assessment and disaster planning and recovery. Network security is like a three-legged stool:
  - It is composed of the *design of the data communications systems,*
  - The *technology* implemented to secure that system, and
  - The *management controls* put in place to ensure that policy is carried out.

Implication

Adding data communications increases both costs and risks but not adding them has even greater risk.

The content of this book provides the basis for understanding the vitally important topic of data communications. As a way to aid the reader, several techniques are used. First, the OSI seven-layer model is presented in the Introduction and each chapter is related to that model or its heir-apparent, the TCP/IP five-layer model; *Clarification* boxes are included on pages to ensure that the text is clear; *Implication* boxes indicate organizational implications and considerations; *margin notes* show emphasis (but not duplication). Included at the end of each chapter are *discussion questions* that relate to the chapter material, *projects* that extend the material and include an *Internet project*, *recommended readings*, and *management* and *technology critical thinking* questions for graduate students and managers.

Since data communications (network administration) is now security administration, this text supplements the underpinning for the technology of data communications with two chapters on security, wired and wireless. It is vital to the health of the organization that students and managers understand the technology of data communications and the importance of incorporating security in design and usage. *Network security is like a three-legged stool;* it is composed of the design of the data communications system, the technology implemented to secure that system, and the management controls put in place to ensure that policy is carried out. This text leads the students and managers on the path toward a real understanding of this environment.

Caveat

The field of data communications is dynamic and rapidly changing. It is doubtful if any one single entity has all the knowledge. In creating this document, the authors relied on sources from the Internet to ensure consistent knowledge. Specifically, the following sources were used and referenced at many times:

- http:www.webopedia.com
- http://whatis.techtarget.com/
- http://searchsecurity.techtarget.com
- http://en.wikipedia.org/wiki/Main_Page

Wikipedia generously places a free use statement on its site. Part of it is reproduced here. We have made every attempt to give credit to any reference we used, especially those listed here. Without this excellent knowledge, few texts on data communications would be possible.

VERBATIM COPYING

Brief Contents

Contents

Chapter 8
Internet Applications   221

PART FOUR
Wireless Networks   243

Chapter 9
Wireless Networks: The Basics   244

Chapter 10
Wireless Networks: Issues and Management   278

PART FIVE
Security   301

Chapter 11
Network Security   302

Chapter 12
Wireless Network Security   350

Introduction

In our rapidly changing world, connectivity via data communications networks continues to expand. Organizations and individuals have grown to expect rapid and dependable access across the globe. There are two aspects of data communications: the technical and the business. In the final analysis, all technology is a cost and installed to solve a business problem or take advantage of an opportunity. This book focuses on the technology in the realm of organizations that must obtain and maintain a competitive advantage.

Data communications is the sharing of data between machines at a distance over electrical, electromagnetic, or photonic channels. The data may be commands for the receiving machine, information as to the status of the sending machine, or the transfer of information that passes on to other machines and ultimately to users, after conversion.

This book is about the many aspects of data communications, ranging from the technical concepts that make it possible to the management concerns for security. The use of data communications requires an environment that protects the messages from noise and intrusion; it involves copper wires, fiber optics, and radio signals; it necessitates control and management; and it provides significant organizational advantages when used properly, but always at a cost.

Implication

As we completed the manuscript for this book, hundreds of thousands of people in the southern United States were recovering from *Hurricane Katrina.* This storm has been as disruptive as any event in U.S. history. Although other events caused the loss of more lives, Katrina tops the list when it comes to destruction of telecommunications facilities and capabilities. It did not take long for the country to realize that there are several aspects of our infrastructure that had to be stabilized in order for recovery to happen: petroleum pipelines, the transportation highway system, the electrical power grid, and voice and data communications facilities and capabilities. While each of these was important it its own right, the voice and data communications systems tied it all together.

The material in this book will give an appreciation of the role that voice and data communications plays in our personal and corporate lives. As you learn how it works, you will learn how organizations depend on this technology and how our country depends on this vital resource.

WHAT'S IT ALL ABOUT?

First, let's set the stage . . . **communications** is a human process that entails moving an idea from one person to another usually via vocal or written language. Therefore, it requires that the persons be in close proximity for the verbal communications or have a delivery system for the written communications and that they share a common language. While this definition of basic communication is correct, we are more interested in machine-to-machine communications, and the various machines may not be in close proximity. The machines may **interface** with a human, such as in **facsimile** transmission of paper documents, but real **data communications** takes place between the machines.

As the persons or machines are moved farther apart, the appropriate term changes from *communications* to *telecommunications. Tele* means "at a distance," so **telecommunications** is where communications requirements are far apart. In this book, we discuss machine-to-machine communications, or *data communications.* From this point on, we will mean machine-to-machine telecommunications whenever the terms *communications* or *data communications* are used, unless otherwise noted. *Business data communications* is the implementation of this technology to achieve business goals.

The various facets of data communications are discussed throughout the book, but first the general characteristics are discussed. The purpose of data communications is the transport of data or information in the form of binary units (bits). These bits represent computer or textual data, audio, video or images. While the initial telecommunications infrastructure was basically for analog voice and video, they are now in the digital realm and will be transported much like other **digital** formats. We are approaching a time when all transported information will be in a digital form. Thus, data communications will become dominant. This evolution makes the study of data communications even more important. (We have included material on analog technologies as an appendix.)

A data communications *network* is a system composed of two or more *nodes,* connected by one or more *channels.* A node might be a telephone (analog or digital), computer, hub, router, TV set, or satellite antenna. In general, the nodes are end-point equipment or switching (i.e., transfer) equipment. The **circuits** are anything that connects them, such as copper wire, fiber strand, coaxial cable, or space as in the case of *wireless (radio) networks.* With technology, the circuits may be divided into **channels,** providing multiple communications paths on a single circuit. Data communications networks connect a group of nodes, for example, computers, so they can pass information. The distance covered may be a few hundred feet, around the earth, or into space. There may be just a few users or millions. *Networks provide* **connectivity,** *allowing machines and humans to share information, share resources, communicate, and operate remotely as if centrally positioned.* (Networks make organizations function).

Distributed Computing

An organization may be spread over several cities, states, countries, or continents. With adequate data communications, it would be possible to have a centralized computing facility with remote entry and reporting. The more likely scenario would be **distributed computing,** that is, placing the computer processing power and data storage at the site requiring it. This structure entails multiple processing sites and data communications to link them for centralized control. Controlling these distributed resources becomes a management challenge. In order to meet the challenge, managers must be aware of the data communications alternatives and their characteristics.

Clarification   **Distributed computing** = local processing with centralized control.

Telecommuting

A different form of distributed computing, or distributed work location, is when employees *work remotely.* Working away from major offices, ranging from **working-at-home,** to working at a satellite office, in a small office home office, or mobile, has advantages for both the employee and the employer, as well as the environment, and requires data communications. With *good connectivity, adequate bandwidth,* and *remote access,* employees can be effective from anywhere while reducing costs for everyone. The remote work using data communications is called *telecommuting* and has become a significant part of many businesses. Of note to our study is that telecommuting requires an underpinning of telecommunications, remote connectivity to the office via voice and data communications that range from an extra telephone line to broadband access such as that provided by cable modems and DSL.

Peer-to-Peer Computing (P2P)

A process that has achieved great visibility is peer-to-peer computing. It started with SETI (Search for Extraterrestrial Intelligence) where people volunteered their computer's idle time to process SETI radio telescope data. The process requires connectivity and a central computer to keep track of the millions of tasks. The central computer sends the tasks to an idle computer where the data are processed. The central computer must then be sure a reply is received. Since SETI, *United Devices* has used the same protocol to process data in the search for possible vaccines for cancer, anthrax, and smallpox by employing formerly unused processing power of over 3.2 million computers (circa July 2005[1]). Data communications makes this all possible and the trend is to use this capability to use the idle processing of an organization's internal resources. The official name for this sort of sharing is evolving to grid computing.

Convergence of Media[2]

Prior to the Internet in the early 1980s, connectivity meant either terminals connected to a **mainframe computer** or modem dialing from one small computer to one acting as a server. Downloads were small, DOS games and programs. E-mail was text only. When the *World Wide Web (WWW)* was created in the early 1990s, the graphical user interface of the browser brought graphics to Internet connectivity as well as adding them to the now widespread electronic mail. **Convergence** means the inclusion of several media types on the bandwidth, resulting in a need for greater and greater bandwidth. **Bandwidth** is driven by convergence; where we once sent text messages, we now include graphics, audio, movies, images, and even executable programs. This trend of *convergence* continues, with no end in sight.

Implication   Convergence (e.g., multimedia) drives bandwidth.

HOW DID WE GET HERE?

Communications and telecommunications have been of vital interest to mankind for as long as family units or social groups have existed. Our history is replete with instances of problems and opportunities resulting from the need to communicate. The legendary Tower of Babel is one well-known instance in project failure directly related to communication difficulties. The development of languages; writing (in all its forms); the need for translating, for example, the Rosetta Stone; and various signals for sending and receiving messages make a rich history in this area.

From ancient times, there has been a need to communicate over a distance. Some early methods were smoke signals, torches or bonfires, mechanical telegraphs, flags, lights, shots (cannons, rifles, etc.), bugles and trumpets, ram's horns, and so forth. Obviously, many of these modes were designed to transmit data—often connected with warfare. As society evolved, so did the need to communicate over greater distances. (Some attribute the defeat of the British military in their war with the American colonies to the failure of global communications.) This need became vital as people developed transportation means such as ships and traveled far from their home territories.

If we examine the United States in the time frame of 1776 to now, we can grasp the rate of change. As the Continental Congress performed its duties, news was spread by word-of-mouth, letters, and the newspaper as the U.S. west was developed. Mail from coast-to-coast took months until the mid-1800s when the Pony Express made the trip from St. Louis, Missouri, to California in nine days. It was not until the third quarter of the 19th century that the (digital) **telegraph** became useful, making communications between stations miles apart possible in seconds. By World War I, **telephone** service was being introduced in the United States. This (analog system) made communications possible in real time. As

---

[1] http://www.grid.org/stats/.

[2] The coming together of two or more disparate disciplines or technologies. For example, the so-called fax revolution was produced by a convergence of telecommunications technology, optical scanning technology, and printing technology.

the Great Depression continued, amplitude-modulated (AM) radio began to be introduced, giving fairly up-to-date news to the few who had them. In the mid-to-late 1930s, the major source of news was from newsreels in theaters, with news a week old for word of the conduct of the war (circa 1942). World War II saw the widespread use of radios in the military; the folks back at home relied on the radio, newspapers, and newsreels.[3] Between World War II and the Korean War, evening news broadcasts on television began. The time delay between world events and their announcement on TV was often proportional to the flight time from the location of the event to the station because film newsreels were transported on airliners—so this might be days. International telephone service during this era was expensive and noisy.

Clarification

24*7 means all the time, every day, every minute.

FNC, CNN, HLN, ABC, CNBC, MSNBC, Weather . . . provide information all the time.

Between the end of the Korean War and the Vietnam conflict, the delivery of news remained static; more people received this information from TV broadcasts and color pictures brought the viewer closer to the action. Events of the Vietnam conflict were brought to the United States and many industrialized nations via television on the nightly news programs. The delay from event to broadcast was only a day or two, as international jet airline service was the norm. Radio news, TV, and the newspapers had replaced newsreels at the movies, but the delay was still present. The real jump in coverage and reduction of delay in time-to-air time occurred between the Vietnam and Operation Desert Storm conflicts. During that time, worldwide networks became available via satellite and full-time news stations began operation on TV. In 1991, the world watched as missiles began to hit Baghdad; the scene was broadcast live via satellite. This event accelerated real-time coverage of world events. Then, in September 2001, millions watched in real time as aircraft, being used as weapons, were flown into the World Trade Center in New York City. Seventeen months later, they again watched, in real time, as the coalition forces liberated Iraq. Over the two-and-a-quarter centuries of the existence of the United States, communications have evolved from in-print to in-your-face, 24*7 on more than nine television news channels, over satellite radio, on the Internet, and via e-mail and instant messenger. All of the systems mentioned above that bring news to viewers rely on data communications.

TECHNOLOGY CAN PROVIDE A COMPETITIVE ADVANTAGE

Competitive advantage—that feature, any feature, of your organization that causes a customer to choose you over a competitor.

The term **competitive advantage** means anything that favorably distinguishes a firm, its products, or services from those of its competitors in the eyes of its customers or end-users in such a way that the consumer chooses to purchase that product or service over another. A monopoly has absolute competitive advantage. A firm[4] operating in pure competition has none. One objective of any firm competing in the marketplace is to have a product, service, or strategy that makes the market choose that firm over all others. A monopoly has no competition and perfect competition has no advantage. In the absence of both, where there is competition but still room to maneuver and create an edge, there is the potential for competitive advantage.

[3] One of the authors, nine years old at the time, learned of the victory in Europe when the telephone operator manually rang the telephones of every home in the county and repeated, "The war is over."

[4] The term *firm* in this book is intended to mean any formal organization, whether educational, governmental, private, public, profit-making, or nonprofit. In like manner, we consider that all organizations, or firms, must consider the idea of competitive advantage. Many would say that organizations such as governmental, religious, philanthropic, military, and other not-for-profit groups do not address competitive advantage. We contend that in these organizations, just as in the profit-oriented firms, there is competition for each budget dollar. Therefore, any project must show an advantage or reason why the dollars should be spent for it and not other projects.

The need for competitive advantage often is a reason for the introduction or upgrade of data communications technology. Data communications can be used to improve the performance or profit of an organization; the technology is not deployed just for its sake alone.

Implication   **Competitive advantage** relies on strategy, performance, and (often) data communications.

Organizations gain and sustain a competitive advantage via four basic strategies: (1) low-cost leadership, (2) focus on a market niche, (3) product and service differentiation, and (4) linkages to partners. If an organization chooses to compete on *cost*, it must reduce its costs, relative to perceived benefits, farther than its competitors. Data communications can aid in this by reducing the time to process transactions; supporting **just-in-time (JIT)** delivery of materials, which reduces inventory costs; and moving information to aid and hasten the decision process. Focusing on a *niche market* means concentrating on a specific part of the total market and doing it better than anyone else. It should not be difficult to see how data communications might be of value here. Organizations that compete on service and product *differentiation* can use data communications effectively in their strategy; Web sites and e-mail are just two examples. Customer service, one form of *linkage to partners*, once made possible by the tried-and-true technology in toll-free telephone numbers and now provided via Internet Web sites, assists the customer in choosing one organization's product over another's.

Clarification   The data communications infrastructure is like the central nervous system for the organization.

In order to compete successfully, as an individual or organization, there is a vital need for data communication. Managers must have a basic understanding of the technologies that are necessary to sustain a networked/connected world and economy. It's not just the store on the corner who is competing with you; it is the competitors from around the world.

COMMUNICATIONS IS A SIGNIFICANT COST FOR ANY ORGANIZATION

Organizations require communications to operate. Present day communication incurs a significant cost, but lack of communications has an even higher cost. As an organization becomes larger and/or dispersed over a wider geographic area, the ability to communicate becomes more difficult and definitely more costly. Even the worst basic telephone service comes at a cost. Installing telephone service, either from the phone company or privately within the organization, even over the Internet, takes an up-front investment in technology. Once installed, the equipment must be maintained, people must be trained, and there are the ever-present long-distance charges. Thus, something as simple as the plain old telephone system (POTS) is an expense to the organization and becomes an even greater expense when the organization is dependent on it and it fails.

Implication   If you think the cost of communications is expensive, consider the cost of the lack of communications!

We have become dependent not only on the wired telephone system and wireless cellular communications but also on written communications, for example, e-mail and instant messenger service. Some people even rely on short message service (SMS), or text messages, on their cell phones. Each of these capabilities requires an initial investment, learning, and maintenance. The information in this book will enable you to learn how to create data communications capabilities for the benefit of the organization.

Clarification

Metcalfe's Law. "The power of the network increases exponentially by the number of computers connected to it. Therefore, every computer added to the network both uses it as a resource while adding resources in a spiral of increasing value and choice." In other words . . . the value of a network increases as its number of **connection** points increases.

A word of caution about relying on technology for competitive advantage: *data communications technology has become a necessity for survival for business*; however, anyone can copy and adopt it, given the requisite financial resources and technical know-how. While an organization can install systems for communications that create a short-term competitive advantage, the underlying technology is easily purchased and such systems can be replicated. When the system is created with the unique use of the technology, it may create a very difficult barrier to duplication by competitors.

Clarification

A modern business has an underpinning of communication; it is dependent on networks. Once voice communication was sufficient; now data communications is dominant. While voice communication was human-to-human, data communications is machine-to-machine to support human activities.

REACH OF ORGANIZATIONAL COMMUNICATIONS

In many organizations, communication is the basis for competitive advantage.

Communications occur within the organization and to external partners. Within the organization, there are department-to-department communications, such as from accounting to finance and marketing to production, and the "chain-of-command" vertical communications. Communications can occur within a country, across national boundaries, and intercontinentally. Just because the communications are internal to the organization does not mean they cover a short distance. The successful organization also has real-time communications with a variety of partners. The first would be the *suppliers* of the material and subassemblies required to make the products of the organization. Once, these vendors and suppliers were held at arm's length. Enlightened organizations now see them as essential partners in order to be more competitive. External communications, therefore, have become important to the extent that partners share data and communications systems with the organization. Interorganization communications now include (a) suppliers, (b) partners, and (c) customers.

An organization that needs to communicate to divisions in different states, provinces, or countries may work with local vendors for local communications. When you must cross state or province lines, federal or national laws may dictate a change of vendors. The distance alone may cause the use of a different system or protocol. Crossing national boundaries may incur taxes as you enter the realm of *transnational information flows* plus other legal considerations. Thus, as the reach and distance of the organization and its partners expand, the considerations expand.

Global competition is now the norm and this means that firms must be prepared for global communications. If a firm in the United States can service a customer in Europe, so can one in Europe, Asia, India, the Far East, or Australia.

In today's competitive environment, *collaboration* is often required. This means partnering with customers, suppliers, and outsourcing vendors as well as others in the organization's business domain. Some collaboration results from regional trade agreements, actions of standards bodies, membership in professional organizations, and communities of interest. Collaboration can take place by written communications, face-to-face meetings, **videoconferencing,** e-mail, instant messaging, and other forms of data communications. With global communications, for example, the Internet and other networks, it is now possible for partners to work on the same document simultaneously, creating true collaboration without the expense of travel and the time delay of the mail. The idea of virtual collaboration has recently taken off and many firms capitalize on the ability of modern data communications networks to offer these environments.

CONSIDERATIONS FOR INTERNAL COMMUNICATIONS

"Moving at the speed of business."

One way to view the need for internal communication is to place it in the context of generalized management. The traditional functions of management include planning, staffing, directing, coordinating, and controlling. Communications are essential to each of these functions. Managers must communicate with others to create plans, ranging from the operational plans that run the business on a day-to-day basis, to the tactical plans that provide resources for the intermediate term, to the strategic plans that determine the long-range direction of the organization. While this planning historically was paper-based, the speed of business requires that near-real-time speed possible with data communications be employed.

Staffing and *directing* may appear to be more person-to-person functions; they often require communications at a distance in the form of telephone and e-mail. The **control** *of operations* encompasses all levels of the organization, thus requiring communications to all levels. The organization communicates to control operations and resources and to *empower* the people who can make a difference. Some of the resources and people will be locally located while others are at remote locations. Communication allows organizations to *coordinate remote resources* and treat them as if they are locally positioned. Control of remote operations has become vital to the survival of the organization.

VOICE OVER IP (VOIP) NETWORKS

With the advent of global connectivity via the Internet, people recognized that the Internet provided a replacement (or supplement) for the telephone system. Thus, *Voice over Internet Protocol* (VoIP), a way to send voice as **data** over networks, has seen significant strides. Not only does this have an important meaning for the telephone companies, it means that internal networks and gateways to the Internet will see added need for bandwidth. Much more on this later.

Clarification   VoIP means voice communications over any network that supports Internet protocol.

INTEROPERABILITY RELIES ON STANDARDS

In order for networks to function, various parties agree to architectures, standards, protocols, and models by which the networks are assembled. *Architectures* define how the equipment is connected. *Standards* are written descriptions of how a piece of equipment will operate. *Protocols* define the rules of communication for machines and networks. *Models* partition the whole system into workable parts and define the functionality of each part. Remember, these topics are expanded later.

A primary model of data communications is the seven layer **open systems interconnection (OSI) model** (see Figure 1.17 in Chapter 1). This model of data communications capabilities was created as a way to partition the **complexity** of equipment and networks so that capabilities could be developed in isolation and still interoperate with other equipment. This interoperation is possible *if* the requirements of the model layer and the boundary conditions are met. While companies continue to develop equipment and systems to proprietary (private) standards, there is great value in engineering to open standards such as the seven-layer OSI model.

Clarification   Inter-oper-ability means that equipment from different sources and vendors will work together.

DIGITAL DIVIDE

The term **digital divide** originally meant the distinction between the Haves and the Have-nots. It addressed the fact that the technology of the computer and Internet access was concentrated in the hands of the more affluent of the population. This is not surprising, as technology is usually purchased with discretionary funds, something lacking for the Have-not population.

The term *digital divide* has three connotations. First, for developed countries, it means the two segments of society, those with resources and those without. In this case, governments saw this as a great concern as a major part of the population would be left behind in the advances in technology. With programs to add computers, and then Internet access, to schools and libraries, 90 percent of the U.S. population has access to such technology. Thus, the distinction is blurred; the people without this technology at home have to make the effort to go to a library or school to access it, but do have access. In addition to the library and school access, Internet cafés, coffee houses (*Starbucks*®), and even small restaurants provide a point for gaining access in many parts of the world.

The second connotation of Haves versus Have-nots refers to developed and developing countries and nations. Some countries such as the sub-Saharan African nations, Mongolia, and Southeast Asia have large expanses of landmass with rural populations widely dispersed. Those who live in cities often do not have the funding and the infrastructure does not exist. Thus, a major portion of the world exists without access to modern data communications technology.

The final view of the digital divide describes the Wants and Want-nots in the population. With the maturity of computers in the home, Internet access, and communications via e-mail and instant messenger, part of our population (upwards to one-half by some estimates) prefer to not use the technology; they choose to be a Want-not. This may seem strange to technology-oriented people, but many prefer to avoid the intrusion of the technology and leave their communications to face-to-face meetings and written correspondence. A recent survey announced on TechTV indicated that 70 percent of people surveyed preferred correspondence, and even advertisements, via snail mail (postal service) versus e-mail. How does that affect us in data communications? We need to understand that not everyone finds data communications friendly or even desirable and that may be the case within organizations. Even though we believe data communications can offer value and efficiency and provide competitive advantage, we will come in contact with some who simply don't want it. This may be especially true for the senior ranks of management and older employees. Don't assume that everyone has or wants data communications technology (see Table I.1), but you can assume most people want to work smarter rather than harder.

Clarification   Have-nots don't possess; Want-nots don't want to possess.

TABLE I.1
Technology
Ownership

	Want	Want-not
Have		
Have-not		

SERVICE AND OPERABILITY 24*7

The term **24*7** is often heard or seen on the news and in advertisements, meaning 24 hours a day, seven days a week, which is all the time. This 24*7 availability is of special interest to us as we consider major enterprisewide systems. This is especially true when the enterprise (e.g., organization) is distributed over multiple time zones with a system availability requirement for service 24 hours a day, seven days a week. This complicates maintenance and upgrades and means expected random down time is significant! Data communications is an integral part of organizations as they adopt enterprisewide systems. As the organization depends on a 24*7 data communications–based system, the networks must be *reliable*, be *available*, *perform* as expected, and have significant *security*. These four characteristics (RAPS) will be of utmost importance as we continue to study data communications.

WE CANNOT FUNCTION WITHOUT DATA COMMUNICATIONS

Technology awareness: Some technologies are so ubiquitous we tend to forget they surround us.

We study data communications because it surrounds us and we are dependent upon it in our personal and professional lives. Data communications and networks are part of nearly every facet of our lives and are everywhere. While we are still approaching an environment of ubiquitous computing and mobility, there are an abundance of systems that have matured to a point of providing a competitive advantage. Some capabilities, such as e-mail, provide widespread capability and great utility for a multitude of uses. Others are more narrowly focused and beneficial for specific groups or industries. Many organizations have implemented enterprisewide systems. These systems were adopted in order to solve business problems and ensure that cross-functional coordination was streamlined. All parties had access to updated data, and systems have **interoperability.** The systems evolved from the manufacturing emphasis of organizations seeking to solve inventory and waste schedule problems. This conservation model spread until the idea of true enterprisewide systems have become popular. We can trace the evolution by exploring the progress from MRP (manufacturing resources planning) to ERP (enterprise resources planning) and CRM (customer relationship management) systems. Many of the enterprisewide ideas were expressed by the proponents of **computer-integrated manufacturing (CIM).** These proponents recognized the need to communicate cross-functionally as well as vertically in their organization. The CIM movement was a true predecessor of the enterprisewide model that would have seamless integration of an organization's functions. This can only be achieved through a coordinated data communications infrastructure. Anyone in a modern organization (manager, employee, end user) is involved in data communications, either directly or indirectly, and must acquire a basic understanding of the features, terminology, and limitation of the networks and their components.

Case I-1

Simplifying Complexity

M. C. Kettelhut began his sporting equipment retail store several years ago in the *North Dallas Forty* mall in Dallas, Texas. He saw his target market as upper-middle-class families in that part of a growing metroplex,[5] one with various recreational and professional sports. Initially,

---

[5] The Dallas, Arlington, and Forth Worth, Texas, metropolitan areas are considered so connected that they are treated as a single, very large metroplex.

he carried a wide variety of goods from all of the major names in sports equipment. With time, and analysis of his customers' buying habits, he culled the inventory to more selective brands. This resulted in increased business and created greater profit per unit of sales. With his success, he needed to either get more space to expand this store or open additional stores.

The first store in Dallas had very basic telecommunications: just three phones and a single phone line. Customers frequently complained that they could not call in, probably at times when the line was being used for a credit card transaction. M. C.'s fame had spread outside of the immediate Dallas calling area and he was attracting repeat business from customers in Arlington, Fort Worth, Frisco, Denton, and Weatherford. To his surprise, he also was receiving regular orders from cities far outside the metroplex including Wichita Falls, Killeen, Waco, Bryan, Texarkana, and Abilene. As M. C. followed up with customers in new markets, he discovered that while his prices and product mix provided a competitive advantage, telephone costs were a major reason that new customers only did some of their business with his company. The phone rates for in-state long distance were not competitively priced as were interstate rates, and local tariffs made it more expensive to call from Waco to Dallas than to call from Dallas to New York or San Francisco. M. C. talked to his Telco about expanding his immediate calling area to include Arlington, Fort Worth, and Frisco. The representative said that would be easy by changing to a metroplex area code that included Arlington, Fort Worth, Frisco, and Weatherford. However, the other cities would require a different solution.

There were actually two business considerations. If he wanted to have customers from distant cities call him without charge, an 800 number would be a simple solution. However, if he wanted to be able to call them, he would have to either use standard long distance or establish a WATS number. As this was just before Thanksgiving and the Christmas buying season, M. C. opted for the metroplex area code and an 800 in-WATS number as a short-term fix, accepting that he would have to pay more for making outgoing calls.

After the hustle-and-bustle of Christmas, M. C. realized that he was ready to expand his business, especially into the cities outside of the metroplex. Frisco was a natural site as it would be easier to access for customers from Fort Worth and Denton. Abilene and Wichita Falls were farther west but were sizeable cities with good recreational sports programs. Killeen and Bryan were south of Waco, between the metroplex and Austin. Texarkana was much to the northeast. M. C. derived a plan to expand, moving to locations where current demand could support a store and where he was closer to his farthest customers. He found a store location in Frisco to serve Frisco and Denton customers. Then he opened a store in Weatherford to serve that community, west Fort Worth, Abilene, and Wichita Falls. As quickly as possible, he followed with a store in Waco to better serve Waco, Bryan, and Killeen. Within six months, he had three new stores stocked and open. He was closer to customers, but he discovered that he was still faced with the same communications problem for many of his customers. Each store was in a different area code. Customers had to know different numbers for each of these stores, and they were still faced with higher telephone costs that drove them to share good portions of their business with other suppliers.

M. C. talked with his Telco rep about a unified telephone plan, one that would reduce his costs and, more importantly, make contact with customers simple and easy. "How can we make it appear to the calling customer that we are all one location? How can they call at no cost? How can I return calls at the lowest cost? How can we make this all work together to and from three locations? What about store-to-store communications?"

QUESTION

How can M. C. find a solution that will answer all of his questions? What system should he choose? Will expandability be an issue?

Summary

Data communications and networks provide connectivity. Connectivity provides opportunities to work at a distance, collaborate, communicate, share resources and information, and merge the various media, all to add value and create competitive advantage for the organization. Communications within and outside of the organization are vital to its survival and success; data communications are now essential. Enterprisewide systems depend on it, coordination with partners and customers depends on it, and the organization's future requires it. Part of management's task is to support the installation and use of technology in general and data communications in particular as an investment, not a cost. While the technology has a cost, it should provide an even greater benefit. With an understanding of the importance of data communications, we now must gain insight about the basics of the technology.

You must have knowledge of voice and data communications to compete. With the basics, you should be able to keep up with the new developments. By the time you finish reading this book, you should realize the need for data communications, understand the technology, grasp it functionality, and appreciate the management principles and concerns that make it a vital resource to the organization.

Key Terms

24*7, *p. xxviii*
audio, *p. xxxii*
bandwidth, *p. xxii*
channel, *p. xxi*
circuit, *p. xxi*
communications, *p. xx*
competitive advantage, *p. xxiii*
complexity, *p. xxvi*
computer-integrated manufacturing (CIM), *p. xxviii*
connection, *p. xxv*

connectivity, *p. xxi*
control, *p. xxvi*
convergence, *p. xxii*
data, *p. xxvi*
data communications, *p. xx*
digital, *p. xxi*
digital divide, *p. xxvii*
distributed computing, *p. xxi*
Dolby Labs, *p. xxxiii*
facsimile, *p. xx*
interface, *p. xx*
interoperability, *p. xxviii*
just-in-time (JIT), *p. xxiv*

mainframe computer, *p. xxii*
Metcalfe's Law, *p. xxv*
monaural, *p. xxxii*
open systems interconnection (OSI) model, *p. xxvi*
stereo, *p. xxxii*
Surround Sound, *p. xxxiii*
telecommunications, *p. xx*
telegraph, *p. xxii*
telephone, *p. xxii*
videoconferencing, *p. xxv*
work-at-home, *p. xxi*

Recommended Readings

Network Magazine (http://www.networkmagazine.com), a monthly managerial and technical publication on data communications.

The Wall Street Journal Reports on Technology (http://www.wsj.com), a periodic insert in the *WSJ* on the fields of technology, often telecommunications.

Communications News (http://www.comnews.com), a monthly periodical covering all aspects of networking decisions.

Information Week (http://www.informationweek.com), a weekly magazine that combines the goals of business with technology to help you make the strategic decisions that affect your company's bottom line.

Wired (http://www.wiredmag.com), magazine emphasizing technologies that impact business.

Management Critical Thinking

MI.1. Consider the implications of conducting a medium-sized business without any modern data communications.

MI.2. If the networked data communications capabilities are destroyed, what functions of business may have to be shut down? Consider the impact on suppliers, customers, and other partners. How could a manufacturing company work around the loss of data communications to a major raw materials supplier?

MI.3. What is the business advantage of having 24*7 availability of access to customers, suppliers, and other partners via telecommunications? What technologies are available for such connectivity?

Technology Critical Thinking

TI.1 Consider the technology of the news delivery in the various wars over history. Can you name an instance in history where a data communications failure changed the outcome of history?

TI.2 What technologies might be deployed to increase accessibility from "have-not" parts of the world to "have" nations?

Discussion Questions

I.1 Compare and contrast the terms *communications, telecommunications,* and *data communications.*

I.2 Examples of data communications that might be found in the home are *AOL®, Earthlink®,* and *CompuServe®* or *MSN®* as they are used with home computers. What capabilities do these services offer? Does noise affect these services?

I.3 Why is it essential for managers to understand data communications? Answer in terms of (a) local, (b) regionally dispersed, and (c) global operations.

I.4 Can you cite examples of how companies have used telecommunications or data communications for competitive advantage?

I.5 How can a decentralized organization operate without telecommunications?

I.6 Compare and contrast the telecommunications equipment and media of Dick Tracy, Maxwell Smart, Napoleon Solo, James Bond, James T. Kirk, and Jean Luc Picard.

I.7 Many individuals and groups start new companies each year. What are the major data communications considerations for such a venture?

I.8 What are the network applications in your work environment?

I.9 Is a mainframe computer–to–personal computer connection a data communications example? Why?

I.10 What were the early means of data communications (prior to the invention of the telegraph)?

Projects

I.1 Bring to class major articles or publication issues that deal with technology in general and telecommunications in particular.

I.2 Bring an article to class that shows how an organization uses telecommunications to carry on its core business.

I.3 Bring an article to class that shows how an organization uses telecommunications for competitive advantage.

I.4 Bring to class examples of technologies being used at points-of-service.

I.5 If you are contracting for a new home, what are the network and data communications specifications that you would incorporate into your design?

I.6 Research the relative "teledensity" of major world regions. What can be done by business, nations, and regional groups to mitigate the "digital divide"?

I.7 Determine the degree to which "convergence" is occurring in the United States, the European Community, Asia, India, Africa, and so forth. What are the implications for global business?

I.8   Trace the way a message might be generated, processed, transmitted, received, and interpreted *without* a data communications network in place.

I.9   Refer to project I.8, above, and trace the same sequence *with* a data communications network in place.

I.10 What kinds of networks do you use at your school, place of business, or organization?

Appendix I-1

Circuits, Channels, and Bandwidth

We will be discussing bandwidth in many forms as we progress through the chapters. As a way to set the stage, let's discuss bandwidth with which you are likely familiar, radio and television. The following assumes only that you listen to FM radio and watch TV; it does not assume a technical understanding of the methods of creating them, which is discussed in later chapters.

MONAURAL MUSIC SYSTEMS ARE SINGLE TRACK

When vinyl records were first introduced in the latter 1940s, they used a *single track,* or *single channel, system.* This means they used one microphone for recording. In playback, we used one speaker for reproduction. In this system, all sounds come from one point because they are recorded with a single microphone and reproduced with a single speaker. Later the recording industry began to use multiple microphones to pick up the various sections of the orchestra better, but they mixed them into a single sound track. The original bandwidth of vinyl sound track was better than AM radio but not as good as FM radio. In the mid-1950s, the companies developed High Fidelity recording with greater bandwidth, but still less than FM radio and using a single track.

Meanwhile, FM radio and television were designed to use a single FM **audio** channel and one speaker. No one believed that "good" sound was a necessity for TV; after all, it was the picture that was important.

This **monaural** capability was used in all systems until the 1960s: radio, TV, vinyl records, and tape-to-tape systems. At this time, records were designed to physically record two channels of sound and music systems were produced that played them as separate channels. This was called *Stereophonic Reproduction.* Why go to the bother?

STEREO IS TWICE AS NICE

Because humans have two ears, they can discriminate from where a sound originates. Monaural, that is, Hi-Fi, has only one origination point. **Stereo** has two origination points; stereo attempts to emulate to the listener's ears the actual performance. To do this, the **stereo**phonic system again used multiple microphones to make the original recordings: the left channel for the left ear and right channel for the right ear. The point was to record the information that would allow the two ears to think they were hearing the original sound. Where there had been a single (though composite) speaker for monaural, there were now two speakers for stereo, about eight feet apart to give musical separation. Again, multiple microphones were used for better sound pickup, but they were mixed into two tracks for final recording. Separation of playback speakers emulates the original sounds; the sound is fuller and richer and stereo contains more information.

LET'S NOT FORGET THE SUBWOOFER

All sound transmission has limits. Humans can hear from 20 Hz to 20,000 Hz. Reasonably priced speaker systems are able to reproduce the mid and high range frequencies well, but the lower frequencies are more difficult. One way to increase the total reproduced frequencies of a stereo system is to add a subwoofer speaker. This is a single speaker that is designed to reproduce only the frequencies in the 30–200 Hz range. Low frequencies are nondirectional, so the placement of the speaker is not important. Also, using a single speaker for both channels works equally well. For this system, the speaker wires are run from the stereo amplifier to the subwoofer and then on to the stereo speakers. The subwoofer contains a crossover circuit that separates out the under-200 Hz signals and sends them to the subwoofer; the remainder of the frequencies are sent to the two stereo speakers.

TV AND FM RADIO

Originally, FM radio had one channel and one small speaker. Stereo was added to FM radio in the 1960s. While vinyl and tape systems record two channels, for example, left and right channels for the left and right ears, FM radio (and eventually TV) had a problem that would not allow this. For radio, the total sound had to be reproduced for either a monaural or stereo receiver. The solution was to transmit right-plus-left music tracks as one "channel," which would satisfy monaural receivers, and right-minus-left music tracks on the other "channel." With this information, stereo receivers could compare the two channels and produce the final right and left music tracks.

Stereo was not introduced to television until later. This was driven, to a great extent, by the advent of MTV; listeners wanted greater bandwidth and dual channels for television broadcast of music videos. When TV stereo sound arrived, it used the same method as in FM radios. TV manufacturers, however, were slow to decode the information at the TV set. However, one could capture the signal at the VCR, where it had been decoded into the two channels, and run it to the stereo system, producing good-quality music. FM radio would perform the separation, but the speakers were very close together; again, running the signal to a stereo system improved the separation and sound. Today, most TVs have two speakers & stereo decoding. The reason for stereo is to provide greater bandwidth, for example, fullness of sound, and a feeling of sensing the location of the individual sounds, for example, the positioning of the parts of the orchestra.

An example of the stereo effect occurred to the author one evening while watching the movie *The Super* on a stereo television. As the plot evolved, he heard a police siren coming from the left side of the room, from the door to the carport. When the TV was muted, the siren silenced. When the TV sound returned, the siren continued, from the left side of the room, apparently 15 feet from the TV. This was impressive.

SURROUND SOUND (DOLBY LABS® HITS THE SCENE)

Dolby Digital 5.1—**Dolby Labs®** Strikes Again.

When stereo information is reproduced, it is desirable to have the singer's, or speaker's, voice coming from the center of the "stage" or from the TV, not from one of the two channels. This is done with **Surround Sound.** The first characteristic of this technology is to combine left and right channels when their signal strengths are equal to produce the center channel. This information is passed to a (new) center speaker. This makes the speaker or singer appear to be coming from the picture or center of stage. Surround Sound's other feature is the additional rear channel that plays as two back channels, making the sound appear to come from behind. This channel is monaural and is based, to a large degree, on time delays. In any case, it increases the effect of Tom Cruise in his F-14 Tom Cat, flying over your head from the back of the room in the movie *Top Gun*.

TABLE I.2
Six Distinct Channels: Surround Sound has 2+

Placement of *Dolby Digital* (5.1) Speakers		
Left	Center	Right
Left rear	Subwoofer	Right rear

After their success with Surround Sound, Dolby Labs developed six-channel recording and reproduction, originally called the AC-3 protocol. Although producing six separate channels of sound, the name of this system is 5.1. (See Table I.2.) This is because the subwoofer channel records and reproduces only the 30–200 Hz, which is about one-tenth of a channel. Again, all six channels are full-frequency and separate; there is no comparison to produce a "synthetic" channel. With these six channels, the full bandwidth of the human ear can be reproduced and the sound emulates the actual performance.

SUMMARY

The purpose of all of this technology for vinyl, cassettes, CD, FM radio, and television is to attempt to faithfully reproduce the location and bandwidth of the original sounds.

Part One

The Basics of Communications

The basics provide the foundation for our exploration of data communications. We outline some basic precepts of communications and discuss the various media employed and the utility of architecture, nodes, and standards.

Chapter **One**

Basics of Communications Technology

There must be a foundation about communications for an understanding of data communications. We begin with the basic communications model before introducing the reader to concepts of circuits, channels, networks, and bandwidth. Also, the technologies of analog and digital signals, circuits, compression, and so forth, are covered.

This chapter covers the basics of the underlying technology used for data communications. Some of this material may seem like old stuff, but it's important to put it into perspective. We begin with definitions, as they are important to later discussions. This evolves to a discussion of the nature of the electrical, **electromagnetic,** and **photonic** signals used for the transport of the **data.** Because of noise, we often have to hide the data in an envelope that is resistant to the noise environment. In data communications, noise is prevalent and becomes our nemesis; it is why we must install much of the equipment in our network. Because the basic data would not travel far, we often have it ride on other carrier signals. Finally, many circuits have a large amount of capacity, but, without special attention, most of the capacity is wasted. We end the chapter showing how to divide circuits into channels in order to take advantage of the total capability.

electromagnetic
Electromagnetic is radio waves.

photonic
Photonic means use of light instead of electricity or radio waves.

Implication

If you think the cost of communications is high, consider the cost of the lack of communications.

THE TECHNOLOGY ORGANIZATIONS USE

The Basic Communications Model

To send data over a distance, you must have a sender (the source), a communications medium (channel), and a receiver (the destination). Figures 1.1 and 1.2 depict these models of communication. Note, in the case of human communications, the sender takes an idea, encodes it into speech, and transmits it over the air-medium channel. The receiver must receive the **message,** decode the words, and receive the original idea or information. Remembering our nemesis, there is an opportunity for noise to be introduced at each step of the communications process.

The basic communications model described in Figure 1.1 is a simple mode of moving data from one source to one or more destinations. The environment and model become more complicated but often more useful when we move to a data-sharing model. This involves moving data back and forth or sharing the data simultaneously between two or more people or devices.

In order to communicate, we must consider the intent of the sender (what is being communicated); the encoding of the message (language, code, etc.); the channel (the medium); the ability of the receiver to decode the message; and the way the receiver interprets the transmitted message. Human-to-human interaction is the concern of "simple communications." For example, two people need to use the same spoken or written language and have

FIGURE 1.1
A Communications Model

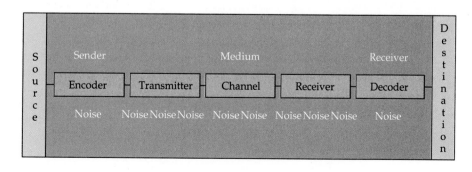

FIGURE 1.2   A (Voice) Telecommunications Process

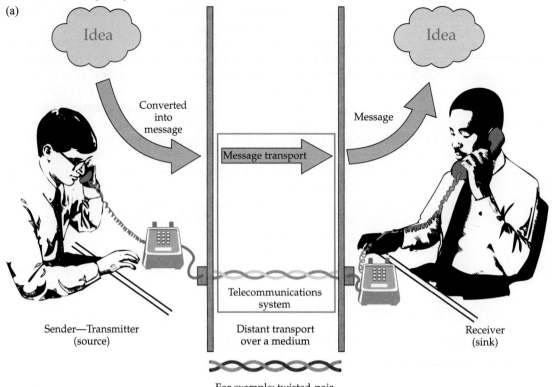
(a)

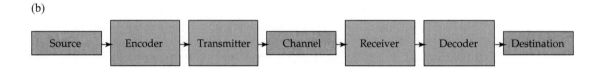
(b)

a similar background of experiences and knowledge to interpret messages in a similar fashion. Additionally, each of the five processes contains noise in many forms. These noises range from incorrect language syntax (*he ain't gonna to do nothing*), to the effect of "bad" telephone circuits, to loud background music.

Implication   Noise permeates our lives and data communications.

In Figure 1.3, we illustrate voice communications by various media. First, we show that even in the simplest case of communications without any devices to augment transmission, a sender communicates to a receiver by transmitting sound waves through the air medium. When we consider cans connected by a taut string, a common communications toy used by children, we see the employment of another medium, the string. In the bottom of the figure, we see simple telephone communications using conducted copper wire, the common medium of twisted pair. Other media are coaxial cable, radio or

FIGURE 1.3   **Voice Communications by Various Media**

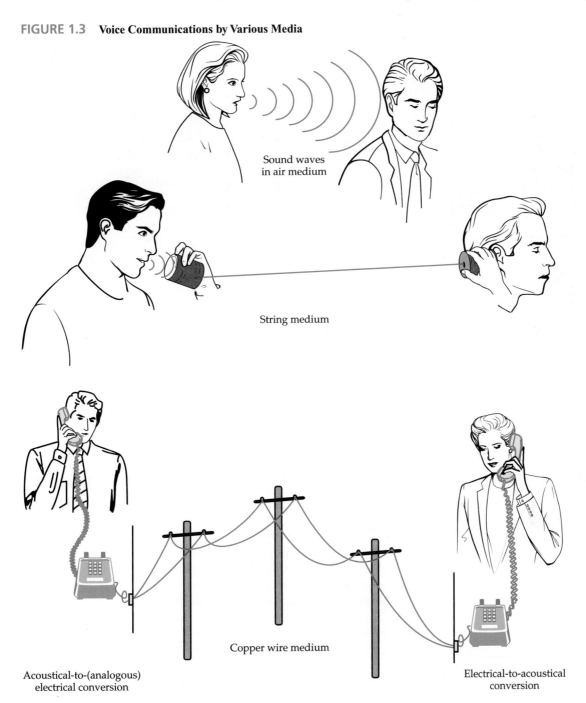

Sound waves
in air medium

String medium

Copper wire medium

Acoustical-to-(analogous)
electrical conversion

Electrical-to-acoustical
conversion

infrared waves through space (generally called **wireless**), and optic fibers. All of these
are discussed in Chapter 2.

Telecommunications versus Data Communications

Fortunately, in data communications, we are not usually as concerned with the intent of the
message but in ensuring the message is accurate and complete. In other words, we generally are concerned that the proper characters were received exactly as sent and in the correct sequence. The receiving machine will be sensitive to any changes made in the data,

FIGURE 1.4 **Data Communications Process**

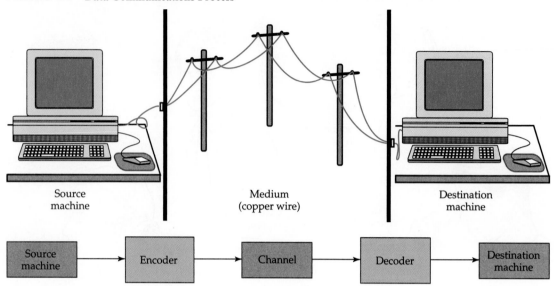

but we people may not able to detect such changes. The data communications process is shown in a simplified illustration in Figure 1.4.

Clarification

Telecommunications is the transfer of data over a distance via electrical, electromagnetic, or photonic means.

Data communications is **machine-to-machine** communications.

Implication

With an understanding of communications in its basic form, we can develop a mental model of its value and use. The better we can understand the process of communications, the more we can use the technology to solve business problems and exploit opportunities.

Circuits, Channels, and Networks

A **channel** is a path, not necessarily a pair of wires, for transmission between two or more **nodes** or points. The channel connects the source to the destination. It is thought of as a one-way communications path, and although the transmission is usually electrical, it also may be photonic. The channel may be all or part of a circuit.

Clarification

You can see a wired circuit, but you cannot see the number of channels on it. You see a single coaxial cable going to your TV set but cannot see the 100+ channels on it.

A **circuit** is a means of connecting two points for communications. It is the *physical connection* between the points; the circuit may be a one-way or two-way communications path and can be divided into multiple channels. A circuit can be either on conducted (wired) or radiated (wireless) media.

There are various data paths over which communications takes place. (See Figure 1.5.) A **path** is the route between any two nodes. A **network,** therefore, *consists of a pattern of paths* and associated equipment that establish connections between nodes. We discuss various classifications and patterns of networks in a later chapter.

Implication

Adding circuits increases both cost and risk.

FIGURE 1.5
Simple and Complex Networks

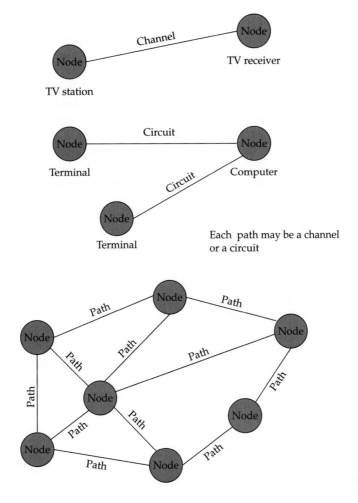

BASIC DATA COMMUNICATIONS LINKS

When sending data from one terminal or computer to another, you are dealing with **data terminating equipment (DTE).** (See Figure 1.6.) DTEs are noncommunications-oriented components of a data communications environment. In order for the data transfer to work, you need some sort of **data communications equipment (DCE).** DCEs are communications-oriented components of a network, such as telephone switching equipment, media, modems, and so on. For a total point-to-point network, you will connect the DTEs via DCEs.

FIGURE 1.6   DTE and DCE

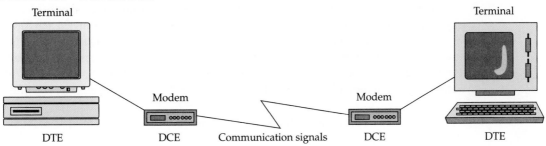

Modes (Directions) of Transmission

In discussing the various physical communications media, we noted that some were **simplex,** meaning one-way communications, all the time, like AM or FM radio, air-broadcast TV, and analog cable television. (See Figures 1.7a, b, and c.) Others were one way at a time but with a bidirectional capability through circuit reversal, which is called a **half-duplex** operation. This is the mode used by many terminals and microcomputers. A noncomputer example of a half-duplex operation would be patrol cars that kept in contact with the base radio station. The third mode is **full-duplex,** or simultaneous bidirectional communications. This mode, for example, makes use of a single twisted-pair circuit and divides the communications channel into two parts, that is, forward and **reverse channels.** An original use of full-duplex is to echo (repeat back) the characters sent from a terminal to a computer to display exactly what the computer received as opposed to just repeating what was transmitted from the source. A more relevant use of full-duplex is to have a small reverse channel to communicate back to a sending node that data were received correctly, without interrupting the forward channel. This is discussed later in the chapter under error detection.

Clarification

Many systems use dual-simplex, using two simplex channels. Fiber circuits do this on two strands.

Serial versus Parallel Circuits

The two circuit forms for sending coded messages are parallel and serial circuits. In a **parallel circuit,** the number of wires from sender to receiver is equal to or greater than the number of bits for a character. Thus, for an eight-bit character, which is what most microcomputers use, the parallel medium from one point to another would have eight or more wires or paths. In such a parallel circuit, each bit of the character travels down its own wire simultaneously with the other bits of the character on other wires. This mode provides fast communications, such as on the data bus[1] within the PC or from computer to the printer or secondary storage, but this mode requires eight or more wires. The alternative, which is used for most data communications, is the **serial circuit.** In this mode, the bits of the character follow each other down a single wire or its equivalent. This form takes eight times as long as parallel communications but requires only one wire-pair. (Oftentimes, the originating system generates the code in a parallel circuit and sends it over a serial circuit. Obviously, this procedure requires a conversion device at each end. This conversion is the task of the modem, to be discussed further later.)

Figure 1.8 shows that, in a given amount of time, a serial circuit will transport a bit while a parallel circuit will transport the total character, which is eight times as much data. Parallel circuits, however, are seldom used for long-haul circuits because of cost (eight wires versus two) and complexity of synchronizing bits.

Bandwidth

In its simplest definition, **bandwidth** *is the information-carrying capacity of a circuit, channel, or medium.* It's like the size of a garden hose and its ability to carry water. A small ⅜-inch plastic hose can carry about five gallons of water per minute over its 50 feet of length at a "normal" level of water pressure. If another length of hose is added, the carrying capacity decreases because of the resistance of the small hose. Exchange the ⅜-inch hose for ¾-inch hose and the quantity of water that can be carried increases substantially to about 12 gallons per minute. Increase the water pressure and, again, the quantity of water increases. Information bandwidth is analogous to the water-carrying capacity of the

---

[1] A data bus is a transmission path or channel where all attached devices receive all transmissions at the same time.

FIGURE 1.7   Communications Channels

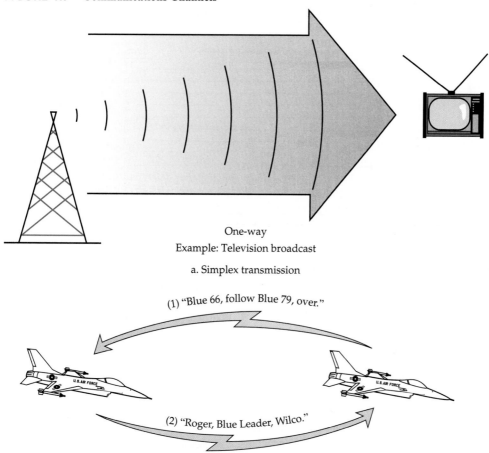

One-way
Example: Television broadcast

a. Simplex transmission

(1) "Blue 66, follow Blue 79, over."

(2) "Roger, Blue Leader, Wilco."

Example: Radio between aircraft

b. Half-duplex transmission (one way at a time)

Example: Telephone with both speaking at once.

c. Full-duplex transmission (both directions simultaneously)

FIGURE 1.8
Serial versus Parallel Transmission

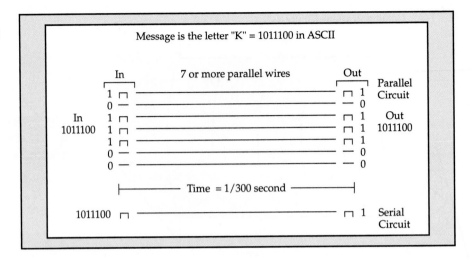

hose: the larger the medium or the higher the pressure, the more data that it can carry. The units of bandwidth vary between the analog and digital environments. The bandwidth for analog is the highest frequency minus the lowest frequency, noted as cycles per second, or hertz; the bandwidth for digital is bits per second.

Clarification

Bandwidth is to communications as water-flow capacity is to a garden hose.

Implication

Moore's Law states that the number of transistors on a chip doubles about every two years.

noise
Noise is any unwanted signal.

Examples of analog bandwidth can be seen in AM and FM radios. AM radio has a listening bandwidth of about 3,000 hertz (Hz), or between 300 and 3,300 cycles per second. This is fine for talk programs but not considered good for music because healthy human ears can hear from below 50 Hz to above 18,000 Hz. FM, on the other hand, has a bandwidth of about 15,000 hertz, which is usually considered very good for music. (Compact discs [CDs]—a storage, not a transport, medium—provide a bandwidth of 20 to 20,000 Hz, which is the maximum that can be heard by a healthy human ear). These are the native bandwidths of the two commercial radio systems. If you add **noise,** then the effective bandwidth could be reduced. One can still hear the music, but the effective pleasure of listening is reduced. If there is a way to remove the redundant bits of data on the transmissions, there would be a greater effective bandwidth. This is **compression.** For example, the MP3 (MPEG3 music) format actually drops some of the sounds, reducing the required bandwidth, which effectively increases the total bandwidth of the channel. This is more evident when in the digital domain, such as direct satellite broadcast TV, satellite radio, high-definition radio, and Internet access.

Dealing with Noise

Most circuits consist of a number of parts, links, connections, and possibly switches (transfer points). The combination of long lines and connections makes the circuit susceptible to noise and loss of signal strength. For these reasons, companies must keep lines and switches of good quality to make analog signal transmission acceptable.

Circuits, especially metal wires, that are not protected from their environment are susceptible to noise from a variety of sources. Some environments are cleaner than others but all host potential sources of error-producing noise. The greatest sources of noise inside a residence or office are the induction motors in refrigerators and air conditioners. Noise is

FIGURE 1.9
Some Wire/Cable Types

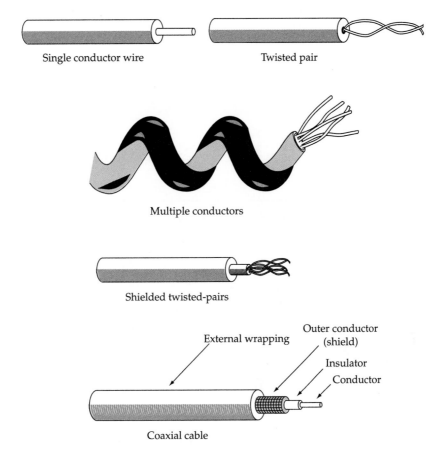

Single conductor wire   Twisted pair

Multiple conductors

Shielded twisted-pairs

External wrapping   Outer conductor (shield)
                         Insulator
                         Conductor

Coaxial cable

also the result of signals on other pairs of wires inside a cable. For this reason, wire-pairs are twisted to reduce the susceptibility to noise, and coaxial cables and some twisted-pairs use an outer, shielded covering. (See Figure 1.9.)

Wire circuits can act as interference transmitters as well as receivers. As the frequency of the signal in the circuit increases, the circuit tends to radiate the signal. Twisting the wire-pair reduces the effect of this transmission, as does twisting the other wires that are potential receivers. It is the tendency of circuits to become radiators or transmitters that limits the bandwidth the circuit can pass as well as increasing attenuation.

Implication

Noise reduces bandwidth and may cause errors, causing an increase in cost due to the need for error detection and correction.

Attenuation

Attenuation means the signal loses strength, becomes weaker, as it travels down the channel.

Amplification means to make the signal larger or stronger.

Figure 1.10 shows three effects taking place in an analog environment: **attenuation** of a signal over distance, *noise* on the medium, and *amplification* of signals to keep them at the desired level. Since the noise is amplified along with the desired signal, and may increase relatively due to noise in the amplifier, the signal-to-noise ratio will tend to deteriorate with distance and repeating amplifiers.

Clarification

Received unwanted signals are noise; so are signals we radiate to others.

Compression

As the amount of data needed to be sent over a channel increases, one must either acquire a higher bandwidth channel or reduce the amount of data to be transferred. Reducing the

FIGURE 1.10   **Three Effects in Analog Environment**

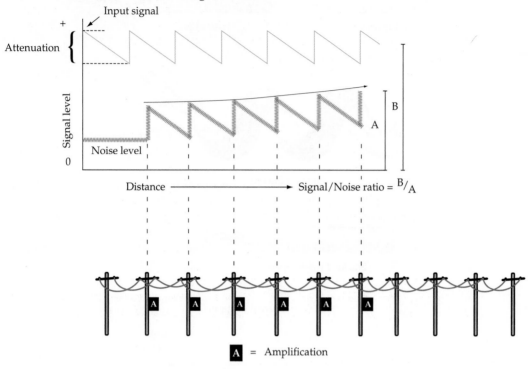

amount of data by *not sending redundant data* is called *compression*. Compression is used to reduce the amount of storage needed and the amount of bandwidth required in transport.

Implication   Compression allows for cost avoidance as it reduces space and bandwidth requirements.

One way to determine redundancy is to ascertain what symbols appear many times in the file or frame. Text files contain many occurrences of blanks and words like "the." It is possible to **encode** the occurrence of a series of blanks as a much smaller amount of data or often-occurring words as single characters. This is called *using a dictionary* and provides compression ratios of up to 2-to-1. At the other end, the process is reversed, recreating the file. With still images or video, a part of each frame or picture is often the same, such as the blue sky. The compression program scans along each line of the frame, from line to line, and frame to frame, and sends only the data that have changed, to reduce by as much as 250-to-1 the amount of data sent.

Implication   Processing to encode and decode compression is generally faster and cheaper than buying more bandwidth.

Compression is becoming a very important issue as an alternative to large storage devices and wide bandwidth channels. A specific benefit of compression technology is in the storing and moving of graphics. The two primary image formats used are GIF (CompuServe's Graphics Interchange Format, .GIF) and JPEG (Joint Photographic Experts Group, .JPEG). Graphic file formats incorporate compression and decompression algorithms as part of the file open/save or import/export operations. The GIF and JPEG formats both use sophisticated compression algorithms, but they differ in that the GIF format incorporates a lossless compression method and JPEG a lossy method. **Lossless compression** preserves all of the visual data in a picture: no information is discarded or altered during the compression/decompression operation. The compression scheme looks for complex repeated patterns within an image. In contrast, JPEG is a **lossy compression** scheme. It achieves very high compression ratios by discarding data. GIF can achieve a 2:1 compression ratio with no loss of quality. JPEG can achieve compression ratios up to 17:1, but at a potential sacrifice of image quality. Compression has one disadvantage for images: compressed data take time to compress and decompress before they can be displayed on the screen. This effect can be experienced on digital cameras.

It is possible to have a compromise. Graphics files can be tweaked to make it seem as if the images are downloading faster than normal through GIF's feature called *interlacing*. This format keeps track of the odd-numbered and even-numbered scan lines in a picture separately. At download time, all of the odd-numbered scan lines are transmitted first. To the viewer, it appears as if the download is much faster, although it takes the same or slightly longer to get the rest of the picture. Netscape Navigator® used this scheme.

Clarification   Cameras allow for variable compression levels, thus determining the size of the file required.

A format with which many are familiar is the Motion Picture Experts Group (MPEG). It was originally designed for compressing digitized versions of movies, resulting in the MPEG-1 and MPEG-2 formats, the latter of which is used widely today, especially in satellite direct broadcast. The extension of MPEG-2 is the MPEG-3 format, used primarily for **audio** (MP3). It is a lossy format but takes advantage of the characteristics of the human ear, meaning that there are times that the ear cannot distinguish certain sounds, so they are discarded. This is unlike MPEG-1 and MPEG-2, which are lossless formats.

Clarification   Compression *increases* the effective bandwidth of a channel.

MPEG-4[2] is a graphics and video lossy compression algorithm standard that is based on MPEG-1, MPEG-2, and Apple QuickTime technology. MPEG-4 files are smaller than JPEG or QuickTime files, so they are designed to transmit video and images over a narrower bandwidth and can mix video with text, graphics, and 2-D and 3-D animation layers. Wavelet technology can compress color images at rates of 20:1 up to 300:1 and grayscale images at 20:1 to 50:1.

Clarification   Overall, MPEG-4 offers a wide range of diversity with a few of the highlights covered here. According to Shelly Sofer, public relations manager of MGI Software (mgisoft.com), "MPEG 4 is designed to deliver video over a much narrower bandwidth. It uses a fundamentally different compression technology to reduce file sizes than other MPEG standards and is more wavelet based. Not only does MPEG 4 offer broadcast capabilities, it offers some interactivity capabilities, meaning than you could click on something in a video and that would launch other things."[3]

Effective Bandwidth

The preceding three topics are individually important and become even more important when combined. Native bandwidth is a characteristic of the circuit on which the signal travels to carry a given amount of data. In reality, this is the most bandwidth that the circuit can achieve. Compression reduces the amount of data being transported, thus increasing the effective bandwidth of the channel. Noise effectively reduces bandwidth by causing the loss of data or the need to retransmit. This brings us to an important definition: the **effective bandwidth**

[2] http://www.webopedia.com.
[3] http://streamingmediaworld.com/video/docs/MPEG4/.

varies based on three things: *the native capacity of the medium, the presence of noise, and the ability to compress the data.* Put mathematically, Effective bandwidth = Native bandwidth minus the effect of noise plus compression ($B_E = B_N - n + C$).

Encryption

Although a different concept than compression, encryption is an allied technology. It does not reduce the size of the data; it disguises the data so it can not be "read." **Encryption** is the process of obscuring information to make it unreadable without special knowledge, generally the knowledge of a key. A simple form of encryption is to substitute one character or number for another. To decrypt the information, the reader must know the key by which the encryption was done. Transmissions over public networks and storage of data in accessible areas require protection through encryption. Because the encryption key is the basis for the process, it must be highly secured.

ANALOG VERSUS DIGITAL SIGNAL DIFFERENCES

Why Consider Analog?

Since this is a book about data communications, it would be natural to ask why we include the subject of analog technology. The reason is *legacy networks*. Data communications outside of mainframe computers was originally carried on the analog POTS (plain old telephone system).[4] Many people continue to access the Internet using analog 56K modems to carry the digital data over noisy telephone circuits. Until we have true digital circuits for universal access, the subject of analog technology will remain important. At the present time, nearly all Internet access to residences is on analog technology; this includes DSL and cable modems. Additionally, all wireless devices, from cell phones to WiFi-enabled PDAs, use analog radio. Analog is widely used and is important for us to study.

Our objective is to move to an all-digital environment because *digital signals are more controllable than analog.* A single digital network, allowing us to handle voice, video, images, music, facsimiles, and data, provides value beyond analog. Noise can be handled in the same manner for all of these modes, as can compression.

Implication

If all the world's communications were digital, all communications could be transported on a single network.

Analog Signals

Analog signals are continuous in wave form; digital signals are discrete, generally binary.

Analog signals vary by time, called *frequency,* and are measured in cycles per second, called hertz (Hz).

The *signal form* of **analog** is that of a continuously varying wave. Consider the simple telephone, as an example of analog technology. The sound coming from the speaker's vocal cords is continuously varying **acoustic** energy. The telephone instrument converts this acoustic energy to analogous electrical energy. If, instead of speaking, we were hearing a single tone from a tuning fork, the signal would be a simple sine wave, as shown in Figure 1.11a. A simple sine wave is a regular repeating wave, represented by crests and troughs around a midpoint. A single repetition of a crest and trough is a cycle. The number of repetitions or cycles that occur within a period (such as second) of time is the **frequency** of the signal and is termed **hertz** [Figure 1.11b]. We think of frequency as the pitch of a signal. The distance from the trough to the top of the crest is the **amplitude** of the signal or its volume; the greater the amplitude, the "louder" the signal. The third parameter of an analog signal is its **phase,** which describes the point to which the signal has advanced in the cycle. Phase will be important to us later in this book as it is used in putting data on analog signals.

---

[4] See material at the back of the book for information on analog voice communications and POTS.

FIGURE 1.11

Amplitude of the sine waves is *measured*. When sound waves have been converted to electrical energy for transmission, we use an electrical measure such as *volts*. A strong wave is relatively high, while a weak wave is relatively low.

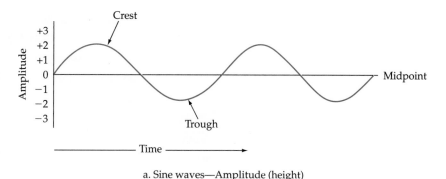

a. Sine waves—Amplitude (height)

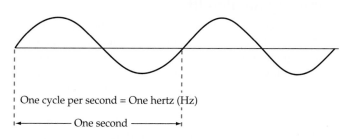

b. Sine waves—Frequency (width)

Clarification   The parameters of an analog wave are amplitude, frequency, and phase.

In the human voice, we have a combination of many simple sine waves. The resultant signal looks like the voice print in Figure 1.12. This voice print is a combination of frequencies from the lowest to the highest that the circuit is to carry. Circuits that have a narrow bandwidth can only carry a small range of frequencies, and, therefore, the richness of the resultant voice print is limited, as is the trueness of the representation of the signal.

We live in an analog world. Even when we use GIF images or MP3 music, the image or music originated as analog waves and must be converted to analog waves before viewing or listening. Where analog is continuously varying signals, **digital,** or more correctly binary, signals are bi-state. The digital data are a collection of a combination of two digits, two numbers: 1s and 0s. Because it is so easy to represent a bi-state occurrence, digital storage and transport is a natural. The digital codes must be converted to analog signals before humans can access them. Machines, on the other hand, exist in a digital world.

Clarification   Human sight and hearing rely on analog receptors; therefore, sound and images recorded, stored, or transported in digital format must be converted before listening or viewing.

FIGURE 1.12
Voice Print

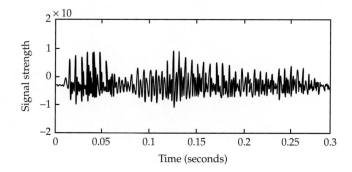

FIGURE 1.13
Acoustic Frequency
Bandwidth of Various
Devices

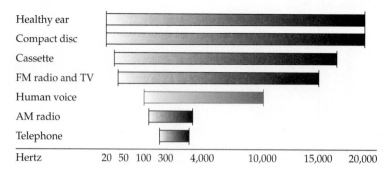

Analog Bandwidth

While electrical signals travel at about the speed of light (186,292 miles per second = 299,654 km/sec), the quantity of information we can pass during a given amount of time is of primary interest. When referring to digital signals, speed is measured in **bits per second.** With analog signals, the primary concern is the ability to pass a given **spectrum** of tone, an amount of frequency, that is, the bandwidth of the circuit (see Figure 1.13). For normal telephone circuits, we need to pass a signal of only 4,000 hertz to communicate the total telephone signal. (The 300–3,000 hertz portion of the frequency spectrum contains the actual voice (see Figure 1.14)). If using one circuit to pass several such telephone signals, a bandwidth several times 4 KHz (kilohertz) wide is needed. As the amount of desirable bandwidth increases, the limits of a given medium are reached. See Figure A.1 in Appendix A in the back of the book.

Noise is *any* unwanted signal. For acoustic signals, noise appears as competing acoustic signals; for example, music or traffic sounds when you are trying to talk. For electrical signals, noise appears as static or electromagnetic pulses. Noise interferes with the true signal and may distort it to a point that it is unpleasant at best and unreliable at worse. Noise can change the original content of the signal.

Clarification

Noise is any unwanted signal. *Spam* is any unwanted electronic mail. Therefore, spam is noise.

FIGURE 1.14
Frequency
Bandwidth for a
Voice channel

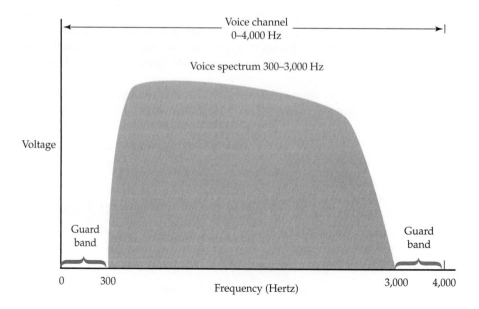

It was previously stated that the effective bandwidth was the result of the native bandwidth, less the effect of noise, plus the effect of compression. Analog signals are difficult to compress and analog circuits are very susceptible to noise. Besides bandwidth—that is, the amount of information a channel can carry—we are interested in the relationship between the signal itself and the noise, called the **signal-to-noise (S/N) ratio.** The greater the S/N ratio, the clearer the signal will appear; the lower the S/N—that is, the greater the noise-to-signal ratio—the more difficult it will be to hear (i.e., detect) the true signal. Some of the noise in analog circuits appears due to our attempt to *keep the signals at a desired level.* As acoustic signals move from the stereo speaker, they decrease in volume; as electrical signals move down a circuit, they diminish in amplitude or strength. To keep the electrical signals at the desired level, repeating **amplifiers** are used. These simple devices take a diminished electrical signal and boost it to a larger signal. The problem arises when the diminished signal contains a small amount of noise; when the signal is boosted, so is the noise. With repeated amplifications of the signal, the noise becomes larger in relation to the signal. Thus, the S/N ratio decreases.

Modulation

Amplitude modulation (AM) and frequency modulation (FM) radios should be familiar. These forms of signal modification are two major examples of signals that carry other signals.

We wish to communicate a signal such as voice or music to a distant point. We **broadcast** from the sending station, but we cannot broadcast the voice or music directly because the signal would degenerate or diminish too quickly. Therefore, we must *superimpose* the desired signal onto a carrier signal. For example, to listen to AM station WREL in Lexington, Virginia, tune the radio to a carrier frequency of 1,450 kilohertz. Superimposed onto this (single frequency) carrier is the music. In the case of amplitude modulation (AM), the music is superimposed, or **modulated,** onto the carrier wave by varying the amplitude of the carrier wave in correlation to the music signal. As shown in Figure 1.15, an analog signal such as music is combined with an analog carrier wave as variations in amplitude to create an amplitude modulated signal. The music signal shown as an example is vastly simplified. Music and voice signals are combinations of hundreds of frequencies, resulting in a wave that is very complex.

In the case of frequency modulation (FM), such as 96 ROCK in Atlanta, Georgia, the analog music signal varies the frequency, not the strength (amplitude), of the carrier wave. Where the AM receiver (550–1,680 KHz carrier frequency range) simply reproduces a signal that is the amplitude variation of the broadcast signal, the FM receiver (88–108 MHz carrier frequency range) reproduces a signal that follows the frequency variation of the broadcast signal.

Clarification

The top and bottom of a frequency modulated signal are clipped to reduce received noise, as noise is generally a variation in amplitude.

Television uses both amplitude and frequency modulation to transmit the picture and sound. Each TV channel has a 6 MHz bandwidth superimposed onto a carrier wave in the range of 50–212 MHz for **very high frequency (VHF)** channels 2–13 and 450–900 MHz for **ultrahigh frequency (UHF)** channels 14–84.[5] The audio signal occurs at 5.75 MHz, using frequency modulation. The black-and-white video signal uses amplitude modulation

---

[5] Originally, television was broadcast over the air. The first 12 channels were 2 thru 13 in a part of the radio spectrum called *very high frequency* (VHF). The next channels added, 14–81, were in the ultra high frequency (UHF) radio spectrum. Since most people now receive television via coaxial cable, these distinctions are not relevant. The cable simply carries a bandwidth of 550 or 750 MHz, divided into the 6 MHz TV channels.

FIGURE 1.15
Modulation

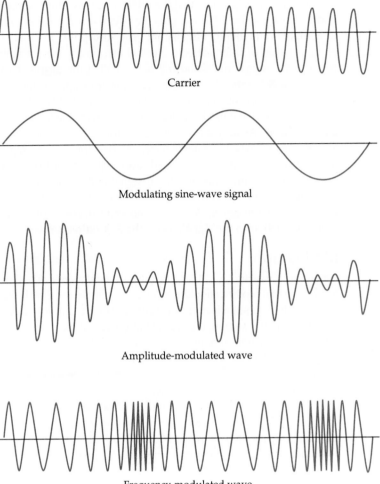

Carrier

Modulating sine-wave signal

Amplitude-modulated wave

Frequency-modulated wave

in the range of 0.5 to near 5.75 MHz, with color information being coded at 3.6 MHz using other AM techniques.

Neither analog nor digital signals can be transmitted directly because of the noise environment or, as in the case of radio and television, the distance required. Therefore, we modulate the desired signals, such as voice or music in the case of analog signals, onto a carrier frequency, which can be transmitted farther. (See Figure 1.15.) The process of modulation bypasses much of the noise **interference.** Because the frequency bands are much wider than required to carry one signal, we divide the frequency into multiple channels. This process is discussed later.

Digital Signals

Analog signals are continuously varying waves of energy. Digital data use discrete levels of **voltage** (e.g., +5 volts for a 1 and −5 volts for a 0). Noise on a circuit can appear like digital data because many devices such as those with motors produce voltage spikes (see Figure 1.16). While there are many advantages in having any signal in a digital form, digital data are much more susceptible to noise; therefore, we must use circuits that avoid noise. The selection of the medium is very important. Fiber circuits are the best as a noise-free environment. Failing a noise-free circuit, we must have reliable error detection and correction. Otherwise, the data are unreliable and suspect.

FIGURE 1.16
Effect of Noise on Digital Data Transmission

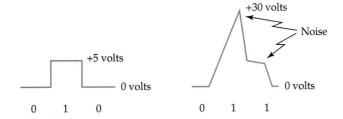

Codes

Codes
Codes represent characters.

Digital computers employ binary (bi-state) codes to represent data: every piece of data that a computer processes or stores is encoded into a series of 1s and 0s in accordance with some standard convention, called a **code.** To transmit digital codes, one must have a channel that does not contain noise that appears as digital data. Table 1.1 shows the decimal, octal, hexadecimal, and binary codes for the numbers zero through nine and binary codes for some characters.

Codes represent characters or collections of characters (alphabetic, symbols, or digits). One may substitute the numbers 1 through 26 for the Latin alphabet, or use a combination of 1s and 0s for the characters. Because data communications is digital, the codes use binary digits.

Code Length

The **Morse code** is variable in length: the number of bits per character (dots and dashes) varies by character. One advantage of the Morse code's variable length is that it uses a minimum of signals or bits (dots and dashes) to send a word, as some letters require only one dot or dash. The code was designed to take advantage of the frequency of occurrence of letters in words; for example, the letter "e" appears often and was given a code of one dot, the shortest signal possible. Since humans were interpreting the signals, this was efficient and effective. Fixed-length codes are preferable for machine-to-machine communications because there is no question when a character is complete.

All communications, whether between humans or machines, require the use of **protocols.** This simply means rules for communications. In the case of the telegraph system, the protocol is the code used, that is, the meaning of each group of dots and dashes, and agreement as to who can transmit and when. There must be a compromise between the number of

TABLE 1.1
Decimal, Hexadecimal, Octal, Binary, and ASCII Conversion Table

Decimal	Hex	Octal	Binary	ASCII
0	0	000	0000 0000	0011 0000
1	1	001	0000 0001	0011 0001
2	2	002	0000 0010	0011 0010
3	3	003	0000 0011	0011 0011
4	4	004	0000 0100	0011 0100
5	5	005	0000 0101	0011 0101
6	6	006	0000 0110	0011 0110
7	7	007	0000 0111	0011 0111
8	8	010	0000 1000	0011 1000
9	9	011	0000 1001	0011 1001
10	A	012	0000 1010	
11	B	013	0000 1011	
12	C	014	0000 1100	
13	D	015	0000 1101	
14	E	106	0000 1110	
15	F	017	0000 1111	
16	10	020	0001 0000	

characters in the code and the amount of signals. An early code system, the **Baudot code,** has five bits to represent a character and, therefore, can represent 32 unique *code points* ($2^5 = 32$), enough only for capital letters and some numbers. Allocating one or more codes as a shift key allows *additional* code points to be designated. (If one character of the 32 is used as a shift character, total characters = $2 * (32 - 1) = 62$ characters. If two characters are used as shift keys, the result is $4 * (32 - 2) = 120$ characters). Baudot uses two escape characters and represents 58 characters. The usual intent is to represent at least all capital letters, numbers, some punctuation marks, and some effector codes that cause the machine to take action, such as go to top-of-page.

Implication   Computer keyboards generally have 104 keys with Shift, Ctrl, and Alt special keys. How many characters can they represent?

The two most used computer codes are the **American Standard Code for Information Interchange (ASCII),** sponsored by the **American National Standards Institute (ANSI),** and **Extended Binary Coded Decimal Interchange Code (EBCDIC),** sponsored by IBM. The seven-bit ASCII code shown in Figure 1.17a can represent 128 characters ($2^7 = 128$) and can represent all capital and lowercase Latin letters, numbers, special symbols such as punctuation, and transmission control characters. Adding an eighth bit to the ASCII doubles the number of code points. The eight-bit ASCII code, Figure 1.17b, is used for most microcomputers to represent the characters noted plus graphic characters. IBM continues to utilize their mainframe-originated eight-bit

FIGURE 1.17

		\multicolumn{8}{First Three Bit Positions (Bits 7,6,5)}
		000	001	010	011	100	101	110	111
L	0000	NUL	DLE	SP	0	@	P	`	p
A									
S	0001	SOH	DC1	!	1	A	Q	a	q
T									
	0010	STX	DC2	"	2	B	R	b	r
F									
O	0011	ETX	DC3	#	3	C	S	c	s
U									
R	0100	EOT	DC4	$	4	D	T	d	t
	0101	ENQ	NAK	%	5	E	U	e	u
B									
I	0110	ACK	SYN	&	6	F	V	f	v
T									
	0111	BEL	ETB	'	7	G	W	g	w
P									
O	1000	BS	CAN	(	8	H	X	h	x
S									
I	1001	HT	EM	)	9	I	Y	i	y
T									
I	1010	LF	SUB	*	:	J	Z	j	z
O									
N	1011	VT	ESC	+	;	K	[	k	{
S	1100	FF	FS	,	<	L	\	l	|
	1101	CR	GS	—	=	M	]	m	}
	1110	SO	RS	.	>	N	^	n	~
	1111	SI	US	/	?	O	–	o	DEL

a. ANSI ASCII 7-bit binary code for telecommunications

000	(Graphics	010	(Graphics
020	characters)	030	characters)
040	( )*+,	-./01	050 23456 789:;
060	<=>?@	ABCDE	070 FGHIJ KLMNO
080	PQRST	UVWXY	090 Z[\]^_`abc
100	defgh	ijklm	110 nopqr stuvw
120	xyz{	}˜Çü	130 éâäàå çêëèï
140	îïÄÅE	æÆôöó	150 ûùÿjÖÜ ¢£¥P▪
160	áíóúñ	Ñ°¨¿⌐	170 ¬½¼i « » ▪▪|
180	╢╡╖╗	╣║╕╜	190 ╛┐└┴ ┌├╞
200	╚╘╒╓	╫╪┘┌	210 █▄▌▐▀ ╪╤ ┤
220	▄ ▌∞	ßΓπΣσ	230 μτ ΦΘΩ δ∞φε∩
240	≡±≥≤⌠	⌡÷≈°·	250 ·√ⁿ² ▪

b. Eight-bit (256) ASCII characters with decimal codes

EBCDIC code in some of their PC-based programs. This difference in code representations necessitates an EBCDIC-ASCII code conversion in many cases. Fortunately, there is a direct relationship between the codes and conversion is quick.

Code Length versus Throughput

When a code is devised, the originator must compromise between the possible number of *code points,* or *characters,* and the number of bits it takes to represent a code point. This relates to speed of transmission, which is **throughput** or bandwidth. The effect of the size of the code is apparent when text is sent faster via five-bit Baudot code than eight-bit ASCII code because of the fewer bits required per unit of time. Organizations are finding that they need to send more and more data, which requires either that there is greater bandwidth or that the data must be made "smaller" by use of a smaller code or by compression.

Implication   **code length** The more bits per characters in a code, the longer the time to transmit or the wider the required bandwidth.

The codes discussed above encode specific groups of characters. If there is a need for more characters than places in the code, such as in European and Oriental languages, a 16-bit code, called the **Unicode,** can represent 65,536 ($2^{16} = 65,536$) characters, believed to be sufficient for any occasion. Because there is backing from a group of significant computer companies in the field of computers, the Unicode may become a standard, replacing ASCII and EBCDIC codes.

Digital Bandwidth

The unit of a digital, or binary, signal is the **bit** (**b**inary un**it**). Digital bandwidth is measured in the quantity of bits per unit of time, specifically bits per second (bps). This is often referred to as the speed[6] of a circuit. The greater the bandwidth of a digital circuit or channel, the more information that can be passed in a specified amount of time. A 10,000 bit-per-second (bps) channel can pass five times as much information as a 2,000 bps channel.

Clarification   Analog bandwidth is noted in hertz; digital bandwidth is noted in bits per second.

Just as with analog signals, digital signals reduce in strength as they travel down the circuit. Just as with analog circuits, noise may be present. For digital signals, noise either changes the bits of the digital code or causes the specific bit waveforms to become misshaped. In the latter case, as with analog signals, devices are introduced into the circuit to put the signal back at the desired level. For digital signals, the bits are *regenerated* to be the exact replicates of the original bits. **Regeneration** leaves the misshaping of noise behind, producing a perfect output; the signal is both reshaped and amplified. If, however, the bits have been totally altered by noise, the new bits will still be perfectly wrong.

Clarification   Digital speed means bandwidth.

Noise and Data Signals

As data communications is the transmission and receipt of binary codes (electrical pulses or light representing 1s and 0s) between terminals and computers, any injected (introduced) signal that appears to be like the binary codes will distort the information being sent.

---

[6] Remember, it was noted that the speed of electrical or photonic signals is the speed of light. In the case of bandwidth, *speed* is used to denote how fast a collection of characters can be transported to its destination.

Injected noise appears much like binary bits and can either add to the transmitted bits or change them. It is for this reason that we must either change the binary signals from terminals and computers to be immune to noise or change the medium to exclude noise. The first situation occurs when using modems, and the second occurs when using shielded wires, coaxial cables, fiber-optic strands, or conditioned[7] data networks. Much of what we will discuss in the area of data communications addresses either of these two situations. In order to understand the nature of data communications, we need to understand the alternative circuit forms and their importance. They are discussed in the next section.

Implication

Although bandwidth comes at a cost, adequate and reserve bandwidth means new capabilities can be implemented quickly, possibly providing a competitive advantage.

Clarification

Two primary codes used are seven-bit and eight-bit ASCII, which can represent 128 and 256 characters respectively.

MOVING DIGITAL DATA OVER ANALOG CHANNELS

We stated earlier that something must be done to or with the signals from a terminal or computer in order to transmit them over existing telecommunications circuits. The reason is that these circuits are susceptible to injected noise signals. The noise introduces errors in the data communications and could potentially render the data useless. A primary reason for the problem is that much of the telecommunications network presently in use is the analog telephone system. Digital data communications networks have been developed for commercial customers, but small nonbroadband users and small amounts of data transmission at slow rates continue to rely on POTS. Thus, we want to change the digital computer signals to analog signals for transmission over the analog network. This will not be a true **digital-to-analog** conversion (as explained elsewhere for digitizing voice) as it is only the **encapsulation** (insertion into a protective envelope) of bits to move them in a noisy, analog environment.

Modems Carry Digital Data on Analog Circuits

The equipment used to make the change from digital data and then the change back to digital form is a **modem** (MOdulator-DEModulator). In its simplest form, the modem (a common piece of DCE) receives a digital bit, generates an analog signal, and adds it to the basic frequency that is carrying the signal. To change the computer signals, the modem will modulate (add) a specific frequency onto a carrier frequency for transmission and demodulate it back to the original signal at the receiving end. In the simplest modem, such as the Bell Type 103 modem using the **frequency shift keying (FSK)** technique, a specific frequency (1070 Hz) is generated for a 0 bit and a different frequency (1270 Hz) is generated for a 1 bit. These frequencies are added to the carrier frequency for a finite amount of time, appearing as a burst of energy at the specified frequency. The receiving modem listens for these two frequencies and decodes this signal by detecting the frequencies (subtracts them from the carrier signal) and generating the 0 and 1 bits accordingly. This process, and the resultant tone that is generated, is very similar to Touch-Tone dialing.

The OSI Model and the Modem

Reference to the OSI seven-layer model, to be explored in Chapter 3, in relation to a modem should give the reader an indication of how the model works. The ultimate

---

[7] Typically, this is a nonswitched voice-grade channel with the addition of equipment to provide minimum values of line characteristics for data transmission; that is, to add devices to a line to reduce the noise environment and/or to increase its bandwidth.

FIGURE 1.18
Seven-Layer OSI Model

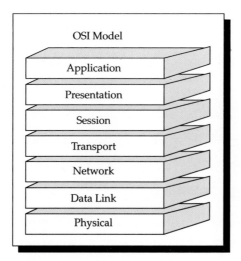

objective of a modem is to send or receive electrical signals, done so in the Physical layer. (See Figure 1.18.) The layer above (Data Link layer) deals with how the bits are grouped into characters, that is, data grouping, and **reliability,** that is, error detection and correction. The Network layer is concerned with addressing and the Transport layer is responsible for a reliable communications channel. These four layers are the ones most used to describe the functionality of a modem; the functions of the remaining three layers of the model may well be handled outside of the modem.

Bits per Second (bps) versus Baud

An asynchronous[8] network that sends 10-bit characters at a rate of 300 bits per second (bps) will transmit 30 characters per second. The term *baud,* which we often use interchangeably with bps, refers to the number of times a change of analog signal occurs in the circuit. If one signal change occurs per bit, then baud and bps measure the same rate. However, as the data transfer rate increases above 600 bps over 4 KHz telephone circuits, the bandwidth limit of the circuit is reached. For FSK modulation to work, the circuit must have a bandwidth of 1.5 times the baud rate. Thus, at 1,200 baud, 1,800 Hz would be required for transmission in each direction, requiring a total analog bandwidth of 3,600 Hz, which exceeds the **voice spectrum** of a POTS line. Therefore, when FSK is used at or above 1,200 baud, the transmission is limited to half-duplex modems, which are 600- or 1,200-baud modems.

Clarification

One has a 56 Kbps modem, not a 56K baud modem.

Caveat

When acquiring a POTS line from the local telephone company, you are only acquiring an analog voice channel that can be switched to another telephone. You are not buying or being guaranteed bandwidth. Because of noise, the shiny new 56 Kbps modem may be communicating at 28.8 Kbps. Some **Telcos**[9] will condition[10] your line, for a fee, giving you higher bandwidth.

Faster Modems

To achieve a greater speed, the modem must use some technique to encode more than one bit per baud. If we can assume that we are actually pushing the envelope and running at

---

[8] An asynchronous network transmits a character at a time. This is discussed in a later chapter.

[9] Telco is a term used to refer to any local telephone company.

[10] Conditioning is the process of balancing the capacitance and induction of the circuit to limit noise.

TABLE 1.2
Example of Phase Changes Possible by Use of QAM

Phase Change (Degrees)	Relative Amplitude	Quadbit
0	3	001
0	5	1001
45	$\sqrt{2}$	0000
45	$3\sqrt{2}$	10000
90	3	0010
90	5	1010
135	$\sqrt{2}$	0011
135	$3\sqrt{2}$	1011
180	3	0111
180	5	1111
225	$\sqrt{2}$	0110
225	$3\sqrt{2}$	1110
270	3	0100
270	5	1100
315	$\sqrt{2}$	0101
315	$3\sqrt{2}$	1101

1,200 baud, we must be able to encode 12 bits per baud at 14.4 Kbps and 24 bits at 28.8 Kbps. To do this, the machine must use one of the three parameters to encode more than one bit per baud. As noted with FSK, we have reached the limit of frequency modulation. Amplitude modulation, by itself, is not often used. The technique most used because the equipment has the highest probability of detecting a change is **phase shift modulation.**

If an analog sine wave starts at zero degrees and continues through its form, a phase shift is said to occur if the instantaneous phase of the wave is set to something new. For example, if the wave entering 90 degrees is abruptly shifted to 180 degrees, a phase shift has occurred and can be readily detected. Plus, this provides a way to place a bit on the waveform due to the shift in phase. If a shift happens every 90 degrees, you have two bits encoded on the baud (e.g., 4 * 90° = 360°). Thus, in this simple case, a 1,200-baud wave, with shifts at 90 degrees, could carry two digital bits and create 2,400 bps. Taking this further, shifting at 45 degrees gives three bits, or 3,600 bps. Adding amplitude shift with phase shift, as in **quadrature amplitude modulation (QAM),** you could have the possibilities shown in Table 1.2, which is four bits. **Trellis code modulation (TCM),** called such because the form of the bits on an oscilloscope resembles the trellis used in rose gardens, uses QAM but marks specific codes as unusable to improve reliability.

Implication

Dial-up communications provides **connectivity** at very low cost, unless one considers the cost of the time transmission takes due to low bandwidth.

Using these techniques, one way to encode data is to shift when there is a change from 1 to 0 or 0 to 1. This is referred to as **phase shift keying (PSK)** (see Figure 1.19a). Thus, the measuring device will detect each time period and assume that the new time slot is the same digit as the last if there is no shift. Using **differential phase shift keying (DPSK),** a shift occurs when a 1 is sent (see Figure 1.19b).

Compression, Again

We have noted how it is possible to have more than one bit represented in each baud. Thus, with these techniques, one can presently have 14.4 Kbps, 28.8 Kbps, 33.6 Kbps, and 56 Kbps. This is the *actual bit rate.* Using compression, the encoding of redundant characters allows for a greater *effective bit rate.* Using V.34bis modem standards, 4:1 compression can be achieved for files that are mostly text. This means that the effective bit rate can range from 33.6 Kbps (actual) to 134.4 Kbps (effective).

FIGURE 1.19 Phase Shift Modulation

a. Phase shift modulation

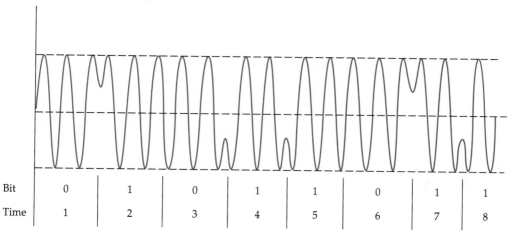

Bit	0	1	0	1	1	0	1	1
Time	1	2	3	4	5	6	7	8

b. Differential phase shift modulation

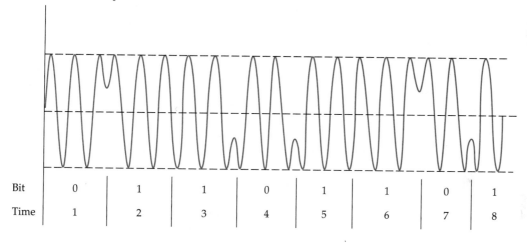

Bit	0	1	1	0	1	1	0	1
Time	1	2	3	4	5	6	7	8

Implication   The effective bandwidth depends on the level of compression.

Caveat   True V.34 modems perform the compression and decompression with hardware on the modem. Some low-priced modems do this with software (on the processor in the computer, resulting in a slower rate). See Case 2-1 for more on compression and modem capabilities.

Clarification   Nodes that process via hardware are faster than more general-purpose nodes that process via software.

Modem Standards

A **standard** is a definition or format that has been approved by a recognized standards organization or is accepted as a de facto criterion by the industry. It is a set of rules covering equipment and software. Recall that protocols are sets of rules governing how equipment and software applications **interoperate.** The capabilities of and standards for modems have evolved greatly over the past 25 years. One standard is based on the way modems communicate with each other; another, the form of compression they use; and the third, and oldest, the instruction set used. This later standard is called **Hayes-compatibility,** meaning it

TABLE 1.3
Modem Standards

Modem Standard	Speed	Features
Bell 103	300 bps	Leased from phone company (300 baud)
Bell 212A	1200 bps	Leased from phone company (600 baud)
v.*22bis*	2400 bps	4 bits per baud
v.32	9600 bps	Full duplex at 9600 bps, error correcting
v.*32bis*	14.4 Kbps	Allowed speed fallback in noise
v.32fast	28.8 Kbps	
v.34	28.8 Kbps	
v.34	33.6 Kbps	Software upgrade from 28.8 Kbps, serial number 40051 or higher
x2 or K56Flex	56 Kbps	Not ITU standard; iIncompatible with each other
v.90	56 Kbps	ITU standard; software upgradeable from x2 or K56Flex
v.92	56 Kbps	fast handshaking, v.44 compression; not upgradeable from x2 or K56Flex

responds to a standard set of commands originated and standardized by Hayes Microcomputer Products, a major modem manufacturer of the 1980s and 1990s. When you purchase a modem that meets certain standards (see Table 1.3), you most likely can communicate with any modem that meets the same standards. For example, all V. 90 modems should be able to interoperate, that is, work with each other.

What you can't ignore is software. The final value of a modem purchase depends on getting a communications program that's both sophisticated and easy to use. Many modems come bundled with simple software. In any case, the software will hide the modem's complexity behind simple commands and screen displays that will let you roam the world by phone from your desk.

Clarification   Modems *encapsulate* digital data in an analog signal to reduce the effect of noise.

Modems in Direct Connection

There are occasions when we use modems but not the telephone system. One case is when modems are used for connection of dumb terminals or PCs emulating terminals to a computer such as a mainframe. The modem is used instead of a higher-speed circuit, the distance is long, and/or the environment is noisy. Legacy mainframe systems historically connected terminals to front-end processors of the mainframe with coaxial cable for bandwidth and noise protection. Imbedded twisted-pair (extra telephone) wire is far cheaper to use and the bandwidth of 19.2 to 33.6 Kbps is sufficient for textual information and many graphics applications.

Clarification   Modems are used to carry digital signals in a (noisy) analog environment.

Modems in Enhanced Connection

With different software, the connection changes from direct "dumb" connection to that of a PC as a node on a network, like the Internet. The software is SLIP/PPP software. **SLIP** or **serial line Internet protocol** is a communications protocol that supports an Internet connection (using TCP/IP) over a dialup line. There is also a common variant of SLIP called compressed SLIP or CSLIP; it can be somewhat faster in operation than standard SLIP. **PPP** or **point-to-point protocol** is a newer protocol that does essentially the same thing as SLIP or CSLIP; however, it's better designed and more acceptable to the sort of people who like to standardize protocol specifications. When using the Internet or online services, especially when graphical data such as interfaces are involved, SLIP and PPP are a must.

DIGITAL REPRESENTATION OF AN ANALOG SIGNAL

Digital Voice

We have noted that the public switched telephone network is an analog network developed to handle voice traffic. As the signal travels down the wire, it loses strength and attenuates. Recall that line amplifiers are placed every mile or so to boost the signal strength, keeping the volume at a predefined level. One problem with boosting analog signals through amplifiers is that they amplify all parts of the signal including any noise that is picked up along the way. A method of avoiding this problem is to *convert* the analog voice signal to a digital form, *transport* it digitally, and then *transform* it back to an analog form at the receiving end. Although this signal also will attenuate over distance, the digital signal is regenerated and amplified instead of just being amplified when required, leaving the noise behind. Thus, transmitting voice as data communications instead of analog voice signals means a cleaner signal at the destination. In addition, we can handle digital voice in the same way as data, with the exception that reception delays are less tolerated for voice than data. In other words, transmitting voice and data together on the same circuit will work well, but, if the voice portion is delayed, the result may be unacceptable. See Figure 1.20 for a diagram illustrating signal regeneration.

Clarification   A-to-D = Convert, transport, and transform

Moving Analog Data over Digital Channels

To convert an analog signal to a digital form, equipment must sample (i.e., measure) the signal many times a second, convert the measured analog signal to a digital (integer) figure, and

FIGURE 1.20   Digital Signal Regeneration

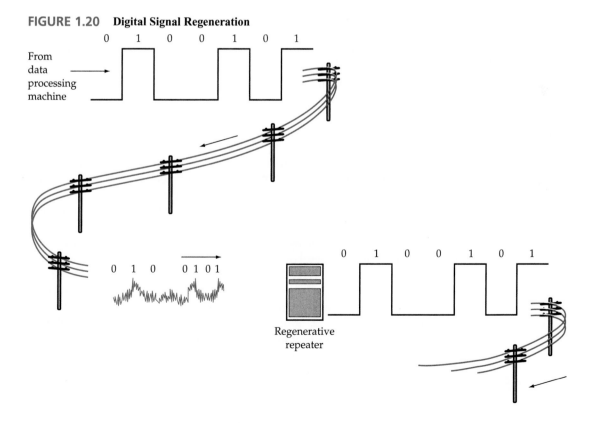

FIGURE 1.21   **Pulse Code Modulation to Convert Analog Signals to Binary (Digital)**

a.

b.

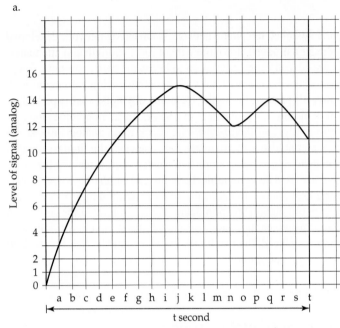

Sample Time	Signal Strength (Analog)	16-Bit Code (Digital)
0	0	0000
1	3	0011
2	5	0101
3	7	0111
4	9	1001
5	10	1010
6	11	1011
7	13	1101
8	14	1110
9	14	1110
10	15	1111
11	15	1111
12	14	1110
13	13	1101
14	12	1100
15	12	1100
16	13	1101
17	14	1110
18	14	1110
19	12	1100
20	11	1011

Sampling rate = 20 samples/second   16 levels = 2^4 = 4 bits/integer

Bandwidth = (4 bits/sample)* (20 samples/second) = 80 bits per second

encode the digital figure as a digital code. In Figure 1.21a, the sine wave represents a simple analog signal, much like a voice conversation. A device samples the signal strength at each point, a through *t*, and measures the values of 0 through 15 volts (analog). The equipment at the receiving end of the digital circuit converts the binary numbers back to analog signals, to voltage levels, using the same timing as the original sampling device. Thus, a spoken analog signal is represented by a digital signal, transmitted, and then converted back to its original analog form for playback (hearing).

Implication

Converting POTS voice to digital is the precursor of **VoIP.**

The fidelity of reproducing the analog signal is dependent on the sampling rate and the number of integer sample points. A rate of 8,000[11] times per second is required for a good representation of a telephone conversation. Also, the system requires 256 integer values of measurement, which can be represented with eight digital bits (2^8 = 256). This combination necessitates a digital signal of eight bits, 8,000 times per second, or a digital speed of 64,000 bps. The standard that employs these rates is called **pulse code modulation (PCM).** Thus, PCM samples the level of a voice conversation 8,000 times per second using a scale of 256 integer points, converts each integer measurement into an eight-bit code, and transmits the resultant information at a speed of 64,000 bps. The receiving unit reverses the process and a voice conversation is reproduced.

A variation of this method is **adaptive differential pulse code modulation (ADPCM),** which measures the signal just like the PCM method but transmits the difference between successive measured values instead of the value itself.

[11] The pulse code modulation scheme was developed at Bell Labs, where researchers found that sampling a signal at twice its highest frequency faithfully reproduced the signal.

Clarification   ADPCM is a lossless form of compression.

ADPCM allows the use of a four-bit code to represent the difference value and reduces the signal requirement to 32,000 bps. The system uses three bits to represent eight integer values and one bit to note whether the change was positive or negative. This reduction from 64 Kbps is a form of *lossless compression,* that is, there is no loss of data with a 2-to-1 reduction in bandwidth. By additional compression (leaving out the silence periods and sending redundant data as multiples) and encoding techniques, the digital rate can be further reduced without sacrificing quality. Equipment is available that reduces voice to 16,000 bps with no discernable loss of quality, and AT&T uses a proprietary method to compress to 8,000 bps[12] in its nationwide long-distance service. The importance of analog-to-digital (A-D) conversion at this point is that almost all long-distance carriers use A-D and D-A conversion and move analog telephone conversations as digital data. One result is noise-free, clear conversations.

Implication   Using the Bell Labs rule of thumb for sampling rate, what is the rate of sampling for a CD?

If we convert real-time analog signals, such as telephone conversations and television programs, to digital signals, we must not delay their transmission and conversion back to analog or a disturbing result will occur. When using the satellite medium, there is a delay caused by the long distance the signal must travel and the transponder's switching time. This is annoying for voice conversations because they are two-way, back-and-forth, but completely transparent to TV broadcast because it is unidirectional and continuous. Delay due to satellite transmission would seem to be irrelevant to data communications. The problem in the latter case, however, is that delays are imposed by responses from the receiving station in noting errors in transmission. This delay translates into significant amounts of wait time and, thus, nonuse of the circuit during the reversal of a half-duplex circuit.

CIRCUIT SHARING (MULTIPLEXING)

Any of the media we use for communications circuits can handle a wider analog bandwidth or higher digital speed than one communications signal requires. For example, a twisted-pair line can accommodate a one-megahertz bandwidth, and a telephone channel requires only 4,000 Hz bandwidth. To make efficient use of circuits, carriers would want to transmit more than one conversation over a circuit. This is especially true with circuits having large volumes of traffic, such as between central offices in the same calling areas and between calling areas (a toll trunk call). Two basic methods of **multiplexing,** frequency division multiplexing (FDM) and time division multiplexing (TDM) of the primary circuit, accommodate circuit sharing. While used in different environments, FDM and TDM perform the same basic function: the sharing of a circuit. FDM is used on analog broadband circuits and TDM is used on digital baseband circuits.

Implication   The nominal analog bandwidth of twisted-pair is one megahertz.

Frequency division multiplexing (FDM) places several signals onto one channel or circuit by placing each at a different part of the (analog) frequency spectrum. For example, 12 continuous telephone circuits are required between Atlanta and Savannah, Georgia, and there is only one twisted-pair wire circuit. Instead of stringing 11 additional

---

[12] Some researchers believe the bandwidth for voice-over-the-Internet will soon be reduced to 4 Kbps.

FIGURE 1.22
Multiplexed Circuit

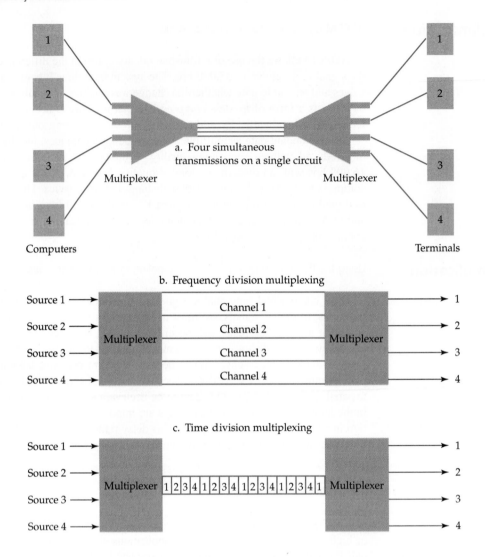

a. Four simultaneous transmissions on a single circuit

Multiplexer   Multiplexer

Computers   Terminals

b. Frequency division multiplexing

Source 1 → Multiplexer | Channel 1 / Channel 2 / Channel 3 / Channel 4 | Multiplexer → 1 2 3 4

c. Time division multiplexing

Source 1 → Multiplexer | 1 2 3 4 1 2 3 4 1 2 3 4 1 2 3 4 1 | Multiplexer → 1 2 3 4

twisted-pair wire circuits, we can use frequency division multiplexing to share the existing one circuit by stacking the conversations on top of each other. (See Figure 1.22a, b.) In a simplistic illustration, the first conversation occupies the frequency between 0 and 4,000 hertz, as it would on a simple wire. Assuming we needed no buffer zone between conversations other than that provided within the 4,000-hertz envelope, the second conversation would be shifted up 4,000 hertz and occupy the frequency between 4,001 and 8,000 hertz. The signal actually being transmitted would occur in the band between 4,001 and 8,000 hertz, but it would be converted back to 0 to 4,000 hertz before being sent to the receiver. The third conversation would be between 8,001 and 12,000 hertz, and so on. These 12 conversations could be accommodated by a single circuit with a bandwidth of only 48 KHz. A normal twisted-wire circuit could potentially handle 250 conversations (250 * 4,000 hertz = 1 megahertz) but is usually limited to 48 4-KHz channels or less. In this example, the Telco could increase the number of telephone channels between two cities hundreds of miles apart 5,000 percent by adding multiplexing equipment at each end.

Clarification

guard bands Also called a frequency guard band, a narrow frequency band between adjacent channels in multiplexing that is kept unused to prevent the channels from overlapping and causing crosstalk among modulated signals.[13]

Implication

FDM and TDM are achieved by technology at the ends of circuits, making the circuits more useful and valuable.

We are accustomed to frequency division in our daily lives. For example, broadcast television transmits 12 VHF channels via the air medium over six megahertz channels (channels 2–13 broadcast on a frequency band of 50–212 MHz with FM radio channels or stations occupying 200 KHz each in the 88–108 MHz frequency range between channels 6 and 7). Many additional channel slots are also available in the 450–900 MHz UHF spectrum, and satellites have several hundred more. We use a filter to select the channel of choice from the large number being received, and the TV converts the signal to video and corresponding audio. With a cable TV (CATV) network, the coaxial cable coming to the home or office can carry over 100 TV channels[14] multiplexed on the one circuit. Once the viewer determines the one channel desired, the receiver selects the signal for that channel alone and the TV converts it to video and audio.

Implication

Multiplexing allows the sharing of a resource, thus reducing costs.

As a different way to view this, the portion of the total electromagnetic spectrum set aside for television programming has a number of discrete channels occupying/dividing the spectrum. The same is true for CATV. In the TV set, the RF (radio frequency) receiver selects the desired frequency bandwidth (channel) and converts the signal by subtracting out the frequency that had been added to move the "real" frequency to the allocated slot. This signal, in turn, is converted to video and corresponding audio.

Examples of Divided Channels

When a channel is established but not totally utilized, the potential exists to further divide the channel for additional uses. Such is the case in **subsidiary communications authorization (SCA)** subchannel transmission over the FM radio spectrum. Although a commercial FM channel is 200 KHz wide (100 KHz useful), the spectrum used for stereo broadcast occupies frequencies of 50–15,000 Hz for the primary channel, 19 KHz for the stereo pilot signal, and 23–53 KHz for the stereo subchannel. The spectrum from 53 KHz to 100 KHz is unused. SCA uses the 53 KHz to 92 KHz band for either analog or digital transmission. For example, an FM music station in Atlanta, Georgia, leased a small bandwidth at the 67 KHz frequency to Lotus Information Service to broadcast (simplex) its stock market quote service, digitally and encrypted at about 8 Kbps. Subscribers in the city can receive updated stock market quotes on specially designed FM radio receiving units. The second leaser of a subchannel from this radio station is a digital paging service that uses the 57 KHz frequency to broadcast a very narrow signal to portable paging units.

Implication

With all the unused bandwidth on an FM radio channel, many signals could be **piggybacked** (carried) on the carrier channel.

[13] http://www.webopedia.com.
[14] The number of channels that a coaxial cable can deliver is dependent on (1) the analog bandwidth of the cable (e.g., 350, 550, or 750 Mhz), (2) the analog bandwidth of the channel (standard or HDTV), and (3) whether the channel is carried as uncompressed analog (6 Mhz) or compressed digital (3–10 channels per 6 Mhz bandwidth).

The background music service Muzak® typically will lease the 67 KHz frequency with a 6 KHz bandwidth for transmitting specially arranged music to subscribers from commercial FM stations. Some grocery stores in Atlanta receive music via another mode at the 92 KHz frequency, called *AM double sideband suppressed carrier*. This technique uses sideband carriers with bandwidths of 10 KHz each to send two channels of high-quality music. All of this activity is occurring in addition to the commercial FM broadcast.

A final use of piggyback signals is the new **MSNDirect**® (www.MSNDirect.com). It uses SCA technologies to transmit data to watches being worn by subscribers who receive real-time data on stock quotes, sports, weather, and news.

Another service using multiplexed transmission occurs when CATV networks transmit news and stock market data in digital form in addition to the usual analog signals. Usually, CATV subscribers connect the cable to a television set or decoder and receive analog TV programs. In some areas, subscribers pay for a special attachment device (modem equivalent) and software for their microcomputers and receive the transmitted text data from the CATV cable. While commercial online stock quote services operate bidirectionally, allowing subscribers such as stockbrokers to ask for specific information, both the CATV and SCA services are simplex. Some CATV systems update the data only three times daily.

Time division multiplexing (TDM) shares the circuit's time allocation instead of the frequency allocation. (See Figure 1.22c.) A simplistic but seldom used form of TDM physically switches from originator to originator to share the time available, and the receiving unit does the same in synchronism. If the signals change slowly or they can be stored, this method will work well. The master controlling unit may mechanically switch among slave units on a scheduled basis, giving each slave unit a predetermined portion of time, or provide a nonequal amount of time based on past need, called *statistical multiplexing* (see below). An alternative method is for the controller to poll the slaves to see if each wishes to be connected. **Polling** provides more control and requires more overhead, but, as with statistical methods, active devices have more access time.

The transmission of digital data, which can be stored, makes the best use of time division multiplexing. The originating sending computers or terminals store their outgoing messages until the originating **multiplexer (MUX)** connects to each for a short time period. Part of the connected sender's message is transmitted, and the MUX switches to the next sender. Since the sending unit can store its message, any timing problem is eliminated. Another logical view of this capability is that the time-division-multiplex sending modem stores the message of each of the terminals or computers to which it is attached and interleaves their messages into one long string, a portion at a time. If you have four slow, 2,400 bps computers communicating over a single 9,600 bps circuit, the computers will notice no delay because the total speed of the four input signals sums to the speed of the communications circuit. In the case of very high speed circuit media such as fiber optic cables capable of a transmission speed of 1,000+ Mbps, a large number of the slower-sending units can share one circuit with no waiting.

A specific use of TDM was in an office that uses video display terminals connected to a remote computer. In the 1970s, organizations installed coaxial cables from a controller near the computer to each terminal and transmitted at a high digital speed (10+ Kbps). In reality, the terminal operators were inputting data at a speed of much less than 10 characters per second, requiring a speed of less than 100 bps (using 10 bits per character). Further, it is unlikely that the operators could sustain even this speed, so the combined speed of 10 operators would average less than 1,000 bps. A circuit capable of maintaining 1,200 bps TDM, therefore, could support these 10 terminals, replacing 10 coaxial cables. If this equipment were installed today, the most cost-effective configuration would be use of a single 1,200 bps modem and a phone line or a statistical multiplexer, to be discussed below. Most offices have extra twisted-pair circuits that are part of the phone line cables and, therefore, require no extra installation cost. Today, TDM speeds are much higher than discussed as modem speeds have increased.

Clarification
A multiplexer (MUX) places several signals onto a carrier, with equal sharing.

Implication
Digital signals can be easily stored, compressed, and encrypted. The same is much more difficult for analog.

A special form of time division multiplexing is **statistical multiplexing (STDM).** The most common use of this technology has been the terminal-host configuration, where the terminals attached to the CPU are not always transmitting. The technology is very cost-effective in an environment where multiple computers must communicate directly to other computers, but only a single communications channel is available (non-LAN environment). The time during which the terminals or computers are idle is a lost resource if using TDM and multiple lines. **Statistical multiplexers** are intelligent devices capable of identifying which terminals are idle and which require transmission and allocating line time only when it is required. This means line time is provided only when a terminal is transmitting, making much more efficient use of the line resource, which is often the scarce and expensive resource. There is no allocation of units to channels; however, the identification of the transmitting terminal must be sent with the data.

Clarification
Statistical multiplexing does not share equally but shares according to the current data requirements.

STDM is the beginning of the **bandwidth-on-demand** capability. For example, suppose you have 10 terminals, each connected via 5 Kbps modems (requiring 10 modems at each end). These modems are not expensive, say, $30 each. Twenty of them, however, cost $600. If you use them at a rate of 10 characters per second input (keyboarding) and 1,000 characters per second output (to display half a page of a document), you would have a lot of wasted time on each circuit. You could replace all modems and lines with two synchronous statistical modems running at 56 Kbps (costing about $200) and one line and have sufficient bandwidth. Efficiency is gained at a lower total cost because less equipment is required and there is less idle time. Newer STDM units provide additional capabilities, such as data compression, line prioritization, mixed-speed lines, host-port sharing, network port control, automatic speed detection, internal diagnostics, memory expansion, and integrated modems. (See Figure 1.23.)

Clarification
STDM is a form of dynamic bandwidth allocation.

Data communications networks can use either of the methods of sharing the bandwidth. *Broadband networks* generally utilize frequency division multiplexing for video, data, and/or audio. They divide the frequency bandwidth into multiple analog channels. *Baseband networks* generally use time division multiplexing, dividing the digital bandwidth into multiple logical channels. Whether the network is a local area network providing service to a small number of units (2–50) over a radius of one-half mile distance or to a large commercial network, the considerations are the same. In all of the instances discussed, we shared the frequency or time allocation of a circuit to increase the circuit's utilization and/or reduce cost.

FIGURE 1.23
Statistical Time
Division Multiplexing

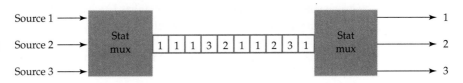

Case 1-1

Data Compression

Len Barker

INTRODUCTION

Data compression consists of reducing storage requirements by eliminating redundancy (repetitive patterns in data). It is used in a variety of applications including sector-oriented disk compression and compression of large data files, data being communicated over networks, backup or distribution files, and sampled voices and graphic data. The technology is advancing quickly, but problems exist, such as the plethora of incompatible and nonstandard compression products, the increased demand on the host processor, and the greater hardware and software complexity it requires. This case helps make sense out of all of this by examining the basics of data compression, data compression applications, and data compression techniques.

THE BASICS

Data compression works because a block of data has an information content substantially smaller than the block itself. For example, you could use an infinite number of bits trying to express p as the decimal number 3.14159265 . . . or you can use just a few bits and represent the value as p. Both the symbol and the decimal number represent the same value, but the character p provides a shorthand for expressing it. You also can use an equation to represent p, once again using far fewer bits than might have once seemed necessary.

The concept of data compression is not new. A familiar example is the system employed in Morse code. Each letter is encoded using a sequence of dots and dashes. The simplest approach would be to specify the 26 letters of the alphabet using a five-bit binary code, with each bit consisting of a dot or a dash. Instead of assigning the same number of bits to each letter, Morse code performs some very simple data compression by using a variable-length code. Samuel Morse knew that the bottleneck to information flow was the speed with which human operators could send messages using this code. By assigning fewer bits to more frequently used letters, he minimized the number of bits that had to be used to encode messages.

Other examples of data compression existing long before computers include

- Secretarial shorthand, where entire words or phrases are condensed to a single symbol.
- Classified newspaper ads, where "DWM seeks SWF" shortens the message.
- Signal flags on ships, where a single pennant can indicate "divers over the side" or "ammunition being loaded."

Data compression has been used for years; however, the techniques and applications used as a result of the computer age are recent inventions.

APPLICATIONS

Data compression is used in many applications, but its use is not always beneficial. Complex hardware and software are required to run the compression algorithms, which also use extra CPU time. Compression is beneficial if a large amount of data is to be communicated over a path that either is very expensive or has low bandwidth. Likewise, if storage space is limited or expensive, compression would be appropriate. Finally, data are often compressed if they either are used frequently or need to be accessed only at human speeds (rather than at database-access speeds). Typical data compression applications in use include

- Sector-oriented disk compression, such as that found in the Double-Space feature of MS DOS 6.0.
- Compression of large data files where appropriate. A good example is the compression of help files in Windows.
- Compression of files for backup or distribution. Programs such as PKZIP often are used to compress files before they are downloaded over phone lines or stored on floppy disk.
- Compression of sampled data in graphics data. Graphics data in particular can effectively overwhelm any system's online storage capability very quickly. Compression techniques designed especially for graphics reduce storage requirements by a factor of 100 and can help with the flood of data.
- Compression of data being transmitted over low-speed networks. V.42bis compression in modems, as well as proprietary techniques used by routers, can help alleviate the bottleneck created when sending data over phone lines.

TECHNIQUES

Data compression techniques are either lossy or lossless. Most compression methods in the past were lossless, which means that the original data can be reproduced entirely from the compressed data. However, the best compression ratio that can be expected from this technique is about 2 to 1. Lossless compression is a necessity for some forms of data, like spreadsheets, where any loss in accuracy is unacceptable.

Lossy compression means that a given set of data will undergo a loss of accuracy or resolution after a cycle of compression and decompression. This fast-growing discipline is usually performed only on sampled voice or graphics data. Lossy compression algorithms typically have a compression factor that can be modified to allow more or less compression of data. By giving up varying degrees of resolution, compression ratios of as much as 250 to 1 are possible. While this generally results in a loss of resolution, the loss of resolution can be adjusted to be completely undetectable.

Entropy Encoding

Entropy encoding is an early lossless technique developed by information theorists. It attempts to do exactly what Morse did by devising coding schemes that use varying numbers of bits depending on the probability that a given piece of data will appear.

The most well-known form of entropy encoding is Huffman encoding. Huffman encoding uses a simple algorithm to develop an encoding scheme for a given block of data, once the probabilities of its various bytes are known. Many variations on Huffman encoding exist, including adaptive schemes that continually modify the encoding scheme based on the character's changing probabilities. Huffman's encoding scheme has been proven to be optimal for fixed-length binary codes. For other types of encoding, better compression methods may exist. But if you are using fixed-length codes for a block of data, the Huffman algorithm will provide you with the best-possible encoding scheme.

Dictionary-Based Methods

Dictionary-based compression was developed in the early 1980s, supplanting entropy encoding as the most popular encoding technique. Lempel and Ziv developed the two most popular techniques, known as LZ77 and LZ78.

Dictionary-based compression takes a completely different approach to compression than entropy encoding does. Instead of trying to efficiently encode each character of a message as entropy encoding schemes do, dictionary-based schemes try to build a code book of commonly seen strings of bytes. Whenever one of these strings appears in a message, the encoder can substitute a simple reference to that string in the dictionary table.

LZSS data compression uses a simple dictionary technique. Its dictionary consists of a window into the previously seen text. Window sizes of 4KB and 8KB are typical. Whenever a string of bytes in a message is identical to a string that appears in the window of previously seen data, a pointer to the previously seen block is passed, along with its length.

LZW data compression operates somewhat differently. Instead of using only the last 4KB or 8KB of the message to form a dictionary, LZW compressors build a dictionary that uses strings of bytes from the entire file or message being compressed. Managing the dictionary may require more work but potentially offers a more comprehensive dictionary of byte sequences.

STANDARDS

Two international standards for lossy graphics compression are known as MPEG, for the Motion Picture Experts Group, and JPEG, for Joint Photographic Experts Group. As the names imply, JPEG is used to compress single graphics images, while MPEG is used for motion pictures. These algorithms are capable of compressing photographic images to ratios of 10 to 1 or greater with no visible loss of resolution. By giving up varying degrees of resolution, compression ratios of as much as 250 to 1 or greater are possible.

Many different schemes have been developed over the past few years by industry associations and individual vendors. The following is a partial list: Motion JPEG, MPEG (I, II, III, IV), Indeo (by Intel), Cinepak, Microsoft Video 1, Fractal (used by Microsoft Encarta), and a revolutionary compression technique by Digital Compression Technology (DCT) that promises to send multimedia data across copper wire at 16 Mbps.

SUMMARY

Data compression will be very important to the future of computing. Increasing hard disk capacity using compression is common as is the compression of commercial software before distribution. However, data compression will play its biggest role in bringing video to the desktop because of the volume of data involved in capturing, storing, and displaying video. The industry is struggling to define standards for compression so that large volumes of information can be delivered to the masses.

REFERENCES

1. "Data Compression, Part Three." *Computer Shopper* 14, no. 2 (February 1994), p. 636.
2. "Fractal Image Compression." *Byte* 18, no. 11 (October 1993), p. 195.
3. *Infoworld* 16, no. 10 (March 1994), p. 82.
4. "Just One Word: Compression." *Windows Sources* 1, no. 10 (November 1993), p. 103.
5. "Small Firm Delivers Big on Digital Compression." *Computer Shopper* 14, no. 5 (May 1994), p. 61.
6. "The Squeeze on Data." *Lan Magazine* 9, no. 5 (May 1994), p. 129.

Summary

This chapter explored the basis and underlying technology that make machine-to-machine communication and networking possible, without requiring a technical background. Based on a concept of what data communications is and the strong effect of noise, the material covers and differentiates between the analog and digital worlds. As much of our world uses analog technology and the analog, but noisy, nationwide and worldwide telephone networks are available for transport of data communications, it is important to understand how we merge the two worlds.

We have introduced the basic models that provide the basis for understanding how communications works. With this understanding of fundamentals, the role of data communications was elaborated. The basic underlying technology that is required for data communications to take place was outlined. Concepts of circuits, channels, and networks were described, as was the importance of coping with noise and its impact on bandwidth and accuracy.

With respect to the concept of channels, it becomes important to be able to create them from the larger resource of circuits. This is only one-half the battle, as we must often superimpose the data of interest onto a carrier to get it through the noisy or distant environment. With these two problems addressed, next is the concern for bandwidth to enable the capacity we desire, whether by native capability or by use of compression.

Transport of digital signals in an analog environment takes technology because of distance and noise. It is, however, worth the trouble as data in a digital form is much easier to deal with, transport, and store. Because of these benefits of the digital environment, it is valuable to understand how to convert and then carry analog signals on a digital channel.

The concepts, technologies, terms, and basics of data communications have been provided. These fundamentals are the basis for understanding the data communications material that follows in subsequent chapters. It is important to build on a vocabulary about data communications as we expand on most of the ideas introduced here.

Key Terms

acoustic, *p. 14*

adaptive differential pulse code modulation (ADPCM), *p. 28*

American National Standards Institute (ANSI), *p. 20*

American Standard Code for Information Interchange (ASCII), *p. 20*

amplifier, *p. 17*

amplitude, *p. 14*

analog, *p. 14*

attenuation, *p. 11*

audio, *p. 13*

bandwidth, *p. 8*

bandwidth-on-demand, *p. 33*

Baudot code, *p. 20*

bit, *p. 21*

bits per second, *p. 16*

broadcast, *p. 17*

channel, *p. 6*

circuit, *p. 6*

code, *p. 19*

compression, *p. 10*

connectivity, *p. 24*

data, *p. 3*

data communications, *p. 6*

data communications equipment (DCE), *p. 7*

data terminating equipment (DTE), *p. 7*

differential phase shift keying (DPSK), *p. 24*

digital, *p. 15*

digital-to-analog, *p. 22*

effective bandwidth, *p. 13*

electromagnetic, *p. 3*

encapsulation, *p. 22*

encode, *p. 12*

encryption, *p. 14*

Extended Binary Coded Decimal Interchange Code (EBCDIC), *p. 20*

frequency, *p. 14*

frequency division multiplexing (FDM), *p. 29*

frequency shift keying (FSK), *p. 22*

full-duplex, *p. 8*

guard bands, *p. 31*

half-duplex, *p. 8*

hardware port, *p. 40*

Hayes-compatibility, *p. 25*

hertz, *p. 14*

interference, *p. 18*

interoperate, *p. 25*

lossless compression, *p. 13*

lossy compression, *p. 13*

machine-to-machine, *p. 6*

message, *p. 3*

modem, *p. 22*

modulation, *p. 17*

Morse code, *p. 19*

MSNDirect®, p. 32

multiplexer (MUX), *p. 32*

multiplexing, *p. 29*

network, *p. 6*

node, *p. 6*

noise, *p. 10*

operating system, *p. 41*

parallel circuit, *p. 8*

parallel port, *p. 40*

path, *p. 6*

phase, *p. 14*

phase shift keying (PSK), *p. 24*

phase shift modulation, *p. 24*

photonic, *p. 3*

piggyback, *p. 31*

polling, *p. 32*

point-to-point protocol (PPP), *p. 26*

protocol, *p. 19*

pulse code modulation
(PCM), *p. 28*
quadrature amplitude
modulation
(QAM), *p. 24*
regeneration, *p. 21*
reliability, *p. 23*
reverse channel, *p. 8*
serial circuit, *p. 8*
serial line internet
protocol (SLIP), *p. 26*
signal-to-noise
(S/N) ratio, *p. 17*
simplex, *p. 8*

spectrum, *p. 16*
standard, *p. 25*
statistical
multiplexer, *p. 33*
statistical time division
multiplexing
(STDM), *p. 33*
subsidiary communications
authorization (SCA), *p. 31*
Telco, *p. 23*
telecommunications, *p. 6*
throughput, *p. 21*
time division multiplexing
(TDM), *p. 32*

trellis code modulation
(TCM), *p. 24*
ultrahigh frequency
(UHF), *p. 17*
Unicode, *p. 21*
universal serial
bus (USB), *p. 41*
very high frequency
(VHF), *p. 17*
voice spectrum, *p. 23*
VoIP (Voice over Internet
Protocol), *p. 28*
voltage, *p. 18*
wireless, *p. 5*

Recommended Readings

ComputerWorld (http://www.computerworld.com/), "America's #1 publication for IT Leaders."

Information Week (http://www.informationweek.com/), "a weekly magazine that combines the goals of business with technology to help you make the strategic decisions that affect your company's bottom line."

For all occasions:

http://www.howstuffworks.com/router.htm, the leading source for clear, reliable explanations of how everything around us actually works.

http://whatis.techtarget.com/, a knowledge exploration and self-education tool about information technology, especially about the Internet and computers.

http://www.webopedia.com/ is a free online dictionary for words, phrases, and abbreviations that are related to computer and Internet technology.

http://searchnetworking.techtarget.com/ is the most active online networking community. It provides networking professionals with information regarding routing, switching, and closely related technologies like security, network/systems management, VoIP, and wireless LANs so they can keep their networks up to date and cope with constant change.

http://en.wikipedia.org/wiki/Main_Page is the free-content encyclopedia that anyone can edit. This English version, started in 2001, is currently working on 648,113 articles.

http://slashdot.org/ is owned by OSTG. OSTG sites provide a unique combination of news, original articles, downloadable resources, and community forums to help IT managers, development professionals, and end-users make critical decisions about information technology products and services.

www.SearchSecurity.com provides IT security professionals with the information they need to keep their corporate data and assets secure. They are the best information source in the IT security market— no other medium has as many e-newsletters or Webcasts dedicated to security. The site offers security-specific daily news, more than 1,000 links to the most useful security sites on the Web, and interaction with leading industry experts.

Management Critical Thinking

M1.1 Millions of modems have been installed in homes and offices. Considering their low speed (bandwidth), what are the implications of continuing them for data communications? Are they just for backup or do they have a primary purpose?

M1.2 Justification for adding capabilities or resources is generally based on a cost-versus-benefit analysis. What are the implications of installing bandwidth without a growth reserve when the analysis cannot justify the cost?

Technology
Critical
Thinking

T1.1 Trace the evolution of modems over the last 15 years. What is the probable next technology for modems?

T1.2 If the organization has company phone lines that will support modems as well as voice, what are the implications of using them for voice-over-data with a modem to share the bandwidth of each phone line?

T1.3 Trace the development of multiplexing. What are the technology advantages and disadvantages for each form?

Discussion
Questions

1.1 Is it practical to use communication devices other than those discussed in this chapter?

1.2 Why is the voice telephone network used for data communications since the data cannot be transmitted in their original digital form?

1.3 If you buy an article at a grocery store and its UPC bar code is read at the checkout point-of-sale (POS) terminal, how is data communications being used?

1.4 List all of the data communications that you have employed in the past week. Separate them into analog and digital modes.

1.5 Does noise seem to be a problem in your world, for example, TV, radio, and so on? What does it come from? How to you reduce it?

1.6 What communication methods were used by George Washington and his British counterpart in the Revolutionary War of 1776?. By the general in the movie *Gladiator*? By the allied field commanders in Iraq in 1991 and 2003?

1.7 Discuss the analog communications used in your school or organization. Could they and should they be converted to digital format? What effect does noise have?

1.8 How would the conversion in 1.7 be done?

1.9 Some telephone sets have music-on-hold built in so that the other party will be entertained when the caller places them on hold. How does this work?

1.10 What is your prediction for the future of analog voice versus digital within the industrialized world? Lay out a time line for your predicted evolution.

Projects

1.1 Check out Radio Shack or the Internet and determine the cost of building an analog-to-digital converter for voice that will work on a computer. Can you now use this over a modem? How much bandwidth does it take? How much security does it provide?

1.2 Research the bandwidth requirements for the following disc protocols: CD, SACD, and DVD-A.

1.3 Investigate the compression method for MP3 music. Is it lossy or lossless?

1.4 Compare and contrast the modulation used in television audio: stereo versus surround sound.

1.5 Determine the multiplexing used in television and radio.

1.6 Call the cable TV provider and determine all of its services. Does it have digital radio, stock quotes, bidirectional communications? What media does it use?

1.7 Find a geographically dispersed retailer using multiple POS terminals and determine bandwidth requirements for collection of sales data.

1.8 Visit a bank and determine how its ATMs are connected to the bank and to the telecommunications networks such as AVAIL, CIRRUS, HONOR, ALERT, and others that provide access to other banks. What type of media, protocol (rules or standards of communications), and speed does each use?

1.9　If your state has a lottery, what type of communications does it use to the ticket outlets? First ask the local ticket seller and then the central lottery agency for the state.

1.10　Put the terms *multiplexing* and *modulation* into the search window of an Internet search engine. How many results did you receive? How many of the first 20 were informative? If using *Google®,* what are the responses on the right side of the page?

Appendix 1-1

Ports on a Computer

In computing,[15] a port (derived from seaport) is usually an interface through which data are sent and received. An exception is a software port (derived from transport), which is software that has been "transported" to another computer system.

HARDWARE PORT

A **hardware port** is an outlet on a piece of equipment into which a plug or cable connects, that is, a connector. For instance, most home computers have a keyboard port into which the keyboard is connected. Hardware ports can almost always be divided into two groups: Those that send and receive one bit at a time via a single wire are called serial ports; those that send multiple bits at the same time over a set of wires are called **parallel ports.**

NETWORK PORT

A network port is an interface for communicating with a computer program over a network. Network ports are usually numbered and a network implementation like TCP or UDP will attach a port number to data it sends; the receiving implementation will use the attached port number to figure out which computer program to send the data to. In TCP and UDP, the combination of a port and a network address (IP number) is called a socket; for example, the list of well-known ports (computing).

Alternatively, in a TCP/IP network,[16] it is a number assigned to a type of service. The port number is included in all transmitted packets to link incoming data to the appropriate application. For example, port 80 is the standard port number for HTTP traffic (Web traffic), and port 80 packets are processed by a Web server. Port numbers range from 1 through 65535. The software that responds to its port number is said to be "listening" for its packets. More accurately, the term should be "looking" for its packets because it is comparing numbers, not "listening" to numbers.

EXAMPLES OF HARDWARE PORTS

Manufacturers place connectors, or ports, on the back, and now also the front, of computers by which we can connect to the systems bus and motherboard from external networks and peripherals. First we discuss the wired ports that are commonly found on computers.

---

[15] http://en.wikipedia.org/wiki/Port_number.

[16] Techweb (http://content.techweb.com/encyclopedia/defineterm.jhtml?term=TCP/IP+port).

FIGURE 1.24
**Rear View of
Computer Showing
Ports**

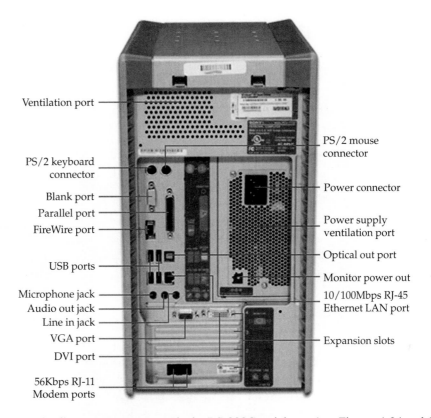

Ventilation port

PS/2 keyboard
connector

Blank port

Parallel port

FireWire port

USB ports

Microphone jack

Audio out jack

Line in jack

VGA port

DVI port

56Kbps RJ-11
Modem ports

PS/2 mouse
connector

Power connector

Power supply
ventilation port

Optical out port

Monitor power out

10/100Mbps RJ-45
Ethernet LAN port

Expansion slots

The first connector or port is the RS-232C serial port (see Figures 1.24 and 1.25). This nine-pin or 25-pin female connector is designed to transfer all data sequentially, or serially, to the device connected to it. Most PCs have serial ports, noted as COM1 and/or COM2, but they and the parallel ports are being replaced with USB ports (discussed later). Although they may be used for peripherals such as the keyboard or mouse, they can be used for external connectivity. For example, although it is preferable to have an internally installed modem, connected directly to the motherboard and bus, we can have an external serial port–connected modem. The good thing about the serial port is that it is on most machines and can be used for a variety of connections. The weak point is that it has a limited bandwidth, originally only 20 Kbps. However, using the Windows (network) **operating system** or other software, one can connect two or more computers via their serial ports and communicate at this slow speed.

Another of the original hardware ports found on PCs is the parallel port. This 25-pin male connector is designed to communicate on eight or nine wires in parallel, thus transferring a full byte at a time. The parallel port, designated as LPT1 or PRN, is generally used to communicate on an N = 2 network to the printer. It was used before the advent of USB and FireWire for multiple tasks concurrently, such as external storage. This port is most often used for peripherals and not for network communications.

Probably the most used port on the computer is the **universal serial bus (USB)** port. This small, rectangle connector in its initial form, USB 1.0, has a speed of 12 Mbps; USB 2.0 has a speed of 480 Mbps. It is designed to overcome the bandwidth restriction of the RS-232C standard serial port. An additional feature of the USB port is that it can be connected to a USB hub, allowing multiple connections to the computer. The connection can be expanded, possibly by daisy-chaining, to a total of 127 lines into one USB connector. The USB protocol is such that when a device is connected to a USB port, the systems automatically recognize one another and begin operation. Because of its expanded bandwidth

FIGURE 1.25
**Ancillary Ports on
a Computer**

DVD+RW/
CD-RW drive

CD-ROM drive

Floppy drive

Memory Stick®

S-Video in   Video in   Audio   Audio   i.LINK 4-pin   USB 2.0 ports
left   right

and popularity, manufacturers include multiple USB ports on the front and back of computers and on printers, cameras, and other user peripherals.

The IEEE 1394, or *FireWire* (also called i.LINK), port supports data transfer rates of up to 400 Mbps. These ports are found on devices that, like USB, wish to communicate at near hard drive speeds. FireWire requires a 1394 port and can, like USB, extend this single port to up to 63 devices. Like USB, FireWire ports are becoming common on portable devices for downloading files such as images and video.

Users can add a network interface card (NIC) that connects inside the PC directly to the motherboard and internal bus; some manufacturers put the NIC directly on the motherboard. This connection provides the maximum bandwidth to the NIC, with the NIC determining the bandwidth to the external world. NICs generally operate at 10 and 100 Mbps even though the network speed may be much faster. Figure 1.24 shows the external RJ45NIC port.

A final set of ports are for the sound card and video card. Sound uses mike in, line in, and speaker out. These allow external inputs such as a video camera to place sound on the sound card and allow external speakers to give fullness to sound. The usual video out is VGA (video graphics adapter), where digital image data are supplied to the video card, which then makes a D/A conversion and provides good-quality analog color signals to the monitor. An upgrade of VGA is AGP (advanced graphics port), which has better analog speed and definition. A further upgrade, the DVI (digital video interface), accepts digital data from the bus to the video card and keeps it digital, passing it on to a DVI port on the monitor. DVI is presently (circa 2005) the highest-quality video signal on PCs.

Figure 1.25 shows additional ports that may appear on select PCs, which range from game ports to digital sound input. S-video-out allows the analog video output to go to a TV set, using the good definition of S-video; S-video-in accepts TV signals from an external source. S/PDIF (Sony/Philips Digital Interface) is a standard audio file transfer format developed jointly by the two companies that allows the transfer of audio signals from one to another without having to be converted first to an analog form. This port, like the Memory Stick provided on Sony machines, is likely to be particular to specific brand equipment.

The previous section described wired ports. We now describe a wireless port, followed by descriptions of other wireless connectivity. They occur in two forms: infrared (IR) light and radio frequency.

The infrared (IR) port is covered with plastic and there is no way to insert a connector. The port first appeared on laptops and other portable devices as a way to connect wirelessly to printers. IR ports have become standard on all portable and moveable devices, allowing users to communicate with and synchronize PDAs, laptops, and desktops at 100+ Kbps.

Just as they can add a modem to the computer for POTS communications and have an IR port for wireless data transfer, users can add a wireless modem, which is really a wireless network interface card. Where a standard NIC uses unshielded twisted-pair wire for the channel, the wireless NIC uses radio frequencies. The IEEE 802.11b wireless modems, also known as WiFi (*Wireless Fidelity*), operate in the 2.4 GHz spectrum and have bandwidth of about 11 Mbps at distances from several dozen to several hundred feet (10–200 meters). Their obvious advantage is that they can roam within their area of coverage and remain connected. The other end of the wireless connection is often a wireless adapter that then connects to a wired network. Thus, the wireless unit is an extension of a standard wired network. The *Intel Centrino*® capability incorporates IEEE 802.11b interior to the laptop computer.

The other primary wireless protocol that will impact us is *Bluetooth,* which is designed to work over 10-meter's distance to replace short wires and synchronize portable devices.

Bluetooth and Wireless Fidelity standards are of great importance as predictions indicate that by the end of 2005 there will be almost one billion handheld devices used by enterprise employees. Employees are expected to gain increased productivity by scheduling, e-mail access, e-m-commerce, GPS, and other applications that enable the "wireless workforce."

Clarification

While all externally powered equipment will have an AC (alternating current) plug, it is not a communications port.

Chapter Two

Media and Their Applications

We outline the electrical, electromagnetic, and photonic circuits and channels. The common wired media and technologies are all detailed. Wireless forms and technologies are covered and recent wireless developments are included as the area continues to evolve rapidly.

In this chapter, we introduce basic telecommunications technology in the form of media used for voice and data communications and some major applications. All transmission must be carried by some **medium,** whether copper, glass, or radio or light waves in space. Each has a capacity and susceptibility to **noise.** (Although much of the technology may seem outdated, to understand the value of the technology in use today, one should understand the historical background).

Since we are discussing the physical media circuits onto which channels can be placed, we are discussing the *physical layer of the OSI model.* (See Figure 1.18 in Chapter 1.) Regardless of the source of the signal, it must be converted to a physical form— for example, electrical, electromagnetic, or photonic—to be transported. This chapter discusses these physical media forms. The OSI model is explored in the next chapter.

We also continue the discussion of the components of the communications model presented in Chapter 1. In Chapter 1, we discussed the nature of the signals from sender to receiver. Here, we concentrate on the channel between sender and receiver and the media used therein.

When we discuss voice transmission in this chapter, the emphasis is on analog. However, today most voice transmissions use some digital circuits. We will discuss digital technology in Chapter 4 and concentrate on analog concepts here.

communications media
The physical paths through which signals are transported.

CIRCUITS AND CHANNELS

communications circuit
A physical path between nodes over which a signal can travel.

communications channel
A logical path between nodes.

A communications circuit is a **path** over which a signal can travel. **Circuits** are generally thought of as something physical, like wire or a radio wave. A **channel** is the actual path for the signal and may occupy the total circuit or a portion of it. For example, a TV **cable** is a circuit (physical coaxial cable) and has many channels for the TV programs. Some media include paths in both directions, either simultaneously or in one direction at a time. A circuit for a voice channel can be established using several media. We discuss each medium fully as we progress through the book, but it is prudent to show the alternative media available at this point.

Electrical

Twisted-pair wire is the simplest **electrical** circuit; it is a pair of wires, lightly twisted to reduce noise. The other electrical circuit, coaxial cables, is logically the same as twisted-pair wire except one (ground) wire surrounds the second wire.

Electromagnetic

Electromagnetic circuits use radio waves in some form. Microwave radio channel circuits are provided over radio channels between line-of-sight transmitters and receivers. Satellite circuits are the same as microwave channels except that the signal is received, repeated, and forwarded to the receiving unit by a receiver-transmitter (satellite transponder) located in space. Omnidirectional radio systems use transmitters that radiate (transmit) in all directions. All wireless systems use either electromagnetic (radio) or infrared (light, see next) as their carrier.

Implication

All wired media require space and security, and, at some point, electrical power. As bandwidth is valuable, the media on which it is transported have a cost.

photonics
The technology of generating and harnessing light and other forms of radiant energy.

Light (Photonic)

Infrared systems use light as the carrier. Fiber optic cable uses glass or plastic, whereas unguided systems use air or space. Each medium has different characteristics and capacity. We discuss photonic media later as they are used primarily as data communications and control systems.

WIRED (CONDUCTED) MEDIA AND TECHNOLOGIES

twisted-pair
Common copper wire medium that is used for most telephone circuits.

Twisted-Pair

A circuit of copper wires is a loop with a single wire going from the sender to the receiver and back. This single loop is seen as a pair of wires, but a complete loop out and back is required for electrical conductivity. If the pair of wires is free from electromagnetic **interference,** it can carry a frequency bandwidth of about one megahertz analog. Thus, a telephone system copper wire circuit would seem to be able to carry a very large number of analog conversations. Noise, however, is a problem that affects the capacity of copper wire circuits, especially as distance increases. One basic method of reducing susceptibility to noise occurring in the form of injected, unwanted electrical signals is to twist the wire, producing the popular term *twisted-pair* or **unshielded twisted-pair (UTP).** The twisting causes cancellation of the injected signal. (See Figure 2.1.) In other words, the noise signal is induced onto portions of wires that twisting has placed in reverse order, canceling the noise signals because they are out of phase with each other. This procedure also prevents the copper pair from radiating signals that would cause interference with nearby circuits.

We constantly encounter wires carrying electrical current. The wire on a hair dryer, the speaker wires on the stereo, the toaster cord, the lamp cord, the long cord on the vacuum cleaner, the power cord on the television, and the cord on the clothes washer, dryer, and iron are just a few. All but one of these carry current at 110–220 volts potential and result in electrical power being conveyed to a device to do work. The amount of power in the devices mentioned ranges from 2 watts for the stereo speakers, to 60–100 watts for the lamp, 1,200 watts for the hair dryer, and 5,000 watts for the electric clothes dryer. These wires, except for those of the stereo speakers, carry sizeable amounts of electrical power, at dangerous **voltage** levels.

Telecommunications and data communications circuits carry information via electrical voltage and current. The levels of voltage and current in the circuit, however, are very small compared with powered devices in the home. The voltage used by communications networks is in the range of 0 to 20 volts. This is the same range used by the telephone system except for a 90-volt ringing signal, and the same as a stereo speaker system and even an automobile battery (pre–42 volt adoption). Unlike stereo and auto circuits, data communications channels carry micro- to milliamperes of current and convey micro- to milliwatts of power. Telecommunications circuits have the same voltage and current characteristics as devices in the home, but the voltage level is not dangerous. Communications circuits are designed to reliably transmit low-voltage signals over a long distance.

FIGURE 2.1
Twisted-Pair Wire

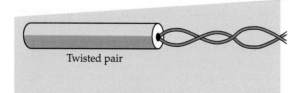

Twisted pair

Implication

There is a large amount of in-place/installed UTP. It is, therefore, either a resource for alternate uses or a sunk cost.

Plain Old Telephone System[1] (POTS)

One of the most-used methods of voice and data communications is the **plain old (wired) telephone system (POTS).** During the nineteenth and twentieth centuries, AT&T and other phone companies ran millions of miles of UTP copper wires from homes and offices to their **central office (CO),** the place where a **switch** connected the calling and called parties. The circuit from the residence/office wall to the central office is UTP and called a **local loop.** If the called party was not on this switch, a cable carried the call to another central office or to the office of a **long distance provider** and then on to the central office connected to the called party. Initially, all of these cables were copper wire; long distance is now handled primary by fiber cables.

Clarification

Plain old telephone service (POTS) is basically hardwired switching systems as opposed to software-controlled systems—often referred to as signaling systems.

POTS is an extensive UTP network that connects almost all home and offices in the industrialized world. Therefore, these paths can be used for data communications. A problem with POTS is that it was designed for analog signals and is susceptible to noise. That is why we must use modems, as we describe in more detail shortly.

Telephone Channel Capacity

The *analog* **bandwidth** *(the range of frequencies) for a telephone channel is 4,000* **hertz** *(Hz),* or 4 kilohertz (KHz). The **voice spectrum** within the telephone channel covers the frequencies from 300 to 3,000 Hz and is 2,700 Hz wide, with the remainder of the channel space being allocated to guard bands[2] that provide separation space. Note that the bandwidth is not related to the actual frequencies used for voice transmission, only with the difference between the upper and lower limits of the range. This range of 2,700 Hz is adequate for voice communications but becomes a limit when we wish to carry more analog or digital information, such as high-fidelity music or digital data at a high rate of speed. While the **public-switched telephone network** provides access to almost every home and office in the developed world, the size of this *information pipeline* is restricted, and, thus, so is the speed or quantity of information flow. (As with the garden hose, one can pump only a limited volume though a small pipe.) In addition, much of the **spectrum** is not available because of equipment constraints and not the medium.

Digital Subscriber Line (DSL)

Unshielded twisted pair (UTP) is still a valuable resource. Its native analog bandwidth is about one megahertz, although POTS uses only a small part of that. Since there is so much UTP deployed in developed countries, there is an opportunity to use the available greater analog bandwidth to create wide digital bandwidth. That's what **digital subscriber line (DSL)** does. DSL, in several forms, takes advantage of the total bandwidth of the circuit and gives great digital bandwidth, often with several channels in that bandwidth.

Clarification

Digital subscriber lines (DSL) are digital services over analog POTS circuits to increase performance on the local loop.

[1] The information herein is a cursory overview of the telephone system. POTS is more fully covered in an appendix at the end of the book.

[2] As noted in the "Introduction" chapter, guard bands separate one channel from interference from another by providing frequency spacing.

How Does DSL Work?

In the DSL environment, modem-equivalent technology is placed at each end of the local loop. If there are no repeating amplifiers on the local loop, we can use the total analog bandwidth and can carry 250 voice channels (1 MHz/4 KHz = 250 channels). Because we don't need the guard band of each telephone channel, we can use the total 4,000 Hz of bandwidth. Therefore, we have 250 channels of 4 KHz bandwidth each. If you think of the technology as placing a modem on each of the channels, you would have 250 times the digital bandwidth of each resulting channel. For DSL, the digital bandwidth of each 4 KHz channel is 60 Kbps, for a total theoretical bandwidth of 250 * 60 Kbps = 15,000,000 bps = 15 Mbps. Because there is noise on the circuit, some channels are reduced in bandwidth or unavailable, with resulting lower bandwidth. Basic DSL **standards** call for no active devices in the circuit, for example, no repeaters, and a distance limitation of 18,000 feet. In practice, DSL providers create three channels at a much lower bandwidth, one for full-time analog telephone and two for data transport, such as Internet access. One data channel is generally slower-speed upstream (from user to provider) and the other is for hi-speed downstream (from provider to user). We discuss the particulars later, but this technology allows a resource that is in place to be used to compete with high-speed cable modem access to the Internet and multiple multiplevoice and data channels to a small or home office (SOHO).

Implication   Using an updated DSL standard with distance limitations of 9,000 feet, a bandwidth of 27 Mbps can be provided.

Integrated Services Digital Network (ISDN)

By placing DSL technology on each end of the local loop, we can turn the analog circuit into a multichannel digital telephone capability and provide two channels for voice or computer connectivity and one for data and control. In the **integrated services digital network (ISDN)** environment, the circuit for the call is converted from analog to digital and is directed by a digital switch. Next, the circuit (e.g., local loop) is time-division multiplexed into three channels. These are two **bearer (B) channels** of 64 Kbps each for voice or data and one delta (D) channel of 16 Kbps for control, resulting in a circuit of 144 Kbps bandwidth. (An additional 16 Kbps of bandwidth are allocated for overhead, bringing the actual total bandwidth to 160 Kbps.) When a B channel is used for voice, the analog voice is converted to an electrical analog equivalent and then converted as described above to its PCM digital equivalent. While this is being sent on one B channel, the other B channel may be used for computer-to-computer communications. This latter capability is as simple as plugging the serial port of a computer into the serial port on the ISDN telephone and bringing up the software to cause communications, all while you speak on the phone. ISDN phones are special; they can receive Caller ID® and can handle Call Waiting®, among a long list of other services.

Clarification   **Integrated services digital network (ISDN)** is a digital form of switched circuits with higher transmission speeds; a form of DSL.

Table 2.1 shows the comparison of analog POTS and digital ISDN telephone communications. In each case, the local loop is a single twisted-pair copper pair from the switch to the premise wall. For POTS, an analog bandwidth of 4 KHz is used, with a voice channel inside of 2,700 Hz. This voice channel, when used for data communications, can be encoded via a modem and carry 56 Kbps. ISDN has a bandwidth of 144 Kbps, which is multiplexed into three channels, two of which have a bandwidth of 64 Kbps. Although there is some overhead, the *two B channels* provide a guaranteed **full-duplex** bandwidth.

TABLE 2.1
POTS versus ISDN Telephone

Characteristic	POTS	ISDN
Local loop	Twisted-pair, copper	Twisted-pair, copper
Channels	1	3
Bandwidth	2,700 Hz analog	160 Kbps digital
Noise level	Medium	Low and controllable
Signaling	Touch-Tone®	Data channel

Coaxial Cable

coaxial cable
Wire surrounded by a shielding sheath to reduce noise.

A **coaxial cable** is similar to a pair of copper wires with the exception that one wire is a braided or solid sheath that encompasses (shields) the other wire. Insulation material separates the two wires, the center wire is at high signal level, and the shield wire is at ground (zero volts) potential. The grounding of the outer **shielding** wire means that interference cannot easily penetrate the coaxial cable and induce noise onto the circuit. Shielded/coaxial cables also do not radiate signals to other circuits as nonshielded circuits may. A coaxial cable provides a much larger analog and/or digital bandwidth. The cable does cost significantly more than UTP, and the signal attenuation per mile is much greater than for twisted-pair circuits. Tables 2.2 and 2.3 show characteristics of various media.

Multiple Channels Using Multiplexing

multiplexer (MUX)
Device that allows a single communications channel to carry multiple transmissions.

Coaxial cables became increasingly common through their use for cable television (CATV), which sends 50 or more full-color television channels to both homes and offices. The channels are created from the available 350, 550, or 750 MHz bandwidth by using frequency division multiplexing (FDM). By dividing the spectrum into 6 MHz channels, the cable TV provider is able to send 50, 80, or 115 channels respectively. The number of channels can be increased further, the resolution of each channel can be increased, and the noise level of each channel can be reduced by converting the transmission from analog to digital signals. By using modems at each end of the coaxial cable, a 6 MHz analog channel can be converted to a 27 MHz digital channel. This allows the carrying of at least three standard-quality digital TV channels per analog channel by using time division multiplexing (TDM) on the digitized channel. This technology provides the potential of over 300 digital channels on a single CATV cable.

Cable TV has historically been forward-only **simplex** transport. With the advent of digital channels, the potential existed to use this technology for Internet access. However, a return channel was required because Internet access requires at least **half-duplex** operation.

Clarification

The analog coaxial cable installed for television was a fixed resource; installing more was very costly. Therefore, cable companies used digital technologies to increase the number of channels, effectively increasing the size of the resource.

TABLE 2.2
Bandwidth and Speed of Telecommunications Media

Circuit Media	Analog Bandwidth (MHz)
Twisted-pair	1
Coaxial cable	350, 500, and 750
Microwave radio	30 (one channel)
Satellite radio	6 (per channel)
Omni-radio	0.010 AM, 0.2 FM
Television	6 (per channel)
Telephone channel	0.004

TABLE 2.3
Characteristics of Circuit Media

Wire

1. 22–26 gauge copper
2. Low bandwidth
3. Used for virtually all local loops
4. Low installation cost
5. Susceptible to noise

Coaxial Cable

1. Ground is shielded (immune to interference)
2. Bandwidth of 350 MHz, 500 MHz, 750 MHz
3. Up to 10,800 voice conversations
4. Amplifiers every mile
5. 50–100 analog TV channels/cable
6. Cable tapped easily; low to medium security

Microwave Radio Terminal

1. 4–28 GHz frequency range
2. Up to 6,000 voice circuits in a 30-MHz-wide channel
3. Line of sight—20–30 miles between towers
4. Mostly used for analog
5. Subject to interference by rain
6. Must have FCC license, regulated
7. No right-of-way permit required; great for building-to-building within city

Satellite Radio

1. Uplink and downlink each 22,300 miles (geosynchronous orbit)
2. Footprint is one-third of earth
3. Propagation delay = 44,600 miles/186K mps = 0.2398 second
4. Most common carriers have left satellite for terrestrial
5. Only security is encryption

Omnidirectional Radio

1. Wireless—replaces wires and cables
2. Passes thru walls
3. Very localized
4. Easy to install; easy to move

Infrared (Photonic)

1. Within a set of walls (room)
2. Omnidirectional
3. Low speed
4. Easy to install; easy to move
5. Not secure
6. Very localized

Fiber-optic Cables

1. Made of glass or plastic
2. Difficult to splice
3. Secure due to items 1 and 2
4. Very high speed
5. Unidirectional strand
6. Difficult to split signal
7. Immune to RFI, EMI, crosstalk
8. Most expensive, greatest bandwidth = low $/bit

Cable Modem for Internet Access

Probably nothing has captured the imagination of people like the Internet and the World Wide Web. We discuss the Internet in more detail in Chapters 9 and 10. When computers were first introduced in the home, the only external access was via modems to *electronic bulletin board systems (BBS)*. These electronic bulletin boards were other computers set up to hold shareable programs and information in small DOS files. Once the Internet and its graphical access through the WWW and the browser became popular, far more bandwidth was required to download the Windows files, text pages, and graphics. Text requires little bandwidth, while graphics and large files are bandwidth hogs. Where a 2,400 bps modem was sufficient for a BBS or text e-mail, Internet users want much higher bandwidth while downloading graphical content so they do not have to wait. A bandwidth 500 times the 2,400 bps (1.2 Mbps) is not unusual for cable or DSL download speed.

> DSL and cable modem speed are provider dependent. Generally, downstream bandwidth is 1.5 Mbps (T-1) or higher, reaching 4.0 Mbps in some areas.

The introduction of cable modems for Internet access in the United States was quite natural because, by mid-2000, of the 105 million homes in North America that are passed by CATV coaxial cable, more than 75 million were cable TV subscribers. Since one in two homes in the United States had a microcomputer, a ready market existed with the addition of the modem and the digital infrastructure.

> Coaxial cables come in four types: 350, 550, 750, and 1,000 MHz.

Cable modem technology includes equipment that replaces or supplements the computer's modem. The allocation of one low-band channel (below 50 MHz) for upload (about 250 Kbps) and one or more high-band (above 900 MHZ) channels for download (1–3 Mbps) of data is provided by the CATV company. The actual downstream speed of Internet access on a cable modem system is dependent on the inherent speed of the download channel and the number of users online because users share this bandwidth. While the speed is superior to a standard modem, sometimes 100 times faster, users may experience much less bandwidth during times of heavy use. Cable modem bandwidth is shared by up to 5,000[3] users per branch on an on-demand basis. Although several hundreds (or thousands) of users may be active, if a request occurs during a lull, very high bandwidth will be available.

Clarification   DOCSIS® (Data Over Cable Service Interface Specification) defines **interface** requirements for cable modems involved in high-speed data distribution over cable television system networks.[4]

Fiber-optic Cable

> **fiber-optic cable**
> Cladded glass or plastic fibers that transmit data via light pulses.

Single Mode

Optical fibers can be used in both the analog **broadband** and digital baseband modes, but, in the telecommunications arena, they are used almost exclusively in the baseband mode for digital communications. Fiber-optic circuits are one-way communications paths, with a light source (laser or light-emitting diode) at one end that pulses on and off and a light detector at the other end. The medium is a very thin, high-quality glass or plastic fiber wrapped with protective coverings (see Figure 2.2). The speed of communications is extremely high, potentially one terabit per second, and the circuit is impervious to electromagnetic noise. Since this medium uses light as opposed to electrical signals, it neither radiates nor receives electromagnetic interference. **Crosstalk** (radiation of signals from one circuit to another) from cable to cable that occurs in twisted-pair circuits is nonexistent. As noted in advertisements for long-distance telephone carrier services, fiber-optic circuits are being installed wherever possible, often replacing microwave and satellite circuits, to improve transmission quality and bandwidth.

Clarification   Broadband is where two or more signals share a medium through FDM or TDM. Baseband in the digital environment means the whole channel is used without multiplexing.

Caveat   The term *broadband*, however, is used for all high-speed communications.

---

[3] A single 6 MHz analog TV channel can be converted into a 27 Mbps digital channel. With 5,000 users (rule-of-thumb) active 0.5 percent of their time online, they would receive about 1 Mbps each.

[4] http://www.cablemodem.com/.

FIGURE 2.2 **Fiber Optics**

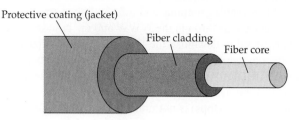

Fiber-optic cable (glass or plastic)

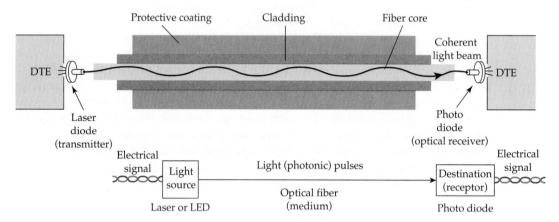

Circuits that use fiber-optic capabilities often begin and end with some other media. For example, if you have a computer or terminal on a network that uses optical fiber as the medium, the signals originate as electrical pulses. They would then be passed as parallel signals on the originator's data bus to a device that transforms them into a serial stream, still electrical. These serial electrical pulses activate a laser or light-emitting diode that sends pulses of light onto the fiber media. At the other end of the fiber, a light-sensitive element detects the presence or absence of light at precise times and generates electrical pulses accordingly. Thus, until the originating and receiving units are photonic devices, there must be an electrical-to-light-to-electrical transformation.

Implication

Fiber strands/cables are inherently secure as the fiber (a) does not radiate signals and (b) must be cut to be shared, thus no eavesdropping; cutting the fiber signals a breach of security.

Fiber-optic cables are small, sometimes being added as a part of larger twisted-pair or coaxial cable bundles. Reduction of production costs, coupled with high quality and fast transmission, makes fiber an excellent medium. The difficulty in splicing to itself and to other media, however, increases cost of use. Fiber optics will become the medium of choice for high-bandwidth requirements of the future as essentially all telecommunications convert from analog to digital modes.

Implication

The native bandwidth of fiber is very large. Out ability to tap into the bandwidth depends on the boxes at the end of the fibers. Thus, the addition of new technologies at the end of the installed cables increases the value of the resource.

Multimode Fiber

We normally assume that there is a single "pipe" of light on a single strand of fiber. With such an environment, the prediction is that we can achieve a bandwidth of about one

FIGURE 2.3
Switch

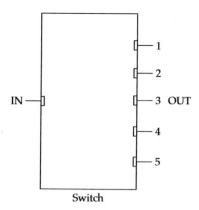

Switch

terabit per second, or one thousand gigabits per second, which is 1,000,000 megabits per second. That seems like a lot of bandwidth. However, in high-traffic channels, the demand actually may exceed this bandwidth. What would you do in this situation? You either have to have spare (dark) strands or install new strands at a cost of about US$330,000 per kilometer. Otherwise you need to use technology to enhance the bandwidth available.

With different technology at each end of the fiber strand, pipes of different light can be added to the same strand, presently up to 64 different light waves, to increase the bandwidth of a single strand by 64 times. This is called *dense wave division multiplexing (DWDM)*, providing a predicted bandwidth of 64,000,000,000,000 bps. DWDM could provide sufficient bandwidth to carry 3.2 million HDTV channels simultaneously. While most might not wish to transport 3.2 million channels, as indicated in the section that follows, some may wish to narrowcast one of 100,000 prerecorded programs to each of 3.2 million viewers.

Clarification

Multimode fiber can carry a greater bandwidth than single-mode but for shorter distances.

Switching of Optical Signals

Historically, optical signals were switched electrically. This required an optical-to-electrical conversion, electrical switching, and electrical-to-optical conversion. If error checking of the data was required at the switch, it was done in the electrical environment. However, when no data checking was required, these conversions just slowed down the switch process. See Figure 2.3.

Clarification

Switching is the process of moving a signal from one incoming line or **port** to one of several outgoing ports or lines.

Because of the worldwide deployment of fiber-optic cables, some of it is "dark," meaning it is waiting in reserve for a use. This wide deployment has greatly reduced the cost of bandwidth. The major costs for using fiber are splicing the strands and the higher-bandwidth terminating equipment.

Texas Instruments pioneered the development of optical switching, producing microelectromechanical (MEMS) mirror arrays. One product on the market notes that "Cost-effective transparent connectivity is possible thanks to an optical switch core module that routes light from any of 80 input fibers to any of 80 output fibers. Developed by *Glimmerglass®*, the palm-sized subsystem is based on 3D microelectromechanical mirror arrays. Called *Reflexion,* it can switch signals within 10 ms."[5]

Hybrid Fiber/Coax (HFC)

Fiber was used initially as a hybrid combination with coaxial cable, called **hybrid fiber/coax(HFC).** Consider that CATV companies had strung coaxial cable to millions of homes in the United States but that the bandwidth of coaxial cable is small compared with

[5] http://www.elecdesign.com/Articles/Index.cfm?ArticleID=5942.

optic fiber strands. The combination of the two media allows for high-bandwidth fiber to the neighborhood and the use of existing coaxial cable from the neighborhood to the home, providing a good trade-off between the cost of installation of new media and the exploitation of present media.

Clarification   **Hybrid fiber/coax (HFC)** is fiber to the neighborhood and coax to the home to bring higher bandwidth and lower noise.

Wire-based CATV systems presently use coaxial cables with frequency bandwidths of either 50–350 MHz, 50–750 MHz, or 50–950 MHz. This limits the number of analog channels that can be delivered to 50, 100, and 150 respectively. To increase the number of channels, there are two choices: (1) use a medium with greater bandwidth or (2) digitize the signals and multiplex three digital channels into one analog channel.

The first choice of medium would seem to be optical fiber cable as its bandwidth, for all intents and purposes, is unlimited. All new replacement fiber cable, however, would have to be installed from the signal source to each premise. This is a very expensive alternative, especially the "last mile." A competing alternative is to install optical fiber from the signal source to neighborhoods of from 200 to 1,200 houses and then distribute the signal to the residence via the existing coaxial cable. This creates a hybrid installation of fiber-optic strands and coaxial cables (*hybrid fiber/coax*). (See Figure 2.4a, b.) When the length of the coaxial cable is limited to a thousand feet or less and is in good condition, a greater bandwidth can be achieved. This is part of what gives HFC greater ability.

Implication   The "last mile" is a term to describe the distance from the neighborhood distribution circuit to the residence/office. This is the most difficult/expensive to install because it generally involves personal property.

Implication   HFC **telephony** can provide integrated support for video and telephony.

The advantage of HFC to the CATV company is lower cost by using the part of the existing system that would be the most expensive to replace while delivering more services and channels via the new fiber distribution cables. Additionally, the newly installed fiber-optic cables negate the need for most amplifiers in the distribution cables and are far less susceptible to disruption by electrical storms. The hybrid configuration provides a higher-quality signal and lower maintenance costs. A bidirectional signal is provided for (interactive) commands from the users.

Clarification   HFC is the best of two worlds. Fiber provides good bandwidth at low noise and coax provides delivery at low cost.

The second method of increasing the number of channels on an existing coaxial cable is to transform the analog channels into a digital environment. This is done by creating "modems" on each end of a specified channel. These modems (the receiving one is in the TV's set-top box) create a digital environment of 27 Mbps, which is enough for three or more NTSC[6] TV channels, one HDTV and one NTSC channel, or Internet **traffic.** Thus, with the addition of technology at each end of the coaxial cable, the number of channels can be increased without the installation of additional media.

Clarification   In many CATV systems, all TV channel and Internet access is brought to the neighborhood via fiber cables. It may branch in the neighborhoods but is eventually converted to coaxial cable for the entry into the residence/office.

[6] National Television Standards Committee, or standard definition, television.

FIGURE 2.4   Hybrid Fiber/Coax

a. Basic HFC for video and voice

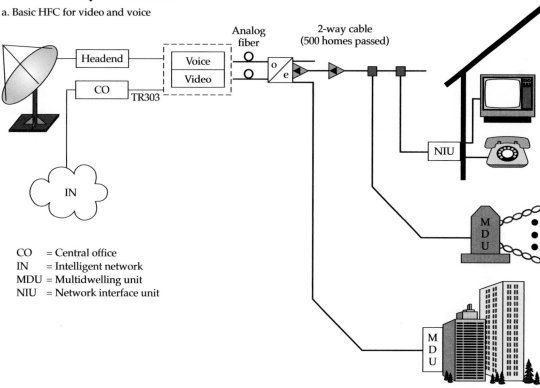

CO   = Central office
IN   = Intelligent network
MDU = Multidwelling unit
NIU = Network interface unit

b. HFC bandwidth

Downstream bandwidth = 700 MHz = 110
Analog channels (50–750)

Upstream bandwidth = 37 MHz (5–42 MHz)

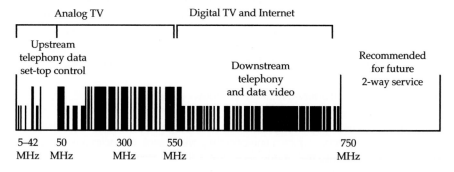

Fiber to the Curb (FTTC); Fiber to the Premise (FTTP)

As HFC and digital channels and compression (discussed later) increase the number of channels in existing systems, the ultimate circuit to the curb is a pair of fiber optic strands from the source to a point very near each residence. Connectivity from that point into the residence will be less important but generally more expensive. This will provide significantly greater bandwidth in both directions and make the services' offerings restricted only by the imagination of the providers and the purses of the receivers. The true ultimate circuit is to provide fiber to the premises because this provides significant bandwidth at present with good growth potential. Several communities have experimented with this medium, but the cost has been extremely high.

Implication	FTTP presents the potential to deliver more services over fixed-cost media, providing new revenue streams.

WIRELESS (RADIATED) TECHNOLOGIES AND MEDIA

Wired connectivity, discussed previously, means a physical electrical or **photonic** channel. **Wireless** is the absence of such physical channels. Wireless generally takes two forms: electromagnetic (radio) waves and light (infrared) waves.

Implication	**Wireless** technologies hold the promise of the "untethered enterprise."

Most are accustomed to wireless communications meaning mobile communications, for example, **pagers** and cellular telephones. While wireless makes mobility possible and offers significant value, the potential of fixed wireless also has great value.

Clarification	People often use the term *channel label* as though it were the medium. For example, broadcast AM and FM radios carry information via radio waves through the air and space between transmitter and receiver; thus, the medium is the air or space. However, most books refer to the radio waves as the medium for simplicity. We will follow this practice. When we use the term *guided media,* this includes the categories of UTP, coax, and fiber strands. **Radiated** media fall into the categories of omnidirectional radio (AM/FM/TV, cellular telephone, Wi-Fi), directed radio (satellite and microwave), and infrared (light).

Modes of Wireless

Fixed wireless is point to point and is not designed to be movable. One specific use is to replace or compete with POTS UTP local loops. While local Telcos have UTP providing service to the home, new providers do not have this access and must rent access from the Telco or install the last mile, generally a very expensive option due to the cost of stringing wire or digging trenches for cable. Fixed wireless offers the potential to bypass this expense. One simply installs an omnidirectional antenna at a central point and a unidirectional antenna on the residence or office. It is a good business decision if the provider can (1) create the network for less than the cost of renting local loops and (2) achieve sufficient bandwidth to provide the desired service. Fixed wireless can be installed much faster than wired alternatives.

Clarification	**Fixed wireless** access is a radio-based replacement for wired circuits.
Implication	Installation of wireless networks generally requires two important considerations: (1) because radio waves at these frequencies require a line-of-sight environment, obstacles are a problem, and (2) wireless is broadcast into the air and security is a special problem.

Companies are using different portions of the spectrum and different technology to provide broadband fixed wireless service to businesses. Connecting a building in a major city with fiber costs about US$400,000, while fixed wireless can cost only US$50,000.

Most data and voice access is bidirectional. Cable modem delivery of entertainment used to be concerned with only one-way delivery, but Internet access now requires bidirectional communications even if the bandwidth is asymmetric, that is, higher downstream than upstream.

Asymmetric circuits have different bandwidth in each direction.

Implication	Wireless not only provides untethered access but provides a way to compete with embedded wired infrastructure, such as CATV, with a much smaller up-front cost.

Forms of Connectivity

The first form of connectivity is **fixed node,** generally wired connectivity. This connectivity would be for desktop machines and servers. LANs are generally implemented via UTP cables due to existing technology, security, and desired bandwidth. Fixed wireless capabilities are just now reaching competitive bandwidth. Radio frequency radio and infrared light (IR) wireless networks (discussed in the following pages) could perform the same function. Advantages of wireless would be the ability of going through walls for multiroom connectivity for *radio frequency* (RF) radio networks and not going through walls for limited-area coverage for IR networks. Disadvantages include a small area of coverage and limited bandwidth.

The second form is **movable node,** that is, laptop computers that are carried from point-to-point. These devices can be supported with wired media, but wireless media allow the machines to be at any of a number of locations and be connected as long as there is a connection point within a few hundred feet. Additionally, the connection does not require any effort; it is made when the laptop computer comes within range and is active.

The third form is **nomadic (moving) nodes,** including devices such as wireless PDAs. Palm Pilot®, Handspring®, and Blackberry®, as well as palmtop computers and connected machines in the factory or warehouse, provide the user with information while actually in motion. Wireless connectivity for these computers means that they provide a wide area of connectivity with multiple access points, there is no user effort involved in the connectivity, and they continuously provide *information exchange* as the user roams.

Radio (Air Medium)

All wireless services, except those using infrared signals, use the same basic technology. The transmitter uses an antenna to transmit a "high power" *analog radio (electromagnetic) carrier wave*. The information to be transported is placed on the carrier wave via **amplitude, frequency,** or **phase modulation.** This is true for AM/FM radio, broadcast and cable TV, Wi-Fi data communications, cordless or cellular telephones, and microwave and satellite transmissions. Thus, when listening to 96 ROCK FM radio station in Atlanta, the **radio** is tuned to an analog **carrier frequency** of 96.1 MHz and is extracting "data" that are frequency modulated onto the carrier. The "data" are then reproduced as **acoustic** energy. With digital cellular telephones, conversation is carried via digital encoding on the analog carrier. Figures 2.5 shows the *radio spectrum* for air and coaxial cable. This is an all-analog spectrum, divided and allocated by governing agencies to avoid conflict of signals.

FIGURE 2.5   The Broad Spectrum

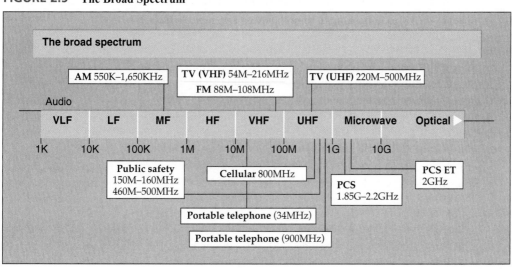

Clarification

Radio transmission is communications channels using radio wave technologies to send data from transmitter to receiver.

Omnidirectional Radio

Many wireless capabilities use *omnidirectional radio* transmitters and receivers, like those used for AM and FM radio and television programs. Although not prime candidates for many data communications systems, they are used for such systems as stock market quotations and personal pagers, along with taxi companies, police and fire departments, and cellular telephones.

Radio transmission and reception in any form, AM/FM/TV, relies on one unit acting as a transmitter and one as a receiver, although these functions may exchange places. A transmitting antenna on a tower broadcasts in all directions **(omnidirectional)** and a receiver, such as a personal AM/FM radio, home TV set, or police mobile radio, receives the signal. In the latter case, patrol cars also have a transmitter than can send signals back to the tower. AM/FM radio and television broadcasters are interested in transmitting from one source to as many receivers within the local coverage areas as possible. Whether the radio/TV is for private or public consumption, the transmitter has a reliable receiving range that is selected at the time the license is granted. Range for radio is a function of the power of the transmitter (from 50 to 50,000 watts of power), the height of the antenna above the terrain (mountain tops are excellent), the terrain itself (mountains are terrible), and weather conditions.

Clarification

Omnidirectional broadcast gives broad coverage over a large area but at lower power or higher cost than unidirectional broadcast over a smaller area.

terrestrial radio link
Common carrier radio link systems carry analog signals via FM for point-to-point communications.

Very-low-power AM radio stations, for example, 250 watts, have a range of a few tens of miles; 10,000-watt transmitters can extend this up to 50–100 miles. Clear-channel stations, using 50,000 watts of output power, can be heard across the United States, partially due to power and partially because they follow the curvature of the earth. At night, they bounce off of the ionosphere.

FM radio and broadcast TV signals, because of their higher frequencies, are line-of-sight and carry only 100–200 or so miles, depending on the receiving antenna.

When transmitting radio or television signals, the source is at a point and the destinations are usually in a large circle around the source. This is called **broadcasting.** Using microwave transmission from one tower to another tower within line-of-sight is **narrowcasting.** The choice of the two modes of transmission is based on the size and location of the audience (receivers).

The frequency **range** for U.S. commercial AM radio is 550 KHz to 1.6 MHz. The variables are the same for commercial broadcast television, except that the frequency range is 50 to 212 MHz for **very high frequency** (channels 1 thru 13) and 400 to 950 MHz for **ultra high frequency** (channels 14 thru 84). Because of the higher frequencies for TV, higher power (100,000 watts) is allowed in the VHF bands, and antennas are usually located on very high structures. UHF stations, on the other hand, are often licensed to provide very localized coverage and may have lower power, for example, 100 to 5,000 watts, so as to minimize interference in the receiving area. FM radio uses the frequency range of 88 MHz to 108 MHz, which falls between TV channels 6 and 7. See Figure 2.5 for the broad spectrum.

An example of the use of over-the-air radio can be found at a local fast-food establishment. McDonald's uses technology to allow the person handling the drive-through window mobility. The person at the window has a mobile radio pack and can walk around the store while answering the requests from the drive-through lane. Each end of the conversation has an FM transmitter and receiver.

Clarification

Terrestrial broadcast radio includes AM, FM, citizens band (CB), shortwave, and various TV transmissions. *Satellite radio* increases coverage and bandwidth.

Cellular (Radio) Telephone

radio telephony
Cellular mobile
communications
systems, using both
analog and digital
modulation, always on
an analog carrier.

A specific use of radio that has gained significant popularity is for mobile telephones. This is due to the use of cellular radio. Formerly, circa 1980, mobile radiotelephone meant that there were one or two transmitters for a medium-size city, limited to only a few hundred FCC-assigned frequencies. Each telephone permanently had one of the frequencies; therefore, you could have only a few hundred vehicles with telephones. (Mobile radiotelephone systems, like cellular systems, connect with the local Telco switch via landlines from base stations.) With **cellular (radio) telephones,** the coverage area is divided into cells of about five miles in diameter, although the size is dependent on the density of telephone traffic and physical terrain. (See Figure 2.6.) Each cell has a base station with a computer-controlled transmitter-receiver, each of which is connected to a master switching center, called a *mobile telephone switching office (MTSO)*. As a vehicle places a call within a cell or enters the cell while talking, the vehicle's phone communicates to the cellular station on a frequency selected at the moment. When the vehicle leaves the cell and enters another cell, or the computer determines that another cell receiver-transmitter can provide better-quality service, the computer, with the aid of the MTSO, will hand off the vehicle transmission to the transmitter-receiver in the new cell and clear the frequency in the old cell for a new caller. If the receiver of the conversation is also a cellular customer and in the same cell as the sender or any cell connected to the sender's MTSO, which is not likely or

FIGURE 2.6   Mobile Cellular Telephone System

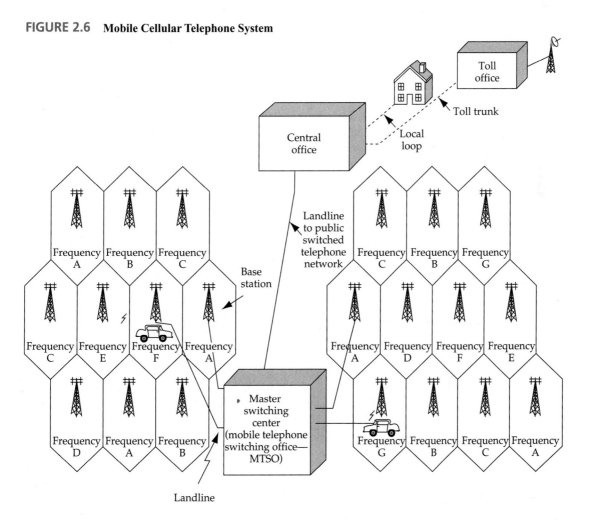

FIGURE 2.7
**Full Communications
Coverage Provided by
Three Geostationary
Satellites**

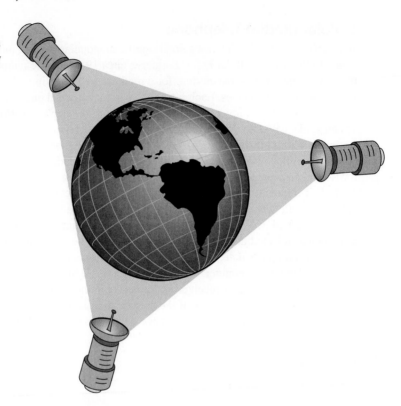

required, the call potentially bypasses the local phone company's CO. Most of the time, the call goes to the local phone company's switch and on to either a **terrestrial** phone or out of another MTSO's cellular transmitter, then to the receiving vehicle phone. The caller can be in Miami and the receiver can be in New York City, and the call can continue as long as each participant is within a cell. However, if one participant enters an area not covered by cellular radio, the conversation stops. The vehicle can be an airplane, and the conversation can continue up to about a half hour's distance beyond the coastline or, using **satellite,** can continue as long as it is in the cone of coverage. (See Figure 2.7.)

In the case of cellular telephones, the cellular switching computer and MTSO act as a central office switch. The local loop is the radio connection between the transmitter tower and the vehicle or portable cellular unit. The switch in this instance not only creates the path, it also allocates radio frequencies. Unlike a local CO, the cellular switch knows a hand-off of the call will occur within a small amount of time.

Clarification

Cellular generations: Cellular communications have progressed from 1G (analog) to 3G (high bandwidth digital data).

Cellular telephones use analog radio as the carrier, initially in the frequency range of 850–900 MHz. In high-density areas, the traffic can become so great that it is not unusual for a call to be blocked as no frequencies are available for a new call or a hand-off. Use of digital technology relieves this in all but the highest-density areas by providing more frequencies, less noise, and a better environment for data communications. Digital cellular protocols are discussed in the following pages.

Implication

Cellular technology provides ubiquitous coverage for voice and a potential for data connectivity with limited investment.

While some contend that the costs of a cellular telephone are high relative to a landline system (monthly costs of US$200 are not unusual), the value to the business often surpasses the cost. People who spend a large amount of time traveling in their cars can now utilize this time to make contacts with customers, receive and give information to their associates, or check their voice mail. An example of the acceptance of the "mobile" market is that the worldwide cellular user base is expected to be 1.9 billion by the end of 2006. This means that penetration of cellular will be at 30 percent. A feature of cellular phone systems is that phone numbers are not generally available via the information directory. This will likely change with the deregulation of the industry and the enforcement of number portability: the ability to move a specific number from switch-to-switch and system-to-system.

Implication

Where cell phones were once just a replacement for POTS, they now incorporate PDA functions, viewing screens, and cameras.

Caveat

This discussion centers around the U.S. deployment of cellular telephones. The United States has different cellular standards than Europe and many other nations, but the concepts remain the same. As service is established in new areas, the choice of equipment will often be standards-based and influenced by the standards adopted nearby.

Cellular Radio Protocols

There are a confusing number of cellular standards, protocols, and technologies. In the next sections, we provide some basics to help in understanding the differences. Table 2.4 showing timelines and generations summarizes the major differences.

Analog Cellular—Advanced Mobile Phone Service (AMPS)

AMPS started as an analog system using the 800 Mhz frequency band. AMPS is widely deployed in the United States, Canada, and Mexico. It is also a de facto standard in Central and South America and is used in the Pacific Rim, Russia, and Africa.

Clarification

Early mobile telephone architecture: The "cellular" concept evolved from the traditional mobile radiotelephone service. Mobile radiotelephone used a single powerful transmitter in a 50-km-radius service area, with a limited number of frequencies in this coverage area. Cellular telephone services use low-power transmitters in 2.5-mile-radius "cells," with frequency reuse in nonadjacent cells.

The analog version was deployed by AT&T in 1983. Early estimates were drastically short of the actual subscriber rates achieved. At the advent of cellular service in 1983, predictions were that total handsets in the United States would reach one million by the year 2000; there were 11.6 million subscribers when 2000 arrived. Part of the market penetration came as a result of technological advances in handsets, the electronic components, and mass market economies of scale that brought prices down.

TABLE 2.4 **Timeline of Cellular**

Generation	Rollout	Data Rate Mbps	Uses	Component Technologies
1G	1981–83	Analog	Early car phones	AMPS, TAC, NMT
2G	1991–95	10 Kbps	Mass-market cell phones	GSM, TDMA, CDMA
2.5G	1999	.064–.144	Wireless net access	Same + GPRS, EDGE, IS-95B
3G	2001–2005	.348–2.0	Location-based service	WCDMA
4G	2010	10–100	Video	OFDM

Digital Cellular Telephone Protocols

The wireless industry began to explore converting the existing analog network to digital as a means of improving capacity in the late 1980s. The two major (competing) digital systems are TDMA and CDMA. **code division multiple access (CDMA)** is a **spread-spectrum**[7] technology that allows multiple frequencies to be used simultaneously. This protocol codes every digital packet it sends with a unique key; a receiver responds only to that key and can pick out and demodulate the associated signal. CDMA lets everyone transmit at the same time; it has a very high "spectral efficiency." It can accommodate more users per MHz of bandwidth than any other deployed cellular technology.

Implication

CDMA is a digital air interface standard that provides 8 to 15 times the capacity of analog.

Time division multiple access (TDMA) does its job by chopping up the channel into sequential time slices, using *time division multiplexing.*[8] Each user of the channel takes turns transmitting and receiving in a round-robin fashion. There are actually three different flavors of TDMA in the PCS market. Each of these technologies implements TDMA in a slightly different way. The most complex implementation is **global system for mobile communications (GSM),** which overlays the basic TDMA principles with many innovations that reduce the potential problems inherent in the system. *IS-136*[9] *(Interim Standard-136)* is the second and crudest form for TDMA. Finally, *iDEN (High Speed Circuit Switched Data)* is a proprietary technology by Motorola for workgroups. iDEN phones can be used as two-way radios but with no need for both parties to be within a certain range. These two-way radio communications use the cellular telephony network. (See Table 2.5.)

Implication

TDMA provides users with time slots so several calls can occur on one frequency.

Because of its adoption as the European standard, the Japanese Digital Cellular (JDC), and North American Digital Cellular (NADC), TDMA GSM and its variants are currently the technology of choice throughout the world. GSM networks are leaders in many typically "digital" services including the **short message service (SMS),** Over the Air (OTA) configuration, and GPS positioning. SMS is a feature available in most modern digital phones that lets users receive and send short text messages (from 150 to 160 characters) to other cell phones, usually limited to phones activated on the same network.

Clarification

Cellular towers have a single data channel for signaling and multiple voice channels, which are allocated to individual users as the need arises. SMS and "dialed phone numbers" are sent over the data channel.

An extension of SMS is **multimedia messaging service (MMS).** This capability provides automatic and immediate delivery of personal messages, just like SMS. However, MMS allows enhanced messages that can be a combination of text, sounds, images, and video to other MMS-enabled handsets. It is also possible to send messages from phone to

---

[7] Spread spectrum is a technology where the signal changes channels (hops) many times per second or spreads the signal over multiple channels simultaneously.

[8] Time division multiplexing divides the space of time into slots and allocates data alternately.

[9] IS-136 was the second-generation (2G) mobile phone system, known as Digital AMPS (D-AMPS). D-AMPS uses existing AMPS channels and allows for a smooth transition between digital and analog systems in the same area. Capacity was increased over the preceding analog design by dividing each 30 KHz channel pair into three time slots and digitally compressing the voice data, yielding three times the call capacity in a single cell. A digital system also made calls more secure because analog scanners could not access digital signals. Calls were encrypted, although the algorithm used (CMEA) was later found to be weak. http://en.wikipedia.org/wiki/IS-136.

TABLE 2.5
Mobile Phone Standards

Source: http://en.wikipedia.org/wiki/IS-136.

Generation	Standard or Nomenclature
0G	• PTT • MTS • IMTS • AMTS • IPTS
0.5	• Autotel/PALM • ARP
1G	• NMT • AMPS
2G	• GSM • iDEN • D-AMPS • cdmaOne • PDC
2.5G	• PRS • CDMA2000 1xRTT
2.75G	• EDGE
3G	• W-CDMA • UMTS • FOMA • CDMA2000 1xEV • TD-SCDNA
3.5G	• HSDPA
4G	

e-mail and vice versa. Some providers make available such services as postcards, animated pictures and graphics, access to news, stock market quotes, greeting cards, video messaging, and so. MMS is an open industry standard.

Implication SMS and MMS messaging are a significant potential revenue stream once the voice market saturates.

A unique and essential component of GSM phones is the **Subscriber Identification Module (SIM) card** carrying the identify of the phone and an address book. Therefore, putting the SIM card into a new phone allows the user to proceed as though no change had been made. If, however, you are switching to or between non-SIM-based instruments, the phone must be identified anew to the system and the entries of the address book must be keyed in again.

Considering its technology and presence through much of the world, GSM is in a good position for global roaming and many new GSM phones are called "global phones," since they can be used in virtually any country. GSM provides eight slots in a channel 200 KHz wide.

Clarification World-class international phones are tri-band (1900, 1800, 900 MHz) for travel in over 100 countries.

3G

The generic term for *third-generation* mobile phone technologies is 3G. The advances in technology bring very-high-speed connections to cellular phones, enabling video presentation and applications requiring digital broadband connectivity, such as e-mail and Internet browsing.

General packet radio service (GPRS) is a distinct change from other digital technologies on cellular. On the other technologies, the voice was encoded continuously on the analog carrier. GPRS takes the digitized voice and sends it as digital **packets** (small blocks of data) on packet-switching technology for GSM networks. It's an advanced data

TABLE 2.6
**Microwave
Frequencies
(IEEE US)**

Sources: http://en.wikipedia.
org/wiki/Radio_frequencies..

Band	Frequency Range
L band	1 to 2 GHz
S band	2 to 4 GHz
C band	4 to 8 GHz
X band	8 to 12 GHz
Ku band	12 to 18 GHz
K band	18 to 26 GHz
Ka band	26 to 40 GHz
V band	40 to 75 GHz
W band	75 to 111 GHz

transmission mode that does not require a continuous connection to the Internet, as with a modem. Instead, GPRS uses the network only when there are data to be sent, which is more efficient. We will explore packet networks in much greater detail later.

TDMA versus CDMA

Imagine a room full of people, all trying to carry on one-on-one conversations. In TDMA, each pair takes turns talking. They keep their turns short by saying only one sentence at a time. As there is never more than one person speaking in the room at any given moment, no one has to worry about being heard over any background noise. In CDMA, each pair talks at the same time, but they all use a different language. Because none of the listeners understand any language other than that of the individual to whom they are listening, the background noise doesn't cause any real problems. These two standards will continue to compete on technological and cost bases.

Microwave

Terrestrial microwave circuits are special, high-frequency, line-of-sight radio systems that send signals from one transmitting station to one receiving station, up to 30 miles apart. They use an analog carrier in the GHz ranges and can carry either analog or digital information. (See Table 2.6.) Where twisted-wire and coaxial circuits could operate bidirectionally by taking turns or using different parts of the bandwidth, the **microwave** radio system is **unidirectional** (simplex) because one station is a transmitter and the other is a receiver. Because of this, a separate microwave radio link is required for the return path. (In reality, all media are one-way due to the need to add directional amplifiers to overcome attenuation. Two paths are required where two-way communications are needed.) Satellite microwave is discussed in a following section.

Implication   Wireless point-to-point and point-to-many-point overcome many problems of **right-of-way** and construction.

Wireless Cable Television (LMDS, MMDS)

The title of this technology seems to be contradictory, that is, how can you have cable-less cable? The terminology may be unfortunate, but the idea was that wireless technology could offer similar functionality to conventional cable.

When the alternative to over-the-air broadcast TV was launched, it used the cheapest, low-noise medium available: coaxial cable. Therefore, it was called cable television. The title has stuck, so when the medium changed, the title was simply extended.

Coaxial cable transport of cable TV uses the same standards and frequencies as on-air broadcast. The signal is modulated onto a carrier wave of the same frequency that would be used for omnidirectional broadcast. In extreme cases, the channels are changed, but the technology is the same.

Wireless cable systems use protocols called **Multipoint Microwave Distribution System (MMDS)** and **Local Multipoint Distribution Service (LMDS)** to place multiple channels of television and Internet access on a frequency spectrum that is wirelessly broadcast. Generally, they use a single omnidirectional broadcast antenna and broadcast to a 36-inch conical microwave antenna located on the wall of a building, much like direct satellite broadcast television. In this case, LMDS and MMDS don't use a satellite transponder; they send their signals line-of-sight from their antenna to yours, MMDS at 2–3 GHz and LMDS at 26–29 GHz. The lesser capability has a total bandwidth less than coaxial cable and satellite, about 25 TV channels, and the distance covered is about a 25-mile radius. The quality competes nicely with cable, it can be installed much faster, and the cost is less because there is no need to run the cables and rent or install utility poles. Newer systems have far greater bandwidth, are bidirectional, and are digital, offering Internet access.

Wireless Internet Access

Wireless Internet access allows getting information from the Internet on cellular telephones, pagers, or PDAs, for example, Palm Pilot® or Blackberry®. There are two aspects of wireless Internet access: (1) information to a portable device and (2) access to a stationary computer.

Wireless Internet access will be competition for wired Internet access whether standard modem, cable modem, or DSL access. With new cellular and other wireless technologies and protocols, it soon may be possible and economically feasible for the local wireless provider to provide a 128-to-767 Kbps wireless channel to the Internet from the home or hotel by way of existing cell towers. Wireless technologies are developing so rapidly that they are beginning to compete effectively with wired technologies.

Implication

We want to be untethered AND we want Internet access, anywhere, anytime. The potential is a virtual office.

Satellite

Satellite radio is logically the same as a microwave radio circuit except that the sending station transmits to a satellite relay station 22,300 miles above the earth. The signal is relayed down to another earth station. In the *high earth orbit* mode, the satellite is in a geosynchronous orbit, causing it to appear stationary above the earth. Three satellites can cover the earth's surface (see Figure 2.7). The major drawback to satellite communications is the time delay in the uplink to the satellite, transponder delay in frequency conversion, and the downlink to earth. The total propagation delay is about half a second for data and three seconds for voice and television. While satellite transmission allows communication across bodies of water and difficult terrain, the delay is generally difficult for people to tolerate in voice communications. An alternative to high earth orbit is to place the satellites in lower orbit. This is discussed below. We continue a discussion of satellite communications used by business in later chapters on data communications.

Microwave and satellite radio systems, which had such great promise for voice communications, have been replaced with other media in many instances. Fiber optic (digital) technology has replaced microwave technology for most long-distance systems. Microwave technology still finds use in local, short-haul environments such as in cities between buildings. Ease of installation and avoidance of right-of-way permits in congested areas remain advantages.

Clarification

Satellite microwave is a line-of-sight transmission to an orbiting satellite where a transponder amplifies and relays the transmission back to earth via downlink.

Very small aperture terminals (VSATs) have smaller (approximately 8 to 16 inches, 20 to 40 cm) antennas than satellite earth station dishes. A VSAT network has many of these terminals linked through the satellite to a central control station on the ground. For example, the next time you pass by a major new car dealer, note the satellite antenna on the roof. If it is less than 1.5 feet in diameter, it is VSAT.

Implication

Because of the high frequencies used for terrestrial microwave and satellite communications, rain can cause sufficient attenuation as to disrupt the signal.

Direct Broadcast Satellite TV

With the advent of low-cost electronics and lower-cost satellite launching facilities, several companies have placed geosynchronous satellites in place for direct broadcast of digital television programming and data. Satellite direct broadcast is not a new idea, but having a small, very-low-cost installation that doesn't take up a lot of space is. This fixed wireless system is the first major competitor to cable TV. The **direct broadcast satellite (DBS)** TV providers chose to send the signals in digital form from the beginning as opposed to the original analog service. Digital transport increases quality and adds enhanced features such as HDTV and AC3 sound while making better use of the frequency spectrum. The cable companies are coming to this same conclusion as they convert upper channels to digital.

Until the early 1990s, satellite TV transmission was analog, using C and K bands, which required ground antennas of at least 6, preferably 10, feet in diameter for home reception. In the early 1990s, the Hughes Corporation allocated US$2 billion to initiate direct broadcast satellite service using three TV broadcast satellites, two for active orbit and one spare. In mid-1996, Hughes began another offering, called DirecPC®, which offered Internet downloads of 400 Kbps to a PC and the capacity for multimedia services. Initial calls for service are made by the modem and then uplinked to the DirecPC Network Operations Center to the Galaxy IV satellite. While more expensive than services by Internet providers, this system provides coverage of the 48 contiguous states of the United States at a significantly higher bandwidth. Hughes now offers DirectWAY, a combination of DirecTV and bidirectional DirecPC, for Internet access anywhere in mid-North America.

Implication

Satellite broadband: The need for high bandwidth wireless has spawned a demand for dedicated broadband service via satellite.

Orbiting Satellites

Direct broadcast satellite television uses geosynchronous satellites that appear to be stationary in orbit located 22,300 miles above the earth's surface with their power dissipated during transmission. Recent entrants into satellite broadcast, both voice and data, are choosing to put up not just one geosynchronous satellite but a number of *low earth orbit (LEO)* satellites; 77 were planned in the case of the original Iridium system. LEOs are in low earth orbit only several hundred miles above the planet. This means they will be in constant motion, being in view of an earth station for only minutes at a time. This requires that the satellite control system switch the user from satellite to satellite as they go out of and come into view. It also means that the amount of power required for broadcast is less for the same reception. While geosynchronous satellites have limited areas for parking to see the optimal portions of the earth and have long expected lifespans, the LEO satellites have almost unlimited orbit locations but very limited lifespans. In addition, Mid Earth Orbit (MEO) altitudes can be between 300 and 22,500 miles. For example, the 24 GPS satellites are at 10,700 miles altitude.

Clarification

Low earth orbit (LEO) satellites have less propagation delay. Twelve LEOs can cover the entire earth.

Infrared—Digital on Light

The other wireless technology used for control, with which you are most likely familiar, is infrared (IR), such as that used in the TV, VCR, CD/DVD player, and stereo remote controls. This technology, like fiber optics, uses light waves instead of electricity. IR devices are used almost exclusively for digital signals. The remote control used to command a TV transmits invisible light pulses that the detector in the TV interprets as commands to change channels, volume, and so on. A very different use of this form of data transmission is in the IR data links used in laptop computers and printers. The laptops send data in IR form at a rate of about 115 Kbps over an IrDA channel, and the printers or any IR receivers accept the data without the use of wiring. The IrDA 1.0 channel is specified at 115 Kbps, whereas the new IrDA 2.0 channel has transfer speeds of 1.152 Mbps, with a new goal of 10 Mbps being researched. Most laptop and notebook computers began to include this IR link as of the end of 1996.

Implication

IR for WIP: Some industries find IR a natural fit for monitoring their work-in-progress in hostile environments.

Clarification

Infrared systems developed by Einstruction Corporation are used as "classroom performance systems" for increased interactivity in the lecture hall. Students employ "clickers" (individual response pads) with buttons A–H for responses. The system is coupled with a database access so that grades are collected and accumulated. Receivers are placed so that line-of-sight is maintained.

Other Wireless

A significant amount of business is being conducted in a wireless environment and the trend appears to be accelerating. Several firms have reported that their agents now have totally substituted wireless palm-held devices for the notebooks and laptops used just a short time ago. This conversion has contributed to the success of *Research in Motion Ltd.,* a formerly obscure Canadian company. During the first two years after they introduced their BlackBerry® device, about 800,000 were sold. The device has been called essentially a pager that can link to corporate e-mail servers to send and receive e-mail. The BlackBerry still has a long way to go to equal the Palm Pilot®'s success: an estimated 10 million handheld computers sold. The European introduction provided another boost in BlackBerry sales and popularity. The impact of such wireless devices on the conduct of business has grown as mobile workers have discovered the advantages of connectivity.

High Definition Radio

A new protocol, now referred to as **high-definition (HD) radio,** carries the information (e.g., music) on the carrier wave in digital form. The use of digital form, **compression,** and digital **modulation** allows a greater bandwidth of useful information. FM will now rival CDs and AM will rival the old FM quality. This is all brought about by changing the modulation format of the signal.

Satellite Radio and Television

The satellite radio market, according to investment bank Merrill Lynch, consists of 200 million motorists, including 3 million truckers, 9 million RVs, and more than 22 million consumers underserved by radio. *Sirius®* and *XM®* each provides 100 channels of high-definition radio from satellites. This new technology is gaining customers slowly, but the equipment costs are dropping rapidly. Now a user can listen to the same music or talk show all around the continental United States, in his/her car or at home. Many of the channels are commercial-free, but this may change.

According to the January 22, 2001, issue of *Autoweek,* the Datron Cruise TV was displayed at the 2001 Consumer Electronics Show, allowing users to receive DirecTV® in their cars. So, along with VHS VCR or DVD in the back for the kids, now you can receive live video, movies, and music via DirecTV.

Implication   **Frequency spectrum constraint:** There is a fixed amount of frequency spectrum that is available for many competing uses.

Wireless Bandwidth

The amount of bandwidth we can achieve on any medium, wired or wireless, is dependent on the native bandwidth available, the congestion of the spectrum, the noise environment, and the amount of compression possible. For UTP, coaxial cable, and fiber strands, the user is in a baseband mode and has all of the native bandwidth unless there are active devices in the line. With IR wireless, a limited portion of the light spectrum is available. With radio wireless, the spectrum is congested and there are many requests for all frequencies. This means that the use of digital modulation onto the analog radio carrier is even more important because significant compression ratios are possible. As noted with the cellular telephone, the various protocols offer different forms of compression and spectrum sharing, thus providing greater bandwidth.

Bluetooth

A person on the move will often have a laptop computer, a personal digital assistant for scheduling, a cellular telephone, and possibly a pager. These devices often carry copies of the same data. Until the advent of *Bluetooth*®, a cable was required to make connections, if connectivity between devices was supported. Next it would be necessary to move or synchronize the data on the different machines. *Bluetooth* is a short-range wireless technology that connects electronic devices, including cell phones, printers, digital cameras, and palmtop computers. *Bluetooth* is designed to exchange data at speeds up to 720 Kbit/s and at ranges up to 10–100 meters.

Clarification   *Bluetooth* protocol supports personal area networks.

Global Positioning System (GPS)

Knowing our exact location and directions to where we need to go via information from satellites has changed the way we travel and find our destination. Airlines, trucking firms, express delivery vans, military aircraft, submarines, soldiers in the field, hikers, and tourists who once referred to maps or used a compass now rely on a small handheld **global positioning system (GPS)** receiver. This technology receives signals from several of 22 orbiting satellites to calculate precise location, altitude, and speed. The signals also are used by other equipment for exact time synchronism. When combined with D-GPS, the location is so precise that mining companies use it to determine boundary locations.

Clarification   **GPS** was installed by the U.S. Department of Defense for military purposes. It has since changed the way military, civilian, and personal location finding is achieved.

Implication   **GPS applications:** Several innovative GPS uses are changing the way business is conducted, such as precise agriculture mapping, tracking, charting, and surveying.

When the location data of the GPS circuitry are combined with charting information, as in the eMap Deluxe® by Garmin, the users have a visual display of where they are, what is around them, and a map of directions. This device, the size of a small, flat calculator, contains a 12-parallel-channel GPS receiver and weighs a mere six ounces.

Implication   **GPS** is a large-fixed-cost, very-low-variable-cost (maintenance) system, providing numerous low-cost uses.

Voice and Data Connectivity from the Air

GTE's *Airfone*® service allows calls to be placed to anywhere in the world while flying over the contiguous United States, southern regions of Canada, Mexico, and within 200 miles of the U.S. coastline. When flying out of range of the ground station network, calls can still be placed on airlines equipped with satellite service. There is also an ability to hook up a computer to a data port and go online. One can check e-mail, send a fax, or download a file while traveling on an airliner.

Traditional Wireless Formats—A Reference Point

Point-to-point links can communicate reliably as long as the two endpoints are close enough to one another to escape the effects of **RF interference (RFI)** and path loss. If unable to achieve a reliable connection initially, sometimes relocating the radios or boosting the transmit power can achieve the desired **reliability.** An example of this is a microwave system.

Point-to-multipoint links (e.g., IEEE 802.11 or *Bluetooth*) have one base station, or access point, that controls communications with all of the other wireless nodes in the network. Also referred to as a hub-and-spoke or star topology, this architecture is similar to wired "home-run" systems in which all of the signals converge on one terminal block. The reliability of these networks is set by the quality of the RF link between the central access point and the end points. In industrial settings, it can be hard to find a location for an access point that provides dependable communications with each end point. Moving an access point to improve communications with one end point often will degrade communications with other endpoints.

Mesh-Enabled Architecture (MEA)

Unlike historical hub-and-spoke wireless networks, the **mesh-enabled architecture (MEA)** produces a point-to-point-to-point, or peer-to-peer, system called an *ad hoc, multihop network.* A **node** can send and receive messages, and in a **mesh network,** a node also functions as a router and may even relay messages for its neighbors. Through the relaying process, a packet of wireless data will find its way to its destination, passing through intermediate nodes with reliable communication links. Like the Internet and other peer-to-peer router-based networks, a mesh network offers multiple redundant communications paths throughout the network.

Wi-Fi, IEEE 802.11b/a/g

IEEE 802.11, refereed to as **Wireless Fidelity,** or **Wi-Fi,** is a standard for **packet-radio networks,** that is, the connectivity of nodes while using radio as the medium. Although there are three wireless standards in use today, Wi-Fi is the predominant one in the area of networks. The other two wireless standards are *HomeRF*® and *Bluetooth*®. *HomeRF* was developed for a home environment and wireless portable phones and ***Bluetooth*** is designed for multidevice communications, to be a "personal area network," with coverage up to about 10–100 meters, depending on the power of the device. IEEE 802.11 is designed for attachment to a network device at a range of 100 meters and above; it is designed as a replacement or complement to a wired network of devices.

Implication   The **IEEE 802.11 (Wi-Fi) standards** are still undergoing rapid evolution as Wi-Fi expands.

FIGURE 2.8
Wireless T-1
(Point-to-Point)

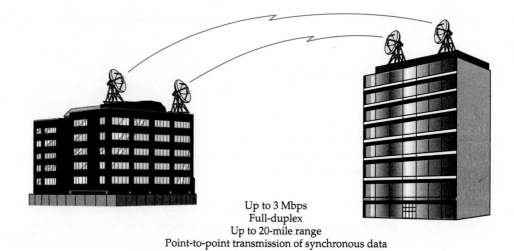

Up to 3 Mbps
Full-duplex
Up to 20-mile range
Point-to-point transmission of synchronous data

Wireless Point-to-Point and Multipoint

Figures 2.8 and 2.9 illustrate two types of wireless connectivity. The first is point-to-point, representing a wireless T-1 circuit. Figure 2.8 shows two simplex circuits, one in each direction, providing good connectivity without requiring right-of-way permits or construction, as long as the two buildings have unobstructed line-of-sight. Figure 2.9 shows the use of point-to-multipoint radio, providing bidirectional, omnidirectional service to several buildings. This instance shows a metropolitan area network providing services to business or SOHOs in an urban area. **IEEE 802.16 (WiMAX)** is the complement to 802.11 for coverage of areas as large of 30 miles in radius and bandwidths up to 256 Mbps.

The Last Mile Considerations

The distance from the junction point of POTS, CATV, or even power service to the home is referred to as the **last mile.** Here is where single circuits must be installed and often it

FIGURE 2.9
Wireless MAN
(Metropolitan Area
Network)

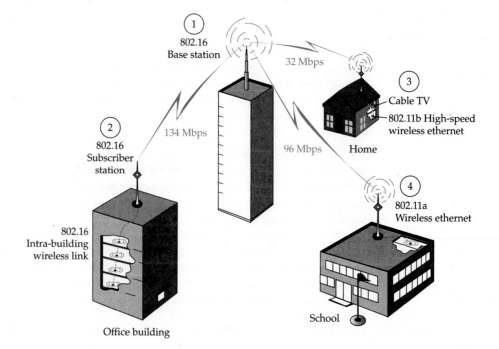

means adding poles in neighborhoods or digging up the street, sidewalks, and yards. It's the most expensive portion of the total circuit.

Telephone companies have local loops of UTP for the last mile to most homes and offices; CATV companies have coaxial cable; power companies have electrical circuits. As the need for greater bandwidth grows, the question arises as to whether the present last mile circuit can transport greater bandwidth or if new media must be installed. For example, pilot installations are providing *broadband over power lines (BPL)*, using the existing power grid. Because data do not go through transformers, the link from the distribution line to the home/office may well be wireless. POTS UTP can be converted to xDSL with only the addition of technology at each end, providing the present voice channel and higher-bandwidth data channels. A question that remains is whether utility companies will install fiber circuits to the home or go wireless.

Implication   Telephone amplifiers have a bandwidth bottleneck in the **last mile** media that is a legacy from the POTS era.

The decision to provide fiber or a wireless environment is dictated by several considerations. First, what is the total bandwidth required? Second, what amount of compression is available; for example, is it enough to use existing channels? Third, will switching be done at a central office or in the home? These last two considerations can be demonstrated by television. Do you want to deliver and receive 300 channels simultaneously, requiring fiber, or just one or two channels at a time, like satellite DBS, with the channel switching done at a central office? This latter is possible via DSL over existing copper UTP.

As a reminder, we discussed LMDS and MMDS in the category of wireless CATV. These are ready candidates for wireless last mile considerations. The new entrant to this arena is the technology of time modulated–ultra wideband architecture. This emerging technology allows for high-bandwidth wireless digital communications over a finite distance to the home from a central neighbor podium.

Case 2-1

Cellular Technology

John Matlock

INTRODUCTION

NTT DoCoMo, Japan's leading mobile operator, launched its commercial 3G (third-generation wireless) service on Monday, October 1, 2001. Known as Freedom of Mobile multimedia Access (FOMA), NTT DoCoMo is using industry standard IMT-2000 to achieve downlink speeds of 384 Kbps, enabling fast and smooth video streaming and wireless data communications for customers. [1] In contrast, companies in the U.S. wireless market are unlikely to launch commercial 3G anytime soon given the current economic slowdown and ongoing military operations worldwide. Negotiations between U.S. Federal Communications Commission, Commerce Department, Department of Defense, Federal Aviation Administration, other government agencies, and wireless providers are on hold pending the "war on terrorism." "As a country, we're all worried about the most basic human need for security," said Blair Levin, an analyst for Legg Mason in Washington, DC. "That's going to take priority over mobile data needs by civilians so we can watch movies on our cell phones." [2]

MOBILE COMMUNICATIONS CONCEPT

Mobile communications embrace two key concepts. The primary concept is that smaller is better. Cellular companies break a large area down into adjacent hexagonal cells of roughly equal area. Radio towers, known as base stations, occupy the center of each cell and communicate with wireless devices. The second concept is that less is better. This concept diminishes transmission power from base station towers. Applying both concepts simultaneously enables reduced transmission power and reduced cell areas to allow for reuse of radio frequencies (in nonadjacent cells). These concepts allow for many more users of the service compared to one frequency over a larger area. [3]

WIRELESS/CELLULAR FUNDAMENTALS

The key to wireless/cellular networks is the mobile switching center (MSC) or mobile telephone switching office (MTSO). The MSC connects smaller cells into a larger networked system.

1G (FIRST GENERATION ANALOG WIRELESS)

Better known as analog cellular, first generation wireless uses technology known as advanced mobile phone system (AMPS) to optimize use of transmission frequencies. AMPS devices connect to a base station using a specific frequency. Approximately 1,000 frequencies are available for use at any particular time and two cells are able to use the same frequency for a voice call if they are not adjacent to each other. [4] Ericsson AMPS uses a frequency range of 850–900 MHz. Digital TDMA standards (known as D-AMPS) provide the evolutionary path to a digital environment. [5]

1.5G (DIGITAL WIRELESS)

During the mid 1990s, providers introduced digital wireless service. These devices were based on a variety of standards including TDMA, GSM, and CDMA. GSM is by far the leading standard with 220 million handsets shipped in 2001, compared to 49 million and 72 million handsets shipped for TDMA and CDMA respectively. [6] This wireless technology provides digital voice service, 9.6 to 14.4 Kbps data service, and enhanced calling features such as call waiting, Caller ID, and voice mail. However, 1.5G does not provide the customer with always-on data connection. Time division multiple access (TDMA) communication technology divides a single radio frequency channel into six unique time slots, allowing a number of users to access a single channel at one time without interference. By dividing the channel into slots, three signals (two time slots for each signal) can be transmitted over a single channel. Thus, TDMA technology, also referred to as ANSI-136, provides a three-to-one gain in capacity over analog technology. Each caller is assigned a specific time slot for transmission. [7]

GSM, or global system for mobile communications, is a digital cellular or PCS network used throughout the world that shares the same time division multiple access transmission method as TDMA. CDMA is a spread spectrum technology, spreading the information contained in a particular signal of interest over a much greater bandwidth than the original signal. [8] TDMA and CDMA are used primarily in the Americas, while GSM is used primarily in Europe and Asia, although GSM has a wide user base in the United States, for example, T-Mobile.

3G (THIRD-GENERATION HIGH-SPEED DIGITAL WIRELESS)

Competing companies worldwide are migrating to 3G wireless as fast as possible. Most firms are preparing to migrate to 3G in various steps. In Europe GSM will migrate to Universal Mobile Telecommunications System (UMTS) using Enhanced Data Rates for

Global Evolution (EDGE), which in essence bumps data rates up to 384 Kbps. EDGE also is used on TDMA networks to migrate to 3G. Another technology, general packet radio service (GPRS), will enable GSM networks to transfer data and wireless Internet content over the network at 115 Kbps. Using a packet data service, subscribers are always connected and always online, so services will be easy and quick to access. Finally, CDMA competitors will migrate to cdma2000, an additional standard under IMT-2000 (International Telecommunication Union [ITU], a UN agency). cdma2000 enables providers to use existing hardware to initially migrate to 3G 1x, providing data rates of 144 Kbps and doubling the voice capacity of the network. The follow-on evolution for cdma2000 is to migrate to 3G 1xEV with data rates capable of 2.4 Mbps.

Other technologies for migration include wideband code division multiple access (WCDMA). In WCDMA, voice, images, data, and video are first converted to a narrowband digital radio signal. The signal is assigned a marker (spreading code) to distinguish it from the signal of other users. WCDMA uses variable rate techniques in digital processing and can achieve multirate transmissions. WCDMA has been adopted as a standard by the ITU under the name IMT-2000 direct spread. [7] This is the technology in use by NTT DoCoMo in the Japanese market.

SUMMARY

Undoubtedly, migration to G3 will be slow and evolutionary here in the United States but potentially very fast and efficient once issues related to frequency management are resolved. Europe has adopted a standard with limited bandwidth in its migration to G3. Finally, Asia is moving quickly to adopt IMT-2000/WCDMA standards to achieve exceptional bandwidth now. Sprint PCS perceives the threat from existing/new entrants versus bandwidth to look like the chart (see Table 2.5 on page 63).

BIBLIOGRAPHY

1. FOMA, NTT DoCoMo. Retrieved October 15, 2001, from http://www.NTTDoCoMo.com.
2. "U.S. Taking the Slow Road to 3G." Retrieved October 15, 2001, from http://www.zdnet.com/zdnn/stories/news/0,4586,2815630,00.html?chkpt=zdnn_rt_latest.
3. R. Costello. *Basic Concepts of Communications: An Introduction.* Retrieved October 15, 2001, from http://www.gartner.com.
4. *Technology A–Z.* Retrieved October 15, 2001, from http://www.ericsson.com/technology.
5. H. Carr and C. Snyder. *The Management of Telecommunications.* Burr Ridge, IL: Irwin/McGraw-Hill, 1997.
6. B. Baxter. *Wireless Update.* Half Moon Bay, CA: Murenove, Inc., 2001.
7. "Frequently Asked Questions." Retrieved October 15, 2001, from http://www.uwcc.org/edge/tdma_faq.html.
8. "Leading the Evolution to 3G." Retrieved October 15, 2001, from http://www. sprintpcs.com/aboutsprintpcs/Cdma_3g/index.html.

Summary

This chapter discussed the channels of telecommunications. It addressed the circuit media of these channels, the physical paths on which the data or voice ride. Whether twisted-pair of the telephone local loop, the coaxial cable of CATV, or space for radio, television, and satellite, all channels must have a circuit medium.

Voice communications have historically used analog signals and the media of copper wire, microwave, and satellite radio. This remains true for the Telcos, but most of the long-distance traffic now rides on fiber optic cables. The use of radio in cellular systems has gained great acceptance and could threaten the existence of wired switched telephone systems. Cellular technology has such value to businesses that the costs are readily accepted as a good alternative for mobile workers. The original analog technology is giving way to the PCS digital mode to provide greater clarity and security.

Recent changes in legislation in the United States have allowed many more operators to provide telecommunications services. Some, like the CATV and phone companies, have large investments in coaxial and twisted-pair cable plants. It is financially prudent to use the installed physical plant, so the CATV companies are working to make use of their coaxial cable in the neighborhood while distributing the signals there via fiber cables. The Telcos are finding that their twisted-pair circuits have the ability to carry significant bandwidth, giving them new uses for this old technology.

Wireless technology, radio, microwave, satellite, and infrared are being used increasingly for data communications. These media allow channels without a need for right-of-way permits and expensive digging. They offer the potential to change the way the business operates through convenient and low-cost connectivity.

The subject of media involves technology, bandwidth, and connectivity. As with all choices, there are trade-offs; one may be low cost but managerially constraining. The question will always arise of whether to use an existing medium or install a new one. The existing one may be lower-cost, but the new alternatives can offer new opportunities. The choice of medium always has management implications.

Key Terms

acoustic, *p. 57*	frequency, *p. 76*	Local Multipoint
amplitude, *p. 57*	full-duplex, *p. 48*	Distribution Service
bandwidth, *p. 47*	general packet radio	(LMDS), *p. 65*
bearer (B)	service (GPRS), *p. 63*	long-distance provider, *p. 47*
channel, *p. 48*	global positioning system	medium, *p. 45*
Bluetooth, *p. 69*	(GPS), *p. 68*	mesh-enabled architecture
broadband, *p. 51*	global system for mobile	(MEA), *p. 69*
broadcasting, *p. 58*	communications	mesh network, *p. 69*
cable, *p. 45*	(GSM), *p. 62*	microwave, *p. 64*
carrier frequency, *p. 57*	half-duplex, *p. 49*	modulation, *p. 67*
cellular (radio)	hertz (Hz), *p. 47*	movable node, *p. 57*
telephone, *p. 59*	high-definition (HD)	multimedia messaging
central office, *p. 47*	radio, *p. 67*	service (MMS), *p. 62*
channel, *p. 45*	high-definition television	Multipoint Microwave
circuit, *p. 45*	(HDTV), *p. 77*	Distribution System
coaxial cable, *p. 49*	hybrid fiber/coax	(MMDS), *p. 65*
code division multiple	(HFC), *p. 53*	narrowcast, *p. 58*
access (CDMA), *p. 62*	IEEE 802.11, *p. 69*	node, *p. 69*
compression, *p. 67*	IEEE 802.16	noise, *p. 45*
crosstalk, *p. 51*	(WiMAX), *p. 70*	nomadic (moving)
digital subscriber line	infrared, *p. 46*	node, *p. 57*
(DSL), *p. 47*	integrated services	omnidirectional, *p. 58*
direct broadcast satellite	digital network	optical fiber, *p. 51*
(DBS), *p. 66*	(ISDN), *p. 48*	packet-radio network, *p. 69*
electrical, *p. 45*	interface, *p. 51*	packets, *p. 63*
electromagnetic, *p. 45*	interference, *p. 46*	pager, *p. 56*
fixed node, *p. 57*	last mile, *p. 70*	path, *p. 45*
fixed wireless, *p. 56*	local loop, *p. 47*	phase modulation, *p. 57*

Recommended Readings

http://www.networkworld.com provides networking news, reviews, opinions, and forums from the leader in network knowledge.

Management Critical Thinking

M2.1 The discussion of media, both conducted (wired) and radiated (wireless), covers much material that would seem to be aimed at the "techies" or cost accountants. How can management use this information to achieve competitive advantage?

M2.2 How does an organization choose how much growth potential (extra) bandwidth to include? That is, how does the organization choose the expense of "extra" capability?

Technology Critical Thinking

T2.1 What are the environments in which conducted (wired) media are a must? What will make it possible to replace conducted with radiated (wireless) technology? What sort of timetable is associated with this technological capability?

T2.2 As radiated media increase, what are the chief technology constraints that can limit deployment?

Discussion Questions

2.1 What effect will cellular phones (or devices with cellular phone capability) have on small business in the next 10 years?

2.2 How feasible do you think it is for every employee of a firm to have a portable telephone?

2.3 What transmission media would you expect to dominate in voice communications in the year 2010? 2020? 2050?

2.4 Are personal communications devices, in general, and cellular telephones, in particular, safe? What are the categories of safety that one should consider?

2.5 The concept of a communications brooch (appliance) was introduced on *Star Trek, The Next Generation.* Discuss what technology this might encompass and how it might be used.

2.6 Discuss some of the cost-benefit categories that a manager must understand in making the data communications media decisions for an organization.

2.7 The telephone companies have a significant installed UTP infrastructure. Would it be better for them to forgo it and move to total implementation of fiber-to-the-home?

2.8 How can a business be operated by employees who remain at home most of the work week?

2.9   What are some privacy issues surrounding the incorporation of GPS in personal communications devices?

2.10 As cellular moves to 4G bandwidth on the data channel, what are the implications, positive and negative?

Projects

2.1   The primary wired competitors for Internet access are DSL and cable modem. Determine the deployment of each in your city and bandwidth versus cost.

2.2   If your city offers DSL and cable modem Internet access, determine who in your class subscribe to which. Compare and contrast the services and cost. Do the users of each do so for convenience or for technical reasons?

2.3   What media are employed in your organization or school? How much has been replaced by wireless? Is the planning calling for mostly wireless in the future?

2.4   Find cities that are using MMDS or LMDS to provide either television service or Internet access. Determine the services offered and compare the costs to coaxial cable services.

2.5   Search on the Internet for your local *Lowes* or *Home Depot* and find comparative prices for the purchase of UTP, coax, and fiber. Call the local telephone company and see if you can find out installation costs. What conclusions would you draw from this exercise?

2.6   Visit a local business and determine the media that they employ. What media have been added in the past 10 years?

2.7   What are the factors that enable POTS to survive in an era of increasing bandwidth demand? Build a matrix to compare POTS and its competition.

2.8   If you were going to create a competitive "cable television service" in your city, what media would you use? What would be the time for installation and how cost competitive would it be?

2.9   Make a table listing various media and attempt to project growth rates for each. Project these for the next 15 years.

2.10 Go to the NetworkWorld Web site and determine how many articles could be considered as about data communications and media.

Appendix 2-1

Visual Radio—Television

The reason to explore television is to consider the bandwidth for this universally popular entertainment mode. Broadcast spectrum and cable bandwidth are limited, users wish more channels, and providers wish to move to higher-quality production. This always results in a need for greater bandwidth or a change in present standards.

Clarification   Each TV channel is described by a **frequency band,** as defined by the FCC in the United States.

STANDARD TELEVISION BANDWIDTH AND SPECTRUM

When television was introduced in New York in 1937, it was via omnidirectional over-the-air broadcast. Since then it has often changed to coaxial cable and direct broadcast satellite. However, the format has remained basically the same. Color video and stereo, as well as surround sound audio, have been added; each requires more bandwidth or a more judicious use of the existing bandwidth. The total bandwidth has remained at six megahertz per channel since inception. The information content, which equates to picture quality, remains at 400 lines vertically, 350 lines horizontally, with a picture rate of 30 frames per second. This standard was set by the *National Television Standards Committee (NTSC),* which is how standard TV is referred to. The European TV standard, PAL, has more definition.

The improved color picture and enhanced sound of present TV are the result of viewers wanting more. MTV and its presentation of stereo music was a prime reason that sound improved. With stable color video came a desire for better definition under all conditions. As personal computers gained sharper definition on their monitors, consumers wanted the same or better picture definition on television. In response to consumer demand, high-definition television was created.

HIGH-DEFINITION TELEVISION (HDTV)

High-definition television (HDTV) can be in an analog or digital environment. Japan chose analog and the United States chose digital for HDTV. The objective in each case is to provide greater picture information and quality and better sound bandwidth. In order to achieve the desired outcome, changes were required. One must either allocate more analog bandwidth, for example, twice as much as an NTSC channel, or convert to digital and use compression to reduce the bandwidth requirements. In either case, the detail of the picture is greater with better clarity and the sound quality has moved from equal that of FM radio to that of compact discs.

Clarification   **High-definition television (HDTV)** delivers a high-quality picture and CD-quality sound via digital 19.2 Mbps broadcast.

The 1080i-format HDTV, from air broadcast, on cable or satellite direct broadcast, will be the U.S. primary standard in 2006. This standard requires new television sets to decode the 1080i format and/or new set-top boxes for the TV. HDTV sets are being produced to also convert NTSC signals into a 480P format. Where NTSC TV in digital format requires about 8 Mbps bandwidth, HDTV requires 19.2 Mbps.

ENTERTAINMENT-ON-DEMAND

Recorded or transmitted entertainment comes in three primary forms or modes: audio, video, and gaming. Entertainment also comes in four forms or schedules: periodic scheduled, special events, near-on-demand, and on-demand.

We are accustomed to periodic scheduled events as that is the general format of radio and television. Special events on TV are becoming more commonplace and are often pay-per-view as opposed to being part of the generalized subscription of cable TV or premium channels. We only get near-on-demand entertainment in places like hotels where there may be a sizable list of movies one can order and the hotel staff inserts the appropriate cassette that is ordered. The format we are moving to is on-demand radio and television.

VIDEO-ON-DEMAND (VOD)

Historically, most people have had to depend on advertised schedules of events, whether on television or at public places, and then bend their schedules to accommodate the fixed schedules of the events. With technology, this dependence on a fixed schedule can change.

Clarification   *Video-on-demand* potentially means viewing any program/movie ever recorded, at any time, from anywhere.

The first way to have scheduled entertainment accommodate the viewer was to video-tape programs (storing) using a programmable video cassette recorder (VCR). This required the viewer to determine when the program was to be aired, place a tape in the machine, and program the machine correctly. Next came pay-per-view movies on premium TV channels that were aired several times a day. The newest entrant accommodating viewers is the *digital video recorder (DVR)*, which uses a hard drive for digital storage and a downloaded schedule of events to find the programs of interest and automatically record them. With DVR, the television viewer's first option is to watch what was recorded that fell in his/her area of interest, as opposed to just watching what is being broadcast at the current time.

Another level of user-controlled entertainment is *video-on-demand (VoD)*. This means that the user orders what s/he wants and it is provided. VoD requires more bandwidth and a server that contains a lot of programs. Ultimately, it means that there will be digital video servers containing all programs of interest, old and new, from which the viewer can choose. It also means a high-bandwidth channel is required to the user's television and, potentially, pay-per-view for everything. If the program is NTSC television, only about 6–8 Mbps is required; if HDTV format is used, 19.2 Mbps is required. Additionally, a **reverse channel** is required so that the selection can be made. (This presents an architecture problem as the server has a constraint in that ports are limited).

Charter Cable offers video-on-demand, allowing rental of a select group of movies, and treats them as if on a DVD or VCR. The rental period is 24 hours, during which the viewer can watch as often as wished, stop, resume, rewind, and fast forward. This implies that Charter is storing the movies on a server and unicasting an individual movie stream to each renter on a common channel. The system is using the upload channel for commands and sharing the VoD channel among all renters, just like Internet access. With this bidirectional action, VoD is truly limited entertainment-on-demand, with the caveat that it is only at your location and only the select set of movies. This requires digital cable and the upload channel.

INTERNET BROADCAST TV (IP-TV OR IP-VIDEO)

The Internet means connectivity; the World Wide Web (WWW) and browsers give a graphical user interface. The uses of the Internet have increased rapidly. First there was special radio broadcast of sports over the Internet and now there is continuous broadcast of hundreds of radio stations over the Internet 24 hours a day, seven days per week (24*7). This means that the audience of a particular radio station is not limited by the reach of its air-broadcast signal; it has a global reach. Just as the small UHF television station WTBS in Atlanta, Georgia, became a major player due to the use of satellite delivery, other radio and television stations can achieve a global presence as the bandwidth of the Internet accommodates their signal. As the bandwidth of the Internet increases and the delays decrease, the use of this "free" resource continues to evolve and can make even a local provider into a global competitor.

Implication   With sufficient bandwidth on the Internet, it is a logical delivery medium for all entertainment.

Digital television is generally delivered in, of course, digital format (after D/A conversion) on a channel that has modems at each end. Whether satellite-based or on CATV, the digital information is encoded as a digital stream in a digital channel dedicated to TV.

A new mode coming to market is *IP-television*. It is based on the increasing number of homes that have broadband access. This access, generally using DSL, carries the Internet protocol (IP). With IP-television, the information is changed from analog-to-digital and encoded in the IP format of packets. It can then be sent over any IP network. Thus, one of the new DSL lines, carrying 8–10 Mbps, can carry four to five TV channels as the compressed signal occupies 2 Mbps/channel.

Additionally, since the IP access has the interactivity of a normal IP channel, users can send commands upstream. Providers are using the DSL conduit to connect to a nationwide, if not global, network that has servers containing many varied programs. For example, TV stations would place their programming on a local server and users could access it using IP-television, download a file or use streaming, and watch a current or old program. Some programs will be free and others fee-based. Subscribers who wish to have access to HBO or other premium channels would arrange a contract with HBO for standard payment and receive the programming via IP-television.

ENHANCED TV

The American Broadcasting Company describes enhanced TV as a live interactive TV experience, converging on-air and online programming. It does not replace on-air programming; the added computer information is available to make the viewer's experience more interesting. Included are quizzes, viewer opinion polls, and chat sessions, adding involvement to the viewing.

To use Enhanced TV, you must have both an online computer and television in the same room. While a program, such as Monday Night Football, is in progress, ETV allows viewers to test their play-calling skills and, during commercials, ETV asks trivia questions. Viewer polls also are displayed when an instant replay is in progress, giving the armchair quarterback much more power. Another form of ETV is used with *Web-TV*®. For example, a chef on the *Food Channel* prepares a specialty item; clicking on the screen brings up the recipe for saving and printing.

Implication   Because television is considered a lean-back (passive) form and the computer is a lean-forward (active) form, their combination may not be as logical as it seems.

TELEVISION OVER DSL

Because DSL technologies provide significant digital bandwidth, the potential exists for telephone companies and others to compete with cable TV by using TV-over-DSL. VHS-quality full-motion, full-color video and audio can be delivered with as little as 2 Mbps, although programming with significant activity requires more bandwidth to reduce smearing. Let's assume we can deliver acceptable-quality television at 6 Mbps. This means a DSL provider could deliver a TV channel to users, assuming the download bandwidth of 6 Mbps is available. This is not competitive with cable TV unless the user has a reverse channel (DSL provides) whereby a user would switch the channel at the DSL central office. DSL could then provide an unlimited number of channels, one-at-a-time. The quality could be increased to the limit of a normal TV unit; HDTV requires more bandwidth than DSL presently offers. In any case, the addition of DSL technology onto analog POTS local loops allows telephone companies and others to provide direct CATV competition with cable companies and satellite providers.

Implication   If switching is done at the central office, Telcos can compete in the cable TV market, offering an infinite number of channels over DSL, one at a time.

Appendix **2-2**

Other Technologies

CONNECTED PALMTOP

As bandwidth to cell phones, laptop computers, and palm devices increases, their use will increase, which will fuel the demand for higher bandwidth. This phenomenon can be illustrated to the expansive use of cell phones with the prediction of one million phones in use 17 years after introduction missed by 1,000 percent. More use stimulates demand and more capability enables and encourages increases in use.

MULTIFUNCTION APPLIANCES

Mobile voice telephone service has been available for more than 40 years by a variety of technologies. Initially, it was provided by a few mobile radiotelephones. The equipment was fairly large and spectrum was very scarce. When miniaturization made cellular telephones feasible in the early 1980s, the bag and car phones took over from the mobile radio, providing many more channels and giving more people mobile voice service. Now that the cell phone is small enough to carry on a belt or in a purse, mobile voice communications capabilities have become essential for a large percentage of the population in advanced and advancing nations. This idea of portable connectivity has evolved with the personal digital assistant (PDA), which ranges in capability from keeping a calendar to being a full-fledged pocket personal computer. Some are GPS-enabled with mapping to show present location and path to destination. PDAs also have become Internet-enabled, giving the user access to his/her e-mail and Web sites. In like manner, cell phones have short message service, are Internet-enabled, and may have digital cameras included. Input to a PDA or cell phone, especially for Internet control, is difficult due to small space for a keyboard or the need to use handwriting translation, even with the use of a stylus or mini-keyboard.

The next evolution point, as any Star Trekkie will tell you, is the brooch as worn by the crew of the *Federation Starship Enterprise*. These computer- and GPS-enabled devices are voice-responsive and small enough to be worn as jewelry on clothing. This negates the use of a display, as on present PDAs, but the input problem is gone. Again, what started with mobile radiophone access has evolved to allow the user to be personally available and connected, all the time, from anywhere. This capability requires more and more bandwidth. Some refer to the evolving devices as communications *appliances*.

Implication   The *Starship Enterprise* communications brooch is the hands-free, wireless model for the future.

PERSONAL COMMUNICATIONS

The first radio device that showed the way for personal communications was the pager. This device began as a small radio receiver carried on the person that gave a beeping sound (hence the pager was often called a *beeper*), notifying the user to call a specified telephone number for a message. This device progressed to the point where a short audible announcement could be sent to the pager. The present digital technology sends textual information that is stored in the receiver. Pagers can be as small as pencils, and some can store pages of textual messages. SkyTel® pagers use a satellite network to locate the receiving pager in thousands of cities and towns across the United States. Their latest pagers feature two-way capabilities, for example, the ability to respond to a page with a preprogrammed or customized response. Meanwhile, newer cellular telephones and laptop computers are including paging capabilities.

While the pager capability has already merged into the cellular telephone, the PDA will be next. It is not likely that the pager will disappear as text messages will likely be a lasting requirement; often we want to be informed but do not want to carry on a conversation.

RADIO AND POTS

The telephone system that Alexander Graham Bell and others developed has traditionally been a wire-based system for most homes. What broke this tradition was the introduction of the portable (or cordless) telephone.

Implication   The telephone that started as place-to-place communications has evolved to person-to-person.

Cordless telephones use two FM radio transmitter-receivers (transceivers) to replace the wire from the handset to the base station. The transmitters operate at different frequencies, to provide full-duplex, just like on a wire. These systems use analog carrier waves and originally carried the voice as analog waves modulated onto the 34 MHz carrier. Therefore, it was easy to detect and extract the voice signal. Five developments have made the cordless telephone of better quality and more secure:

1. The first change seen on the original 34 MHz phones was the use of multiple channels. The user could change the frequency of the channel of transmission from the base station to the handset to get away from noise or an eavesdropping neighbor. These phones have from 2 to 40 channels.

2. Next came the movement to the 900 MHz portion of the radio spectrum. This is a cleaner (less noise) frequency than the lower frequency. Operation in the 900 MHz radio band allows the signal to penetrate walls and other physical barriers more easily. The mode is still analog, but techniques are added to either change channels, offset channel frequencies, or partially encrypt the voice for security.

3. When the providers digitized the voice, they carried it as digital data on the analog carrier. This makes interception of the voice much more difficult and provides better voice quality.

4. Fourth, some phones use spread spectrum, which is a technology that constantly changes channels (frequency hopping) between the handset and the base station on a predefined frequency shift. If the eavesdropper does not know the allocations and timing, the results will appear as noise, making detection almost impossible. Thus, the change of frequency, digitizing the voice, and adding spread spectrum give a higher-quality signal that is very secure. Encryption adds even more security.

5. The latest change in the cordless systems is the movement to the 2.4 and 5.8 GHz frequencies. While the phones are advertised as having greater reach, the primary advantage is clarity and less interference from other phones.

Chapter **Three**

Architecture, Models, and Standards

We use concepts of architecture to provide for building a logical structure for data communications systems. Included in the establishment of architecture are the standards that make for interoperability. To facilitate the building of a logical infrastructure, we employ layered models, including the OSI model.

This chapter provides an important foundation for the provision of the systems that make both voice and data communications possible and coherent within an organization. We begin by defining the term *architecture* as applied to the systems' overall design, configuration, and constituent components. Next we provide models that enable the system's functional parts to be separated so that the resulting layers enable standards to be applied within the particular layer without impacting the other layers.

The role of standards in enabling systems to communicate is explained and some major standards are discussed. Without standards, we would have islands of capability, unable to share with others; that is, communications would not take place.

ARCHITECTURE

Architecture, like a city map, shows the buildings on each block, not what's inside.

An **architecture** is an overall system plan that is implemented in a set of hardware, software, and communications products. It should be expressed in terms of logical or functional and physical configuration and design. The architecture specifies components and interfaces that make up the systems, to include protocols, formats, and standards to which all hardware and software in the network must conform. Architectures should include documentation of system usage, functionality, performance parameters, and so forth, of all components. The physical connection, location, identification of nodes, circuits, networks, and standards should be specified. **Network architectures** attempt to facilitate the operation, maintenance, and growth of the communication and processing environment by isolating the user and the application program from the details of the network. This was necessary to support distributed data processing with multiple computers in a network; the architecture should be open so that it can accommodate new technology. The purpose of network architecture is to

- Provide an orderly structure for the communications network that ensures a specific level of compatibility.
- Provide isolation of the application systems from the physical hardware.
- Support faster development and easier maintenance of application systems by using system software utilities to perform communications functions.
- Be reliable, modular, and easy to use.
- Accommodate new devices and software for the network without changing the application systems.
- Replace individual pieces of the system without affecting other pieces.

In order to support our discussion and to assist in the definition of architecture, we typically use models that deal with layers. A layer or level evokes an orderly or logical grouping of functions. As long as the rules of the boundary between layers are honored, the intra-layer architectures are independent. In addressing a model for open systems, we consider first the generic three-level model, the OSI model, and then the TCP/IP model.

Standards

A **standard** is a set of rules or descriptions about a specific piece of equipment, software, or service. For example, commercial (entertainment) FM radio in the United States has a spectrum standard that set the carrier frequency between 88.1 and 108.1 MHz. Other countries may establish different ranges of spectrum, but this is the standard for the United States.

Standards can come about in several ways. First, as in the present case of DVD (digital versatile discs) used for recording movies and computer data, individual companies may set standards as they develop the technology. In the case of the DVD, there are DVD-R, DVD-RW, DVD+R, DVD+RW, DVD-ROM, and DVD-RAM standards. The problem with the various standards is that a machine from one company may not be able to play back a DVD recorded on the machine of a different company or to a different standard. That is, the equipment does not have **interoperability.** (When the 56K modem was developed, Motorola and Lucent had different standards that were incompatible.) While each company believes its standard should be universally accepted because that company then is able to charge others for use of its **proprietary standard,** the market prefers a single, **open standard** so that equipment from different suppliers is interoperable.

interoperability means that equipment from different sources and vendors will work together.

The 56K problem of noninteroperability of two proprietary standards was solved by software upgrades to the V.90 standard.

Implication
The Importance of Interoperability

As the impact of the destruction caused by Hurricane Katrina was still being assessed, some telecommunications issues became clear. Among them was the great problem caused by lack of interoperability of the communications systems used by early responders to the disaster, for example, police, firefighters, homeland security, U.S. Coast Guard, and others. Because of the incompatibility, responses were much slower and often uncoordinated or duplicated. When the lack of interoperability is coupled with widespread loss of infrastructure, the needed resources are not dispatched where they are most required, resulting in loss of life and further destruction.

de facto
Actually exercising power, although not legally or officially established.

If there is a single company developing a standard and it has sufficient power in the industry or the technology is simply the first to be accepted, it may create a **de facto** standard. That is, the standard that the company sets for its equipment is accepted by the rest of the industry and other companies then make equipment to that standard. Such was the case for IBM for the early decades of the computer industry. When companies cannot agree to a single standard, *standards bodies* enter the picture.

Clarification

A **standard** is a definition or format that has been approved by a recognized standards organization or is accepted as a de facto standard by the industry.[1]

Some standards bodies, such as the **Federal Communications Commission (FCC),** are created by U.S. law; states create equivalent agencies. Others are a cooperative and collaborative effort by the companies interested in the equipment. The *ATM Forum* is an example of the latter for the development of the asynchronous transfer mode technology. A third form of standards body is exemplified when a professional organization chooses to set standards, as in the case of the **Institute of Electrical and Electronics Engineers (IEEE).** This international organization of thousands of engineers creates committees who negotiate among members and companies to arrive at accepted and open standards.

Clarification

IEEE is an organization composed of engineers, scientists, and students. The IEEE is best known for developing standards for the computer and electronics industry. In particular, the IEEE 802 standards for local-area networks are widely followed.[2]

When a *manufacturer develops a new capability,* it is often based on that manufacturer's view of technology, architecture, and unique or proprietary standards. Proprietary standards may allow that manufacturer's equipment to communicate with others in the product line, but these standards generally preclude *interoperation* of "foreign" equipment, often

---

[1] http://www.webopedia.com.
[2] http://www.webopedia.com.

by design. If telecommunications managers buy into a set of proprietary standards, they may preclude the use of other manufacturers' equipment. If the proprietary standard is kept secret by implementing it in firmware within the equipment, other manufacturers cannot replicate it and are kept from communicating with the original equipment except as the original equipment allows or through conversion equipment.

Several approaches exist for establishing a network of computers. If the proposed network is homogeneous (all the same equipment), a single vendor may provide all the network capability, including the electrically dependent and the application-dependent services. Also, the latter services may be obtained from vendors who will provide host-compatible software. On the other hand, if the network hosts are heterogeneous, there are two networking alternatives. One is to buy networking software from vendors who provide (middleware) software to bridge from the one **proprietary** system to the other. These packages are usually limited to connect the major vendors. Another option is to buy software that provides a bridge into a standards-based environment. The open architecture, to be discussed next, lets the nodes interoperate, basing that interoperation on standard protocols.

The decision maker should be concerned with the way that a vendor chooses to provide interconnectivity. It is a decision-making situation that has far-reaching implications; it falls into a "pay me now or pay me later" category, that is, the **total cost of ownership.** If one buys into a manufacturer's proprietary scheme, it is a long-term decision and commitment. It means you must work within that vendor's scheme or standard, often to the exclusion of others. If an open systems standard is chosen, you will have many more choices later as the competition for your business will be far greater. This is the tenet of interoperability: the ability to provide connectivity between equipment of different vendors without any special hardware or software interfaces.

Electronic Data Interchange (EDI)

A **system** *is a group of interrelated and interdependent parts working together to achieve a common goal.* A *cooperative system* is a specialized type of system that requires at least two parties with different objectives but common goals to collaborate on the development and operation of a joint system in support of these common goals. Typically, partners in a cooperative system develop and operate their own specific portions of the common system environment. Standards usually play a significant role in cooperative systems, and third-party facilitators are often involved.

Electronic data interchange (EDI) is, by definition, a cooperative system. EDI is the process of direct computer-to-computer communication of information in a standard format between organizations or parties (companies) that, as a result of this communication, permits the receiver to perform a specific set of business functions (e.g., purchasing, invoicing). The primary purpose of EDI is to provide a *communications standard* for the electronic transfer of common business documents between the respective computer systems of individual and diverse trading partners. That is to say, EDI uses existing communications technology and takes data from an existing computer-based information system and places it into another. Generally, no new equipment is required, just data standards and EDI management software. EDI provides a significant opportunity to lower the cost of doing business between trading partners, strengthens the partnership, and has other effects on the parties involved. As a result, EDI offers significant competitive advantage to those organizations utilizing the technology. EDI, however, requires not only the use of standards *but also* contractual agreements among the parties because the electronically transmitted information is to be used as if contained within a written contract.

In computing, **middleware** consists of software agents acting as an intermediary between different application components.[3]

Total cost of ownership (TCO) considers *all* costs of using a specific set of equipment, including purchase, installation, maintenance, and upgrade.

EDI
A set of standards for electronic transfer of business documents.

[3] http://en.wikipedia.org.

Clarification   EDI allows inventory management systems at one firm to communicate with a manufacturing requirements planning system at another without knowing the file structure.

Implication   EDI requires contractual agreements among the participating parties. Each agrees to accept the contents of the message as if negotiated by humans, even though the system will often not require human intervention.

Benefits of the Use of EDI

1. Improved customer service.
2. Improved accuracy of data.
3. Reduced clerical errors.
4. Faster access to information.
5. Decreased administration costs.
6. Reduced delivery times.
7. Improved cash flow.

The discussion of EDI continues in the appendix to this chapter.

Governmental Agencies

Most data communications services are unregulated; many voice services are regulated. Regulated services require the approval of tariffs.

The FCC has the ability to force the following of standards via establishing regulations. FCC approval may be required when operating across state lines in the United States. A state agency counterpart to the FCC that establishes rules that a group with which users of equipment may have to comply is the **public service commission (PSC),** also called a **public utilities commission (PUC).** Although PSCs and PUCs do not generally set or approve standards, they do set **tariffs,** statements of telecommunications services and prices to be charged. Organizations must be aware of such agencies and their tariff-approval ability when offering products and services within a state or province. Additionally, governmental agencies can influence standards through their imposition of regulations and rules.

Implication   Governmental agencies often have oversight for technology and services that require approval of related tariffs.

LAYERED MODELS

prototyping
The use of one capability to emulate another, generally in a inexpensive manner and short time frame; an early example.

simulation
The process of imitating a real phenomenon with a set of mathematical formulas. It may be achieved on a working system by use of equipment to simulate real conditions, such as network traffic.

Models have long been used to assist in the understanding of complex concepts, systems operation, behavior of organizations, construction of structures, and so on. There are several taxonomies of models. Most of us are familiar with some classifications, for example, analog, iconic, verbal, physical, mental, mathematical, descriptive, and so forth. It has become common to use models of systems before placing them into production—a process called **prototyping.** We also find **simulation** models of networks to be useful before fixing the final design configuration.

Many children and some adults enjoy building model aircraft. These models are both iconic (they look like the reality they represent) and analog (they act like the reality they represent). When Richard Morris Hunt was commissioned to build a country home for George Vanderbilt, he created a model of the house in order to better explain his ideas to the new owner and the newspaper. Some have attributed the historical success of England on the high seas to their practice of creating ship models by which the admiralty could better understand the vessel's capabilities. Models in data communications are narrative and graphic descriptions of the product or system to be created. Using a layered model allows the total entity to

FIGURE 3.1
Generic Three-Layer Model

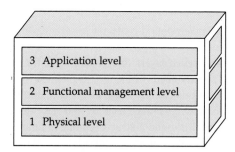

Layered models are all about boundaries; honor the boundary and live in freedom within the layer.

be partitioned so that developers and manufacturers can create products that comply with one **layer** or across layers. The one rule of a layered model is that different developers can create different products within a layer as long as the *boundary conditions* are met.

Generic Three-Layer Model

In the computer field, the generic model consists of input, processing, and output. In the generalized layered model of Figure 3.1, the total system is partitioned into the physical or electrical domain at the bottom, the application or user interface level at the top, and the management of network functions in the middle. Other larger models expand on this basic model.

The OSI Model

A groundswell of interest in connecting heterogeneous computer systems has spurred demand for a set of standards that allow communication between dissimilar equipment. The International Organization for Standardization (www.iso.org) (ISO) and the Consultative Committee on International Telephony and Telegraphy (CCITT), subsumed by International Telecommunications Union (ITU), promulgated the Reference Model for Open Systems Interconnection in 1983. This model has become widely known as the **open systems interconnection (OSI) model.** OSI was developed as a **de jure** (legally derived) standard to replace the IBM SNA de facto (commonly accepted) standard. OSI has currently evolved into de facto usage with fewer than seven layers. (See Figure 3.2.)

de jure
According to law; by right.

FIGURE 3.2
OSI Layered Model

Source: http://en.wikipedia.org/wiki/OSI_model.

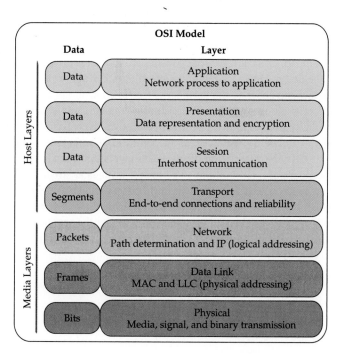

Clarification

Open systems interconnection is an OSI standard for worldwide communications that defines a networking framework for implementing protocols in seven layers.[4]

The OSI model consists of seven layers that contain specific standards for each control level. The uppermost levels specify needs of the user as planned for the application, while the lower three levels concern communications in the physical realm, such as electrical parameters, circuit buildup and breakdown, message routing, and error detection and correction.

The layers in the OSI model provide a structured or modular approach to allow modification and development to be done at a particular level without impacting other layers. The layers are hierarchical in that lower layers provide services such as control information to the layer just above it. The layers are, however, specialized according to functions.

Business managers should be aware that the OSI model facilitates control, analysis, modification, replacement, and management of communication network resources because it provides reference standards. The adherence to OSI model standards makes software and hardware development far easier. This model has become important as the need for "enterprise computing" has increased. Systems managers have become much more concerned about interoperability. This occurs without the user taking an intervening step; s/he does not have to "tweek" the devices. Increasingly, users expect to connect various devices and have them communicate without having to change hardware and software settings. Protocols, standards, and models such as the OSI seven-level model provide for interoperability and, ultimately, lower costs.

Clarification

ISO-OSI model: The primary objective is to provide a basis for interconnecting dissimilar systems for the purpose of information exchange.

The OSI seven-layer model is a plan by which communications software is designed. The model facilitates control, analysis, upgradeability, replacement, and management of the resources that constitute the communications network. The seven-layer OSI model is put into practice as software that handles the transmission of a message from one terminal or application program to another distant terminal or application program. Use of this open standards model makes it easier to develop software and hardware that link incompatible networks and components because protocols can be dealt with one layer at a time. The use of such a model and its layers in design of network software and applications provides the developer with a clear roadmap. Often a module is constrained to the protocol of a specific level. This means that if the design meets the requirements of the interfaces between that level or layer and the ones below and above it, the designer is free to concentrate on only that task.

Like most things in telecommunications, the OSI model was designed around the analog world, and as digital transmission becomes even more dominant, analysts find this model too unwieldy. Because of its value as a reference model, we will continue to describe it. In the description of the seven layers that follow, remember that the layers are like the floors of a tall building. The lowest layer is at street level, while the highest level is in the penthouse, or executive level. Ideas take place at the highest level and work their way down the floors until they come out the front door as messages that can be transported to another building. These physical messages enter the second building and make their way up the floors, being transformed along the way until they reach the highest (executive) floor as ideas.

Clarification

Most of the functionality in the OSI model exists in all communications systems, although two or three OSI layers may be incorporated into one.[5]

[4] http://www.webopedia.com.
[5] http://www.webopedia.com.

TABLE 3.1
Physical Layer—
Level 1

What the Layer Does	What's Going On
• Physically establishes a connection when requested to do so by the data link layer (layer 2) • Physically transmits data as bits • Concerns itself with establishing synchronization of bit flow—duplex, half-duplex, point-to-point, multipoint, asynchronous, or synchronous transmission • Defines quality-of-service parameters	• Defines electrical standards and signaling required to make and break a connection on the physical link to allow bit stream from the DTE onto the network • Specifies the modem interface between DTE and the line • Concerned with voltage levels, currents, simplex/half- or full-duplex

Additionally, remember that in each layer, and on each floor of our buildings, there are different and specific tasks that take place. For example, messages are received physically on the ground floor, where they are handled by a department dedicated to receiving and transmitting physical messages. On each succeeding higher level, different groups or programs carry out unique functions to ensure the reliability of the ultimate message as it is delivered to the correct destination in its complete form. Figure 3.2 displays a concise view of the seven layers of the OSI model.

Clarification

OSI-compliant X.400 and X.500 e-mail and directory standards are among the few widely used OSI capabilities.[6]

Layer 1—The Physical Layer

The objective of digital transmission is to send bits across the network. Regardless of where a message originates or which data code is used, a message must be reduced to an electrical, electromagnetic (radio), or photonic signal of specified voltage or frequency and characteristics for transmission. The protocol in the **physical layer** ensures that when one node transmits a signal representing a logical 1, it is received and interpreted as a logical 1 at the next node. That is, this layer specifies the electrical or photonic connection between the transmission medium and the computer system. (See Table 3.1.) This layer, the lowest in the hierarchy, is basically a computer-to-computer protocol describing the conventions of the electrical or photonic circuits and mechanical system. If the network is circuit-switched, the physical layer also includes the procedures for establishing the circuit. This layer typically deals with connection cable standards, pin assignments, voltages, current levels, impedances, and timing rates. The major functions and services performed by the physical layer are[7]

physical layer
Defines all electrical and physical specifications for devices.

- Establishment and termination of a connection to a communications medium.
- Participation in the process whereby the communication resources are effectively shared among multiple users; for example, contention resolution and flow control.
- Modulation, or conversion between the representation of digital data in user equipment and the corresponding signals transmitted over a communications channel. These are signals operating over the physical cabling—copper and fiber optic, for example.

Layer 2—Data Link Layer

The **data link layer** concerns itself with the actual characters and the sequence in which they are transmitted. It establishes and controls the physical paths of communications to the next node. This layer attempts to deliver error-free data over the circuit between adjacent computers that is established by the physical layer. (See Table 3.2.) Within this layer are such things as error-detection methods, error-recovery methods, error-correction methods,

The physical layer is electrical/optical; the **data link layer** is reliability.

[6] http://www.webopedia.com.
[7] http://en.wikipedia.org/wiki/OSI_seven-layer_model.

TABLE 3.2
Data Link
Layer—Level 2

What the Layer Does	What's Going On

• Segments bit stream into frames
• Coordinates data flow split over multiple physical connections
• Is responsible for error detection and correction
• Monitors flow of data frames and compiles statistics
• Ensures that data frames are in sequence
	• Defines standards for structuring data into
	  • How machine knows where a frame starts
	  • How transmission errors are detected, corrected
	  • How polling and addressing are handled
	  • How machines are addressed
	• Data link protocol used is High-level Data Link Control (HDLC). Normally located in host FEP
	• Whereas modems ensure that bits are accurately sensed for the communication line at the receiver, this layer groups bits into characters for processing
	• Responsible for error checking VRC, LRC, and CRC codes
	• ACK/NAK for receipt of good frames/blocks, and requesting retransmission if error found

methods for resolving competing requests for a shared communication link, and framing (data grouping) requirements for the data.

Layer 3—The Network Layer

network layer
Performs network routing, flow control, segmentation/ desegmentation, and error control functions.

The third layer defines message addressing and routing methods. When there are a number of nodes in the network and multiple paths from the source to the destination, the **network layer** routes the data from node to node. This layer also controls congestion if the network is overloading certain computers. (See Table 3.3.) It does end-to-end routing of packets or blocks of information; collects billing, accounting, and statistical information; and routes messages.

Layer 4—The Transport Layer

transport layer
Provides transparent transfer of data between end users.

The **transport layer** is responsible for maintaining a reliable and cost-effective communications channel from a user's application software process in one computer to a user process in another. It is the highest layer concerned with the world external to its processor. The transport layer views the intervening network as a transparent entity that simply provides a service. The transport layer does not determine the route (the temporary or virtual

TABLE 3.3
Network Layer—
Level 3

What the Layer Does	What's Going On
• Routes packets • Establishes connections between transport layers on two computers that will communicate with each other • Controls flow; stops the flow of packets upon request from the transport layer • Maintains correct sequence of packets • Provides transport layer with acknowledgment that packet has been received correctly • Maintains quality of service • Is responsible for detecting and correcting errors • Is responsible for multiplexing several network connections to a single data link connection for maximum productivity • Handles internetworking, the movement of data from one network to another network	• Provides network addressing and routing • Generates ACK (positive acknowledgment) that entire message was received correctly • Breaks message from level 4 into blocks for level 2 • Decides over which communications circuit to send packet

**TABLE 3.4
Transport Layer—
Level 4**

What the Layer Does	What's Going On
• Ensures that transport data protocol units arrive in proper order • Specifies grade of service, including acceptable error rates • Monitors status of connection • Establishes connections with another transport layer on a second computer • Detects and recovers errors	• Selects the transmission route between DTEs, comparable to postal system for letters • Handles user addressing; controls form of messages to compensate for speed mismatch • Prevents loss and duplication of entire message • Multiplexes several streams into one physical

channel chosen from multiple segments to create the virtual, or temporary, channel), but it ensures that a reliable channel exists between the computer processes.

This layer may break a long message or file into smaller segments at the source and reassemble it at the destination if it is required because of a network packet-size limitation. (See Table 3.4.) The transport layer provides addressing to a specific user process at the destination, message reliability, sequential delivery of the data, and flow control of data between user processes.

Layer 5—The Session Layer

session layer
Provides the mechanism for managing the dialogue between end-user application processes.

The **session layer** deals with the organization of a logical session. It organizes and synchronizes a period of interactive communication between user processes. That is, it establishes the connection between applications, enforces the rules for carrying on the dialogue, and tries to reestablish the connection if a failure occurs. In particular, it authorizes the transport connection between user processes and maintains the continuity of the connection, for example, the order in which the applications are allowed to communicate and the pacing of information, so the recipient is not overloaded. For example, if the transport connection fails, the session layer provides a logical synchronization point so the activity can be resumed. (See Table 3.5.) This layer provides access procedures, rules of half-duplex or full-duplex dialogues, rules for recovering if the session is interrupted, and rules for logically ending the session.

Layer 6—The Presentation Layer

presentation layer
Handles encoding, encryption, and similar manipulation of the presentation of data.

The **presentation layer** is responsible for formatting and displaying the data to and from the application layer and deals with the transmission format of the data. The presentation layer may define how to compress the user data to improve data transfer rates or how to encrypt it for security. (See Table 3.6.) The presentation layer provides transmission syntax, message transformations and formatting, data encryption, code conversion, and data compression.

A major component of this layer is the concept of the **virtual terminal.** This concept allows the programmer at the application layer to send data to a universal virtual terminal where the layer will reformat the data to meet the presentation needs of the actual terminal.

**TABLE 3.5
Session Layer—
Level 5**

What the Layer Does	What's Going On
• Establishes and maintains transport connection with a certain designated quality of service • Organizes and synchronizes dialogues between presentation entities • Ensures the reliable transfer of data • Reestablishes transport connection if there is a transport connection failure • Expedites data transfer for high-priority items	• Sets terms and conditions of session, such as who transmits first, how long, and so on • Establishes accounting functions for charging

TABLE 3.6
Presentation Layer—Level 6

What the Layer Does	What's Going On
• Selects the syntax to be used • Negotiates the syntax with its corresponding presentation layer on a second computer • Requests the establishment of a session • Encapsulates data in protocol data units	• Determines way data are formatted and presented to user at terminal • Application program talks to virtual terminal in this layer and this layer transforms to the real device

TABLE 3.7
Process or Application Layer—Level 7

What the Layer Does	What's Going On
• Establishes authority to communicate • Identifies intended communication partners • Agrees upon level of privacy • Determines whether resources are adequate for communication • Decides upon an acceptable quality of service • Agrees upon responsibility for error recovery • Agrees upon responsibility for data integrity	• Services to user applications • Layer for data editing, file update, user thinking • Source and destination for data

application layer
Interfaces directly to and performs common application services for the application processes.

The following mnemonics may help you remember the layers:
"Peter Dances Near The Soft Pink Apples."
"Programmers Dare Not Throw Salty Pretzels Away."
"People Design Networks To Send Packets Accurately."
"All Pizza Seems To Need Diet Pepsi."

Such a capability relieves the application programmer of having to account for many possible terminal types.

Layer 7—The Application Layer

Applications, from the user perspective, are functionally defined by the user and include such things as banking through ATM machines and airlines reservations. When the application involves cooperation between separated computers, the OSI **application layer** focuses on unique communications services, such as file transfer, electronic mail, and remote terminal-entry protocols. (See Table 3.7.) This layer, the highest in the hierarchy, provides interfaces to user-oriented services such as determining the data to be transmitted, the message or record format for the data, and the transaction codes that identify the data to the receiver. (See Table 3.8 for the entire seven-level model.) This layer does not define what the user sees or how the user interfaces with the computer or network; the application client does that. This layer is the intermediator between the client program and the presentation layer.

TABLE 3.8   Summary of the Functions and Services Provided by OSI Model Layers

What the Layer Does	What's Going On
Level 7—Process (Application) Layer—Exists in Host Computer	
• Establishes authority to communicate • Identifies intended communication partners • Agrees upon level of privacy • Determines whether resources are adequate for communication • Decides upon an acceptable quality of service • Agrees upon responsibility for error recovery • Agrees upon responsibility for data integrity	• Services to user applications • Layer for data editing, file update, user thinking • Source and destination for data
Level 6—Presentation Layer—Exists in Host Computer	
• Selects the syntax to be used • Negotiates the syntax with its corresponding presentation layer on a second computer • Requests the establishment of a session • Encapsulates data in protocol data units	• Determines way data are formatted and presented to user at terminal • Application program talks to virtual terminal in this layer and this layer transforms to the real device

TABLE 3.8 (Concluded)

Level 5—Session Layer—Session Control—Exists in Host Computer

- Establishes and maintains transport connection with a certain designated quality of service
- Organizes and synchronizes dialogues between presentation entities
- Ensures the reliable transfer of data
- Reestablishes transport connection if there is a transport connection failure
- Expedites data transfer for high-priority items

- Sets term and conditions of session, such as who transmits first, how long, and so on
- Establishes accounting functions for charging

Level 4—Transport (Control) Layer—Exists in Host Computer

- Ensures that transport data protocol units arrive in proper order
- Specifies grade of service, including acceptable error rates
- Monitors status of connection
- Establishes connections with another transport layer on a second computer
- Detects and recovers errors

- Selects the transmission route between DTEs, comparable to postal system for letters
- Handles user addressing; controls form of messages to compensate for speed mismatch
- Prevents loss and duplication of entire message
- Multiplexes several streams into one physical

Level 3—Network (Control) Layer—Exists in Front-End Processor (FEP)

- Routes packets
- Establishes connections between transport layers on two computers that will communicate with each other
- Controls flow; stops the flow of packets upon request from the transport layer
- Maintains correct sequence of packets
- Provides transport layer with acknowledgment that packet has been received correctly
- Maintains quality of service
- Is responsible for detecting and correcting errors
- Is responsible for multiplexing several network connections to a single data link connection for maximum productivity
- Handles internetworking, the movement of data from one network to another network

- Provides network addressing and routing
- Generates ACK (positive acknowledgment) that entire message was received correctly
- Breaks message from level 4 into blocks for level 2
- Decides over which communications circuit to send packet

Level 2—Data Link (Control) Layer—Exists in FEP and/or Controller

- Segments bit stream into frames
- Coordinates data flow split over multiple physical connections
- Is responsible for error detection and correction
- Monitors flow of data frames and compiles statistics
- Ensures that data frames are in sequence

- Defines standards for structuring data into
  - How machine knows where a frame starts
  - How transmission errors are detected, corrected
  - How polling and addressing are handled
  - How machines are addressed
- Data link protocol used is High-level Data Link Control (HDLC); normally, located in host FEP
- Whereas modems ensure that bits are accurately sensed for the communication line at the receiver, this layer groups bits into characters for processing
- Responsible for error checking VRC, LRC, and CRC codes
- ACK/NAK for receipt of good frames/blocks, and requesting retransmission if error found

Level 1—Physical (Link Control) Layer—Exists in Terminal Modem

- Physically establishes a connection when requested to do so by the data link layer (layer 2)
- Physically transmits data as bits
- Concerns itself with establishing synchronization of bit flow—duplex, half-duplex, point-to-point, multipoint, asynchronous, or synchronous transmission
- Defines quality-of-service parameters

- Defines electrical standards and signaling required to make and break a connection on the physical link to allow bit stream from the DTE onto the network
- Specifies the modem interface between DTE and the line
- Concerned with voltage levels, currents, simplex/half- or full-duplex

FIGURE 3.3
OSI Model

Layer	Title	Actions
1	Physical	Delivers bits from node to node
2	Data link	Delivers error-free data
3	Network	Defines message addressing and routing
4	Transport	Maintains reliable and cost-effective communications channel
5	Session	Organizes logical session
6	Presentation	Formats and displays the data, provides security, encodes, and encrypts
7	Application	Functionally defined by the user

Implication

The seven-layer model has often been extended in a humorous manner, to refer to nontechnical issues or problems. A common joke is the nine-layer model, with layers 8 and 9 being the "financial" and "political" layers. Network technicians will sometimes refer euphemistically to "layer-eight problems," meaning problems with an end user and not with the network. Carl Malamud, in his book *Stacks,* defines layers 8, 9, and 10 as "Money," "Politics," and "Religion." The "Religion layer" is used to describe nonrational behavior and/or decision making that cannot be accounted for within the lower nine levels. (For example, a manager who insists on migrating all systems to a Linux platform "because everyone else is doing it" is said to be operating in Layer 10.)[8]

OSI Summary

The subject of open systems architecture is somewhat removed from the decisions of the manager, except that it provides an underpinning for his/her decisions. Realizing that upgrade and interconnect are inevitable, managers must choose equipment that will readily connect to that of other suppliers and that is designed for expansion and added features. (See Figure 3.3 for an OSI model summary.) Choosing a design that inhibits either of these attributes not only commits your organization to one vendor, it locks you into this one vendor's view of expansion and upgrade. Figure 3.4 also gives an analogy of data moving down and up the seven layers. In this case, a telephone call with language translation is used. If the situation had involved data, the appropriate layers would have employed error detection and correction, **security,** encryption, compression, code/protocol conversion, communication channel continuity, and response in the event of a disruption.

The TCP/IP Model

The **transmission control protocol/Internet protocol (TCP/IP)** was developed by a Department of Defense (DoD) research project begun in the 1960s to connect a number of different networks designed by different vendors into a network of networks. The project was initially successful because it provided a few basic services everyone needed, including file transfer, electronic mail, and remote logon (see Chapter 7 on the history of the Internet). Because TCP/IP is a file transfer protocol, consisting of the two layers TCP and IP, it can send large files of information across sometimes unreliable networks with great assurance that the data will arrive in an uncorrupted form. It allows reasonably efficient and error-free transmission between different systems, and has become the standard of choice when using the Internet.

[8] http://en.wikipedia.org/wiki/OSI_model.

FIGURE 3.4 **Multiple-Layer Model Showing Translation of Communications**

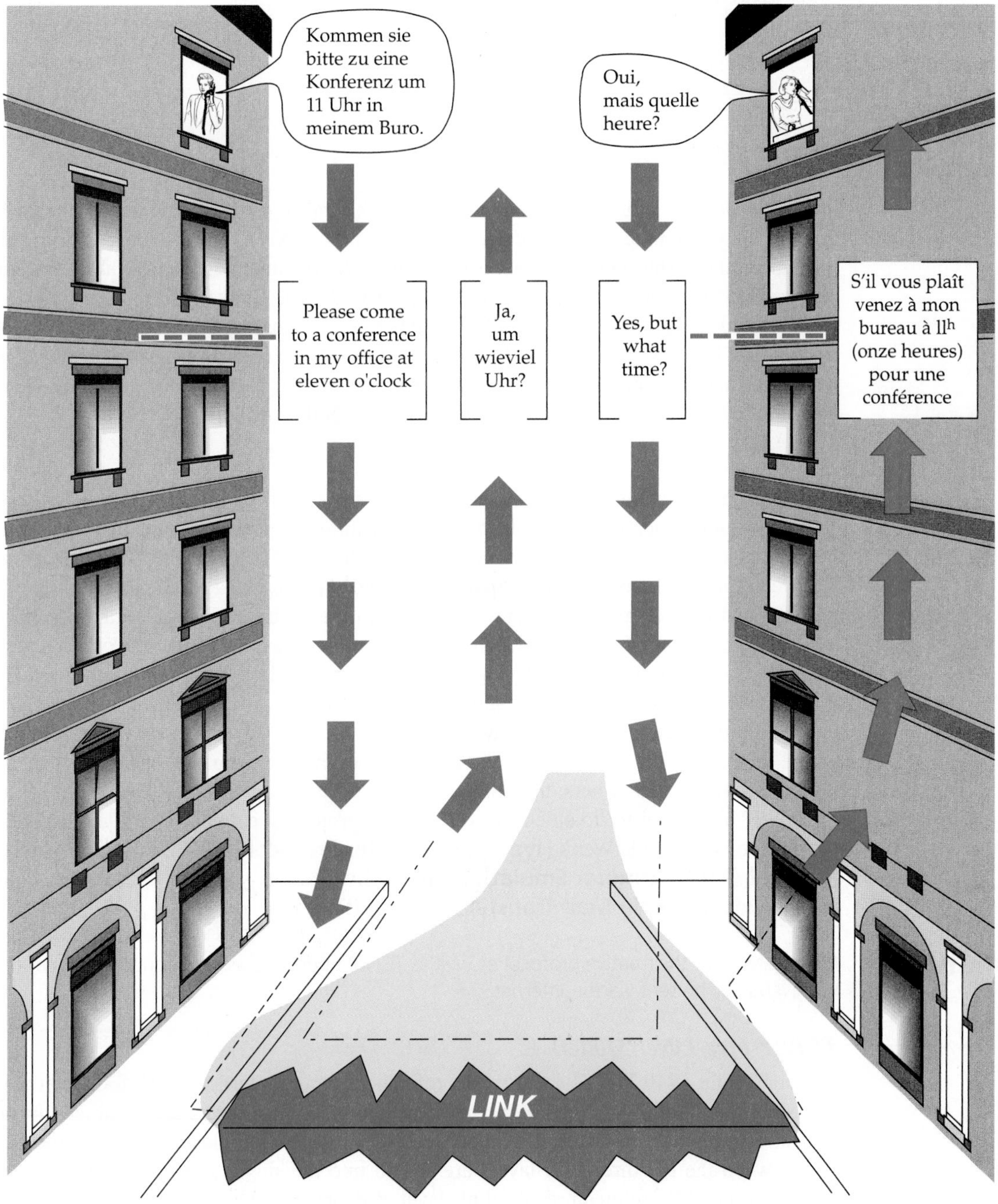

Implication   The **TCP/IP** is the most used protocol on public networks and generally co-resides on private networks.

The Internet protocol suite, commonly known as TCP/IP, is a de facto standard that is used to express the details of how computers communicate with each other. It is also a set of conventions for interconnecting networks and routing traffic. TCP/IP is independent of network hardware, carrier systems, and interconnected networks. All it needs is a corresponding suite

TABLE 3.9
**TCP/IP Five-Layer
Model**

TCP/IP Layers

Application
Transport
Internet
Data link
Physical

of TCP/IP details at the far end, and it will pass the traffic as needed. It adapts for every form of long-haul data transport currently in use such as ATM, frame relay, X.25, private line, and dial up. The Internet is based on TCP/IP. Network and machine operating systems such as MS Windows (9X, ME, NT, 2000, XP), Novell, Banyan, IBM (OS/2, VMS), Hewlett-Packard, and SUN all support TCP/IP because it transcends operating systems such as DOS, UNIX, or Windows. TCP/IP utilizes a simple basic rule: "Act only as an envelope with a destination address and a return address. Ensure that the envelope gets to its destination safely and sealed. Don't worry about what is IN the envelope."

What Is TCP/IP?

TCP/IP is the basic communication language or protocol of the Internet. Basically, it is a set of two communication protocols that an application can use to package its information for sending across a private network or the public Internet. TCP/IP is a two-layer program. The higher layer, **transmission control protocol (TCP),** manages the assembling of a message or file into smaller packets that are transmitted over the Internet and received by a TCP layer at the receiving end, which then reassembles the packets into the original message. TCP is responsible for verifying correct delivery of data from client to server. The lower layer, **Internet protocol,** handles the **address** part of each packet so that it gets to the right destination. Each gateway computer on the network checks this address to see where to forward the message. Even though some packets from the same message are routed differently than others, they'll be reassembled at the destination. (See Table 3.9.)

TCP/IP also can refer to an entire collection of protocols, called a TCP/IP suite. These include the World Wide Web's **HyperText Transfer Protocol (HTTP); the File Transfer Protocol (FTP); Terminal Emulation (Telnet),** which lets you log on to remote computers; and the **Simple Mail Transfer Protocol (SMTP).**

Clarification

Most LANs have their native protocol as well as TCP/IP under the assumption that the packets will eventually be sent via the Internet.

TCP/IP Five-Layer Model[9]

Although eight of 10 TCP/IP layer model references on the Internet define four layers, we will show the five-layer model, realizing that the bottom two layers are combined in the four-layer model. An advantage of the five-layer model is the close relation to the OSI model when the top three OSI layers are folded into the single application layer. In the four-layer TCP/IP architecture, the link layer and physical layer are normally grouped together to become the **network access layer.** TCP/IP makes use of existing data link and physical layer standards rather than defining its own.

The Physical Layer   The physical layer describes the physical characteristics of the communication (e.g., conventions about the nature of the medium used for communication such as wires, fiber optic links, or radio links) and all related details such as connectors, channel codes and modulation, signal strength, wavelengths, low-level synchronization

[9] This discussion relies heavily on http://en.wikipedia.org/wiki/TCP/IP.

and timing, and maximum distances. The Internet protocol suite does not cover the physical layer of any network.

The Link Layer The **link layer** specifies how packets are transported over the physical layer, including the framing (i.e., the special bit patterns that mark the start and end of packets). Ethernet, for example, includes fields in the packet header that specify for which machine or machines on the network a packet is destined. This layer is sometimes further subdivided into logical link control and **media access control (MAC).**

Clarification

The data link control (DLC) layer of the OSI reference model and sometimes the TCP/IP model is divided into two sublayers: the logical link control (LLC) layer and the media access control (MAC) layer.

Internet Control Message Protocol (ICMP) supports packets containing error, control, and informational messages.

Internet Group Management Protocol (IGMP) establishes host memberships in particular multicast groups on a single network.

A companion to the TCP is UDP (User Datagram Protocol), a connectionless protocol that is less reliable except for small amounts of data.

The Network (or Internet) Layer The network layer uses the Internet protocol to send blocks of data **(datagrams)** from one point to another. IP is the major protocol of TCP/IP because each piece of data is sent over the network as an IP packet. All upper- and lower-layer communications must travel through IP as they are passed through the TCP/IP protocol stack. In addition, there are many supporting protocols in the network layer, such as ICMP, to facilitate and manage the routing process. IP performs the basic task of getting packets of data from source to destination. IP can carry data for a number of different higher-level protocols. Some of the protocols carried by IP, such as ICMP (used to transmit diagnostic information about IP transmission) and IGMP (used to manage multicast data), are layered on top of IP but perform network layer functions, illustrating an incompatibility between the Internet and OSI models.

The Transport Layer

The transport layer is where the transmission control protocol (TCP) is used by most Internet applications such as FTP, HTTP, and Telnet. TCP is connection-oriented and, therefore, the sender and receiver must establish a connection before data can be transferred. The transport layer provides a reliable byte stream, which makes sure data arrive complete, undamaged, and in order. TCP tries to continuously measure how loaded the network is and throttles its sending rate in order to avoid overloading the network. Furthermore, TCP will attempt to deliver all data correctly in the specified sequence.

The Application Layer

The application layer is the layer that most common network-aware programs use in order to communicate across a network with other programs. Processes that occur in this layer are application specific; data are passed from the network-aware program, in the format used internally by this application, and are encoded into a standard protocol.

Some specific programs are considered to run in this layer. They provide services that directly support user applications. These programs and their corresponding protocols include HTTP (the World Wide Web), FTP (file transport), SMTP (e-mail), SSH (secure remote login), DNS (Name <-> IP address lookups), and many others. Once the data from an application have been encoded into a standard application layer protocol, they will be passed down to the next layer of the IP stack.

Clarification

UDP (User Datagram Protocol) is a communications protocol that offers a limited amount of service when messages are exchanged between computers in a network that uses the Internet protocol (IP). UDP is an alternative to the transmission control protocol (TCP) and, together with IP, is sometimes referred to as UDP/IP. Like the transmission control protocol, UDP uses the Internet protocol to actually get a data unit (called a datagram) from one computer to another. Unlike TCP, however, UDP does not provide the service of dividing a message into packets (datagrams) and reassembling it at the other end. Specifically, UDP doesn't provide sequencing of the packets that the data arrive in. This means that the application program that uses UDP must be able to make sure that the entire message has arrived and is assembled in the right order. Network applications that want to save processing time because they have very small data units to exchange (and therefore very little message reassembling to do) may prefer UDP to TCP. The Trivial File Transfer Protocol (TFTP) uses UDP instead of TCP.

UDP provides two services not provided by the IP layer. It provides port numbers to help distinguish different user requests and, optionally, an error-checking (checksum) capability to verify that the data arrived intact. In the open systems interconnection (OSI) communication model, UDP, like TCP, is in layer four, the transport layer.[10]

IP Addressing and Routing

IP addressing is the backbone of what TCP/IP sets out to accomplish. This allows for a node that has an IP address to interact with other nodes connected to a network. TCP/IP also supports **routing,** which allows for computers across different networks to communicate. Any computer that has an IP address is allowed to connect to other computers that have IP addresses.

Clarification

Routing is the act of moving information across an internetwork from a source to a destination. Along the way, at least one intermediate node typically is encountered. Routing is often contrasted with bridging, which might seem to accomplish precisely the same thing to the casual observer. The primary difference between the two is that bridging occurs at layer two (the link layer) of the OSI reference model, whereas routing occurs at layer three (the network layer). This distinction provides routing and bridging with different information to use in the process of moving information from source to destination, so the two functions accomplish their tasks in different ways.[11]

IP Addressing and Subnetting

Currently there are two types of Internet protocol addresses in active use: IP version 4 (IPv4) and IP version 6 (IPv6). IPv4 was initially deployed on January 1, 1983, and is still the most commonly used version. IPv4 addresses are 32-bit numbers often expressed as four octets in "dotted decimal" notation (for example, 192.167.32.67).

This IP address[12] is referred to as a hierarchical address, as opposed to a flat or non-hierarchical address. A flat address would be a Social Security number; that is, it does not show parental relationships. An example of a hierarchical addressing scheme is the telephone system. For example, the phone number 251.887.9999 can be broken down. The first three digits, the 251 prefix, shows that the phone number trying to be reached is in lower Alabama, that is, the area code for the phone system. The next three digits are the exchange within that area code (Mobile County), and the final four digits are the user code. The IP hierarchical numbering scheme allows an organized way to route information across the Internet. The IP address is broken into two parts: the *network address* and the *node address*. These two parts are what gives IP addressing its layered structure.

The **network address** uniquely identifies each network. Every machine on the same network shares that network address as part of its IP address. For example, all IP addresses on Auburn University's campus have the network address of 131.204.

The node address is assigned to each machine or node on a network. This part of the address must be unique because it identifies a single machine. This number also can be referred to as a *host address*. So to continue with our example, in the complete IP address of 131.204.65.245, 131.204 is the network address and 65.245 is the node or host address.

[10] http://searchwebservices.techtarget.com/sDefinition/0,,sid26_gci214157,00.html.

[11] http://www.cisco.com.

[12] Adolfo Rodriguez, John Gatrell, John Karas, and Roland Peschke, *TCP/IP Tutorial and Technical Overview,* 7th ed. (Upper Saddle River, NJ: Prentice Hall, 2001).

TABLE 3.10
Network Addressing

Class	1st Octet	Number of Networks	Number of Nodes
A	1–126	Small	Very large
B	128–191	Medium	Medium
C	192–223	Large	Small

router
A device that forwards data packets along networks. A router is connected to at least two networks, commonly two LANs or WANs or a LAN and its ISP's network. Routers are located at gateways, the places where two or more networks connect.[13]

The designers of the Internet created classes of networks based on network size. (See Table 3.10.) For the small number of networks possessing a very large number of nodes, they created the rank Class A network. At the other end is, alternately, the Class C network, which is reserved for the numerous networks with a small number of nodes. The Class B networks fit in between. The classes of networks are determined by the first octet of numbers in the IP address. Class A networks occupy 1–126, Class B networks occupy 128–191, and Class C networks occupy 192–223. For the IP address of 24.17.123.145, in a Class A network, 24 is the network address and the node address is 17.123.145. This allows for a small number of networks with a large number of nodes or computers. However, in a Class C network, with the IP address of 204.12.134.245, the network address will be 204.12.134 with a node address of 245. This division of networks allows for organized routing of information. For example, if a router gets a packet that is slated to go to the IP address of 131.204.65.245, the router will know that, by looking at that first number, 131, the packet will be routed to a Class B network. This cuts down on inefficient routing. (See Table 3.11.)

A problem arises if an organization has several physical networks but only one **IP network** address. This can, however, be handled by creating subnets. **Subnetting** is a TCP/IP software feature that allows for dividing a single IP network into smaller, logical subnetworks. This trick is achieved by using the host portion of an IP address to create a subnet address. It also can be thought of as the act of creating little networks from a single, large parent network. An organization with a single network address can have a subnet address for each individual physical network. Each subnet is still part of the shared network address, but it also has an additional identifier denoting its individual subnetwork number. For example, take a parent who has two kids. The children inherit the same last name as their parent. People make further distinctions when referring to someone's individual children, like "Kelly, the Joneses' oldest, left for college." Those distinctions are like subnet addresses for people.

[13] http://www.webopedia.com.

TABLE 3.11 **IP Addressing Classes**

Class	1st Octet Range	Number of Networks/ Number of Hosts	Highest Order Bits	Characteristics (Number of Hosts versus Number of Networks)
A	1–126.xxx.xxx.xxx 7-bit network, 24-bit host	126 /16,777,214	0	Small / very large (50% of IPv4 unicast address space)
B	128.0–191.255.xxx.xxx 14-bit network, 16-bit host	16,384 /65,534	1-0	Medium / medium (25% of IPv4 unicast address space)
C	192.0.0–223.255.255.xxx 21-bit network, 8-bit host	2,097,152 /256	1-1-0	Large / small (12.5% of IPv4 unicast address space)
D	First octet 224–239		1-1-1-0	Used for IP multicasting

Back to the Auburn University example, the IP address is 131.204.65.245. Now, the network address is 131.204 because the first number, 131, falls within the Class B network. So, 65.245 is left. On Auburn University's campus, there are more divisions between the different colleges. The Auburn College of Business has three subnets, 64, 65, and 66. So we know that in 131.204.65.245, the 65 is the subnet address and the 245 is the node address, which is identifying a specific computer.

Methods of Delivery Based on IP Addresses

The majority of IP addresses refer to a single recipient; this is called a **unicast** address, a one-to-one relationship between a single source and a single destination. **Broadcasting** can take four forms. *Limited broadcast* addresses use all one bits in all parts of the IP address (255.255.255.255). This refers to all hosts on the local subnet. Routers do not forward this packet. *Network-directed broadcast* addressing is used in an unsubnetted environment. The network number is a valid network number and the host number is all ones, referring to all hosts on the specified network. An example IP address under this condition would be 190.109.255.255. The third form is *subnet-directed broadcast* addressing, where the network number is valid, the subnet number is valid, and the host number is all ones, for example, 192.109.54.255. This then addresses all hosts on the specified subnet. *All-subnets-directed broadcast* addressing is next, with a valid network number, the network subnetted, and the local part all ones (e.g., 192.109.255.255); the address then refers to all hosts on all subnets in the specified network.

After unicasting and broadcasting is **multicasting,** sending out data to multiple destinations. For large amounts of data, such as audio and video programs, IP multicast is more efficient than normal Internet (streaming) transmissions because the server can broadcast a message to many recipients simultaneously. Unlike traditional Internet traffic that requires separate connections for each source-destination pair, IP multicasting allows many recipients to share the same source. This means that just one set of packets is transmitted for all the destinations. Multicasting is achieved by use of Internet Group Management Protocol (IGMP). It is used to establish host memberships in particular multicast groups on a single network. The mechanisms of the protocol allow a host to inform its local router, using Host Membership Reports, that it wants to receive messages addressed to a specific multicast group.[14]

Finally, **anycasting** is somewhat of reverse notation. It provides an IP address that would specify a number of hosts; the first host to respond would then be used for service. This might be used for downloading files from multiple sites, with all sites notified; the first to respond, potentially the least busy, would be the one to provide service.

Internet Protocol Version 6 (IPv6)

As noted previously, IPv4 is four sets of three octets (192.168.65.12) and supports about 4 billion ($4 * 10^9$) addresses, while **IPv6 addresses** are 128 bits long; this corresponds to 32 hexadecimal digits, supporting about $3.4*10^{38}$ (340 undecillion). IPv6 addresses are composed of two logical parts: a 64-bit network prefix and a 64-bit host-addressing part, which is often automatically generated from the interface MAC address. They are normally written as eight groups of four hexadecimal digits. For example, 2001:0db8:85a3:0000: 1319:8a2e:0370:7344 is a valid IPv6 address. If a four-digit group is 0000, it may be omitted; for example, 2001:0db8:85a3::1319: 8a2e:0370:7344.

IPv4 addresses are easily converted to IPv6 format. For instance, if the decimal IPv4 address was 135.75.43.52 (in hexadecimal, 0x874B2B34), it could be converted to 0000:0000:0000:0000:0000:0000:874B:2B34 or ::874B:2B34.

Reusable-IP Addresses—NAT

To get around the IP address shortage problem, it is increasingly common for networks—ranging from large corporate networks to small home

---

[14] http://www.webopedia.com.

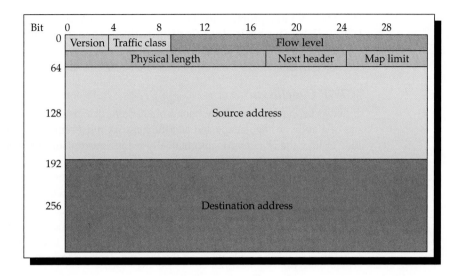

FIGURE 3.5
Structure of IPv6
Packet Header

networks—to be deployed using reusable-IP addresses. By connecting a reusable-IP network to the IP Internet through a **Network Address Translation (NAT)** gateway, unidirectional connectivity to the IP Internet is provided. That is, in general, reusable-IP hosts can initiate connections to IP hosts but not vice versa. Moreover, between two reusable-IP hosts belonging to different networks, there is generally no connectivity. Also NAT makes it difficult or impossible to use some peer-to-peer applications, such as VoIP and multi-user games. Currently, the big drive for IPv6 is new uses, such as mobility, **quality of service (QoS),** privacy extension, and so on.

IPv6 Packet The IPv6 packet (see Figure 3.5) is composed of two main parts: the header and the payload. The header is in the first 40 bytes of the packet and contains both source and destination addresses (128 bits each), as well as the version (4-bit IP version), traffic class (8 bits, packet priority), flow label (20 bits, QoS management), payload length (16 bits), next header (8 bits), and hop limit (8 bits, time to live). Next comes the payload, which can be up to 64K in size in standard mode, or larger with a "jumbo payload" option.

Limitations of TCP/IP

Any device that connects to the Internet must have a unique IP address or be behind a device that has one.

Finite Space The major problem with the current TCP/IP suite is the total number of unique IP addresses that can be created. Using the bit numbering system, there can be a maximum of 2^{32} (roughly 4 billion) unique numbers. This sounds like quite a large number, but it is pretty small when you look at everything that is connected to the Internet. Soon, there will be refrigerators and other home appliances connected to the Internet for such services as automatically ordering milk when the carton is low. The Internet is growing too quickly to support the number of unique addresses. To combat this problem, IPv6 was created. Compared to the IPv4, IPv6 will be 128 bits long. That is the old IPv4 addressing squared twice!

Maintenance

The use of TCP/IP on smaller networks is another drawback. TCP/IP requires a lot of overhead and therefore can slow up the speed of the network. "Overhead" in this context refers to the additional network control information, such as routing and error checking, that the protocol adds to data that the application layer needs to send across the network. It would be easier to set up another type of protocol on a small to medium-sized network. Another problem with maintenance is the configuration of the devices that use TCP/IP. Someone will have to be in charge of distributing the IP addresses and staying in charge of who-got-what.

This poses a large problem for IT managers. To counter this, there is another feature of TCP/IP called **Dynamic Host Configuration Protocol (DHCP).** DHCP allows for the user to just log onto the company network and the server will automatically assign that computer an IP address. This eliminates the hassles of keeping track of the IP address.

TCP/IP Conclusion

IP networks, having a military background, were designed to be robust; for example, the loss of a node or line is expected and the network must remain operational. With loss of a node or line, the IP network automatically reconfigures itself, because redundancy is built in. A congested node will attempt to reroute traffic; failing that, it discards the packets, relying on TCP to request retransmission until the total message is completed.

TCP/IP has had a dramatic effect on the Internet. It allows for an organized way of communicating across networks with different operating systems and hardware platforms. TCP/IP also has special features such as HTTP, FTP, Telnet, and SMTP. The benefits of TCP/IP completely outweigh the costs and disadvantages. These benefits include the ability to provide what people want on the Internet, leading to growth over the last few years.

Comparing TCP/IP and OSI Models

The TCP/IP model does not exactly match the OSI model. There is no universal agreement regarding how to describe TCP/IP with a layered model, but it is generally agreed that there are fewer levels than the seven layers of the OSI model. Most descriptions present from three to five layers. (See Figure 3.6.) The layers of the TCP/IP model are defined in the following sections.

Application Layer

In TCP/IP, the application layer also includes the OSI presentation layer and session layer. In this document, an application is any process that occurs above the transport layer. This includes all of the processes that involve user interaction. The application layer determines the presentation of the data and controls the session. In TCP/IP, the terms *socket* and *port* are used to describe the path over which applications communicate. There are numerous application level protocols in TCP/IP, including Simple Mail Transfer Protocol (SMTP, port 25) and Post Office Protocol (POP, port 110) used for e-mail, Hyper Text Transfer Protocol (HTTP, port 80) used for the World Wide Web, and File Transfer Protocol (FTP, port 20). Most application-level protocols are associated with one or more port numbers.

FIGURE 3.6
TCP/IP versus OSI Model Layers

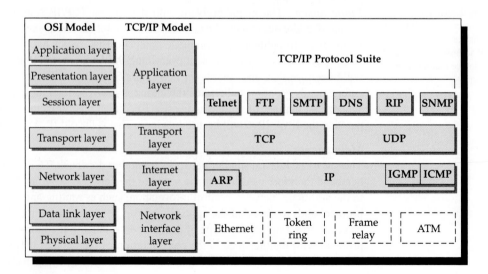

FIGURE 3.7
Use of Headers in TCP/IP

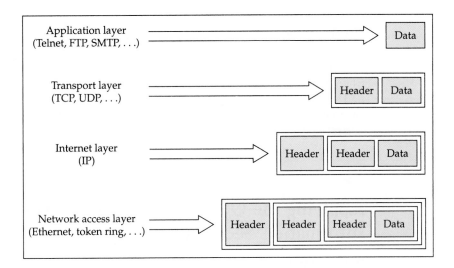

Transport Layer

In TCP/IP, there are two transport layer protocols. The transmission control protocol (TCP) guarantees that information is received as it was sent. The user datagram protocol (UDP) performs no end-to-end reliability checks.

Internet Layer

In the OSI reference model, the network layer isolates the upper-layer protocols from the details of the underlying network and manages the connections across the network. The Internet protocol (IP) is normally described as the TCP/IP network layer. Because of the internetworking emphasis of TCP/IP, this is commonly referred to as the Internet layer. All upper- and lower-layer communications travel through IP as they are passed through the TCP/IP protocol stack.

Network Access Layer

In TCP/IP, the data link layer and physical layer are normally grouped together. TCP/IP makes use of existing data link and physical layer standards rather than defining its own. The characteristics of the hardware that carries the communication signal are typically defined by the physical layer.

The four-layer structure of TCP/IP is built as information is passed down from applications to the physical network layer. When data are sent, each layer treats all of the information it receives from the layer above as data and adds control information to the front of those data. This control information is called a **header,** and the addition of a header is called *encapsulation*. When data are received, the opposite procedure takes place as each layer removes its header before passing the data to the layer above. (See Figure 3.7.)

Summary

To communicate and share complex ideas, man has used models, abstractions of the reality. At the lower levels, this may be a standard by which some item is produced so that it will operate with other objects. Eli Whitney and Henry Ford, for example, were among the first to use standardization as a way to speed up manufacturing and lower the cost of production, using a method where parts were interchangeable between weapons and vehicles respectively. At a high level, architecture lays out how a group of components is related and will operate. A data communications architecture establishes the basis for the installation and connectivity of the parts of the system.

When dealing with a complex system, it is often necessary to partition the capability into layers. With a layered model, the parties agree that what happens on the interior of a layer is up to the designer but what happens at the boundary is set in the layered model. Such a layered model allows for interoperability from and across components provided by a numbers of vendors.

The primary layered models for data communications today are TCP/IP for the Internet environment and OSI for the macro environment. Understanding these models and their interrelationship allows the reader to follow processes that occur within and across them.

Case 3-1

Applying the OSI Seven-Layer Network Model to Information Security

Damon Reed

ABSTRACT

Data networking is a critical area of focus in the study of information security. This paper focuses on reviewing a key area of data networking theory—the open systems interconnect (OSI) seven-layer network model. This paper demonstrates the application of the model's concepts into the context of information security. This paper overall presents the perspective that common information security problems map directly to the logical constructs presented in the OSI seven-layer network model and seeks to demonstrate the seven-layer model's usefulness in evaluating information security problems and solutions. The OSI model is presented by way of both formal definition and practical terms that affect information security on a layer-by-layer basis. For each layer, examples of common information security threats and controls are evaluated by how they fit into the OSI seven-layer model's layers of classification, with notes on exceptions and special cases. Once the seven layers have been covered as a basis for the discussion, it is presented that the seven-layer model's scheme for interaction between the layers gives insight to some of the problems faced by focused, "single-layer" security solutions. To answer these problems, a multilayer "defense-in-depth" approach is examined by example, taken from the viewpoint of network model layers rather than discrete solutions and logical or physical hardware layers. This paper concludes with some proposed extensions to the model that complete the model's application to information security problems.

To read this excellent article on the OSI layered model and network security, please refer to http://www.sans.org/rr/whitepapers/protocols/1309.php

---

Key Terms			
	address, *p. 96*	de jure, *p. 87*	Federal Communications
	anycast, *p. 100*	Dynamic Host	Commission
	application layer, *p. 92*	Configuration Protocol	(FCC), *p. 84*
	architecture, *p. 83*	(DHCP), *p. 102*	File Transfer Protocol
	broadcasting, *p. 100*	electronic data interchange	(FTP), *p. 96*
	data link layer, *p. 89*	(EDI), *p. 85*	header, *p. 103*
	datagram, *p. 97*	electronic funds transfer	HyperText Transfer
	de facto, *p. 84*	(EFT), *p. 109*	Protocol (HTTP), *p. 96*

Recommended Readings

Network Magazine (http://www.networkmagazine.com/), periodical about networks and network management.

Computerworld (http://www.computerworld.com/) "brings you the business issues, news and technology advice you need to succeed in today's electronic, global communications."

Management Critical Thinking

M3.1 Why is it important for the telecommunications manager to understand the various layered models?

M3.2 Is it important for other senior managers to understand these layered models? Why or why not?

M3.3 What is the purpose of a data communications architecture? Should all businesses have an architectural plan? What facets should be incorporated in such a plan?

Technology Critical Thinking

T3.1. The choice of an architecture has far-reaching consequences. How can the choice of a particular architecture be made a part of organization thinking so it can guide technical decision making?

T3.2. What are the major technology issues involved in adopting OSI to TCP/IP and vice versa?

Discussion Questions

3.1 What are the implications of buying noninteroperable (proprietary) equipment?

3.2 Does your organization/school have an architectural plan for the campus? How does it relate to the telecommunications architecture plan?

3.3 What is TCP/IP? What does TCP do? What does IP do?

3.4 Which is the most important of the seven OSI layers?

3.5 What is the value of using IP subnetting?

3.6 Why do we employ layered models if there is lack of agreement on a standard?

3.7 Would there be a difference in the architecture plan for a wired and wireless network?

3.8 What are the major advantages offered by EDI?

3.9 Does it appear that your organization/school has a building architecture plan? What are it's objectives?

Projects

3.1 Determine the architecture used in your school or organizational LAN.

3.2 Interview telecommunications managers and determine their beliefs as to the relevance of the OSI layered model.

3.3 Call an IBM representative and ask if the company is moving from their SNA model to the OSI model.

3.4 Determine the components of the IP address of your school or organization <network: host>>. Does the organization use subnetting?

3.5 Research telecommunications/data communications literature and the Web to find predictions about the future of TCP/IP.

3.6 Interview a telecommunications manager and determine the role of standards in preparing purchasing documents for data communications systems.

3.7 Create a figure that compares OSI and TCP/IP layering to indicate how the layers match up with one another.

3.8 Go to the Web and determine the standards agencies that manage the various layered models.

3.9 Create an IP address for a new firm; specify each component. Do it with IPv4 and IPv6.

3.10 Determine the standards that are required to make the Internet. Where can you find this information in summary form.

Appendix 3-1

Electronic Data Interchange (EDI)

BACKGROUND

To demonstrate the potential for EDI to improve the information and paperwork flow within and between organizations, consider the following scenario. The purchasing department at Company A decides to procure a particular product from supplier Company B. In order to accomplish this task, the Company A purchasing department creates a purchase order. Typically, this process consists of either manually typing the necessary information on the company's unique purchasing form or computer generating the form through some mode of online entry. The purchase order document is then forwarded to Company A's mail room for subsequent pickup by a mail service (Postal Service or other carrier). The document is taken to the local mail service office where local mail is removed and the residual forwarded to a central sorting site for routing to the mail service office closest to the supplier. At the local mail service office, the document is placed in a vehicle for transport and delivered to supplier Company B's mail room. The mail room at the supply company sorts the inbound mail and delivers the purchase order to the order department, where it is likely to be keyed into an order processing system. The order processing system triggers the creation and/or delivery of the product or service and the generation of a corresponding invoice. Whether the invoice is manually generated or computer printed, it requires distribution to the mail room for pickup. Some invoices may require envelope stuffing either before or at the mail room. The invoice is then transported to the local mail company pickup office and to the central sorting site, and again routed to the office closest to Company A for delivery. The invoice is then received by the mail room of Company A, sorted, and distributed to the ordering department for verification of receipt of service or merchandise.

The invoice may be keyed in and is then marked for payment and forwarded back through the mail room to the accounts payable department and keyed for payment. The notification of recently received merchandise also might be input into some computer-based inventory or other asset management system.

In this example, it is important to note the number of opportunities for the document to be delayed, mishandled, or lost. In addition, the same basic information has to be keyed and verified (or potentially corrected) a number of times. Of great importance is the time required to complete the process.

EDI SCENARIO

EDI is the direct computer-to-computer communication of information in a standard format that permits the receiver to perform a specific business function. Consider the foregoing purchasing example within an EDI framework. Figures 3.8(a) and 3.8(b) show the points of the transaction in question. In an EDI environment, the purchasing agent at Company A would simply enter the purchasing information into Company A's micro- or mainframe computer, or a computer-based inventory or scheduling system would generate a purchase automatically, an action unlikely to occur without EDI. The system would electronically forward the purchase order on a predetermined schedule to supplier Company B's computer, after EDI management software or software created by the MIS department had converted the stored format data to EDI format. The receipt of the electronic purchase order after conversion by Company B's EDI management software would create an automatic update of the order entry system, triggering the generation and/or delivery of the product or service by the supplier. As a result of the rendering of the product or service, Company B would electronically generate, translate, and transmit a corresponding EDI invoice to the purchasing company's computer. Verification of merchandise received becomes a check-off process and inventory and accounts payable files or databases are automatically updated.

FIGURE 3.8(a)
The Integrated Electronic Business Cycle

The integrated electronic business cycle
*ASN = Advance Ship Notice

FIGURE 3.8(b)
An EDI Model

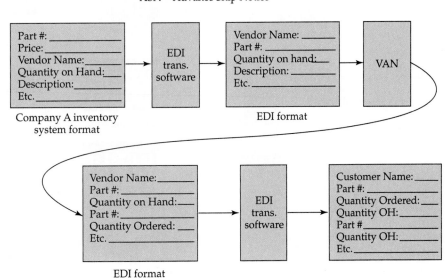

Although each partner company may be fully automated within their own company environment, each step in the manual intercompany transfer process takes a significant amount of time and has an opportunity for error (e.g., either lost paper or incorrectly keyed information). The old adage that "time is money" is appropriate, particularly if the purchased item is to be utilized as a component in a subsequent product offered for sale. A new adage might be "delay is at least expensive, if not a lost sale." Figure 3.9a shows some of the uses of EDI in business transactions.

FIGURE 3.9
Financial EDI versus EDI

a. Example standard business transactions that can use EDI

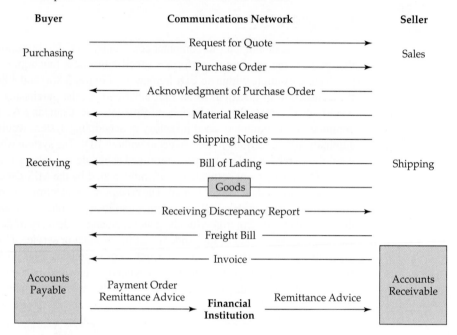

b. The payment process using EDI and EFT

- Single format
- Single bank
- Multiple payment types

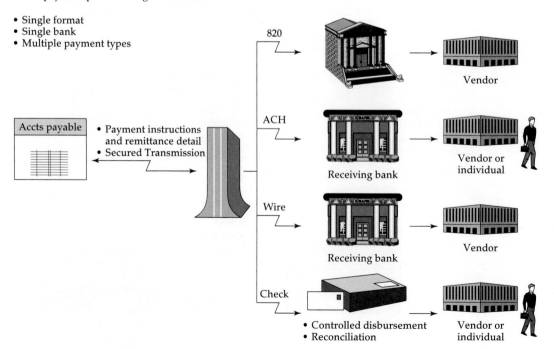

EDI, along with the use of **electronic funds transfer (EFT),** allows two or more organizations with differing computer systems to communicate their needs in real time, without much, if any, human intervention. (See Figure 3.9b.) Just-in-time inventory methods rely heavily on EDI, allowing a manufacturing system to request delivery of parts and make them available on the factory floor when needed. This negates the need for warehouses and stockpiles of expensive goods. It all relies on common standards and a telecommunications network.

Clarification

EDI-formatted files place the order and acknowledge receipt; EFT pays for the order.

While EDI appears to offer great advantages, there are several problems that must be addressed. For example, the standards used vary both by industry and internationally. While the X.12 ANSI standard is designed to be universal, industries such as grocery and transportation began developing a standard before X.12 came into being. Meanwhile, EDIFACT was developed in Europe and has differences from U.S. standards. There is also a need for EDI translation (management) software for both sender and receiver(s). The value-added network (VAN), while considered an added expense, provides the interface between differing operating systems. The VAN also provides **store-and-forward** capabilities. The VAN can be replaced by a direct connection, including the Internet; however, they will not have store-and-forward capabilities.

Clarification

A value added network (VAN) is a privately owned network, as opposed to the Internet, providing transport and other services. If only transport is provided, this is called a *common carrier*. VANs are not regulated but have been replaced to a great extent by use of the Internet.

There are three levels of players in the arena of EDI. As Figure 3.10 shows, relationships exist between sources, destinations, and third-party providers. The alliances made in this environment change the nature of business. For example, the means of transportation change with many small packages sent just-in-time by UPS instead of one large shipment by a freight carrier.

FIGURE 3.10
Parties and Their Interactions with EDI

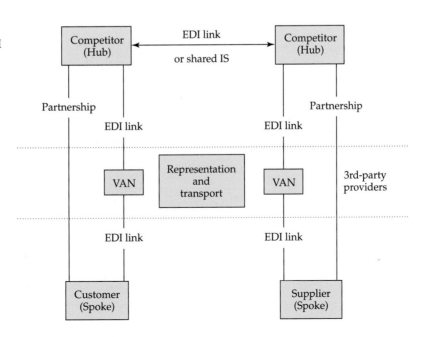

Part Two

Network Basics

In order to establish networks, there must be an understanding of a myriad of concepts and terms. Networks are classified by topology, protocol, forms of connectivity, reach, and the equipment components that make them work. Issues of codes, signal reliability, protection and security, and storage are all important. While most of the network basics refer to wired networks, wireless technology is introduced.

Chapter Four

Building a Network: Topology and Protocols

FIGURE 4.1
Point-to-Point (*n* = 2)
Network

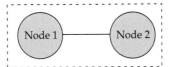

In order to build a network, we need an understanding of topology options and their features. The role of protocols is important in the network design and configuration. There are several equipment technologies that need to be understood if our networks are to operate efficiently and with effectiveness for the organization.

In discussing networks, we need to describe how they are physically constructed, physically joined, and electrically connected. The arrangement of the nodes is called the *topology* of the network. Additionally, all networks must have *rules* that control communications. The sets of rules are called *protocols*. While it is useful to separate the concepts of topology and protocols for initial discussion, we combine them later as they are interrelated.

The arrangement of the chapter is to discuss network form first and then describe the equipment that makes it up. Reversing this order is, obviously, an option. In either case, the discussion of wireless topology and protocol is presented in Chapter 9.

NETWORK CONNECTIVITY BASICS

network
Consists of two or more nodes, connected by one or more channels.

The *n* in *n = 2* means the number of nodes in the physical network.

An *n* = 2 network requires either a special cable or use of a connecting device, making it an *n* = 3 network.

The way that networks are arranged, constructed, and connected is called **topology.** The simplest network is one where there are only two computers (called an *n = 2 network,* the *n* meaning the number of **nodes**). If both nodes are computers, it is a **point-to-point topology.** (See Figure 4.1.) Most likely the medium connecting the machines (nodes) remains UTP. This is the **physical topology,** simply meaning how the wires are run from node to node. **Electrical topology** would be the communications path between machines and how each machine can share the other's resources or can send messages and data to the other. Networks are generally run out from a central telecommunications closet (forming a physical star) and connected electrically at the central node, which will determine the electrical topology. See Table 4.1 for potential benefits of networks.

The alternative to point-to-point topology is *multipoint topology.* In this case, many computers are connected to a common communications **channel.** (See Figure 4.2.) This means that the nodes are connected locally. When discussing widely dispersed nodes, the view is different than multipoint connectivity because messages will be sent differently. Multipoint access, one of the primary aspects of the network protocol, provides rules for multiple nodes to access the network channel and determine how they communicate.

Relationships between Nodes

There are two relationships between connected nodes. One is where (in an *n* = 2 network) both computers are of the same standing and neither is a master or server nor a slave or client. This would be a **peer-to-peer** network. *Windows®* operating systems include peer-to-peer networking. Different relationships are discussed later.

Implication

When you dial into your account on the bank's computer, you have a point-to-point but not peer-to-peer connection.

A way to enhance the use of resources is **cooperative processing,** the sharing of computer processing tasks between two or more processors, either within or between computers.

TABLE 4.1
Potential Benefits of Networks

I. Cost Savings
   A. *Reduced number of peripherals*, such as printers, application and file storage, and facsimile modems, as the network allows sharing of peripherals among users on the network.
   B. *Less storage space required per workstation*, as only one copy of software is stored on the publicly accessible file server.
   C. *Reduced software costs for workstations*, since the LAN keeps the only copy of the software. This generally means that although there must be a license and a number of copies to cover the number of users who use the software at any *one time*, the number of copies is generally less than the total number of users.
   D. *Single-network communications*, reducing the cost of redundant networks. Thus, as more users are connected to one network or as smaller networks are interconnected, communications and sharing increase.
   E. *Savings in installation and maintenance*.
      1. New stations are easy to install and existing stations can be moved without substantial cost.
      2. *Only one copy of each software package must be installed or updated*.

II. Sharing of Resources
   A. *Reduced cost per resource item* as peripherals, applications, and files are shared.
   B. *Potential availability of more expensive peripherals* that would not be justified for single users (for example, plotters or expensive printers).
   C. *Greater use of resources* as they are shared.
   D. *Reduced redundancy* as only single application resources are required to be maintained/ updated.

III. User Interface
   A. *Interface with the network* is transparent.
   B. *Less or no involvement by users in operational tasks* such as backup and recovery. This is true only for files and applications stored on shared devices on the network. Generally, this does not include private workstation storage.
   C. *Capability of sharing data and text for joint work* or review by management, where software exists that allows for joint access to a file or application.
   D. *Added functionality* provided by the network in terms of electronic mail. One of the prime values of telecommunications is facilitation of human communications; networks enable more effective communications.

IV. Encourages Management Control
   A. *Control of the introduction of new technology* by imposition of network standards.
   B. *One network management environment* may be established versus managing multiple network environments at the same time.
   C. *Equipment compatibility* by requiring a single protocol.
   D. *Uniformity of communication and resources* as all users are accessing and sharing the one network.

A similar operation, **distributed processing,** is the dispersion of data processing among computers in multiple sites.

The second relationship is when the nodes on the network are not equal. For example, Figure 4.2 shows a network of six computers. All but one of the computers are user workstations and one provides services to the rest. This is called a **client-server architecture (C/SA).** As it implies, the workstations are clients of the server. The clients request services from the server, such as shared file space, common printers, and shared processing.

Shared resources
Any devices such as a hard drive or printer that can be accessed and used by all nodes on the network.

FIGURE 4.2
Multipoint (*n* > 2) Network

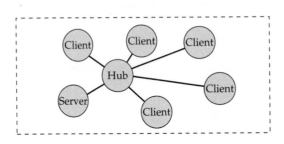

This requires the connectivity of a network and creates an environment where updates for all clients can be as simple as updating a single application on the server. While this avoids going to each client for a software update, licenses for all users are required.

The basic construct behind client-server architecture is a client end and a server end that are distinguishable, yet interact with each other. Basic C/SA consists of two independent processing machines, in separate locations, each capable of performing operations independently, but operating with greater efficiency together than alone. A single **server,** the base machine around which the architecture is designed, can have multiple **clients.** The C/SA environment differs from its predecessor, the mainframe-centered (often called *legacy*) environment, in that much of the processing may now be performed by the client. Previously, the client was a **dumb terminal,** that is, one with neither storage nor processing capabilities. Figure 4.3 presents four typical client/server architectures that could be used in a company's implementation of the technology.

FIGURE 4.3   Models of Client-Server Implementation

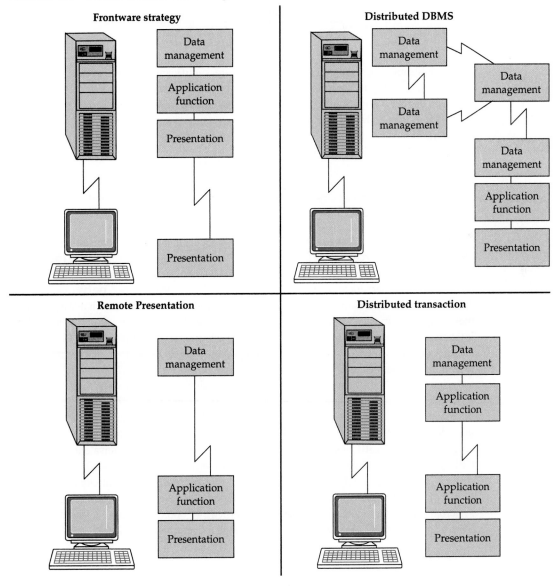

FIGURE 4.4
Client-Server
Architecture

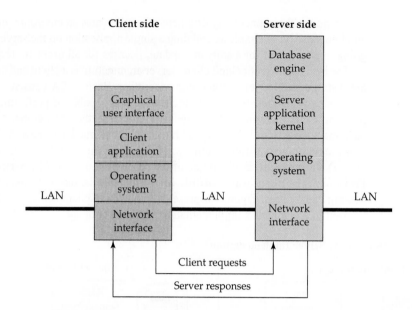

The advantages of client-server operations are significant. The ability to interconnect computers, not only to share information but also to share the processing of information, creates a distinctive synergistic relationship transcending the capabilities of either machine in isolation. By combining the processing capacity and speed of a high-end micro-, mini-, or mainframe computer as a server with the cost benefits and end-user support of lower-cost microcomputer-based applications, clients can quickly and comfortably access data from remote sources.

Clarification

A **server** is a dedicated resource (microcomputer or mainframe) that shares disk space, processing, and other resources to serve the other nodes on the network.

local area network (LAN)
Connectivity for homogeneous computers closely located.

The foundation of C/SA is the network that connects the clients and the server(s). A simple example of this architecture is the software that attaches a printer server to a **local area network (LAN)** to serve all users (clients) on the LAN. A more complex configuration could have these servers and clients connected by way of a **wide area network (WAN).** The extreme example is the use of a Web browser on the Internet. With such a Web environment, a query is made and the Web browser (a client) seeks out numerous servers to find the information, without the user knowing where the information came from. (See Figure 4.4.)

wide area network (WAN)
Connectivity for heterogeneous computers, remotely located.

An example of C/SA and distributed computing is **peer-2-peer** or **grid computing (P2P),** first made practical with SETI (*Search for Extraterrestrial Intelligence*). This environment has a server at its center that loads a data processing program on a very large number of voluntarily connected but independent computers. It then sends data files to be processed to the clients; in the case of SETI, there are 5,000,000 PCs worldwide that perform the processing when they would otherwise be idle. SETI began processing worldwide in 1999. As of August 2005, 5,436,301 members had returned 1,956,938,775 results, equivalent to 2,347,122 CPU years of processing.[1] In like manner, *United Devices* of San Antonio, Texas, created a similar P2P system of 3,186,267 idle computers to process mathematical chemistry data in search of medicines for cancer, anthrax, and smallpox. As of August 2005, the *United Devices* system of Internet-connected computers has returned 341,066,660 results by way of an equivalent of 428,558 CPU years of processing[2] since its inception in April 2001.

[1] http://setiathome.ssl.berkeley.edu/totals.html.
[2] http://www.grid.org/stats/.

Implication

When monolithic software programs can be partitioned into autonomous transactions, grid computing will take advantage of the unused computer resource in the organization.

C/S Fat and Thin Clients

The clients in C/S architecture have come to be called "fat" and "thin." **Fat clients** have large storage and processing power; therefore, they contribute to the processing capability but also to the high costs of C/S computing. **Thin clients,** on the other hand, frequently have no local storage, limited processing power, and reasonable connectivity to the server. Some of them are completely server controlled; some allow users to download applications to the client. The ultimate thin client is the **network computer (NC)** or **network appliance** that is little more than a keyboard, monitor, and network connectivity and depends on the connected server to do all the processing. This emulates the legacy mainframe-terminal environment except this CS/A environment may have access to multiple servers.

Implication

Thick clients offload processing and storage from the centralized server, but at the expense of hardware and software maintenance, upgrade, and replacement. Thin clients require less maintenance, no upgrade, and infrequent replacement. Both will require licenses for use of software.

Clarification

Thin clients can adequately use browser-based applications; some programs, such as *Novell's GroupWise*® and *IBM's Lotus Notes*® require thick clients.

WHAT IT TAKES TO MAKE A NETWORK

A **hub** is a **passive device** that provides a central point of contact for a network; it connects the nodes with shared bandwidth. A hub is like a three-way electrical plug; it shares the power on a single circuit breaker.

Clarification

A **switch** is an **active device** that provides a central point of contact for a network; it connects the nodes and provides dedicated bandwidth. Nodes cannot see other nodes. A switch is like adding a new circuit breaker panel with multiple circuit breakers; it distributes the power of the incoming line.

Clarification
Continuous Data versus Packets

Ever since 1993, *Microsoft's Windows*® operating system (version 3.11) has included networking capabilities. A simple, and slow network could use UTP to connect the serial ports for an $n = 2$ network, using a serial cable. As the network increases in size, that is, more *nodes,* and to achieve greater speed (aka bandwidth), you must install **network interface cards (NICs)** in each node, use a connecting **hub** for networks with more than two nodes, connect the nodes and the hub (also a node) using some **medium** (e.g., **10BaseT,** CAT 3 wiring), and install a **network operating system (NOS)** to make it all work. All local area networks (LANs) with multiple nodes must have all of these elements.

CATegory	Definition	Means	Usage
3	10BaseT	10 Mbps on baseline, using UTP	Ethernet
5	100BaseT	100 Mbps on baseline, using UTP	Fast Ethernet
6	1000BaseT	1000 Mbps on baseline, using UTP	Giga Ethernet

The medium provides connectivity, for example, **100BaseT** CAT 5 cable. Networks of all sizes have limits related to the length of the medium from a node to its central attachment point. Most networks use UTP for bandwidth up to one giga-bps and distances of up to 100 meters. Beyond 100 meters, switching to fiber allows for even higher bandwidths over longer distances.

When data communications first began with the telegraph system, the dots and dashes of the Morse code were continuously sent once a message began. In today's networks, data are organized into **packets.** The format of the packet is discussed in a later chapter, but it is important to realize that the data being sent over wired or wireless networks are first divided into standard-sized groupings (128 to 1,024 bytes) and organized into a packet along with other overhead information. It is this packet that is sent over a network.

The *network interface card (NIC)* operates at the physical layer of the OSI model and exchanges packets with the network as bits. It sits on the node's internal bus and provides connectivity to the outside world. A NIC has sufficient intelligence so that incoming traffic that is not destined for its node does not pass to the node's interior to avoid taking up resources and processing time.

Clarification

The computer's internal bus provides communications among the various devices on the computer, such as the CPU, main memory, hard drive, and CD drive, plus all ports to the external world. Inserted cards attach to this bus to transfer data at a high speed.

Network operating systems contain the protocol used on the network.

The *network operating system (NOS)*[3] is to the network as the *operating system (OS)* is to the node. The NOS contains the **protocols** that institute the *rules for communications* and control access to the network. A NOS is software that (a) controls a network and its message (e.g., packet) traffic and queues, (b) controls access by multiple users to network resources such as files, and (c) provides for certain administrative functions, including security. It provides printer sharing, common file system and database sharing, application sharing, and the ability to manage a network name directory, security, and other housekeeping aspects of a network.

Peer-to-peer network operating systems allow users to share resources and files located on their computers and to access shared resources found on other computers. However, they do not have a file server or a centralized management source.

Client-server network operating systems allow the network to centralize functions and applications in one or more dedicated file servers. The file servers become the heart of the system, providing access to resources and providing security. Individual workstations (clients) have access to the resources available on the file servers. The NOS provides the mechanism to integrate all the components of the network and allow multiple users to simultaneously share the same resources irrespective of physical location, when authenticated by the security function and authorized access.

Windows for Workgroups 3.11 in 1993 started the OS-resident NOS to support peer-to-peer computing. *Apple* and DOS computers had client-loadable NOS as early as the mid-1980s. Networks do not necessarily depend on servers to house the NOS. The two prevalent network-loadable NOS are *Novell Netware* and *Windows NT*.

The final part of an $n = 3$ network is one or more connecting devices. (These are discussed in greater detail later.) In its simplest form, the central connective device provides connectivity of all its ports to a common channel. Nodes connect from their NIC to the device, which then ties all nodes together. Various forms of devices perform added services, such as switching of bandwidth and bridging across networks, to be discussed later.

Implication

A **hot key** is one or more keystrokes in combination that cause a macro command to run. The combination often includes a CTRL or ALT key. An example is CTRL-S to save the current file.

Many people have an $n = 2$ network via the printer attached to their computer. What about a larger network?

When you have a single computer, most of its control and input come from the keyboard and mouse and much of the output goes to the monitor. What happens if you must routinely use a second computer? Does this mean there must be two keyboards, two mice, and two monitors? With keyboard-video-mouse (KVM) technology, the answer is no. *KVM switches* allow two or more computers to be connected to a single monitor, keyboard, and mouse. Thus, changing computers means only pressing a key (such as the Scroll Lock "hot" key) and control changes from one computer to the other. KVM switches have physical cables from each computer to the switch and can control from 2 to 64 computers over a network, with prices ranging from $20 to $18,000. Simple KVM switches use standard cables, whereas larger switches use CAT 5 cables and network addressing for control of a large number of computers. Imagine the alternative in a server farm, that is, a room full of servers.

[3] http://en.wikipedia.org/wiki/Network_operating_system, http://searchnetworking.techtarget.com/sDefinition/0,290660,sid7_gci214124,00.html, http://fcit.usf.edu/network/chap6/chap6.htm.

TOPOLOGY

topology
The electrical or channel layout.

physical topology is how the wiring is laid out.

An **uninterruptible power supply** is a device to protect from power surges and continue providing power in the case of short-term electrical outages.

Physical security means controlled access and protection from theft, fire, and water damage.

Electrical topology is how the nodes use the paths

Most commonly, topology refers to the geographic arrangement of a network's components. In reality, topology comes in two forms: physical and electrical (logical). **Physical topology** is the way the connecting media are laid out. For local networks, the physical topology is *almost always a star*. This means the cables from the end node, the user workstation client or the server, run from that computer to a central point to attach to a hub. Thus, they *star out from the hub,* where it is generally located in a telecommunications closet.

The reason for the physical star network is simple; it is much easier to make the connections and troubleshoot a network in a single place than to go to the locations of the end nodes (usually scattered computers). Also, there is often a need for environmental control (i.e., cooling) and stable electrical power to run the equipment of the network, for example, a large connecting hub or switch and its attendant **uninterruptible power supply (UPS).** This equipment is located in the telecommunications closet with appropriate power, cooling, and physical security.

Electrical Topology

Electrical topology is the way the wires or cables are actually connected. One of the simplest electrical topologies, beyond the point-to-point $n = 2$ network, is the **star network** (see Figure 4.5(a) and (b)). The premise of an electrical star is that the nodes all connect to a single, central, *active node* that controls all communications. An example is a POTS network, where all channels terminate into a computer (or switch). To make a call, one phone (a node) must get the attention of the central switch (switching node) and have the node make the connection. In the star electrical topology, the nodes at the end of the channels acquire the attention of the central node before communications can take place, a

FIGURE 4.5   (a) Electrical and Physical Star Topology **(b)** Electrical and Physical Star Network

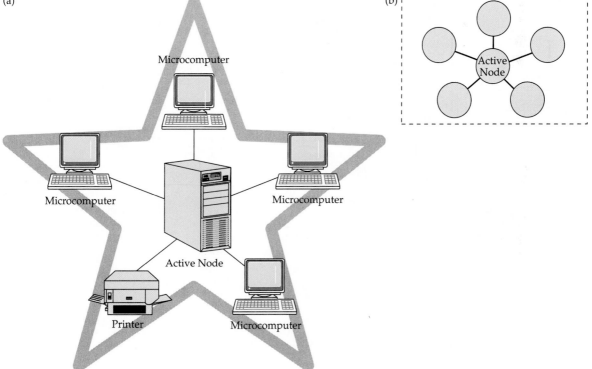

(a)

Microcomputer

Microcomputer

Microcomputer

Active Node

Printer

Microcomputer

(b)

Active Node

FIGURE 4.6   (a) Electrical (Physical) Network Using a Hub (b) Logical Network Using a Hub

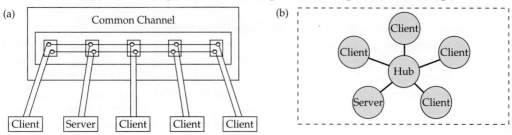

process often called **polling.** A specific problem with electrical star networks is that congestion or failure of the central node will congest or cause total network failure.

The second way to connect the nodes electrically is to make them all *multipoints* on a single, shared channel. As Figures 4.6(a) and (b) show, we use a hub to electrically connect each wire to a central wire, making a **bus network.** The physical topology is still a star, but the electrical connectivity provided by the hub makes it into an electrical **bus.** Figures 4.7(a) and (b) illustrate the physical and electrical topology of a bus network.

The third kind of local connectivity is the **ring network.** As shown in Figures 4.8(a)–(d), each node or computer is electrically connected to one computer above (or to its left) and one below (or to its right). This means that all messages pass around the ring and each transited node will see part of each message. Again, the physical topology is a star, but the hub, called a **multistation access unit (MAU or MSAU),** will connect the wires as a ring.

FIGURE 4.7   (a) Bus Topology (b) Logical Bus Network

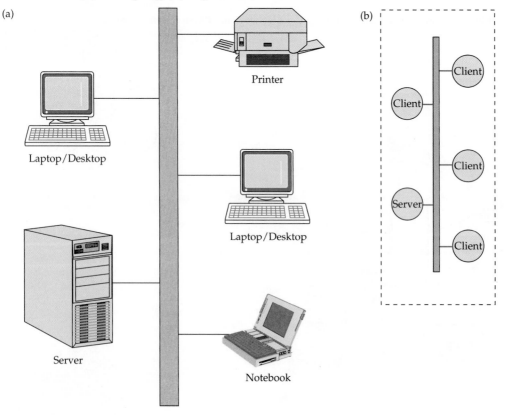

FIGURE 4.8 **(a)** Ring Topology **(b)** Logical Ring Network **(c)** Electrical Ring Network **(d)** Physical (Star-Connected) Token-Passing Ring Architecture

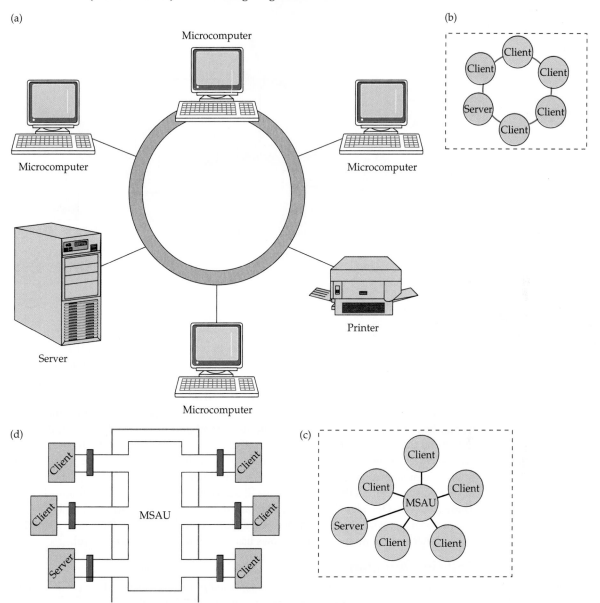

A **hybrid network** is a combination of the three basic and dissimilar network forms. A primary feature of this form is that it generally provides redundancy of paths, providing alternate paths in case of disruption of some links. (See Figure 4.9.)

In the **small office home office (SOHO), broadband** access is usually provided by either a CATV cable-based network or a DSL-based network. A *cable-based network* works as an electrical bus network with each modem connected to a common, shared channel in the form of a multiple-access network. The *DSL-based network* has the connectivity of a star network. Each POTS UTP circuit is dedicated to a single residence or premise and the bandwidth is not shared from the residence to the central active node. At that point, the packets of the multiple lines are accumulated and forwarded in TDM.

FIGURE 4.9
LAN/WAN
Interconnectivity

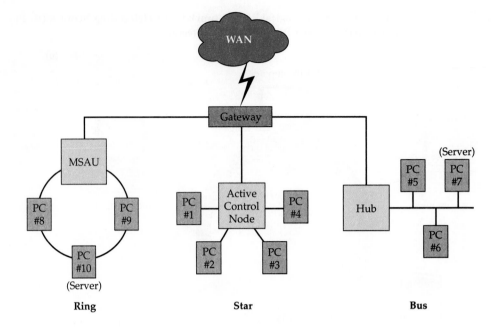

PROTOCOLS

protocol
Determines how data
move on the network.

The purpose of a protocol is to control access to a network and to control interference
between nodes, that is, **collisions.** There is a different protocol for each configuration of
network. Each of the electrical arrangements requires different access and communications
control. This applies for both local connectivity and wide area connectivity. The simplest
protocol is for the local $n = 2$, point-to-point network, well represented by a computer and
its attached printer. When the computer needs to send information to the printer, it begins
to transmit over either its parallel port or a USB port, through the cable, to the printer. The
printer buffer (memory) receives data until the buffer is full, at which time the protocol controlling communications sends an X-OFF message to the computer, signaling the computer
to stop transmitting. As the buffer empties, it sends an X-ON signal, and data flow begins.
These **X-ON/X-OFF** signals continue until all of the message is transmitted to the printer.
All protocols contain some method of data flow control. X-ON/X-OFF is the simplest
method. Each of the protocols that follows has its own method of flow control.

Collision Avoidance

Collisions occur when
two nodes transmit at
the same time.

More complex networks require more sophisticated protocols. Such is the case in a multipoint network such as a star, bus, or ring. In a star network, all outlying nodes are connected
directly to the active central node through multiple hardware ports in the central node. The
star network has the feature of avoiding data collisions by not allowing them to occur. The
central node controls communications by requiring that the outlying node get the attention
of the central node. This can occur in two fashions of polling: (1) the central node can continuously poll, that is, query, each outlying node to see if it incorporates something to send
or is ready to receive data at the central node destined for it, or (2) the outlying nodes can
signal the central node that they wish to send data and then wait until the central node is
ready. This latter mode is the manner used by the telephone system, a star network.

Collisions result in all
communications being
unintelligible.

The second method of **collision avoidance** is the one used in the ring topology. One
way to connect nodes is to make them into a ring, with each node connected to one computer above (or to the left of) it and one below (or to the right of) it. Any data are passed

from one node to the next; thus, everyone receives all messages. A node, however, only reads the messages destined for it. This is multipoint connectivity that, like polling, is *collisionless.*

A node must have permission of the token on a ring to send data.

In the ring topology, a data packet is continuously passed; this packet is called a **token.** It is passed around the ring from one node to the next following the **token-passing protocol.** When a node wishes to transmit, it must first take possession of the token. If another node has possession of the token, it must wait until the token is free. The node with the token then attaches the message, such as a portion of a file, to the end of the token and sends it on its way. Only one node can possess the token at a time, thus avoiding collisions. Think of a relay runner who must have a baton before starting. The token is analogous to the baton.

The IEEE specification for the token passing protocol is 802.5.

Collision Detection

Overhead is necessary for networks but takes up time that can be used for transport.

The alternate method of collision control is to react to them instead of avoiding them. In the systems that choose to react to collisions, some method must be established to detect or recognize that a collision has taken place and then there must be rules for how to react. This is a primary method used in the **Ethernet** bus network.[4] The network allows multiple computers to access the common channel (multiple access). Any node may transmit when it thinks it is permissible. This means there is no recurring overhead, as in the case of polling or token passing.

Ethernet
The brand name for the protocol used on bus networks.

CSMA/CD
The protocol for an Ethernet bus network.

With a bus network, such as an Ethernet bus using **IEEE 802.3 carrier sense multiple access/collision detection (CSMA/CD) protocol,** multiple access of many nodes is provided to a common channel, generally via a hub. This all works because each single node uses **carrier sense** to determine that the common channel is quiet. If no signals are heard, no one else is transmitting, so the node needing to transmit broadcasts a message on the network that all nodes hear. (This is called *broadcasting,* as opposed to narrowcasting or unicasting). Under normal conditions, every node receives and reads the header of the message and only the named recipient reads the actual message. If a second node broadcasts at the same time, a *collision* occurs, making all transmission unintelligible. Each node is intelligent enough to provide **collision detection** and stops transmitting for an individual random amount of time. When the timer for each offending mode expires, they try to transmit again, using the same quiet channel rule.

Collision avoidance or collision detection; you make the call.

The use of carrier sense is a primary way of gaining line access in a bus topology, though it is not the only way (token passing is another bus protocol). *Carrier sense multiple access/collision avoidance (CSMA/CA),* like token passing and polling, precludes collisions. However, the greater possibility of collisions during periods of high use causes **throughput** to decrease. Thus, in choosing between a collision detection and a collision avoidance protocol, the level of busy conditions must be considered. For example, a 10 Mbps CSMA/CD bus will be faster than a 4 Mbps token-passing ring in light loads, but the reverse may well be true in heavy traffic due to the overhead of a bus network.

The message format of the protocol defines the location and amount of true data contained in the message as well as the overhead necessary to ensure that the destination receives the data as they were sent. To do this, the message format of synchronous protocol usually has header, text or data, and trailer sections, and possibly synchronization characters. Figure 4.10 :shows the composition of an Ethernet frame and the function of each part.

---

[4] The original Ethernet was developed as an experimental coaxial cable network in the 1970s by Xerox Corporation to operate with a data rate of 3 Mbps using a carrier sense multiple access collision detect (CSMA/CD) protocol for LANs with sporadic but occasionally heavy traffic requirements. Success with that project attracted early attention and led to the 1980 joint development of the 10-Mbps Ethernet Version 1.0 specification by the three-company consortium Digital Equipment Corporation, Intel Corporation, and Xerox Corporation. http://www.cisco.com/univercd/cc/td/doc/cisintwk/ito_doc/ethernet.htm.

FIGURE 4.10   IEEE 803.2 / 802.2 Ethernet Frame

7 bytes	1 byte	2 or 6 bytes	2 or 6 bytes	2 bytes	\(Data/Padding\) 4–1,500 bytes				4 bytes
Preamble	Start Frame Delimiter	Destination MAC address	Source MAC address	Length	DSAP	SSAP	CTRL	NLI	FCS

Preamble This is a stream of bits used to allow the transmitter and receiver to synchronize their communication. The preamble is an alternating pattern of binary 56 ones and zeroes. The preamble is immediately followed by the *Start Frame Delimiter*.

Start Frame Delimiter This is always 10101011 and is used to indicate the beginning of the frame information.

Destination MAC This is the MAC[5] address of the machine receiving the data. When a network interface card (NIC) is listening to the wire, it is checking this field for its own MAC address.

Source MAC This is the MAC address of the machine transmitting data.

Length This is the length of the entire Ethernet frame in bytes. Although this field can hold any value between 0 and 65,534, it is rarely larger than 1,500 as that is usually the maximum transmission frame size for most serial connections. Ethernet networks tend to use serial devices to access the Internet.

Data/Padding (aka Payload) The data are inserted here. This is where the IP header and data are placed if you are running IP over Ethernet. This field contains IPX information if you are running IPX/SPX (Novell). Contained within the data/padding section of an IEEE 803.2 frame are four specific fields:
  DSAP: Destination Service Access Point
  SSAP: Source Service Access Point
  CTRL: Control bits for Ethernet communication
  NLI: Network Layer Interface

FCS This field contains the *frame check sequence* (FCS), which is calculated using a **cyclic redundancy check (CRC).** The FCS allows Ethernet to detect errors in the Ethernet frame and reject the frame if it appears damaged.

ACCESS AND CONNECTIVITY

A **network segment** is simply a portion of the connecting circuit; a length of the UTP.

Network devices are the equipment used to create the connection between three or more nodes. This section discusses the technology of network devices and examines how they are used to link network segments and create networks. Some of the equipment is used on local area networks and some on wide area networks.

We define a **network** as two or more nodes connected by one or more channels. If there are only two nodes (i.e., $n = 2$), then only a medium is required. Therefore, the nodes are connected by directly connecting them with a cable, such as a 10BaseT CAT 3 cable. As soon as n goes above two or the distance between nodes becomes too great, something must be added in order to have an effective network. That something is one or more of the pieces of equipment discussed next.

NETWORK EQUIPMENT[6]

The equipment discussed works at various levels of the OSI model. Table 4.2 shows this connection.

---

[5] The **media access control (MAC) address** is a hardware address that uniquely identifies each node of a network. In **IEEE 802** networks, the data link control (DLC) layer of the OSI reference model is divided into two sublayers: the logical link control (LLC) layer and the media access control (MAC) layer. The MAC layer interfaces directly with the network medium. Consequently, each different type of network medium requires a different MAC layer. http://www.webopedia.com.

[6] Assistance on this part was provided in part by Bliss Bailey of Auburn University and makes use of information at http://www.webopedia.com and http://searchnetworking.techtarget.com/sDefinition/.

TABLE 4.2
OSI Level of Device

Network Device	OSI Layer
Repeater	Physical
Hub	Physical
Switch	Data link
Bridge	Data link
Routers	Network
Gateway	All

Repeaters

One reason that digital signals are so perfect is that the bits are perfectly regenerated each time they become weak.

A **repeater** is the simplest hardware that can be used to provide connection for multiple network segments. In the analog environment, a repeater receives a signal and *replicates* it to form a new signal that matches the old one at the appropriate amplitude. In the digital domain, a repeater is a device that receives a signal and *regenerates* it. Since it creates a new signal, distortion or attenuation that the old signal carried is removed as the strength of the signal is improved. Repeaters of differing design are used for electrical, electro-magnetic, and photonic signals.

Repeaters are "line stretchers."

Repeaters operate at the first layer (physical layer) of the OSI model (see Figure 1.17 in Chapter 1) and are transparent to data flow. The repeater is restricted to linking identical network segments. For example, a repeater can connect two Ethernet segments or two token-ring network segments. It cannot connect an Ethernet to a token-ring segment.

Repeaters are used principally to extend the coverage of a network by extending the length of any one segment. Although several network segments can be connected via repeaters, there is a maximum distance limitation. The protocol used will determine the segment limitation; often it is between 100 and 500 meters. Photonic repeaters are capable of extreme distances: presently over 100 km.

Hubs

A **hub** is a device that physically connects multiple cables, providing a common connection point for devices in a network. Hubs are commonly used to connect segments of a LAN, providing a port[7] in the hub for each LAN segment. Hubs can be passive or active: If passive, a hub simply provides a conduit for the data among the ports. Hubs share all bandwidth among the nodes; thus, nodes see all traffic. If active, the hub can provide noise filtering, amplification, isolation, and switching.

Basic hubs are single boxes that have 4, 8, 16, or 32 ports. If more ports are required, and the hub is expandable or *stackable*, a cable connects another hub to the original hub, extending the total number of ports. Stacking eliminates the need to replace a smaller hub with a larger one.

Switches

Once called switchable hubs, these devices are a special type of active hubs that forward packets to the appropriate port based on the packet's address. This is in contrast to a standard or conventional passive hub, which simply rebroadcasts every packet to every port. By forwarding the packet to only the required port, **switches** provide better overall performance, for example, better allocation of bandwidth, even dedicated bandwidth. This provides for *load balancing* by dynamically assigning ports to different LAN segments based on traffic, something like statistical multiplexing.

[7] A port is generally an interface on a computer (or node such as a hub) to which one can connect a device. In TCP/IP and UDP networks, a port is an endpoint to a logical connection. The port number identifies what type of port it is. For example, port 80 is used for HTTP traffic. http://www.webopedia.com.

Clarification

In networks, a **switch** is a device that filters and forwards packets between LAN segments. Switches operate at the data link layer (layer two) and sometimes the network layer (layer three) of the OSI reference model and therefore support any packet protocol. http://www.webopedia.com.

Switches allow ports to be reassigned or turned on/off through software control. In the past, whenever ports needed to be reassigned, someone would walk to the wiring closet and physically change the cables to the new port settings. Software reassignment is a form of *remote monitoring (RMON),* a form of **simple network management protocol (SNMP)**[8] allowing remote devices to be managed from a central office. This technology allows users and resources to be placed in logical groups, even if physically separated. Switching allows networks to be separated into small, independent segments. Each segment can run at full network speed, which gives specific ports dedicated bandwidth.

Implication

Switches require power and potentially cooling because they are active devices.

There are two basic types of *switching architecture* or techniques. The first one is called *on-the-fly switching* (or cut-through switching). This type of switch begins to forward a frame before it receives the entire packet. Although it is quicker, this type of switch is more likely to suffer from errors because it detects errors by reading the end of each frame. The second type of switch is called *buffered* (or store-and-forward). As the name implies, it places the data in a buffer (temporary storage) until the entire packet is received. This allows it to make better routing decisions as well as have a better chance of detecting errors. *The tradeoff for these two techniques is speed versus reliability.* As memory (for buffers) has become cheaper and processors faster, most switch vendors have moved away from cut-through switching to store-and-forward. This gives them the flexibility to do a lot more management and shaping of the traffic.

Hubs operate at the physical level of the OSI model; switches operate at the data link layer. The switch looks at each packet or data unit and determines from a physical address (the MAC address) for which device a data unit is intended and switches it to that device. Some newer switches also perform routing functions (network layer) and are called *IP switches*.

Again, the purpose of a hub or switch is to make connectivity and, in the latter case, allocate bandwidth. The switching of bandwidth is logically separating portions of the network into subnetworks. This also can be done with software, for example, addressing. The value of separating networks into partitions is that the traffic of one subnetwork will not interfere with that of another subnetwork. Thus, with the addition of minor hardware or software, it would be possible to reduce the congestion of the network without changing major portions of the network or its medium.

Bridges

Bridges connect networks, thereby allowing separation for security and congestion control.

A **bridge** is a device that connects two local area networks (LANs), or two segments (e.g., subnets) of the same LAN that use the same data link layer protocol (e.g., Ethernet, token ring, etc). It examines each message on a LAN, "passing" those known to be within the same LAN and forwarding those known to be on the other interconnected LAN (or LANs). It does this by reading data frame addresses and uses this information to perform transmitting or

---

[8] The simple network management protocol (SNMP) is an application layer protocol that facilitates the exchange of management information between network devices. It is part of the transmission control protocol/Internet protocol (TCP/IP) protocol suite. SNMP enables network administrators to manage network performance, find and solve network problems, and plan for network growth. http://www.cisco.com/univercd/cc/td/doc/cisintwk/ito_doc/snmp.htm.

translating functions. A bridge uses a combination of hardware and software to connect LANs that have the same or different data link layers, but the upper-level protocols must be the same. A bridge operates at the second layer (data link) of the OSI reference model, (see Figure 1.17 in Chapter 1) and is more powerful than a hub. However, bridges are slower and more expensive to use than switches. A switch is nothing more than a multiport bridge, and the function of bridges has been pretty much taken over by switches.

In bridging networks, computer or node addresses have no specific relationship to location. For this reason, messages are sent out to every address on the network and accepted only by the intended destination node. Bridges learn which addresses are on which network and develop a routing or forwarding table so that subsequent messages can be forwarded to the right network. There are two general types of bridges: transparent and translating.

Transparent bridges connect two LANs that use the same data link protocol. The bridge reads the *source address* of each frame and compares the address to a table of local addresses that it keeps separately for each network. If the address is not in the table, then the bridge will insert it (a process called *learning*). The bridge also examines the *destination address* of each frame and compares it to the table it has built. If the destination address matches an entry in the local address table, then the bridge simply repeats the frame because it belongs on the current network. If the destination address does not match, then the bridge transmits the frame to the other network (a process called *forwarding*). If a bridge compares addresses and realizes that a station has already received the message, then the frame will be discarded (a process called *filtering*).

Translating bridges connect two LANs that use different data link protocols. For example, a translating bridge can be used to connect a token ring network to an Ethernet network. This is much more complicated because of frame and transmission rate conversion. It can become tricky because the networks have different format restrictions.

Bridges, routers, and so forth are computers that contain specialized intelligence for data handling.

A translating bridge performs the same functions as a transparent bridge. However, instead of forwarding the data to the other network, the bridge must provide translation. This means that the frame must be converted to a different format and run at a different speed. The bridge will provide a buffer and negotiate the frame size and speed.

Clarification

Bus networks generally run at 10 and 100 Mbps, dependent on the hub or switch and the NIC cards. Newer standards support 1,000 Mbps with 10,000 Mbps coming out of the lab into practice. Costs follow speed, as does productivity.

For example, Ethernet networks run at 10 or 100 Mbps while token ring networks typically run at 4 or 16 Mbps. This means that the data packets will be traveling at different speeds between the two networks and the bridge must provide a buffer. The second issue involves the size of the frames. The maximum frame size on an Ethernet is 1,518 bytes, while a token ring has a maximum frame size of 5,000 bytes (4 Mbps speed) or 18,000 bytes (16 Mbps speed). Bridges do not have the capacity to split and reassemble frames. Therefore, the only solution is to use software to configure workstations to use the smallest maximum frame size of each network. In the previous example, frames would be restricted to 1,518 bytes since that is the smallest frame size available.

Routers

A **router**[9] forwards data packets between networks. This device is connected to at least two networks, commonly two LANs or WANs or a LAN and its ISP's network. Routers are located at *gateways,* the places where two or more networks connect. Routers use headers and forwarding tables to determine the best path for forwarding the packets, and they use

[9] http://http://www.webopedia.com/.

protocols such as ICMP *(Internet Control Message Protocol,* an extension to the Internet protocol) to communicate with each other and configure the best route between any two hosts. Routers are capable of filtering.

A router examines frames that are specifically addressed to the router. It looks at the logical address, which is typically an address assigned by a network administrator. Frames that are not addressed to the router are ignored.

A router chooses the best possible path for data frame transmission and uses software control to help prevent traffic congestion. Operating at the third OSI layer (network layer), routers must be able to communicate using the same protocols at both the data link and network layers. (See Table 4.2.) However, it is possible for networks with different operating systems to use multiprotocol routers that are able to provide address translation. Of course, the networks must have similar upper-level protocols because it would be useless to route data to a destination point if the upper-layer protocols conflicted.

Routers are very flexible because they have algorithms and protocols that allow them to select the best possible path to transmit data. If necessary, routers have the ability to split packets into fragments that will be reassembled at the destination point. This sometimes results in packets arriving out of order, but instructions are included for reassembly. Routers use flow control to try to prevent congestion. They have the capacity to monitor the bandwidth of the path and will inhibit transmission as well as notify other routers to inhibit transmission if congestion occurs.

The two basic types of routing are called *static* and *dynamic.* The routing table is constructed by network personnel for *static routers.* Once the table is configured, the paths do not change. If a link is disabled, the router will issue an alarm. However, it will not be able to reroute traffic automatically.

Dynamic routers are able to make changes in the routing table and find an alternate path if a link should become disabled. These routers can even rebalance the traffic load. Dynamic routers are constantly communicating with other routers, and any new information is used to update the address routing tables.

Routers are more powerful than bridges because they operate at a higher level in the OSI model. Unlike bridges, routers have the capacity to split and reassemble frames, as well as choose the best possible path for transmission. Bridges must examine all frames (promiscuous mode), while routers selectively examine frames (nonpromiscuous mode). Of course, routers are slower than bridges and more expensive.

Storage routers are used to help build and manage storage area networks (SANs). The storage router enables users to connect switched storage environments on a standard data network. They build a separate network infrastructure for the storage network. Cisco Systems has introduced a storage router that features optical *Fibre Channel*® and **Gigabit Ethernet** ports and 10/100 BaseT ports. The devices support connection to Fibre Channel switches, SCSI devices, or both and feature a **gigabit** interface. These storage routers should facilitate the building of enterprise storage.

Clarification

An **edge router** is a device that routes data packets between one or more local area networks (LANs) and an ATM (asynchronous transfer mode) backbone network, whether a campus network or a wide area network (WAN). An edge router is an example of an edge device and is sometimes referred to as a boundary router. An edge router is sometimes contrasted with a core router, which forwards packets to computer hosts within a network (but not between networks).

ATM (asynchronous transfer mode) is a dedicated-connection switching technology that organizes digital data into 53-byte cell units and transmits them over a physical medium using digital signal technology. Individually, a cell is processed asynchronously relative to other related cells.[10]

[10] http://searchnetworking.techtarget.com/sDefinition/0,,sid7_gci212031,00.html.

Gateways

Generically, a **gateway** *is a node on a network that serves as an entrance to another network.* Specifically, it is a device that uses software to connect networks with different architectures by performing protocol conversion at the application level. A gateway is the most complex device available that will connect networks. It operates at all seven layers of the OSI model, connecting networks that have different protocols (operating at all levels of the proprietary architecture), and performs an actual protocol conversion. A gateway can provide terminal emulation so workstations can emulate dumb terminals. It also provides for file-sharing and peer-to-peer communications between LAN and host. Gateways also provide error detection on transmitted data, as well as monitoring of traffic flow.

In the network for an enterprise, a computer server acting as a gateway node is often also acting as a proxy server and a firewall server (discussed later). A gateway is often associated with both a router, which knows where to direct a given packet of data that arrives at the gateway, and a switch, which furnishes the actual path in and out of the gateway for a given packet.

Gateways are very sophisticated and complicated to install. As capability increases, for example, switch to router to gateway, cost increases. Gateways normally are used to connect LANs to mainframes or connect a LAN to a WAN. It is possible to connect LANs with gateways, but typically other methods are used since gateways are both slower and more expensive than routers.

Network Connections Summary

The various devices that link networks are summarized in Table 4.3 and the characteristics of hubs are summarized in Table 4.4. The list starts with the simplest level and goes to the

TABLE 4.3   **Comparison of Linkage Alternatives**

	Connect	Function	OSI Model	Speed	Cost
Repeater	Similar segments	Regenerates signal Extends network length	Layer 1	Fastest	Least expensive
Hub bridge-transparent	Nodes of LANs with similar protocols, similar data link layers	Makes connection Filters signal Decides if signal should be repeated or forwarded	Layers 1 and 2	Slow/medium Loses speed because extra layer is added	More expensive
Bridge-translating	LANs with similar protocols, different data link layers	Filters signal Forwards signal or translates	Layer 2	Loses speed because extra layer is added	More expensive
Router	LANs with similar protocols, similar data and network layers (unless multiprotocol routers used)	Chooses best path Provides flow control Segments frame, transmits, and reassembles	Layers 1-3	Loses speed because of third layer that was added	More expensive as level of complexity increases
Brouter	LANs with similar protocols, similar or different data link layers	Performs both routing and bridging functions	Examines protocols at layers 2 and 3 Performs bridging at layer 2	Depends on filtering and forwarding rate Can be faster or slower than bridges and routers	More expensive as level of complexity increases
Gateway	Networks with different protocols	Performs protocol conversion	Layers 1-7	Slowest	Most expensive

TABLE 4.4 **Characteristics of Hubs**

	Links	Function	Pros	Cons
Modular hubs	Multiple protocols	Physically connects cables	Can mix FDDI, Ethernet, etc.	Must physically change connections at the wiring closet
Stackable hubs	Single protocol	Physically connects cables, can add units by stacking	Able to fit more units in wiring closet because of smaller size	When connecting the stackables together, interference sometimes occurs (bleed over from the lines) Cannot mix different network segments
Switchable hubs	Multiple protocols	Uses software control to reassign ports Segments network into individual segments	Do not have to physically change connections in wiring closet Allows individuals to have full network speed	More expensive than modular or stackable hubs
ATM	LAN-to-ATM technology	New products allow ATM to be utilized more efficiently	Allows full performance of ATM to be realized	More expensive; ATM requires extra devices to utilize full potential

most complex. Of course, there is always a trade-off. Each time a device adds a layer of complexity, it suffers a loss of speed. The benefits of adding flexibility are costly. Therefore, when building a network, administrators must keep in mind that the more flexible networks will also be more expensive, as well as slower. However, sometimes the extra flexibility is a necessity.

Case 4-1

Building a Home Network

Brandon Billingsley

It is estimated that there are more than 820 million PCs in the world today and there will be over 1 billion by 2007. Over half of these are connected to the Internet, with half of those being connected by broadband. Now that network exposure at the office and at educational institutions is a forgone conclusion, many want to extend this connectivity to the home environment. In the past, this need for connectivity led computer manufacturers such as Dell and HP to include 56K modems in their computers and has now led to the inclusion a network interface card (NIC) as a standard feature. The modem did two things: it provided connectivity out of the box and circumvented the need for an external modem that relied on the bandwidth of the serial port.

The introduction of universal serial bus (USB) ports and internal networking cards further revolutionized home networking. The cable or DSL modems can now be connected directly to the computer via the USB ports or the network cards. The network card and the USB ports send a message to the user that the computer was meant to be connected to the Internet and to other computers.

The simplest way to connect to the Internet is through a dial-up modem connecting to an Internet service provider (ISP). This is a nonbroadband solution and is typically seen as very slow today at 56K. This requires an open phone line and an account with an ISP. Basically, a phone line is run from a home phone jack to the modem in the PC and then you use PC software to dial your ISP's phone number to connect to the Internet. This method is generally slow and will tie up the phone line while the user is connected to the Internet.

For a faster connection, a DSL or cable modem is needed along with service from a telephone or cable television provider. Both of these connections are broadband and run anywhere between 1 and 5 Mbps speeds and are increasing in speed every year. The cable modem is connected to your existing cable television coaxial cable via a splitter. The modem is designed to accept coaxial cable and output to either a USB cable or an Ethernet cable, though Ethernet is by far the most popular today. The DSL service is much like the cable service except a phone line is run to the DSL modem and then a USB or Ethernet connection is made to the computer. Because the DSL modem shares the bandwidth of the telephone line, a filter must be installed on every other phone in the house to separate the two different types of service.

As Figure 4.11 shows, the **home network** with broadband connectivity becomes available when a router is included. Instead of being directly connected to the Internet through the modem, the router is connected to the modem and the computers connect through the router. In the figure, the router receives an IP address from the ISP and then uses DHCP to assign each client device its own IP address. This splits the bandwidth of the connection among all of the clients that are connected to the router.

FIGURE 4.11 **Home Network with Broadband Connectivity**

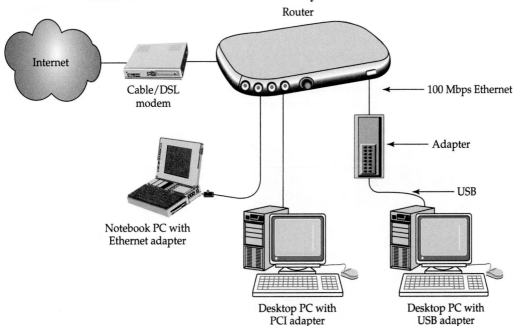

The use of a router brings an element of security to the home network in the presence of a firewall. The firewall is a hardware tool that is the first line of defense in keeping unwanted traffic away from your home network. To further improve the security of the home network, a software firewall should be installed and configured on each PC connected to the network. Additional software that should be added to each computer is antivirus software and spyware removal software. Spyware has become a major problem in today's computing environment as it uses up system resources and can slow down a network connection or even cut a computer off from a network entirely. These protections are more concerned with privacy than security.

The network shown allows four computers to share the bandwidth. With file-and-print sharing activated (default), all computers could share the file space and printers available on other machines. File sharing is not a good idea in that it allows easy entry for intruders that make it through the firewall. It can be provided, however, with password protection. Print sharing is a natural, as it shares the best printers among all nodes.

Setting up a home network does not have to be a complicated process. The steps involved are very simple and can be completed by almost anyone. These steps are

1. Purchase a router and broadband connectivity from a cable or phone service provider.
2. Connect the router to a power source and connect each computer to the router with Category 5 networking cable.
3. Log in to the software on the router through your Web browser and change the default password to something you can remember. Each router manufacturer has its own IP address to access the router software, but Figure 4.12 shows the Linksys IP address and what the login screen looks like. Figure 4.13 shows where to change your password.
4. Connect the DSL or cable modem to the uplink port on the router.

These steps will allow you to connect to the Internet and to any other computers that are connected to the router. To make sure that your computer has pulled an IP address from the router, you can go to the command line by going to **Start | Run.** In the Open box, type **cmd.** Once at the command line, type in **ipconfig.** See Figure 4.14.

If you are having problems connecting, you can try the commands in Figure 4.14. The command **ipconfig /release** will release the IP address that your computer has and **ipconfig/renew** will poll your router and grab a new IP address.

FIGURE 4.12
IP Address for
Linksys Router

Source: Linksys

FIGURE 4.13   **Checking IP Address on Linksys Router**

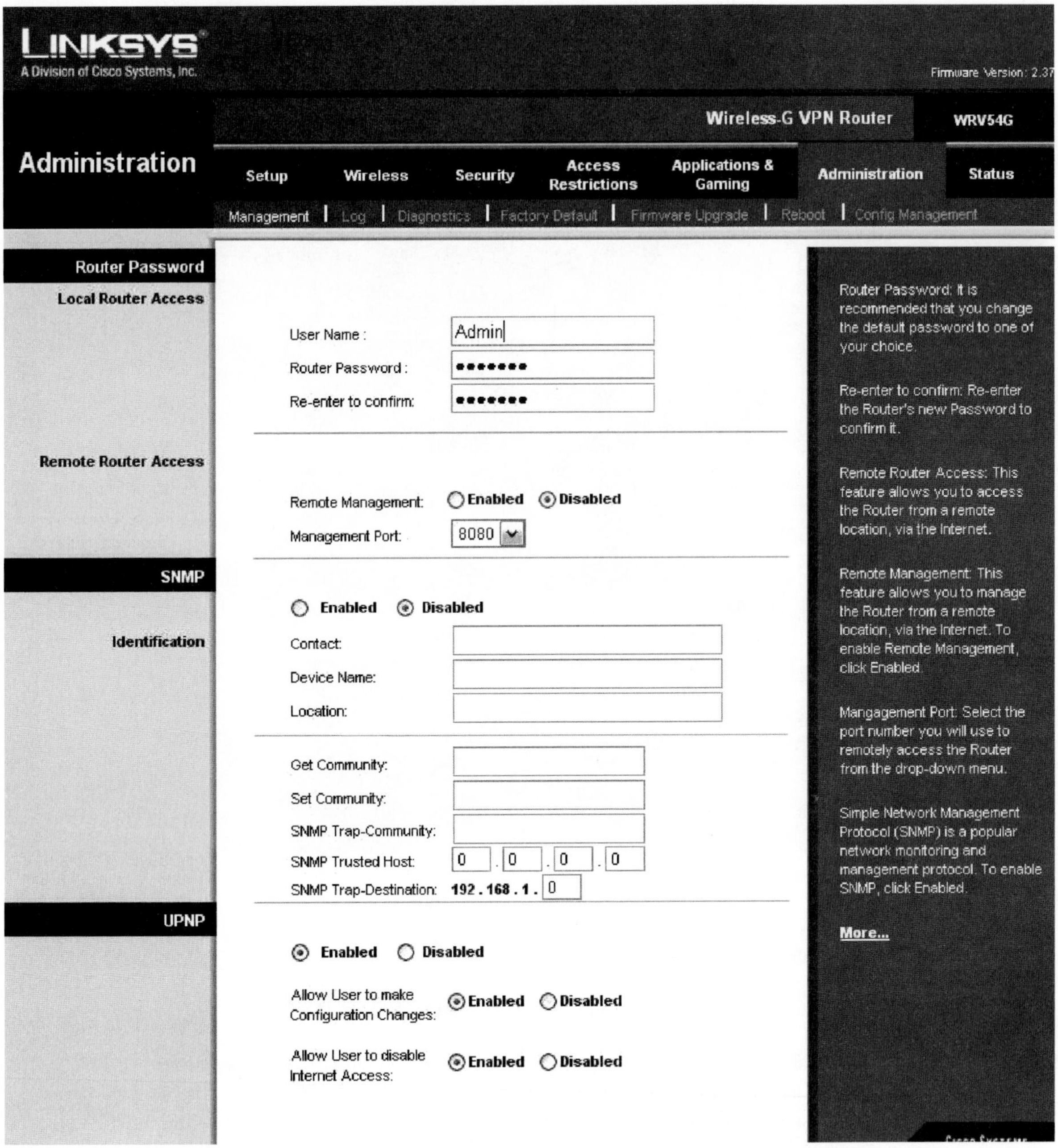

Source: Linksys

FIGURE 4.14 **IP Address Release and Renew**

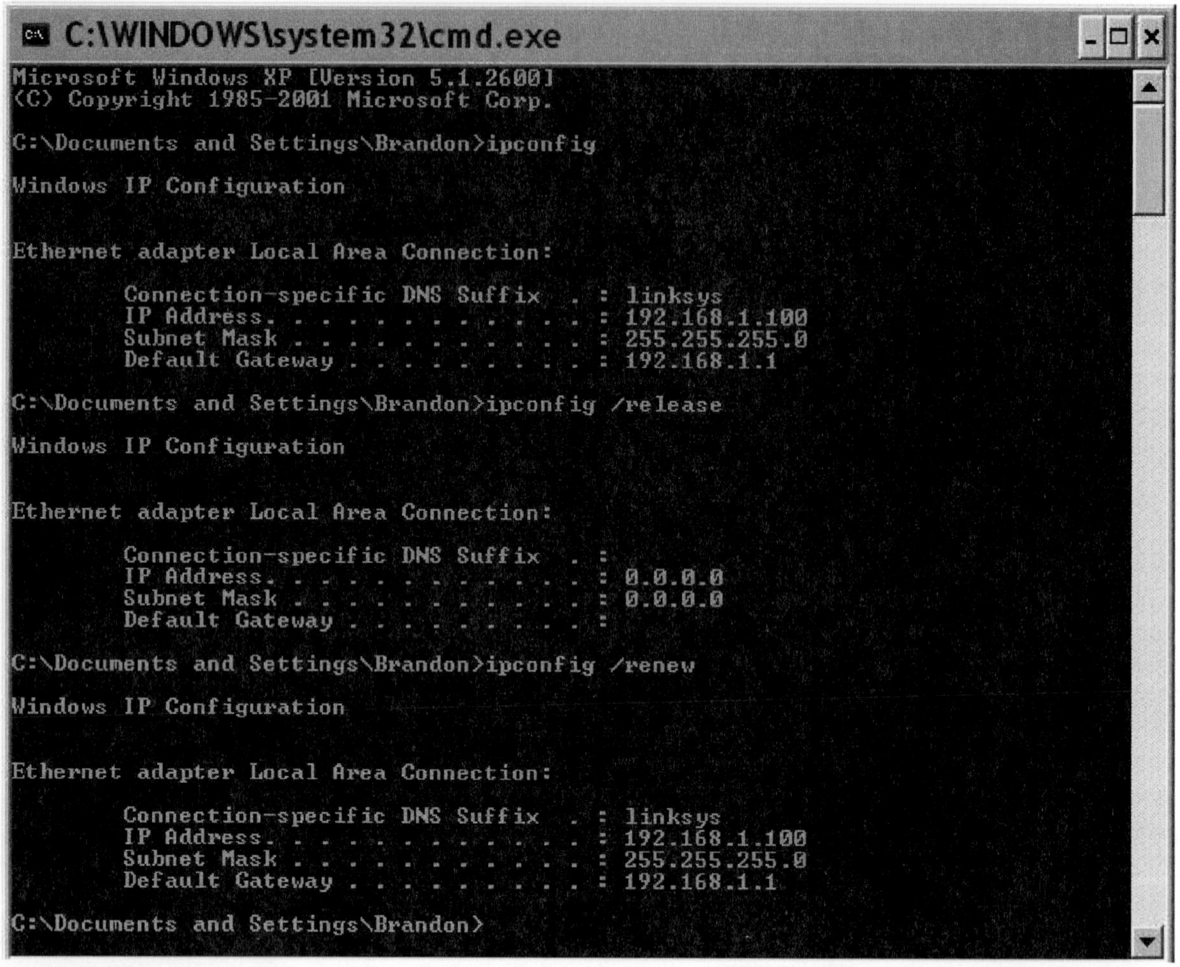

Source: Linksys/Microsoft

Case 4-2

Consolidation through Networking

Dealing Doug and Johnnie Clarence were pleased with their partnership (circa 1989). With the award of the F-16 contract to General Dynamics (formerly Convair), Johnnie's used car business had picked up quite a bit. He wasn't getting just the hand-to-mouth clientele who had been his stock-and-trade; he seemed to be moving up as the economy moved up. The bank told him he was doing well enough to enlarge his car business, so he rented the vacant lot next door and got more vehicles through Doug.

Doug's financing business was flourishing, not only due to Johnnie's success but due to the economy in general. However, Doug could increase his customer base several hundred percent without added facilities. Sometimes, he would hire an additional person to help with the paperwork, but Doug made all of the decisions. He played golf with most of the new car dealership owners, his friends, and began to explore their wholesaling more of their used cars to him to make room for the new model year vehicles the automobile companies were pushing on them. Doug realized, as he sunk a 23-foot putt on hole 18 at Colonial Country Club, that perhaps he was being too successful. If he took advantage of all of the possibilities of all of the six new car dealers, he would have a lot of vehicles, not only to sell but to store. Johnnie could handle a maximum of 63 vehicles and Doug's lot was small, holding no more than 10. If he expanded, he would need much more space.

Doug's finance company was in the middle of the automobile sales area of West 7th Street in Ft. Worth, Texas, while Johnnie's was some nine miles away in White Settlement (named as the nearest white settlement in Indian country in the early 1800s). Johnnie had no more room to grow and his landlord was talking rent increase as the economy, and Johnnie's business, increased. While playing golf, again, with the Pontiac dealer across the street, Doug was told that the owner planned to move from his present location to a new location on the southwest side of town and that his location would be for rent. This facility had the usual new car capabilities, including a large sales building, service department, used vehicle office building, and several acres of parking lot for new and used cars. Doug saw this as an opportunity to merge his and Johnnie's businesses, bringing Johnnie's business to a more visible part of town, one that seemed to appeal to a higher class of customers. It was close enough to his old location that his customers would likely follow him and it certainly gave Johnnie a great way to expand. Doug saw this as a way to not only see more financing but to keep an eye on Johnnie. While the large lot was a great asset, Doug thought there were too many buildings and could not see how he could use most of them. He wanted to keep his present office as a way to be able to escape the hustle and bustle of the expanded business. Perhaps he could rent all but the sales building to other people, and even some of that one, leaving the rest of the space for him and Johnnie.

Doug discussed the changes with Johnnie and soon they were the proud leasers of *Texas Used Cars and Financing*. As he thought, Doug rented the buildings he did not require to others, at a fee that just about paid for the total lot, leaving him with two buildings, about 200 feet apart. Johnnie prepared to move his cars to the lot and Doug was ready to receive his first shipment of wholesale used cars from the new car dealers. To handle the new enterprise, Doug had quickly realized he would need at least one full-time person to help with the financing plus a receptionist/bookkeeper and Johnnie would require three full-time and three part-time salespersons. They had the office space for the people, but communications would be a problem. Each of the 10 people would have an office; Doug would have two, in one building or the other; each would need a phone with at least three lines. Doug saw this as a time to bring in computers for each person to do the paperwork. While these were not powerful desktop computers, they did make filling out the myriad of forms much easier. The computers were about $1,500 each, but the printer was special, costing $5,500. To keep his costs down, he thought a printer in each sales building would be sufficient but didn't know how to make this all work. Besides the printing of paperwork, he wanted to be able to fax documents to the courthouse, two miles away, and let the receptionist send an e-mail to the recipient if they didn't answer a page.

QUESTIONS

How can Doug create a voice and data communications system for the new lot? Describe and design the new system(s), indicating any special considerations and limitations.

Caveat   You are constrained to the technologies thus far covered in the text.

Summary

Because we are able to arrange networks in various ways, it is necessary to understand the basics. The basics include a framework for the components of a network, that is, nodes, a hub, media, a network operating system, and protocols. Given the basic parts of a network are made available, a decision about the topology is required. This decision impacts the choice of protocols and equipment that must be assembled in order to build a network. We can't simply order the connection of various devices at different locations without deciding on a number of parameters that will determine network functionality. Some of the decisions may appear trivial, such as the NIC for the connected computers; however, all of the decisions about various facets of the network are interrelated and often interdependent. There is little hope of achieving a reliable, adaptable, scalable, and effective network without incorporating complementary components. The choices should be made logically so that the data communications system will be effective for the organization.

Networks are created by establishing special equipment connected by channels of media. The media can be wired or wireless; therefore, the nodes must accommodate wired and wireless connectivity and pass-through. The equipment in question begins with inexpensive repeaters and hubs, which are, for the most part, passive devices. The nodes quickly move to active nodes and the complexity of the network increases. A primary concern is the movement into the wireless environment as it brings great mobility but greater risks, for example, easier eavesdropping and intrusion.

Networks cannot be created without the equipment discussed in this chapter nor without budgets and organizations to support them. The nodes and channels can be created at both ends of the budget spectrum, with consequences.

Key Terms

10BaseT, *p. 117*
100BaseT, *p. 117*
bridge, *p. 126*
broadband, *p. 121*
bus, *p. 120*
bus network, *p. 120*
carrier sense multiple access/collision detection (CSMA/CD) protocol, *p. 123*
channel, *p. 113*
client, *p. 115*
client-server architecture (C/SA), *p. 114*
collision, *p. 122*
collision avoidance, *p. 122*
collision detection, *p. 123*
cooperative processing, *p. 113*
cyclic redundancy check (CRC), *p. 124*
distributed processing, *p. 114*

dumb terminal, *p. 115*
electrical topology, *p. 113, 119*
Ethernet, *p. 123*
fat client, *p. 117*
gateway, *p. 129*
gigabit, *p. 128*
Gigabit Ethernet, *p. 128*
grid computing (P2P), *p. 116*
home networks, *p. 131*
hub, *p. 117, 125*
hybrid network, *p. 121*
IEEE 802, *p. 124*
IEEE 802.3, *p. 123*
local area network (LAN), *p. 116*
media access control (MAC) address, *p. 124*
medium, *p. 117*
multistation access unit (MAU or MSAU), *p. 120*
network, *p. 124*

network appliance, *p. 117*
network computer (NC), *p. 117*
network interface card (NIC), *p. 117*
network operating system (NOS), *p. 117*
node, *p. 113*
packet, *p. 117*
peer-2-peer (P2P) computing, *p. 116*
peer-to-peer, *p. 113*
physical topology, *p. 113*
point-to-point topology, *p. 113*
polling, *p. 120*
protocol, *p. 118*
repeater, *p. 125*
ring network, *p. 120*
router, *p. 127*
server, *p. 115*

simple network management protocol (SNMP), *p. 126*	thin client, *p. 117* throughput, *p. 123* token, *p. 123*	transparent bridges, *p. 127* uninterruptible power
small office home office (SOHO), *p. 121*	token-passing protocol, *p. 123*	supply (UPS), *p. 119* wide area networks
star network, *p. 119*	topology, *p. 113*	(WAN), *p. 116*
switch, *p. 125*	translating bridges, *p. 127*	X-ON/X-OFF, *p. 122*

Recommended Readings

Network Magazine (http://www.networkmagazine.com/), a periodical about networks and network management.

Management Critical Thinking

M4.1 An organization that wishes to build a network connecting its branches and divisions so that each node can exchange real-time sales and inventory status with the central headquarters has several choices. Given that the organization has 25 geographically dispersed sites, what are the categories (e.g., topologies, protocols, etc.) of decisions that should be considered? What are the important parameters that the decision makers must determine in order to support the choices.

M4.2 The budget for network equipment can seem excessive. How is such a budget established and how much uncertainty (e.g., risk) is in the value?

Technology Critical Thinking

T4.1 If a fast-food restaurant chain employs a central computer for sales data processing and wishes to collect data from its local clients, what topology would a network likely employ?

T4.2 Networks can be created with a low-cost option or a growth option. How does the technical community justify which one to use?

Discussion Questions

4.1 What peripherals can be shared?

4.2 Discuss the pros and cons of the three major network topologies.

4.3 How can a 4 Mbps token-passing ring outperform a 10 Mbps CSMA/CD bus?

4.4 Discuss the means used on various networks to ensure collision avoidance and minimization of interference.

4.5 Why have LANs become so popular?

4.6 Differentiate between a repeater, bridge, router, and gateway.

4.7 How does a switch add value over a hub? Why would one purchase a hub over a switch?

4.8 How many in the class have home networks? What protocol does each use? What are the limits?

4.9 What are the present trends in routing?

4.10 Which topology involves the most overhead in collision detection/avoidance?

Projects

4.1 Go to several businesses in your city and find the percent of revenues that are spent on telecommunications. What type of industry do you expect to have the larger budgets?

4.2 Design a telecommunications network (analog and digital) for a state lottery system, local election, state election.

4.3 Find an organization that has a complex network. What are the protocols that govern the network? Are there problems involved because of differing protocols?

4.4 Determine various data communications services and characteristics available from your telephone company.

4.5 Develop a list of industries that have become dependent, in a competitive sense, on data communications for their survival. Can you make a list of industries that *are not dependent* on data communications?

4.6 Can you determine the firms that have data communications as the basis for their business? Is there an industry that exists because of data communications?

4.7 Determine the networks used in your organization or school. List the uses and the transmission media; for example, twisted-pair wire, coaxial cable, microwave, fiber optics, and so on. Although we have not discussed wireless, how significant is this technology in your school?

4.8 Divide the class into groups of two or three people. Each group is to find a local company that uses MIS and telecommunications. Write a three-page report on this company, describing its product or service and the use of MIS and, especially, telecommunications capabilities.

4.9 Look up the limitations of link distances for Ethernet and determine how many repeaters could be added to a link.

4.10 Design an $n = 11$ network that has connection to the Internet. Draw and price out the network.

Chapter **Five**

Network Form and Function

We must assess the desired reach of networks before deciding on the connections between nodes. There are a number of choices in terms of technology and bandwidth that offer characteristics to be evaluated. In addition, we need to consider forms of data transmission and signal reliability as we build our networks.

We have described telecommunications as the transfer of information over a distance by electrical, electromagnetic, or photonic means. This applies to both analog and digital communications. Both forms have networks, that is, two or more nodes connected by one or more channels. The purpose of networks is to communicate, share information, share resources, and operate at a distance as if centrally organized. A primary concern for data communications is high throughput with near-perfect reliability.

Data communications moves digital data from machine-to-machine, point-to-point. Because of noise, precautions must be taken to ensure that the data received were the exact data sent. In addition to this error prevention, error detection and correction are vital along with ensuring that our servers and clients have reliable and continuous power. These precautions provide reliability, the most important feature of a network.

WHERE IT ALL STARTED

The evolution of modern data communications equipment began with the first electrical means of communications over a distance: the telegraph (circa 1837). This device consists of a long-distance loop of wire that has an electric storage battery for energy, a key to open and close the circuit, and a sounding unit that responds to the electrical current with a sound (click). A home door buzzer is an equivalent system. The telegraph operator sends the message in a code that bears the name of the equipment's originator, Samuel Morse. The capability requires an operator who understands how to mentally translate (encode) the message into Morse code (Figure 5.1) and tap the code on the key at a reasonable speed. In this case, the medium is copper wire that is strung from the sending station to the receiving station and back (although the return "wire" or path may be the earth). Potentially, a number of receiving stations can be connected along the route, and each would hear all messages sent on the wire. (We will see later that a more effective way to communicate to multiple receiving nodes on a circuit is to place the address of the receiving station on the message so that all other nodes will ignore it.) The agent at the receiving station must be present and able to follow the clicks (dots and dashes) of Morse code, decode the clicks into the receiver's language (expanding abbreviations), and write down the message. The agent must accomplish all of these tasks at the speed of the sender's transmission (real time) since there is no storage capability built into the system.

What follows is machine-to-machine communications; the human operator has been removed from the system except as a possible recipient of the final message from the end machine. The networks created to support this machine-to-machine communications range from small and local to large and global. The range, or reach, of the network is related to the reach of the organization.

REACH OF NETWORKS

Networks connect the organization and give it a cohesive quality.

When the topic of **topology** is discussed, it means the layout of a particular network, not the geographical coverage of the network. Topology is simply the physical connectivity pattern, for example, the arrangement of nodes and their connection to other nodes. We have become accustomed to referring to the networks by labels that reflect their relative size and, to some

FIGURE 5.1
Morse and Computer Codes for Letters, Numbers, and Punctuation

■ is a short signal/sound or "dot"
— is a long signal/sound or "dash"

Letter	Morse telegraph code	Equivalent digital bits	7-bit ASCII (computer) code
A	■ —	0 1	1 0 0 0 0 0 1
B	— ■ ■ ■	1 0 0 0	1 0 0 0 0 1 0
C	— ■ — ■	1 0 1 0	1 0 0 0 0 1 1
D	— ■ ■	1 0 0	1 0 0 0 1 0 0
E	■	0	1 0 0 0 1 0 1
F	■ ■ — ■	0 0 1 0	1 0 0 0 1 1 0
G	— — ■	1 1 0	1 0 0 0 1 1 1
H	■ ■ ■ ■	0 0 0 0	1 0 0 1 0 0 0
I	■ ■	0 0	1 0 0 1 0 0 1
J	■ — — —	0 1 1 1	1 0 0 1 0 1 0
K	— ■ —	1 0 1	1 0 0 1 0 1 1
L	■ — ■ ■	0 1 0 0	1 0 0 1 1 0 0
M	— —	1 1	1 0 0 1 1 0 1
N	— ■	1 0	1 0 0 1 1 1 0
O	— — —	1 1 1	1 0 0 1 1 1 1
P	■ — — ■	0 1 1 0	1 0 1 0 0 0 0
Q	— — ■ —	1 1 0 1	1 0 1 0 0 0 1
R	■ — ■	0 1 0	1 0 1 0 0 1 0
S	■ ■ ■	0 0 0	1 0 1 0 0 1 1
T	—	1	1 0 1 0 1 0 0
U	■ ■ —	0 0 1	1 0 1 0 1 0 1
V	■ ■ ■ —	0 0 0 1	1 0 1 0 1 1 0
W	■ — —	0 1 1	1 0 1 0 1 1 1
X	— ■ ■ —	1 0 0 1	1 0 1 1 0 0 0
Y	— ■ — —	1 0 1 1	1 0 1 1 0 0 1
Z	— — ■ ■	1 1 0 0	1 0 1 1 0 1 0
,	— — ■ ■ — —	1 1 0 0 1 1	0 1 0 1 1 0 0
.	■ — ■ — ■ —	0 1 0 1 0 1	0 1 0 1 1 1 0
1	■ — — — —	0 1 1 1 1	0 1 1 0 0 0 1
2	■ ■ — — —	0 0 1 1 1	0 1 1 0 0 1 0
3	■ ■ ■ — —	0 0 0 1 1	0 1 1 0 0 1 1
4	■ ■ ■ ■ —	0 0 0 0 1	0 1 1 0 1 0 0
5	■ ■ ■ ■ ■	0 0 0 0 0	0 1 1 0 1 0 1
6	— ■ ■ ■ ■	1 0 0 0 0	0 1 1 0 1 1 0
7	— — ■ ■ ■	1 1 0 0 0	0 1 1 0 1 1 1
8	— — — ■ ■	1 1 1 0 0	0 1 1 1 0 0 0
9	— — — — ■	1 1 1 1 0	0 1 1 1 0 0 1
0	— — — — —	1 1 1 1 1	0 1 1 0 0 0 0

extent, their use. Thus, we have local networks that link work groups composed of relatively homogeneous users and wide networks that connect the world. Figure 5.2 illustrates several levels or layers of data connectivity. We will use this reference of layers to show coverage and usage.

Clarification

Figure 5.2 is not a network model; its purpose is to show the onion-type nature of networks. (*Shrek*® said, "Onions have layers.") That is, some networks are designed for use internal to the organization; this is also known as the access layer. The intermediate layer, for example, the distribution layer, provides connectivity within a region, and the wide area, or core, layer reaches out beyond.

At the heart of the layers of connectivity is the machine that connects to other machines. Inside of the computer is its own high-speed network, the *internal communications bus of the machine*. Since it is the lowest or smallest geographic coverage network, we seldom

FIGURE 5.2
**The Onion Model
of Network
Connectivity**

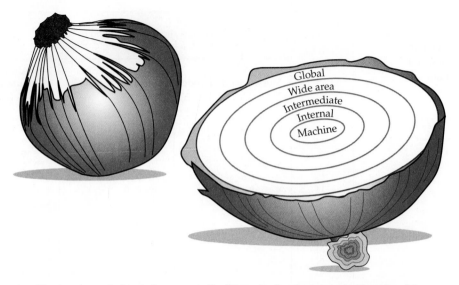

consider it a network, but it does meet all of the criteria of a network. Outside of the computer, we deal with several levels of organizational connectivity, of organization usage of the network. In one sense, the labeling is artificial. We use the terms to give us a human reference.

Outside of the individual computer, we would likely encounter the connectivity to a printer. Because **peripheral equipment** is often considered an extension of the primary device, it is not viewed as a network, although it has all of the characteristics, such as medium and operating system.

The smallest labeled network is the **personal area network (PAN).** This takes on two forms: (1) networking to peripherals on the person and (2) synchronizing of multiple personal devices. For the first type, consider the use of an *iPod*® music player and the earphones used with them. While wired earphones would be seen as a peripheral to the iPod, wireless earphones would more likely be viewed as a network. In the second definition, many people have PDAs, cell phones, and laptop and desktop computers that contain address books and appointment calendars. **Synchronizing** them is valuable and a personal area network using the *Bluetooth*® wireless networking capability means all devices have the same data.

Internal Network Level

The lowest level of organizational connectivity is called the **internal level** in Figure 5.2. Here we are concerned with limited distances and cohesive groups of people and machines. **Local area networks (LAN)** cover an area of less than one kilometer in radius and connect people within an organization, such as one or more buildings, departments, floors, or work groups.

Local area networks have essential components, such as NICs, cables, connecting devices, servers, and operating systems.

Even though all telecommunications should be transparent to the users, the LAN has the greatest visibility and potentially greatest influence on them. During the short time since their introduction, microcomputers (Apple in 1976, IBM in 1981) have found their greatest value when connected to each other and to organizational data sources. In the early days of the 1980s, users went to great lengths to acquire microcomputers, often hiding them in organizational budgets when official approval to purchase was denied. As they discovered the value of this new resource, they realized quickly that the stand-alone microcomputer was missing something very valuable: organizational data. As the number of microcomputers in organizations increased, users searched for connectivity. An early form of resource sharing was via **circuit switching** of two or more microcomputers to a common printer. This is a non-LAN solution.

Clarification

A circuit-switching network, like the telephone system, switches two physical circuits together for the duration of the transmission. The circuit generally is not shared during this time.

Users seeking connectivity first consider either point to point, or modem to modem, as a simple and inexpensive way of getting files from one machine to another. Then they move up to the local area network. The LAN seems like a dream compared to the simple point-to-point connection since even the inexpensive ones communicate at over 1 Mbps. Novice microcomputer users readily see the value of connectivity to communicate and share resources. Speed was not considered important in the beginning, partially due to a lack of experience and partially because 1 Mbps seemed so fast. With experience, speed becomes more important to users. This is more evident as LANs support hundreds of users vying for the same data path and set of resources.

Electronic mail was a driving force for the installation of LANs and then connectivity to the Internet. Now, e-mail is just one form of sharing information and resources.

As more people work at home or remote officers, they create **small office home office (SOHO)** networks. This is just a small LAN and allows connecting computers in this smaller location and, with connectivity such as DSL, provides access to the main office. SOHO networks, as with all LANs, provide for the sharing of resources and support that primary application, e-mail.

LANs tend to be internal to an organization. LANs may potentially connect to partners and customers, creating an **extranet.** This environment is simply the extension of the organization network to include those participants who are outside of the organization. As enlightened organizations realize the importance of including vendors, suppliers, and customers in decisions, they create LANs with external connectivity, extranets. Extranets and intranets are discussed in detail in Chapter 7.

Clarification

Intranets embody the use of Internet technologies within a LAN environment, for example, the browser interface.

A campus area network is, potentially, just a LAN with greater reach.

Backbone networks
One way to distribute networks; the other is subnetting, to be discussed later. Like distributed processing and storage, distributed networks put the dedicated connectivity close to the user.

Backbone networks use bridges and routers to join but keep separate.

An extended form of the LAN is a **campus area network (CAN),** which tends to cover a fairly wide area. This may be several business buildings, called a campus, or an actual educational setting. The only characteristic of interest is that the CAN may be a large LAN, resulting in a greater number of nodes and, thus, need for administration. The CAN may be a single LAN or a connection of LANs by way of a **backbone network.** The backbone network is analogous to the backbone of a human in that it acts as a central structure of the organization. The point of separating areas of a campus or building into separate LANs and then connecting them by a backbone network is to keep local data flow on the local network, resulting in less congestion on other networks.

An example of a campus area network is the fiber optic backbone network (AU-NET) at Auburn University that connects all buildings at 1,000 Mbps (i.e., **gigabit**) to a central building, to the IBM mainframes, and to the Internet. With the backbone in place, departments can attach directly to it and communicate to the mainframes, other LANs, and servers, and have a gateway to the Internet. The College of Business at Auburn University has a 10/100 Mbps network for communications among the 250 staff, faculty, and doctoral students in the college, one gigabit fiber to communicate between wings of the building, and a gateway to AU-NET for communications to the rest of the campus and the world (through the Internet).

A local area network (LAN) can generally be defined as a broadcast domain.[1] Hubs, bridges, or switches in the same physical segment or segments connect all end-node devices. End nodes can communicate with each other without the need for a router. Communications with devices on other LAN segments requires the use of a router. As networks expand, more routers are needed to separate users into broadcast and collision domains and

[1] http://net21.ucdavis.edu/newvlan.htm.

provide connectivity to other LANs. One drawback to this design is that routers add latency, caused by the process involved in routing data from one LAN to another. One solution is the use of virtual LANs.

Virtual LANs (VLANs) can be viewed in one of two ways: (1) as a group of devices on different physical LAN segments that can communicate with each other as if they were all on the same physical LAN segment or (2) by logically segmenting the network into different broadcast domains so that packets are only switched between ports that are designated for the same VLAN.[2] This may be done for either of two purposes: (1) to allow for ease of entry to a LAN and support physical movement or (2) to provide separation of nodes from general broadcast, even though they are on a common channel. In this latter use, VLANs perform traffic separation within a shared network environment.

Implication

The development of VLAN tagging technology has truly changed the way LANs and campus networks are managed. The flexibility added by the use of VLANs had made things possible that would never have been practical just six or eight years ago.

With the implementation of switches in conjunction with VLANs, each network segment can contain as few as one user (approaching private port LAN switching), while broadcast domains can be as large as 1,000 or more users. If implemented properly, VLANs can track workstation movement to new locations without requiring manual reconfiguration of IP addresses.[3]

Many VLANs use the IEEE 802.1Q standard, which functions by adding a VLAN identifier to the header of LAN packets, functioning at the data-link layer. The 802.1Q header bears a 12-bit VLAN header allowing 4,095 ($2^{12} - 1$) different VLANs on an Ethernet network. For LAN segments configured together as a VLAN, packets originating from stations attached to these LANs acquire an 802.1Q header carrying the appropriate VLAN ID as they are forwarded onto the shared backbone network. The receiving router or switch then performs a VLAN ID match against the VLAN it is configured to support to determine whether it should remove the 802.1Q header and forward the original packet to any ports that belong to the same VLAN.

Clarification

Ethernet is the predominate standard for wired LANs. It is a packet-based system, sending out packets on a common (wired) channel for all connected devices to hear. The packet is composed of a header, indicating the source and destination, and a payload, or data portion. All devices listen to the common channel and read the header of all packets. A device will read the data portion of the packet only if it is destined for that device. The protocol for Ethernet is CSMA/CD, allowing common multiple access and detection of collisions and reaction to them.

A **metropolitan area network,** wired or wireless, can be public or private. If created by the city, it can offer a competitive advantage for attracting businesses. **Metro Ethernet networks** are Ethernet access and services across a metropolitan area network.

Intermediate Network Level

Remembering that the division between layers is to distinguish use, the next layer of Figure 5.2's scope is the **intermediate level.** In Figure 5.3 showing a number of LANs, the intermediate layer would be the connection of the LANs into a corporate network, for example, a backbone. Each LAN has its own server for the operating system and shared resources for its groups of users, but each LAN is bridged to the other LANs for connectivity across a campus, division, or corporation.

At the intermediate layer level is the **metropolitan area network (MAN).** As it implies, this is connectivity within a metropolitan-sized area, possibly providing services to many companies. For instance, the financial district of Denver or Chicago could take

---

[2] http://www.cisco.com/warp/public/614/11.html.

[3] David Passmore and John Freeman, *The Virtual LAN Technology Report* (Sterling, VA: Decisys, Inc, 1996), http://www.3com.com/other/pdfs/solutions/en_US/20037401.pdf.

FIGURE 5.3 **Geographic or Layer Network Concept**

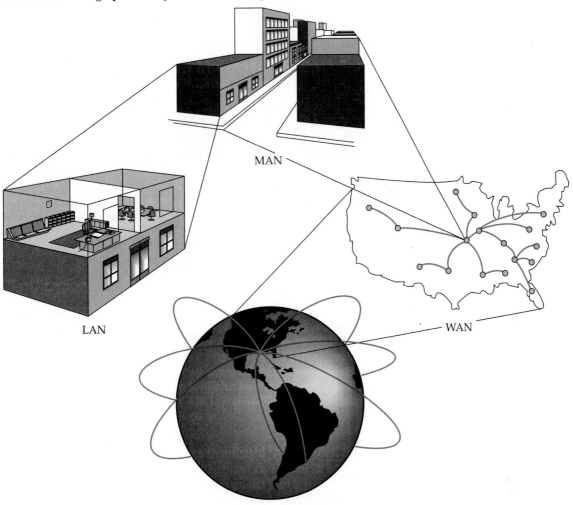

MAN

LAN

WAN

Global network

advantage of such a network to access commonly used resources. Such a network might be a fiber strand providing 10 to 10,000 Mbps connectivity for access to financial services or even to long-distance providers for voice service. A major form of MAN is the **fiber-distributed data interface (FDDI)** network based on optical fiber media operating at 100 Mbps speed. An extension to FDDI, called FDDI-2,[4] supports the transmission of voice and video information as well as data.

Wide Network Level

Wide area networks Connect widely dispersed machines, possibly requiring intermediary machines to make conversion from one protocol to another.

Just as a local area network provides internal connectivity to a small geographic area, and a metropolitan area network extends intermediate coverage to a wider area, *wide area level* coverage is provided by **wide area networks (WANs).** WANs provide connectivity to larger and larger geographic areas, including the world. They can be privately owned and

---

[4] http://www.webopedia.com.

operated for internal use only (such as Southtrust Bank or Wal-Mart), a common carrier resource offered publicly, or the global public Internet. Privately owned WANs often have excess **capacity** that is sold or leased (brokered) to other organizations. WANs are long-haul, broadband, generally public access networks with wide geographic area, crossing rights of way, and are provided by common carriers. **Value added networks (VANs)** are a special form of WAN. They generally provide this wide area coverage and offer services in addition to connectivity in the form of added intelligence such as speed translation, store-and-forward messaging, protocol conversion, data handling, and packet assembly and disassembly. Their circuits may use switched and leased services on terrestrial circuits like wire or optic fiber or air media like microwave and satellite radio. They may offer speeds for basic connectivity of 56 Kbps to multimegabit rates for larger volume and constant traffic.

A final form in the evolution of the hierarchy of networks is the **galactic area network (GAN),** which is a WAN that extends beyond the confines of the earth. Although a WAN may use satellite links, a GAN uses radio to extend the network to other bodies, such as the Mir Space Station, weather satellites, Earth's moon, or Mars. As mankind probes the heavens, connectivity with the exploratory crews and vehicles is needed for control and information. Having links on the moon or Mars should help speed long-distance communications beyond our solar system.

Implication

As the number of satellites transmitting public information such as weather increases, they will likely be connected to the Internet, creating a GAN.

We generally think of WANs as privately owned networks, although they may be owned by a common carrier and made available for public traffic. With the introduction of the Internet, everyone has access to a wide area network. This **packet data network (PDN)** provides connectivity from almost anywhere to almost anywhere, at a moderate speed. Please see Chapter 7 for more detail on this technology and the organizational companion, the intranet.

Securing Networks

Organizations may wish to extend the reach of a LAN or create a temporary WAN by use of a public network, such as the Internet. **Virtual private networks (VPNs)** include software that is intended to make users of the network secure. The VPN acts as a tunnel inside the larger network. By using encryption techniques called **tunneling,** the data travel over a public network but cannot be read by it. VPNs utilize strong encryption, authentication, and access control technologies to connect remote offices, business partners, and mobile employees via low-cost public Internet connections. The subject of network security is covered in detail in Chapter 11.

Clarification
VPN Made Simple and Personal

Virtual private networks (VPNs) don't have to be the private domain of the large organization. According to http://www.onsystems.com/, their product *Tijit*® is a simple virtual private network (VPN) alternative for individuals and businesses. Tijit allows friends and family to network their computers together over the Internet. A Tijit user is able to share a picture or any other file on his home computer with his friends, family, or the entire Internet. Features include share-transfer of files, chat and instant message, secure private connections, and joining of communities to share files with others who share your same interests. VPNs may be established with MS Windows® operating systems.

Implication

VPN technology provides security when connecting to public networks. VPN technology is even more important for wireless networks as all communications there are public.

CONNECTIVITY IN NETWORKS

Networks have evolved since the mid-1980s and the media used also have changed. Two media were commonly used: UTP and coaxial cable of varying size. The standard speed for Ethernet in the early 1990s was 10 Mbps. The common channel medium for this was **category 3 UTP wiring.** Category 3 is a standard for wiring that specifies the gauge size and number of twists per foot, determining the native bandwidth and the radiation to be expected. This wiring allowed networks with specified link lengths to be constructed with two pairs of wires for a 10 Mbps bandwidth. This was called **10BaseT** (bandwidth, mode of transmission, and medium). With different electronics on the ends of the links and using four wire pairs, it is possible to extend the bandwidth to 100 Mbps, called *Fast Ethernet.*

The predecessor of 10BaseT networks has bandwidth of 1 to 2 Mbps, called *10Base2.* With the addition of nodes and the use of networks for more than text-based e-mail, the need for greater bandwidth became evident. In addition, there was a need for more sophisticated access and control. The same thing happened as 10BaseT networks matured; nodes used up all of the bandwidth and many organizations moved to 100BaseT networks. Historically, installing new media in a building that anticipated high-bandwidth usage or growth called for use of **category 5 wiring** for a **100BaseT** network. With the consideration of adding large file transfer, images, and even video conferencing to local networks, the need for bandwidth is apparent. This calls for **Gigabit-Ethernet,** designated as *1000BaseT* running on **category 6** wiring **(1000BaseT)** or fiber **(1000BaseFX).**

IEEE 802.3 specifies a series of standards for telecommunication technology over Ethernet local area networks. Table 5.1 details the different Ethernet flavors and how they differ from one another.

TABLE 5.1 **Ethernet Designations**

Source: http://www.webopedia.com/quick_ref/EthernetDesignations.asp.

Designation	Description
10Base2	10 Mbps baseband Ethernet over coaxial cable with a maximum distance of 185 meters. Also referred to as *Thin Ethernet* or *Thinnet* or *Thinwire.*
10Base5	10 Mbps baseband Ethernet over coaxial cable with a maximum distance of 500 meters. Also referred to as *Thick Ethernet* or *Thicknet* or *Thickwire.*
10Base36	10 Mbps baseband Ethernet over multichannel coaxial cable with a maximum distance of 3,600 meters.
10BaseF	10 Mbps baseband Ethernet over optical fiber.
10BaseFB	10 Mbps baseband Ethernet over two multimode optical fibers using a synchronous active hub.
10BaseFL	10 Mbps baseband Ethernet over two optical fibers. It can include an optional asynchronous hub.
10BaseFP	10 Mbps baseband Ethernet over two optical fibers using a passive hub to connect communication devices.
10BaseT	10 Mbps baseband Ethernet over twisted-pair cables with a maximum length of 100 meters.
10Broad36	10 Mbps broadband Ethernet over three channels of a cable television system with a maximum cable length of 3,600 meters.
10Gigabit Ethernet	Ethernet at 10 billion bits per second over optical fiber. Multimode fiber supports distances up to 300 meters; single-mode fiber supports distances up to 40 kilometers.
100BaseFX	100 Mbps baseband Ethernet over two multimode optical fibers.
100BaseT	100 Mbps baseband Ethernet over twisted-pair cable.
100BaseT2	100 Mbps baseband Ethernet over two pairs of category 3 or higher unshielded twisted-pair cable.
100BaseT4	100 Mbps baseband Ethernet over four pairs of category 3 or higher unshielded twisted-pair cable.
100BaseTX	100 Mbps baseband Ethernet over two pairs of shielded twisted-pair or category 4 twisted-pair cable.
100BaseX	A generic name for 100 Mbps Ethernet systems.
1000BaseCX	1,000 Mbps baseband Ethernet over two pairs of 150 shielded twisted-pair cable.
1000BaseLX	1,000 Mbps baseband Ethernet over two multimode or single-mode optical fibers using longwave laser optics.
1000BaseSX	1,000 Mbps baseband Ethernet over two multimode optical fibers using shortwave laser optics.
1000BaseT	1,000 Mbps baseband Ethernet over four pairs of category 5 unshielded twisted-pair cable.
1000BaseX	A generic name for 1,000 Mbps Ethernet systems.

FORMS OF DATA TRANSMISSION

Data communications requires the generation of streams of characters, composed of bits. The two forms of data communications are *asynchronous* and *synchronous*. During **asynchronous** communications, each character is considered by itself with no relation to the character sent before or after it. All timing and error checking are within the bits of the character. In **synchronous** communications, a block of data (many characters) is sent at one time with significantly greater reliability.

Asynchronous Communications

The data in data communications are composed of **BI**nary digi**T**s (bits) that are combined into bytes, generally seven-bit or eight-bit groupings as determined by a code. For example, the letter A, in eight-bit ASCII would be 01000001. In **asynchronous communications,** data are sent one character at a time. (See Figure 5.4(a).) It is usually assumed that placing digital data onto an analog (e.g., POTS) line will encounter enough noise so as to make the data unusable. Therefore, extra bits are added to each eight-bit character to create a one-character frame to help identify changes during transmission. Thus, the *asynchronous character* is a group of 10 bits, the eight bits of the data and two extra bits for timing and error detection. Following the data may be one bit for error checking (called **parity checking**) and one bit for character **stop bit,** making 10 bits in all. (The reference to a "start" bit actually means the first bit of the character; the line is quiet and the first bit starts the receipt of data. See Figure 5.4(b).) The modems operate on one 10-bit character at a time with no apparent control between characters: No timing is maintained. At the time communications is established, a process called **handshaking** (see Table 5.2), the sending and receiving machines agree on the bit pattern, for example, seven or eight bits for data; even, odd, or no parity; 1 or 2 stop bits; and speed of communications. The slower device will set the speed. If the line is noisy, the two DTEs will slow down to a speed of transmission that is reliable.

Asynchronous communication was originally chosen because of low cost of equipment, but it resulted in a low-speed system with low reliability. Parity bit checking has a low reliability. These problems can be overcome by grouping the data and sending them as a block, or **packet,** of data.

FIGURE 5.4(a)
Asynchronous Data Transmission

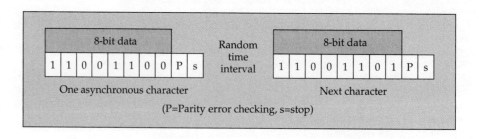

FIGURE 5.4(b)
Asynchronous Data = One Character

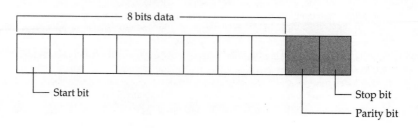

TABLE 5.2 **Some Considerations of Device-to-Device Handshaking, a Function of Protocol**	**A. Communications startup**—communications initiated **B. Character identification and framing**—determine text characters and which are control characters **C. Message identification**—separates characters into messages **D. Line control**—receiving unit says data were good, by turnaround, and requests next block **E. Error control**—what to do in case of error (retransmit), what to do when communications breaks and is reestablished **F. Termination**—normal and abnormal

Synchronous Communications

With synchronous communications, the unit of control is a block of data: a group of text, numbers, or binary characters. See Figure 5.5(a). The total block of data, often called a *packet,* contains a header, payload (i.e., data), and a trailer. (See Figure 5.5(b).) Header characters, which include timing and possibly address information, begin the block and are added in front of the real data being sent. The data group, that is, the payload, is a number of characters (more than 1 but generally fewer than 1,025 characters) that are sent with minimum or no error checking bits among them. Finally, the modem software adds trailer characters to the end, after the data, that include comprehensive error checking (called the **block check characters [BCCs]**) for the total block of data and an end-of-block and/or end-of-message indication. With synchronous communications, we must have a clock at both sender and receiver ends so that the sender will know when to put the data bits onto the line and the receiver will know how often to sample the line.

To send a large amount of data from one computer to another, we usually use synchronous communications to achieve rapid and reliable transfer. This mode results in greater accuracy and speed, and more efficient use of the network. The two forms of synchronous communications are continuous and packet switching. In *continuous synchronous communications,* the

FIGURE 5.5(a)
Synchronous Data Transmission

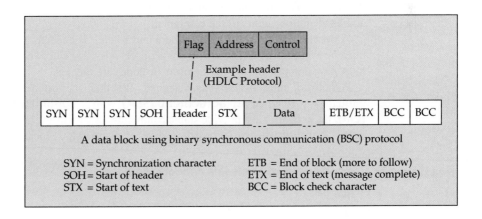

FIGURE 5.5(b)
Synchronous Data = One Block

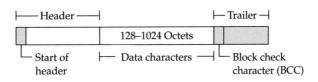

sender or receiver establishes a dedicated (dial-up or leased) point-to-point connection and the source sends data to the destination. The sender transmits a number of blocks of data, containing 128, 256, or 1,024 characters each. For the transmission of an 85-page document (85 pages * 2,000 char/page = 170,000 text characters), the system might transmit 1,024 bytes of data at a time, with no overhead information. This would be 166 blocks at 1,024 bytes/packet for a total of 170,000 bytes. This transmission would take about 30 seconds on a 56 Kbps circuit. This form assumes no error detection is required and is not often used.

The alternative form of synchronous communications is called **packet switching.** Like continuous communications, this mode may use the ASCII or EBCDIC coding scheme and sends the data as blocks of characters. Packet-switched networks are covered in detail later in the book.

Protocols for asynchronous and synchronous communications provide a feature called *transparency*. This characteristic prevents the receiving unit from reading executable code and acting on it.

The ultimate objective of a data communications system is effective throughput of useful information. As we add signal control, designation addressing, and error control information to communications units and blocks, the number of useful bits relative to total bits transmitted decreases. Consequently, our objective of accurate communications is in conflict with rapid communications, and we must have a trade-off. The use of synchronous (block-oriented) communications is the method to increase the throughput over that possible with asynchronous (character-oriented) communications while providing exceptional reliability.

Implication   As we strive for reliability and accuracy, we sacrifice speed.

PARAMETERS OF A VALUABLE NETWORK

Above all, a network must be reliable. **Reliability** encompasses the traits of constancy, dependability, stability, and durability. The network is either error free or has error detection included that will detect **errors** with great certainty. The network is, in human terms, dependable. This includes the reliability of the equipment components, the power supplied, and bandwidth conduits to the equipment.

A network should be *available* where and when it is needed. Increasingly, the network must be available **24*7** if the network supports *mission-critical* capabilities. For access, it must have entry points where the users or nodes are located.

A network must *perform* as designed. Whether supported in-house or outsourced through a **service level agreement (SLA)** with a **committed information rate (CIR),** the network must provide stable bandwidth under varying conditions, such as bursty traffic, **bandwidth-on-demand,** or dynamic bandwidth allocation.

Clarification   The **parameters of a valuable network (RAPS)** are

1. **R**eliability
   a. Error free
   b. Error detection system
2. **A**vailability
   a. 24*7
   b. From where wished
3. **P**erformance
   a. Fixed bandwidth
   b. Dynamic bandwidth

c. Bandwidth-on-demand
d. Bursty traffic
4. **S**ecurity
a. Physical
b. Virtual
c. Data

Finally, a network must have **security.** It must be safe from intrusion and protected from attacks. The network must not allow itself to be used to access attached resources, especially the data it stores and transmits. Part of the security system is monitoring and logging of events; authentication; audits; and software to protect against viruses, hackers, and spam. This subject is covered extensively in Chapters 11 and 12.

Power for Network Equipment

One of the criteria for a network is availability; one form is provided by continuous and suitable electrical power. Computers and network equipment require power, continuously. If the voltage goes too high (e.g., has a spike) or too low, or stops, the equipment may have (serious) problems. There are several ways to counter this.

In networks, we want the attributes of reliability, accessibility, performance, and security. For electrical power, we want similar attributes: *continuity, accessibility, regulation,* and *consistent waveform. Continuity* means power must be available as needed. There must be *access* to power wherever it is needed. The power should be *regulated* so that the voltage levels are between specified limits. If the voltage is too high, the equipment may not operate and may cause a fuse or circuit breaker to actuate. When the voltage falls below the lower limit for an extended period of time, referred to as a *brownout,* equipment may be damaged or even destroyed. Finally, the *waveform* of the power should follow a sine wave format as some equipment is sensitive to this parameter.

Many countries require in-home filtering of line voltages in order to power personal computers and peripherals. Without filtering, the power fluctuations would drastically shorten the useful lifespan of the devices.[5]

Providing Continuous Reliable Power

Dirty power is a term used for electrical power that has too many variations, such as spikes, high voltages, low voltages, and brownouts. Because the local power company may get their power from a variety of sources and has many connected customers, the stability of voltage (the opposite to dirty) cannot always be assured. This possibility must be addressed or equipment could be damaged. The following are progressively better ways to protect equipment from these problems.

Surge protectors are power strips containing electronics that short out high-voltage spikes. These are the first line of defense for home and office equipment and are low cost. However, they only protect from this one threat and may do it only once.

Implication

"We have these problems at many of our austere locations overseas (Turkey is notoriously bad). We have to put regulators on all the lines coming in to the base and then put diesel generators on all buildings containing mission essential equipment due to inconsistent power supply. It's quite a headache!!!!"
Capt J. S. Stewart, Chief, Software Support

Line conditioning is the use of electronics to take the incoming power and change it to conform to more rigid specifications. The equipment has no storage capacity; it just runs the power through transformers and other components that keep the voltage level within a specified range. For the less-expensive models, the output voltage may only approximate a sine wave. Better models produce good sine waves, which is important to

[5] http://www.innovatia.com/software/papers/kalman.htm.

some equipment. Many line conditioners are made to a specification of 89 to 147 volts input to produce an output of 120±10% output voltage.

Uninterruptible power supplies (UPS) are devices that operate much like line conditioners, except they have storage capabilities. Specifically, the incoming (nominal) 110 volts AC power is converted to DC power and stored in batteries. The DC power is then taken from the batteries and run through an inverter, producing 110 VAC, with good regulation and reasonable wave formation. If the primary source of power fails, the batteries continue to supply power until they are drained. If a spike is carried to the UPS, the internal electronics will stop it, a reason many UPS suppliers give a large insurance warranty with the equipment. If a brownout occurs, the UPS takes the power it can and keeps generating good power. The cost of the UPS unit increases as the amount of power required continuously and the length of time power is required during a **power outage.** For the home or organizational setting, a UPS is the lowest-cost reliable protection on the market. Additionally, a better UPS can even shut down equipment gracefully when a specified level of remaining power is reached.

Clarification

The device used by one of the authors is a 1100VAR (volt-amps-regulated = 660 watts) UPS with voltage regulation. It accepts voltages in the range of 90–128 volts and produces an output of 120±5% while providing up to 80 minutes of backup power and constant surge protection for power, phone, and network lines. He uses it for two computers, a monitor, and telephones, producing a backup time of 44 minutes. The onboard software causes a graceful power down of the primary computer via USB after 25 minutes.

Motor-generator (MG) sets are devices historically used to protect mainframe computers. The design connects an AC motor directly to the public power source. The motor is physically connected to an AC generator that produces the desired voltage. Between the motor and the generator is a large flywheel. This flywheel does three things. First, it has sufficient inertia to run continuity for a short time during a power outage, thus acting as a short-term UPS. Second, this inertia prevents spikes from reaching the end equipment. Finally, the generator, being a rotating device just like the generator at the source, produces a good waveform. For the organization with continuous high-power needs, good waveform, and protection from the variations in the public (dirty) power, a motor-generator set, possibly connected to a (smaller) UPS, is an excellent solution.

Clarification

Whether you need line isolation with harmonic filtering, power factor correction, phase conversion, frequency conversion, or voltage conversion, TEMCo's Standard Industrial line of Motor Generator Sets are built to meet your power needs. See http://www.electricpowergenerator.com/motor-generator-sets.html.

Motor-generator (MG) sets provide load isolation for computers and other sophisticated electrical systems. They also can *perform frequency and phase conversion*. MGs offer many advantages over solid-state systems. See http://www.katoeng.com/motorgenset.html.

Backup generators are gasoline-, diesel-, or natural gas–powered engines with generators attached. Like the MG set, they can provide regulated power with good waveform. Unlike the MG set, they provide power in times of extended power outages, as long as fuel is available. When used in combination with a UPS or MG, the best probability exists for continuous power for any eventuality, as long as the backup generator is maintained and tanks are kept full of fuel.

Clarification

The electrical power attributes are

- **C**ontinuity
- **A**ccessibility

- **R**egulation
- **W**aveform
- **P**rotection:
- Surge protectors
  - Line conditioning
  - Uninterruptible power supplies
  - Motor-generator sets
  - Backup generators

SIGNAL RELIABILITY

Performance, accessibility, reliability, and security (PARS): of these the more important is reliability.

When using data communications to transport digital data, there is a probability that the digital bits of the data message will be changed by the noise environment of the circuit. Because of the potential for change, error checking and notification, and preferably error correction, are required to ensure that the message received is the same as the message sent. In all but the simplest cases, this procedure involves putting extra bits or characters in the data stream that can be used by the receiving unit to detect the presence or absence of changed data.

Implication

What is the implication of noise changing a single bit of a character? If you received e-mail to meet a friend/spouse at 6 p.m. (seven-bit ASCII = 0110110) and noise changed a single bit, you might receive the message of "meet me at 7" (0110111) or "meet me at 4" (0110100). In each case, a single bit was changed.

Error Detection

The simplest mode of **error detection** is called **echo checking.** This process involves echoing each transmitted character back from the receiving unit to the sending operator, generally via a full-duplex capability. The system relies on the terminal operator visually checking the returned data and confirming that they are what were sent. This obviously requires a human at the terminal to see the echo and detect errors. It also assumes transmitting only humanly recognizable characters.

Implication

Error detection is a reliability function; it ensures that the data received are the same as the data sent.

Another simple form of error checking for blocks of data is the use of a checksum. A **checksum** is a count of the number of bytes in a transmission unit that are included with the unit so that the receiver can check to see whether the same number of bytes arrived. If the counts match, it's assumed that the complete transmission was received. If the two numbers do not match, a negative acknowledgment (NAK) is sent to the transmitting unit to retransmit the data block. Checksum is a simple calculation and is the method used first in the blocked asynchronous communications called XMODEM. It is also included in TCP and UDP protocols. Most communications software can handle checksum.

Parity checking the simplest error detection scheme for asynchronous data, but it has low reliability. It will not detect even numbers of errors.

Because characters in data communications are generally unintelligible to humans, a different form of error checking is used. The simplest form used in *asynchronous communications* is called **vertical redundancy checking (VRC)** or *parity checking* as it adds a **parity bit** in each character to validate that the character was received correctly. See Figure 5.4(b). When the system uses even-1s parity, the sending modem sums the number of 1s in the data field being sent and places a 0 or 1 in the parity bit position to make the total number of 1s an even quantity. (Odd parity does the same to create an odd number of 1s.) The receiving modem performs the same count and compares the result with the value of the parity bit. Noise often changes an even number of bits at the same time and therefore the error condition is not detected. While this method of error checking is used almost universally, reliance on its use alone carries some risk, that is, about 50 percent, because an even number of changes are not detected.

An error detection scheme similar to VRC that is used on blocks of data is **longitudinal redundancy check (LRC).** Visualize this as stacking all of the data bytes in a vertical stack. This protocol adds a byte of data at the bottom of the stack, the *block check character*. Each horizontal data byte has its own parity bit. LRC software puts as "bit one" of the BCC the parity bit of the first bits of all bytes of the data in the block. Bit two of the BCC is the parity bit of all of the second bits of all bytes of the block of data, and so on. LRC increases the probability that an error in a block of data will be detected to about 98 percent. Still, two errors in 100 get through undetected.

Cyclic redundancy checking (CRC) is the most complex, the most common, and the most reliable error checking mode used for synchronous communications. This method is based on dividing the block of data by a polynomial of order K (such as (0x04C11DB7) and producing a value and a remainder. The CRC character(s) transmitted as the BCC code(s) is the remainder computed. (See Figure 5.5(b).) The receiving unit performs a similar division procedure and compares the remainder it produced with the remainder transmitted. If the two remainders are the same, the system assumes no errors were introduced. Induced noise often appears as a burst of bits, and the check method must be able to detect a number of changes that seem to escape detection in simpler schemes. This method will reliably detect error patterns of K or fewer bit changes. Since the coefficient $K = 16$ is often used, CRC will detect up to 16-bit changes in a block of data. This gives an overall reliability, using $K = 16$, of 0.99999997 percent, one part error in 10^8 data bits.

> **Cyclic redundancy checking** (CRC) is the error detection scheme of choice. CRC-32, four byte, is better than CRC-16, two byte.

Error Correction

If transmitting data in a purely asynchronous mode and an error is detected via VRC parity checking, the receiving modem could potentially notify the ultimate receiving unit that an error occurred. To do this for character-oriented data transmission, the sender would have to wait until the receiver checked the data and replied. In most cases, the receiving station will continue processing data and just discard the bad character. If the receiving unit attempted to correct a single corrupted character through character retransmission, as we will explain in synchronous communications, the network would suffer an intolerable amount of idle time. With modern software and hardware, asynchronous data are usually sent as groups so that block error checking ensures a more reliable result.

For synchronous communications, the nodes are working with blocks of data. When the receiving unit detects an error in the block or packet, it requests that the sending unit retransmit the total data block in question. Commonly, retransmission involves an **automatic repeat request (ARQ).** The receiving unit checks the block of data and informs the source whether the block was good or bad. The data being transmitted are held in a buffer at the source unit until the receiving unit confirms, by reverse communications, that the block of data was good (positive **acknowledgment, ACK**). If an error was detected and a **negative acknowledgment character (NAK)** sent, the data are retransmitted. To speed up the **error correction** process, blocks of data are marked even and odd. The odd packet is sent and received, and stored at the sending station. While the odd packet is being checked, the even packet is sent. If the acknowledgment for the odd packet is ACK, the packet is deleted from the sender and another odd-marked packet is sent while the even-marked packet is checked. If, however, a NAK was sent, the same odd packet is retransmitted.

> Full-duplex circuits greatly speed the ARQ process. Half-duplex circuits use a significant amount of time in switching the channel.

Some systems operate at such a distance that the time to retransmit is unacceptably large. Such systems that have little or no chance of retransmission include additional data that can determine which bits were changed. Such codes include forward error correction (FEC) and are called **hamming codes.** An example system using this code would be a space probe at the outer reaches of our solar system, where the time for the data to reach the receiver is hours to days. Therefore, not only will the data transfer rate be low, to deal with noise it will contain such redundancy so that true data can be created from the corrupted data.

Clarification

Forward error correction (FEC) is a system of error control for data transmission wherein the receiving device has the capability to detect and correct fewer than a predetermined number or fraction of bits or symbols corrupted by transmission errors. FEC is accomplished by adding redundancy to the transmitted information using a predetermined algorithm.[6]

Implication

Network reliability is like a three-legged stool; the legs are *prevention*, *detection*, and *correction*. Detection and correction have been discussed. Prevention is achieved by using a noise-free medium or coding that is more resistant to system errors, such as timing drift. Noise-free circuits can be achieved by use of **fiber optics** (light) or shielding on UTP.

NETWORK STORAGE

When large computer storage devices became available in the mainframe days, they were contained in environments separate from but directly connected to the mainframe CPU. This was a nonnetworked environment; the storage was a peripheral. In today's networked environment, data storage is either provided on individual clients or on servers, for example, a server-centric data storage model. Storage requirements, in general, tend to grow at a rate in excess of 80 percent per year. In the case of data-centric companies like Lithonia Lighting (now Acuity Brands Company) in Conyers, Georgia, the need for storage is growing at a rate in excess of a terabyte per year. This requires something in addition to server-centric storage.

The importance of large data storage areas is that they must be readily accessible from a variety of places. This means that access to data is dependent on data communications. Storage was historically attached to a specific processor and that processor had to be accessed to access the data. All users of a modern network may well need immediate access to data storage. Methodologies have been devised to house data storage in a protected area that is immediately accessible via multiple network connections.

The primary mass storage used on networks is within or attached to servers; a secondary method is to use remote storage-based networks with good bandwidth.

The rise in importance of readily accessible, large data storage has given rise to network-attached storage (NAS). The increasing requirements fueled by growth of data stemming from the Internet, intranets, e-commerce, e-mail, videoconferencing, voice recognition, Web-TV, and so on, have caused additional attention to be focused on traditional server-based storage. New ways to provide networked storage have been developed. There are two major storage management technologies evolving to provide organizations the ability to store and manage their valuable data: network-attached storage (NAS) and storage area networks (SANs). As organizations increasingly consider data as a strategic resource, the ability to manage and store it has gained in importance. New business initiatives that create and use massive amounts of data have started to outgrow existing infrastructure for storing and managing those data. These business initiatives include data warehousing, data mining, customer relationship management, supply-chain management, and e-business.

Storage Area Networks (SANs)

A **storage area network (SAN)** is a high-speed subnetwork of shared storage devices. A storage device is a machine that contains nothing but a disk or disks for storing data. A SAN's

6 http://en.wikipedia.org/wiki/Forward_error_correction.

architecture works in a way that makes all storage devices available to all servers on a LAN or WAN. As more storage devices are added to a SAN, they too will be accessible from any server in the larger network. In this case, the server merely acts as a pathway between the end user and the stored data. Because stored data do not reside directly on any of a network's servers, server power is utilized for business applications, and network capacity is released to the end user.

SANs can provide several benefits for businesses. They help satisfy the explosive demand for storage and use networking to enhance access to data. In addition, there is a means to manage more data with existing human resources. Many are now forecasting that IP networking will accommodate the greatest portion of the world's storage networking requirements. Storage has become the biggest line item in IT budgets of many firms. We have had the emergence of a new player, the storage service provider (SSP), that will manage, store, and maintain customer's data via the network.

Network-Attached Storage (NAS)

The **network-attached storage (NAS)** is basically a system for file sharing that should have the software and hardware preconfigured to make the network storage system. This concept and platform were invented for the Internet and therefore many see NAS as tailor-made for eBusiness.

A NAS[7] device is a server that is dedicated to nothing more than file sharing. NAS does not provide any of the activities that a server in a server-centric system typically provides, such as e-mail, authentication, or file management. NAS allows more hard disk storage space to be added to a network that already utilizes servers without shutting them down for maintenance and upgrades. With a NAS device, storage is not an integral part of the server. Instead, in this storage-centric design, the server still handles all of the processing of data but a NAS device delivers the data to the user. A NAS device does not need to be located within the server but can exist anywhere in a LAN and can be made up of multiple networked NAS devices.

The NAS is basically a system for file sharing; SANs consolidate storage requirements into a common repository. Usually the data come from a variety of platforms and go to a common platform. This requires SAN software, such as **fibre channel,[8]** a serial, high-speed, optical transfer architecture. The multiple connections and multiple devices mean that SANs create additional complexity.

Case 5-1

Bandwidth Resources

"Wild Bill" Wichlei, while on his Texas tour, saw the success of *Billy Bob's Texas*® in Fort Worth and *Gilly's*® in Houston. He figured that this sort of entertainment spot would work well in his home state of Florida. In fact, he thought of three locations where there seemed to be a lot of vacationers but there was a lack of entertainment and "watering holes"; these were in west Orlando, Clearwater, and Destin. Bill found empty buildings, actually former warehouses, and made arrangements to lease them. With a good line of credit at the bank, he set about designing the inside and outside decor, using an architect who had aided with *Billy Bob's Texas*. Following the suggested Fort Worth arrangement, he had large parking

---

[7] http://www.webopedia.com.

[8] *Fibre channel* is a serial computer bus intended for connecting high-speed storage devices to computers. It started for use primarily in the supercomputer field but has become the standard connection type for storage area networks in enterprise storage. Despite its name, fibre channel signaling can run on both twisted-pair copper wire and fiber-optic cables. http://en.wikipedia.org/wiki/Fiber_Channel.

lots plus valet parking at all three locations. The buildings were old, so he enhanced the rustic exterior look and concentrated on upgrading the interior to western and sports themes. He outsourced the restaurants, having two in each location: one for typical western fare and one upscale or semi-gourmet. Each location had a stage for entertainment with a large area in front for the audience to stand or to have temporary seating on the occasion when more refined entertainment came to town. A touch that added much of the atmosphere was the placement of the bars, shoeshine stands, dart boards, and even a mechanical bull. He installed a larger bar with high-definition production television for either closed circuit or PPV satellite broadcasts. "Will Bill" thought that a combination of country-and-western and sports atmosphere would draw customers from two populations. He hoped that the different patrons would help to spread demand over the entire week.

"Wild Bill" knew he could not be at all three spots every night. While he believed he could hire competent, and mostly honest, managers and employees, he needed a way to centralize management of all cash transactions. This would involve connected cash registers and computer-based drink dispensing so he could control inventory and cash/credit receipts. While the point-of-sales cash registers' connectivity seemed straightforward, the credit card transactions would be more difficult to correlate with inventory depletion. Most bars simply took a credit card and opened up a tab account for a customer, summing it at the end of the evening as a single credit transaction. This left room for discrepancy.

He also considered putting TV monitoring for recording in each location with local multicamera recording and centralized on-demand viewing, camera-by-camera. Bill wanted to be able to put a security manager at the home office who could watch all areas of all three establishments on an on-demand basis. The security manager had the ability to remotely control the cameras. This required both the transport of TV data to the home office and transmission of control information from the home office to the individual TV cameras.

"Wild Bill" approached the Telco about service from the three bars to the home office. They told him that he would need continuous communications from each bar to the home office during business hours and that bandwidth would have to accommodate the POS data, the control data for cameras, and the images from the cameras.

QUESTION

Can "Wild Bill" evaluate other than the Telco-recommended solutions for the transport of data? List his options. How does "Wild Bill" determine what "adequate" bandwidth is for his applications? What are the anticipated costs for centralized monitoring and control versus having control only at the individual sites?

Caveat   You are constrained to the technologies thus far covered in the text.

Case 5-2

Network-Attached Storage (NAS)

Steve Guendert

INTRODUCTION

Storage devices that optimize the concept of file sharing across the network have come to be known as network-attached storage, or NAS for short. NAS architectures utilize the mature Ethernet IP network technology of the LAN. Data are sent to and from NAS devices over the LAN using TCP/IP. NAS technology began as an open systems technology in 1985, when it was introduced by Sun Microsystems as NFS, or the Network File System. NFS was an integral element in the growth of network computing, as it allowed Unix systems to share files over a network.

HOW DOES NAS WORK?

By making storage devices LAN addressable, the storage is freed from its direct attachment to a specific server, and any-to-any connectivity is facilitated using the LAN fabric. This may sound very similar to the previous discussion on SAN. The primary distinction between NAS and SAN rests on the differences between data files and data blocks. NAS transports files; SAN transports blocks. NAS uses file-oriented delivery protocols such as NFS for Unix servers and CIFS (Common Internet File System) for Microsoft servers, whereas SAN uses block-oriented delivery protocols such as SCSI. File I/O is a high-level type of request that, in essence, specifies only the file to be accessed, but does not directly address the storage device.

A file I/O specifies the file. It also indicates an offset into the file. For example, the I/O may specify "Go to byte 1567 in the file (as if the file were a set of contiguous bytes) and read the next 256 bytes beginning at that position." Unlike block I/O (SAN), there is no awareness of a disk volume or disk sectors in a file I/O request. Inside the NAS appliance, the operating system keeps track of where files are located on disk. The OS issues a block I/O request to the disks to fulfill the file I/O read and write requests it receives.

Because data blocks are the raw materials from which files are formed, NAS also has a block component; however, this is typically hidden in the NAS enclosure and, to the outside world, the NAS device is a server of files and directories.

NAS accomplishes the primary goal of storage networking: the sharing of storage resources through the separation of servers and storage over a common network. Much like SAN-based storage, NAS overcomes the many limitations of parallel SCSI and enables a more flexible deployment of shared storage. When designed with redundant configurations/architectures, NAS also can provide highly available, nondisruptive storage access. In principle, any user running any operating system can access files on the remote NAS storage device. In addition, a task such as backup to tape can be performed across the LAN, enabling sharing of expensive hardware resources such as automated tape libraries between multiple servers.

NAS BENEFITS

NAS offers a number of benefits, which address some of the limitations of directly attached storage devices (DASD) and also overcome some of the complexities associated with SANs.

- **NAS exploitation of the existing infrastructure:** Because NAS utilizes the existing LAN infrastructure, there are minimal costs of implementation. Introducing a new network infrastructure such as a fibre channel, SAN can incur significant upfront hardware costs. In addition, new skills must be acquired and a project of any size involving a SAN will need careful planning and monitoring to bring to a successful completion.

- **Simple implementation:** Since NAS devices attach to mature, standard LAN infrastructures and have standard LAN addresses, they are typically (according to vendors) extremely simple to install, operate, and administer. This "plug-and-play" operation results in low risk, ease of use, and far fewer operator errors. This equates into a lower **total cost of ownership (TCO).**

- **Reduced total cost of ownership (TCO):** Because of its use of existing LAN network infrastructures, and of network administration skills already employed in many organizations, NAS costs may be substantially lower than for directly attached or SAN-attached storage.

- **Resource pooling:** A NAS appliance enables disk storage capacity to be consolidated and pooled on a shared network resource at a great distance from the servers and clients that will share it. Consolidation of files onto centralized NAS devices can minimize the need to have multiple copies of files spread across distributed clients. Thus, overall hardware costs can be reduced.
- **Improved manageability:** By providing consolidated storage, which supports multiple application systems, storage management is centralized. This enables a storage administrator to manage more capacity on a NAS appliance than typically would be possible for DASD.
- **Scalability:** NAS appliances can scale in capacity and performance within the allowed configuration limits of the individual appliance. However, this may be restricted by considerations such as LAN bandwidth constraints and the need to avoid restricting other LAN traffic.
- **Connectivity:** LAN implementation allows any-to-any connectivity across the network. NAS appliances may allow for concurrent attachment to multiple networks, thus supporting many users.
- **Heterogeneous file sharing:** Remote file sharing is one of the basic functions of any NAS appliance. Multiple client systems can have access to the same file.
- **Enhanced choice:** The storage decision is separated from the server decision, thus enabling the buyer to exercise more choice in selecting equipment to meet the business needs.

NAS DRAWBACKS

On the flip side, NAS is not the perfect storage network solution. Below are some of the "others" to counter the benefits outlined above:

- **Consumption of LAN bandwidth:** Ethernet LANs are tuned to favor short burst transmissions for rapid response to messaging requests, rather than large continuous data transmissions. Significant overhead can be imposed to move large blocks of data over the LAN. This is due to the small packet size used by messaging protocols. Because of the small packet size, network congestion may lead to reduced or variable performance. So the LAN must have plenty of spare capacity (and a willing network manager!) to support implementation.
- **Data integrity:** The Ethernet protocols are designed for messaging applications, so data integrity is not of the highest priority. Data packets may be dropped without warning in a busy network and have to be resent. Since it is up to the receiver to detect that a data packet has not arrived and to request that it be resent, additional network traffic can be created.
- **Proliferation of NAS devices:** Pooling of NAS resources can occur only within the capacity of the individual NAS appliance. As a result, in order to scale for capacity and performance, there is a tendency to grow the number of individual NAS appliances over time, which can increase both hardware and storage management costs.
- **Suitability for database storage:** Given the fact that their design is for file I/O transactions, NAS appliances are not optimized for the I/O demands of some database applications. They do not allow the database programmer to exploit "raw" block I/O for high performance. As a result, typical databases such as Oracle and IBM DB2 do not perform as well on NAS devices as they would on DASD, SAN, or iSCSI.
- **Software overhead impacting performance:** TCP/IP is designed to bring data integrity to Ethernet-based networks by guaranteeing data movement from one place to

another. The trade-off for reliability is a software-intensive network design that requires significant processing overheads that can consume more than 50 percent of available processor cycles when handling Ethernet connections. This is not normally a drawback for applications such as Web browsing, but it is an issue for performance-intensive storage applications.

- **Impact of backup/recovery applications:** One of the potential downsides of NAS is the consumption of substantial amounts of LAN bandwidth during backup and recovery operations. This may impact other user applications. NAS devices may not suit applications that require very high bandwidth.

SUMMARY

One of the major phenomena of the early 1990s was the stabilization and wide acceptance of the Internet. What started as a research tool and evolved into a communications medium for academics and governmental agencies became a major commercial vehicle. Companies have found ways to ensure security on the worldwide conduit, as they established home pages for visibility, shopping malls to purchase from, and databases of their catalogues for total access to the latest literature. This has been fueled in part by the World Wide Web and its user-friendly browsers, led by Netscape and Microsoft's Internet Explorer. Other companies have added plug-in applications to Netscape, making the home pages of greater interest and amusement.

In the mid-1990s, the Internet concept spread to the internal organization, and the concept of the intranet received wide and rapid acceptance. Companies realized that an Internet presence was invaluable and that these same tools could be of great value inside the organization as a single repository for company literature in an easy-to-use format. Now there could be a telephone directory that is always up-to-date and not dog-eared. Organizations could place operating procedures and training manuals in a place so authorized people could pass through the firewall from the outside, or access them internally, and get access to organizational documents they need. This has exploited the great potential for knowledge transfer and training and side-stepped political battles. Information that once was the property of a department (with access at their control) now can be viewed and used by any authorized employee. The United Kingdom, like the United States, has taken large strides to network the country. In the case of JANET and SuperJANET, the plan was overt, starting with academe and research agencies, much as in the United States with ARPANET and Bitnet.

Summary   We now know something about the way data communications developed and evolved. While the advance in technology may seem quite profound, many of the basics remain the same or are very similar. From the dots and dashes of Morse code to the gigabit transmission of EBCDIC or ASCII, we still must use codes to represent data. Regardless of the code employed, there must be adherence to protocols in order for data to be communicated. After choice of code and protocol, we must decide whether to send one character at a time or large blocks of characters. The concept of packets allows a rationale for sending blocks of data. Of course, there is always the need for dealing with possible errors—how can they be detected and corrected?

When we deal with networks, there are many considerations about the media used to connect nodes. Also, the scope of networks, the topology or topologies employed, and the geographic area come into play as we categorize the various networks. Most people are very familiar with local area networks (LANs) and wide area networks (WANS) because most organizations deploy LANs and most people in the United States have access to the Internet (a WAN). Because of the increased vulnerability of data traveling over a network, the topic of security is extremely important. This topic is of such importance, we detail it later.

Key Terms

10BaseT, *p. 147*
100BaseT, *p. 147*
1000BaseT/FX, *p. 147*
24*7, *p. 150*
acknowledgment
(ACK), *p. 154*
asynchronous, *p. 148*
asynchronous
communications, *p. 148*
automatic repeat request
(ARQ), *p. 154*
backbone network, *p. 143*
backup generator, *p. 152*
bandwidth-on-
demand, *p. 150*
bits, *p. 148*
block check character
(BCC), *p. 149*
campus area network
(CAN), *p. 143*
capacity, *p. 146*
category *x*
wiring/cable, *p. 147*
checksum, *p. 153*
circuit switching, *p. 142*
committed information
rate (CIR), *p. 150*
cyclic redundancy
checking (CRC), *p. 154*
dirty power, *p. 151*
echo checking, *p. 153*
error, *p. 150*
error correction, *p. 154*
error detection, *p. 153*

extranet, *p. 143*
fiber-distributed data
interface (FDDI), *p. 145*
fiber optics, *p. 155*
fibre channel, *p. 156*
galactic area network
(GAN), *p. 146*
gigabit, *p. 143*
Gigabit Ethernet, *p. 147*
hamming code, *p. 155*
handshaking, *p. 148*
intermediate level, *p. 144*
internal level, *p. 142*
line conditioning, *p. 151*
local area network
(LAN), *p. 142*
longitudinal redundancy
check (LRC), *p. 154*
metropolitan area
network (MAN), *p. 144*
motor-generator (MG)
set, *p. 152*
negative acknowledgment
character (NAK), *p. 154*
network-attached storage
(NAS), *p. 156*
packet, *p. 148*
packet data network
(PDN), *p. 146*
packet switching, *p. 150*
parameters of a valuable
network (RAPS or
PARS), *p. 150*
parity bit, *p. 153*

parity checking, *p. 148*
peripheral
equipment, *p. 142*
personal area network
(PAN), *p. 142*
power outage, *p. 152*
reliability, *p. 150*
security, *p. 151*
service level agreement
(SLA), *p. 150*
small office home office
(SOHO), *p. 143*
stop bit, *p. 148*
storage area network
(SAN), *p. 155*
surge protectors, *p. 151*
synchronizing, *p. 142*
synchronous, *p. 148*
topology, *p. 140*
total cost of ownership
(TCO), *p. 158*
tunneling, *p. 146*
uninterruptible power
supply (UPS), *p. 152*
value added networks
(VANs), *p. 146*
vertical redundancy
checking (VRC), *p. 153*
virtual LAN
(VLAN), *p. 144*
virtual private network
(VPN), *p. 146*
wide area networks
(WANs), *p. 145*

Recommended Readings

Network Magazine (http://www.networkmagazine.com/), a periodical about networks and network management.

Management Critical Thinking

M5.1 If your firm is making a decision to set up a local area network and current forecasts indicate 10BaseT is adequate for the next three years, should you (as telecommunications manager) recommend it as a low-cost option or consider the likely demand for greater bandwidth to make your recommendation?

M5.2 Why is total cost of ownership (TCO) an important consideration in the choice of NAS and SAN? Where can managers get help in performing evaluations of TCO?

Technology Critical Thinking

T5.1 Building in future growth capacity increases the cost of a project. As a technical project manager, are you concerned about the future or just getting the job done on time and within cost? Does it cost more to add capacity and capability to the system as it is being developed or to add to the apparitional system later?

T5.2 What are the technology trade-offs between one large network for a medium-sized organization versus several smaller networks and a backbone network?

Discussion Questions

5.1 If you are planning a new house LAN (personal or SOHO), what sort of wiring would you specify and why?

5.2 Is it cost-effective for a small, for example, 10,000 population, town to install a MAN to attract business?

5.3 If you are on a project team to install network wiring in a new building, how would you justify the use of a specific type or category cable, and which one would you recommend?

5.4 What PANs do the members of the class use? How many are wireless?

5.5 How practical is a galactic area network? Who is using them now?

5.6 What is the likely future of the twisted copper wire infrastructure in an age of high-bandwidth demand?

5.7 Since the public Internet is basically free, when and why should an organization go to the effort to install a private WAN?

5.8 How important is reliable and continuous power? What should a firm such as a bank do to ensure adequate and continuous power?

5.9 Is cooling of network equipment a problem? How is this solved?

5.10 What electrical power protection do you have at home? With the options discussed in this chapter, what should you be considering and why?

Projects

5.1 Look up the cost per foot of all of the categories of network cabling on the Internet. How well is Gigabit Ethernet supported; how about 10,000BaseFX?

5.2 Determine the cost differential between creating a one-mile-radius LAN using 100BaseT, 1000BaseT, and fiber cabling. Is the cost significant enough to drive the decision?

5.3 Draw a network diagram showing PC-attached servers, LAN-attached servers, and SAN-attached servers. How is each used?

5.4 Interview the network administrator of an organization with a medium-to-large network and find out how much of a problem they have with noise. Is it significant enough to affect data throughput, that is, effective bandwidth?

5.5 Using the Internet or a local electrical supply store, price out the cost of protecting 100 PCs with surge protectors, 10 servers with UPS, and one backup generator to power all of them during an outage. Define the outage periods to determine the capability of each device.

5.6 Determine if any local organizations use motor-generator sets? Do they know what they are and why there are valuable? Do any use them for 1/phase-to-3/phase transfer?

5.7 Interview the network manager of a local organization and determine their use of electrical protection of clients and servers and method of continuous power during outages. Have they tested their plan?

5.8 In project 5.7, determine the use of LANs, backbone networks, and WANs in the organization. Get diagrams if possible. What do you think of their plan?

5.9 In project 5.7, discuss the network parameters of RAPS with the manager and ask his/her opinion. How does it match this chapter?

5.10 What software can you find on the Internet to diagram LANs and WANs. Get a 30-day copy and draw the diagram in project 5.3.

Part Three

Wide Area Networks: The Internet

Wide area networks extend the reach of the organization from near to far. The Internet, that network of networks, is the ultimate network and often connects and extends the organizational private networks. The Internet, like the Interstate Highway System, has changed the conduct and formation of business in the United States and the world.

Chapter **Six**

From LANs to WANs: Broadband Technology

In this chapter, we extend our discussion of networks, moving from LANs to WANs, and the related issues, such as packet switching and other techniques for improving bandwidth use. The important WAN technologies are covered.

We mentioned in previous chapters the hierarchy of networks, based on geography and, to some extent, the scope of coverage. To this point, we have discussed fairly localized networks. We now examine larger, geographically dispersed, and more complex networks. The new topologies and protocols will require different technologies and standards, but as some of the technology remains the same, we will revisit it.

As the need for organizational bandwidth increases and as the reach of the organization increases, new technologies over LANs will be required. One of the features of broadband technologies as found in wide area networks is the ability to expand the capability. A network may start with wide connectivity but low bandwidth and grow, that is, scale, the bandwidth as the need arises. This scalability is a very important aspect of networking and many other technologies when making the initial decisions.

BANDWIDTH IS CONVERGENCE DRIVEN

Convergence is driving bandwidth: the more you add various media to the message, the more bandwidth you need.

For networks, **convergence** is the combination of two or more media, such as text/data, graphics/images/fax, audio, or video. The more information placed on the channel, the greater the bandwidth of that channel must be or the longer the time it will take to move the information. So, as in our definition of effective bandwidth, we must use compression, counter **noise,** or increase the native bandwidth to overcome the added time to transmit.

As an example of convergence, consider the offering of a few Bell Operating Companies; they offer a service called *D1AO,* which is a single ISDN D-channel of 9.6 Kbps that is always connected. This provides a conduit of almost 10 Kbps 24*7, without modems or manual connectivity, by providing an open packet-based channel. What could one do with this capability? How does this relate to convergence?

When **electronic mail (e-mail)** was introduced, it was a text-based phenomenon, using only ASCII characters. Given the premise that an average e-mail is 400 characters in size, then an e-message would be about 4,000 bits (at asynchronous 10 bits/character or less for packets). The D1AO could transmit an average e-mail in less than one-half second (4,000/9,600 = 0.42 second), completely adequate for our needs. (See Table 6.1 for more examples.) The problem arises with *convergence*; e-mails now contain graphics and have attachments that range from executable programs to movies. However, provided with enough time, the D1AO bandwidth could be adequate. For example, if one wished to send an e-mail of one megabyte in size, it would take 17 minutes [(1,000,000 * 10)/10,000 = 1,000 seconds = 16.6 minutes, asynchronous]. (Using **synchronous** packets and assuming 3 percent overhead = (1,000,000 * 8 * 1.09)/9,600 = 858 seconds = 14.3 minutes.) Downloading this e-mail while sleeping might be very acceptable.

TABLE 6.1
Examples of E-mail Growth

E-mail	Characters with Spaces	Size
Text only	300	0.9K *GroupWise*
Text only	5,018	20K *Outlook*
Text only	2,618	10K *Outlook*
Text with .DOC attachment	.DOC (19K)	22K *Outlook*
One-page graphic ad		17K *Outlook*

E-mail is a light load; when supplemented by art graphics, videos, executable upgrades, and database files, the demand for bandwidth swells. For e-mail *with convergence,* a 56K modem could generally suffice, being about nine times faster than the D1AO channel.

Even when just text, e-mail has evolved to be more real time; the extreme example is *instant messenger (IM)*. While IM text messages would work equally well on D1AO, they would slow down if graphics were a factor.

The convergence that really drives the process is Internet access and upload and download of music, graphics, executables, and so forth. *We want higher bandwidth not for e-mail, but for Internet access.* Add also that many homes or SOHOs have installed multiple computers on LANs, all of which require Internet access and use e-mail.

It is for this need for greater bandwidth, greater reach on the network, and connectivity of heterogeneous users that we address technologies beyond the local area network. Just as the usage on the Internet is drastically different than usage on a LAN, widely dispersed networks of heteroecious users require different topologies and protocols than for LANs or even backbone networks.

PACKET OR FRAME FORMAT

In Chapter 1, we discussed the conversion of analog signals to a digital representation. The standard used was *pulse code modulation,* which produces a standard 64 Kbps digital signal. The signal is continuous and both ends are synchronized. Thus, every 64 kilobits is a second of PCM-converted analog signal. This standard is referred to as **Data Services Zero (DS0).**

The Bell System developed the *Data Services* categories of digitized analog signals. PCM samples voice at a rate of 8,000 samples, or frames, per second. For a DS0 voice channel, each frame is an eight-bit code, resulting in a bandwidth of 64,000 bps. Extending the categories produces **Data Services One (DS1)** or **Terrestrial One (T1)** service, which is defined as 24 DS0 channels. In the T1 environment (see Figure 6.1), the equipment sends, not one

FIGURE 6.1
T1 Frame Format

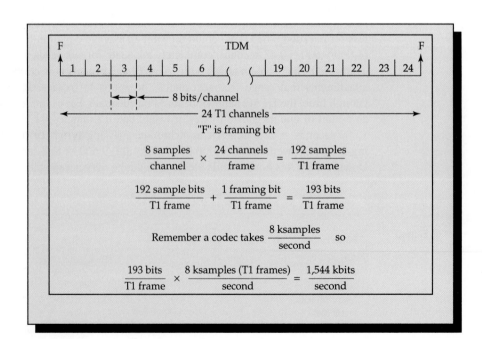

TABLE 6.2
Bandwidths of
T-Carriers

Service	Bandwidth	Number of DS0 Channels
DS0	64 Kbps	1
T1	1.544 Mbps	24
T2	6.312 Mbps	96
T3	44.736 Mbps	672
T4	274.176 Mbps	4,032

channel of eight bits per frame, but 24 channels, for a total of (24*8) = 192 bits/frame. Each frame is separated from the next by a single bit, making a 193-bit frame. This produces a T1 bandwidth of (193 bits/frame *8K frames/second) = 1,544,000 bps. Alternately, this is ((24 * DS0) + 8K) = 1.544 Mbps. Table 6.2 shows the bandwidth of other **T-carrier system** services. The **T3** service has been the workhorse backbone of the Internet.

The DS0 and T1 format are simple standards for data communications. They are simple because the network for them is point to point; thus, no destination or source information or error checking is required. In the multipoint access networks, the packet must carry additional information to support its journey from sender to receiver. The DS0 rate may support 20 2.4-Kbps channels, 10 4.8-Kbps channels, five 9.67-Kbps channels, one 56-Kbps channel, or one 64-Kbps clear channel. In its primary format, a T1 circuit is channelized into 24 DS0 channels.

NETWORK TYPES REVIEWED

Homogeneous usage means that the user community has similar tasks and similar applications on the network to support these tasks. Heterogeneous usage means that the users either do not have similar tasks or are more remotely located.

For illustration, we choose to divide the networked world into two domains: near and remote. Nearby networks include *local area networks* to which users attach and *backbone networks* that connect multiple LANs. For example, our College of Business has an **Ethernet** bus LAN with three subnets (supporting about 250 faculty and staff and 3,500 students) that is connected to the rest of the campus via an optical backbone, itself running Ethernet. This begins the process of *internetworking*. The usage is, for the most part, homogeneous, as is expected on a LAN.

An intermediate connection for a home, office, or campus is a *metropolitan area network (MAN),* which may act as or connect to an **Internet service provider (ISP).** Many CATV providers have MANs to carry analog and digital TV and Internet access. Because DSL is point to point, we don't consider it a MAN or even a backbone unless we have LANs running in the connected home or office. Many metropolitan area networks are constructed as counter-rotating rings, for redundancy, using token-passing protocol.

Clarification

Wide area networks provide communications among widely dispersed, heterogeneous users and networks. WANs are much like the road system for vehicles. Each provides a variety of paths from one point to another; some are congested and low speed and some are high speed. Navigating roads and WANs takes instructions, ranging from road maps and driving instructions for vehicles to information in the packet header for data to the final end point.

Digital Subscriber Line (DSL)—Local Broadband

If your organization is the owner of an extensive twisted-pair copper network, such as the Telco's extensive POTS local loop network, you would want to take advantage of it to deliver **broadband** services, ranging from Internet access to video-on-demand. **Digital subscriber line (DSL)** is a technology that provides digital communications over twisted-pair copper wire from 64 Kbps to 6.0 Mbps and beyond over repeaterless local loops of up

Designation	Bandwidth Limit (one way)	Distance Limitation
DS1 (T1)	1.544 Mbps	18,000 feet
E1	2.048 Mbps	16,000 feet
DS2	6.312 Mbps	12,000 feet
E2	8.448 Mbps	9,000 feet
¼ STS-1	12.960 Mbps	4,500 feet
½ STS-1	25.920 Mbps	3,000 feet
STS-1	51.840 Mbps	1,000 feet

to 18,000 feet in length. (See Table 6.3 for a comparison of bandwidth versus distance for 24-gauge wire.) The initial DSL technology in use was **integrated services digital network (ISDN),** using a bandwidth of 160 Kbps for BRI and delivering two 64 Kbps bearer channels plus a 16 Kbps delta channel for control and packet data. This was the original digital local loop, giving the users a digital domain and ISDN switching, and providing a switched digital communications network. As with all DSL capabilities, ISDN provides digital channels over an analog circuit. That is, ISDN and all DSL technologies each uses its own specific modemlike equipment at each end of the circuit, referred to as *terminators*.

Downstream data rates depend on a number of factors, including the length of the copper line, its wire gauge, and cross-coupled interference. Line attenuation increases with line length and frequency and decreases as wire diameter increases. DSL will perform as shown in Table 6.4. It is predicted that 80 percent of the residences and offices in the United States can be accommodated via DSL due to the presence of 24-AWG- and 26-AWG-gauge twisted-pair wire in the local loops.

Asymmetric Digital Subscriber Line (ADSL)

The specific ADSL bandwidth provided to a customer is determined by the Telco; DSL is capable of high bandwidth, but equipment costs rise as the bandwidth increases.

Asymmetric digital subscriber line (ADSL) converts existing twisted-pair telephone lines into access paths for multimedia and high-speed data communications. ADSL can transmit more than 6 Mbps to a subscriber (downstream) and as much as 640 Kbps upstream. Such rates expand existing access capacity by a factor of 50 or more without new cabling. ADSL can literally transform the existing public information network from one limited to voice, text, and low-resolution graphics to a powerful, ubiquitous system

TABLE 6.4   Copper Access Transmission Technologies

Source: http://www.adsl.com/adsl/dsl_tut.html.

Name	Meaning	Data Range	Mode	Application
V.22, 32, 42	Voice band modems	1.2 Kbps to 28.8 Kbps	Duplex	Data communications
DSL	Digital subscriber line	160 Kbps	Duplex	ISDN service, voice and data communications
HDSL	High-data-rate digital subscriber line	1.544 Mbps 2.048 Mbps	Duplex Duplex	T1/E1 service feeder plant, WAN, LAN access, server access
SDSL	Single-line digital subscriber line	1.544 Mbps 2.048 Mbps	Duplex Duplex	Same as HDSL plus premises access for symmetric services
ADSL	Asymmetric digital subscriber line	1.5 to 9 Mbps 16 to 640 Kbps	Down Up	Internet access, video-on-demand, simplex video, remote LAN access, interactive multimedia
VDSL, BDSL, or VADSL	Very-high-data-rate digital subscriber line	13 to 52 Mbps 1.5 to 2.3 Mbps	Down Up	Same as ADSL plus HDTV

capable of bringing multimedia, including full-motion video, to everyone's home this decade.[1]

Description of ADSL

ADSL allows the use of the installed POTS network to provide high-speed Internet access and limited video-on-demand offerings. As the lines of the network are the most difficult and expensive portion of the total network to install, any scheme that provides digital bandwidth to the home with only the addition of black boxes at each end means that service could have widespread usage.

An ADSL circuit connects an ADSL modem on each end of a twisted-pair telephone line, creating three information channels: a high-speed downstream channel, a medium-speed upstream channel (thus the name *asymmetric*), and a POTS channel. The POTS channel is split off from the digital modem by filters, thus guaranteeing uninterrupted POTS even if ADSL fails, that is, a *lifeline*. The high-speed channel ranges from 1.5 to 6.1 Mbps, while upstream rates range from 16 to 640 Kbps. Each channel can be submultiplexed to form multiple, lower-rate channels.

How does ADSL work on an analog POTS line? ADSL delivery methods divide the downstream channel logically into 256 4-KHz channels, each of which is modulated and can carry as many as 60,000 bps. Noise will make some of these channels inoperable or the **signal-to-noise ratio** will limit the bit rate, giving an effective bandwidth that is dependent on the distance traveled, ranging from 1.544 Mbps at 18,000 feet to 51.840 Mbps at 1,000 feet (see Table 6.1). The upstream arrangement is 32 channels of 4 KHz each. The primary use of ADSL is for Internet access. This environment generally requires a 20:1 or better factor of downstream to upstream data rates. Therefore, it would be possible to provide the full ADSL circuit to the home with 6 Mbps downstream and 640 Kbps upstream.

Caveat

STS-1 = Synchronous Transport Signal
1 = 51.84 Mbps rate

As noted, DSL is dependent on distance and wire size. The defined **customer service area (CSA)** for DSL for good speed is 18,000 feet. If the distance is reduced to 9,000 feet, the speed in the new protocol can approach 30 Mbps. At 1,000 feet, it can achieve STS-1 speeds of 51.85 Mbps. Thus, as the attached central office moves closer to the user, for example, local micro central office, (unshared) bandwidth will increase.

ADSL2—The Next Step[2]

The evolution of ADSL has produced two new capabilities, called *ADSL2* and *ADSL2+*. Meeting G.992.3 and G.992.4 standards, the extension improves data rate and reach performance; ADSL2+ (G.992.5) more than doubles the downstream data rate of ADSL to 25 Mbps. ADSL2 and ADSL2+ provide the following benefits:

- **Better rate and reach:** ADSL2 increases downstream data rates to more than 12 Mbps, as compared to between 8 Mbps and 10 Mbps for original ADSL. ADSL2 extends reach by approximately 600 feet.
- **Diagnostics:** Real-time performance-monitoring capabilities provide information regarding line quality and noise conditions at both ends of the line.
- **Channelization**: ADSL2's channelization capability provides support for Channelized Voice over DSL (CVoDSL), a method to transport derived lines of TDM voice traffic transparently over DSL. CVoDSL transports voice within the physical layer, letting derived voice channels ride over DSL bandwidth while maintaining both POTS and high-speed Internet access.

---

[1] Tutorial from ADSL Forum, Kim Maxwell, chair, available at http://www.adsl.com/adsl/adsl_tut.html.
[2] http://www.nwfusion.com/news/tech/2003/0120techupdate.html.

- **Bonding for higher data rates:** ADSL2 chipsets can bind two or more copper pairs in an ADSL link. The result is fiberlike data rates over existing copper lines.
- **All-digital mode:** An optional mode allows for transmission of data in the POTS portion of the phone line. This adds 256 Kbps to the upstream data rate, which can be an attractive option for businesses that have voice services on different phone lines and value the additional upstream bandwidth.
- **Packet-based services:** Packet-based services such as Ethernet can be transported over ADSL2.
- **ADSL2+:** The ADSL21 standard doubles the maximum frequency used for downstream data transmission from 1.1 MHz to 2.2 MHz. This effectively provides *downstream data rates of 25 Mbps on phone lines as long as 5,000 feet.*

High-Data-Rate Digital Subscriber Line (HDSL)

DSL depends on the use of a pair of matched modems. **High-data-rate digital subscriber line (HDSL)** initially required four modems and four wires (two lines) to deliver up to 2,048 Kbps (E1) speed and is often used in the delivery of T1 services, such as connecting PBXs, cellular antenna sites, routers, and IXC access points. HDSL is often referred to as a repeaterless T1 line, though the actual bandwidth may be between 56 Kbps and T1. Unlike ADSL, HDSL is symmetric with the rated speed in both directions.

HDSL requires two lines. SDSL (symmetric digital subscriber line), which is HDSL over a shorter length, uses only one line, so it will run over POTS and offers T1 symmetric services for the residence with bandwidths of up to 3 Mbps.

Clarification   DSL technology continues to develop and new capabilities are emerging. For example, we describe ADSL as having a bandwidth of 64 Kbps to 9 Mbps downstream and 640 Kbps upstream at distances of up to 18,000 feet. As noted in Table 6.4, VDSL (very-high-data-rate DSL) permits data rates over much shorter lines, transmitting up to STS-1 (51.84 Mbps) speeds downstream and proposing to send between 1.6 Mbps and 2.3 Mbps upstream. Further variants address a 19.2 Mbps upstream, as well as a symmetric bidirectional data rate.

MOVING TO NETWORKS COVERING LARGE AREAS

Wide area networks (WANs) are at the opposite end of the network continuum from LANs and are the primary focus of this chapter. They are networks that cover wide geographical areas. They go beyond the boundaries of cities and may extend globally. The extreme of the WAN is the *global area network.*

Clarification   A **wide area network** is a group of heterogeneous users, widely dispersed.

Packet Data Network (PDN)

One form of WAN is a set of permanent circuits and channels onto which users are connected or into which they dial on an as-needed basis. This WAN provides connectivity between specified points by use of a temporary physical circuit. The connectivity is made possible via a switch at the center of this star network. To create an end-to-end, point-to-point network, the system uses **circuit-switching** technology. This technology has been used for decades in the POTS and Switched-56 networks and has the problem of dedicating the total circuit to a single communication. While this was previously the only way possible for voice networks, and was a logical solution, it is a very wasteful solution for data transmission.

An alternate form of connection for a WAN is the **packet data network (PDN),** which also can provide connectivity to many points geographically. The PDN appears like a cloud with entry/exit points in multiple locations. The internal workings of the network are obscured from the users because of the way in which it provides connectivity. The first, and presently almost exclusive, way is by providing a **virtual circuit (VC)** upon request. In this mode, commands from the requestor cause the network to establish a temporary *physical channel* composed of a series of *physical linkages* for the duration of the communications. Each such setup may be different, based on the availability or busy state of the various linkage paths in the network, although there may be a preferred path, for example, the most direct route. The only time the specific arrangement of the virtual circuit is changed is when a link is disabled, at which time an alternate route is quickly established. This method is called a *virtual circuit* because other communications' virtual circuits may share some of the linkages that make up this circuit. Since the "circuit" is shared by multiple VCs, its bandwidth is shared and more efficiently utilized than circuit-switched circuits.

With PDNs, the total file to be transferred is divided into predetermined-sized packets by a **packet assembler/disassembler (PAD).** The PAD receives the total data file from the sender, breaks it into packets, adds the appropriate overhead (addressing, administrative, and error-checking data), and places the packet onto the network. The PAD at the receiving end reverses the process, handing off the total file to the receiver. A second function of the PAD is to assemble the packets in the correct order upon receipt and ensure that no packet is missing.

Clarification

A **packet data network** is viewed as a cloud with a PAD at the entry and exit points.

PDNs are viewed as clouds since their inside circuitry is hidden from the users.

The second form of connectivity, or data transfer, in the PDN is the **datagram.** In this mode, the data are packetized via the PAD and sent over the network, much as a letter through the Postal Service. The major difference in this analogy is that the Postal Service letter has the total message therein, whereas the datagram has only a portion of the message. (Also, the Postal Service uses virtual circuits although they appear to be using datagrams.) The intelligence of the network routes each packet (datagram) to its destination, based on availability and busy state of each linkage at that moment. The virtual circuit is a circuit-switched network while datagrams use a **packet-switched network.** (See Figure 6.2.) The PDN switches the datagram at each point in the network, with no continuous channels being established. Circuit-switched networks take less overhead after setup, that is, no destination address is required, whereas each datagram must have a destination address.

FIGURE 6.2   Packet Switching

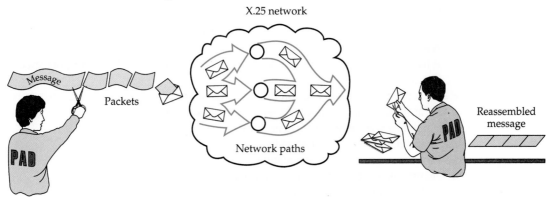

Clarification

Traffic through a PDN is in the form of packets that may travel over specified links as defined by a virtual circuit or proceed on their own as datagrams. Datagrams have a time-to-live indicator so that they will not circulate continuously.

Clarification

- **Circuit switching** establishes a path (temporary circuit) that remains fixed for the duration of a connection.
- In **packet switching** (datagrams), each packet includes source and destination address information so that individual packets can be routed through the internetwork independently.
- **Message switching** transfers the complete message from one switch to the next, where the message is stored before being forwarded again (store-and-forward). The path is determined at the time of switching. It is commonly used in e-mail.

WAN TECHNOLOGIES

WANs may offer broadband services (in excess of T1) or subbroadband speeds (9.6 Kbps to 1.544 Kbps). The **protocol** and distance, not the speed/bandwidth, separate a LAN from a WAN. WANs have two primary uses: connecting disparate users and interconnecting LANs.

Implication

POTS uses circuit switching and allocates the physical circuit to one conversation for the duration of the call. VoIP, telephony over Internet protocol networks, uses virtual channels and shares each link with other users.

X.25 Packet Data Network

One of the earliest WAN protocols is **X.25,** a virtual circuit-based network in its most-used form. It is a connection-oriented, packet-switching system that interconnects noncollocated local area networks or user terminals. X.25 is a publicly based network switching system run by most telephone companies. It acts the same way as the telephone system (see Figure 6.2): it makes a temporary physical connection (virtual circuit) between two locations, allows the conversation (in this case, the transference of data) to occur, and upon completion breaks the physical connection so the middle carrier trunks can be used by someone else. And the good thing is, you only pay for the time you use (much like a normal telephone bill).

X.25 is packet switched at each junction. The system checks each packet at each switch point, resulting in its ability to carry data over noisy circuits but imposing significant time delays. The value of this feature of X.25 is that it only has to return to its predecessor switch point when a packet has to be retransmitted, making it noise-resistant. The negative side of this same feature is that it does not support time-sensitize data, such as voice.

In referencing the OSI standard stack, X.25 is incorporated in the first three layers (refer to Figure 6.3).

1. The physical layer incorporates specifications such as
   a. What type of wire medium (UTP, TP, STP, coax, etc.).
   b. What type of connections (RJ, V.35, RS-232, etc.) and end link transport speeds (2.4 56 Kbps).
   c. What the electrical voltages on the pins will be (spread of ±5 to 25 volts DC).
   d. What the signaling will be (master, slave, clocking and synchronization, etc.).
2. Data link layer describes specific data link protocol efforts such as
   a. Delimiting pointers for start and end of X.25 formatted packets.
   b. Packet numbering and accountability.
   c. Packet acknowledgment, error control, and flow control.
   d. SDLC or HDLC synchronous protocol parameters.

FIGURE 6.3
X.25 Layers and Frame Structures

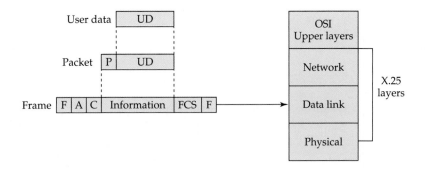

3. Network layer controls
   a. Virtual network establishment between network switches.
   b. Network interface and DNIC (Data Network Identification Connection) addressing.
   c. Routing.
   d. Carrier interconnection flow control.

Packet-Layer Protocol (PLP)[3]

Packet layer protocol (PLP) is the X.25 network layer protocol. PLP manages packet exchanges between DTE devices across virtual circuits. PLPs can run over Logical Link Control 2 (LLC2) implementations on LANs. The PLP operates in five distinct modes: call setup, data transfer, idle, call clearing, and restarting.

Link Access Procedure, Balanced (LAPB) is a data link layer protocol that manages communication and packet framing between DTE and DCE devices. LAPB is a bit-oriented protocol that ensures that frames are correctly ordered and error-free. (See Figures 6.4(a) and (b).)

Three types of LAPB frames exist: information, supervisory, and unnumbered. The *information frame (I-frame)* carries upper-layer information and some control information. I-frame functions include sequencing, flow control, and **error detection** and recovery. I-frames carry send- and receive-sequence numbers. The *supervisory frame (S-frame)* carries control information. S-frame functions inc lude requesting and suspending transmissions, reporting on status, and acknowledging the receipt of I-frames. S-frames carry only receive-sequence numbers. The unnumbered frame (*U-frame*) carries control information. *U-frame* functions include link setup and disconnection, as well as error reporting. U-frames carry no sequence numbers. LAPB frames include a header, encapsulated data, and a trailer. Figure 6.4(a) illustrates the format of the LAPB frame and its relationship to the PLP packet and the X.21bis frame.

Some of the following discussions show a greater level of network detail than we have done so far. These descriptions should give the reader a feeling for **frame formats** and other details as enticement to investigate such information further if desired.

Broadband Integrated Services Digital Network (BISDN)

Broadband generally means bandwidths of T1 and above.

ISDN, in its **basic rate interface (BRI)** form, refers to digital telephony that includes two DS0 (64 Kbps) **bearer (B) channels,** one 16-Kbps D data channel, and overhead to the retail touch point. As a method of transporting higher quantities of data or voice between central offices, *primary rate interface* ISDN may use this technology on a T1 circuit, providing 23 B channels and one **delta (D) channel.** This is ISDN in its "narrow" form, or NISDN. In the broadband form of ISDN, the services provided increase and the bandwidth is raised to T1 and above. This is the generic form of **broadband integrated services digital network (BISDN).**

[3] http://www.cisco.com/univercd/cc/td/doc/cisintwk/ito_doc/x25.htm.

FIGURE 6.4(a)
The PLP Packet Is
Encapsulated within
the LAPB Frame and
X.21bis Frame LAPB
Frame Format

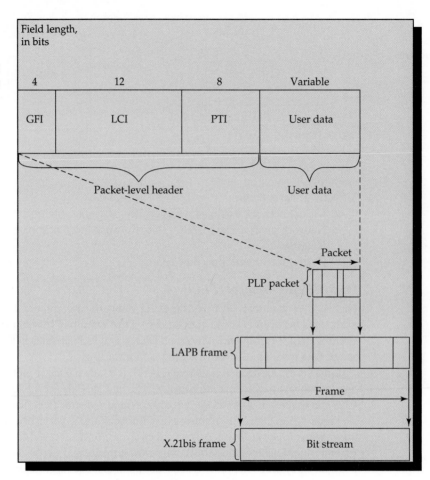

FIGURE 6.4(b)
LAPB Frame Format

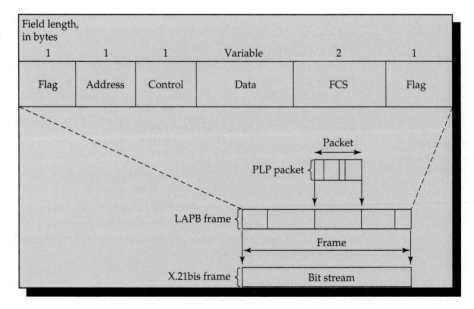

Clarification

BISDN will have two definitions. First is ISDN above T1 that has specific services in the ISDN family. Second is the use of two broadband technologies (SONET and ATM) to provide high bandwidth connectivity.

We introduced WANs with a discussion of X.25 capabilities. The next discussions provide upgrades to this technology and continue to capabilities with greater and great bandwidth. As WANs increase in scope and reach and as organizations need to connect more and more LANs, broadband networks are important.

Clarification

Most references to T1 are to speed/bandwidth, not a specific channelized protocol. When a circuit or network is defined as comparable to T1, it means its bandwidth is about 1.5 Mbps.

Frame Relay

Frame relay was initially introduced in the late 1980s as an additional packet-mode bearer service for NISDN. **Frame relay** can be defined as an ISDN frame-mode service based upon fast packet switching. In simplistic terms, it can be thought of as relaying variable-length units of data, called *frames,* through the network. It is a connection-oriented technology that supports variable-length packets at medium- to high-speed data rates.

Since frame relay provides connection-oriented point-to-point service, it is offered as a **permanent virtual channel (PVC)** service. The PVC must be reliable and mostly error-free because error correction and flow control functions are only performed at the end users' **customer premises equipment (CPE).** Each node in a frame relay network only has to perform three tasks: ensure the frame is valid, ensure the frame possesses a known destination address, and relay the frame toward the destination. If any problem develops while processing a frame, the frame is discarded. The intended receiver must detect any missing frames, notify the sender, and wait for retransmission. Currently, frame relay access speeds range from 56 Kbps to 1.544 Mbps (T1).

Frame relay uses variable-length frames. The frame format (as shown in Figure 6.5) consists of a beginning and ending flag, a frame relay header, a user data field, and a frame check sequence (FCS), referred to as block check characters (BCC) in previous discussions of packet formats. The flags are used to delimit the beginning and end of the frame with the bit pattern 01111110. The frame relay header (two to four octets) contains the address as well as the congestion control bits. The user data field consists of an integral number of octets of information and conforms to the ISDN's Link Access Procedure on the D channel (LAPD), which establishes a maximum frame size of 262 bytes. Finally, the frame check sequence (FCS = two octets) contains the remainder from the cyclic redundancy check (CRC) calculation that is used to detect bit errors.

Advantages of Frame Relay

First, frame relay provides higher performance than traditional X.25 packet switching because error correction and flow control are not performed at every node, as in the X.25

FIGURE 6.5
Frame Relay Frame Format

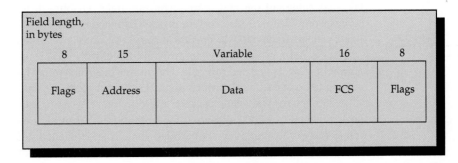

Field length, in bytes				
8	15	Variable	16	8
Flags	Address	Data	FCS	Flags

network. This reduces the amount of packet-handling equipment in the network, which controls communication costs. Frame relay service has the potential to be an economical alternative to private lines offering the same bandwidth. In addition, frame relay achieves higher performance because the nodes spend less processing time looking for errors. As a result, frame relay provides lower delays, higher throughput, and better bandwidth utilization than the X.25 packet-switching network. The single access line with multiple permanent virtual channel features seems a big technical advantage as well as a potential cost advantage.

Second, frame relay is generally a simple software upgrade from most X.25 devices (proven technology), so the investment in currently used equipment is protected. If more than a simple upgrade is necessary, the start-up cost for frame relay equipment is relatively low. Thus, frame relay has the advantage of not only possessing a low technical risk, but a low implementation cost as well.

Third, frame relay supports numerous applications, such as block-interactive data applications, file transfer, multiplexed low-bit rate, and character-interactive traffic. Block-interactive data applications consist of high-throughput, low-delay applications such as high-resolution graphics, videotex, and CAD/CAM files. Multiplexed low-bit-rate applications multiplex multiple low-speed channels onto a single high-speed frame relay data channel. Character-interactive traffic describes low-throughput, low-delay, low-volume traffic, such as text editing.

Implication   Frame relay is a cost-effective upgrade from X.25; the initial time delays of variable-sized frames have been overcome so voice-over-frame-relay is usable.

Disadvantages of Frame Relay

Even though performing error correction and flow control only at endpoints is an advantage of frame relay, these savings can create some potential disadvantages. If an error occurs, the time to correct the error is longer than in an X.25 network operating at the same speed. Also, frame relay requires a mostly error-free transport network. This bases frame relay on the use of PVCs, which limits flexibility.

Another problem is that frame relay suffers from longer delays at the switching points than cell-based systems (discussed later). A receiving switch waits until the entire data unit is received before forwarding it to the next switch. This delay made frame relay, in its original form, unsuitable for voice or steady-flow traffic that requires real-time processing. The problem is compounded by frame relay's low switching rate; its current top switching rate is 1.544 Mbps (T1), while other technologies have rates in the gigabit-per-second range.

Finally, frame relay is more difficult to manage as the network grows. As it uses permanent virtual channels, the number of PVCs grows with the net, causing management problems.

Frame Relay versus X.25

X.25 was designed to provide error-free delivery using high-error-rate links. Frame relay takes advantage of the new, lower-error-rate links, enabling it to eliminate many of the services provided by X.25. The elimination of functions and fields, combined with digital links, enables frame relay to operate at speeds 20 times greater than X.25.

X.25 is defined for layers one, two, and three of the OSI model, while frame relay is defined for layers one and two only. This means that frame relay has significantly less processing to do at each node, which improves throughput by an order of magnitude.

X.25 prepares and sends packets, while frame relay prepares and sends frames. X.25 packets contain several fields used for error and flow control, none of which is needed by frame relay. The frames in frame relay contain an expanded address field that enables frame relay nodes to direct frames to their destinations with minimal processing.

Finally, X.25 has a fixed bandwidth available. It uses or wastes portions of its bandwidth as the load dictates. Frame relay can dynamically allocate bandwidth during call setup negotiation at both the physical and logical channel levels.

Switched Multimegabit Data Service (SMDS)

Local access and transport areas (LATAs) are "equal-sized" population or geographic areas for wired telephone service. Service within a LATA is provided by a **local exchange carrier (LEC),** that is, the local telephone company; service across LATA boundaries is provided by an **interexchange carrier (IXC),** that is, a long-distance provider.

Switched multimegabit data service (SMDS) is a high-speed, **connectionless** (datagram), cell-oriented, public, packet-switched data service developed to meet the demands for broadband services. Bellcore (later Telcordia) standardized SMDS in 1989 as a metropolitan area network (MAN) construction plan in order to accommodate high-speed data-switching services inside the **local access and transport areas (LATAs)** of the **regional Bell operating companies (RBOCs).** SMDS offers six data rates (ranging from 1.17 Mbps to 155.52 Mbps) and uses fixed-length packets or cells. SMDS refers to a service that is aimed at the growing market of interconnecting LANs in a metropolitan area. For example, the STAR Consortium, a group of businesses and Samford University in Birmingham, Alabama, uses SMDS provided by BellSouth to connect the LANs of the member organizations.

Clarification

Prior to 1984, AT&T included the Bell System of local telephone companies and Bell Labs, the research arm of AT&T. This single company, affectionately called *Ma Bell* to retail customers, provided most of the local and long-distance (wired) voice services in the United States. As a result of the agreement with the U.S. Department of Justice, AT&T divested the Bell System, which became seven separate regional Bell operating companies (RBOCs), composed of 22 Bell operating companies (local telephone companies). This is, of course, all in the wired world of voice communications. Bell Labs stayed with AT&T and has fallen on hard times due to lack of funding that came from the original combination of companies within AT&T. The RBOCs created Bellcore as their research arm, later to change the name to Telcordia.

SMDS uses Distributed Queued Dual Bus (DQDB) as an access protocol between the subscriber and the network. DQDB technology, defined in the IEEE 802.6 MAN standard, provides all stations on the dual bus with knowledge of the frames queued at all other stations. This eliminates packet collisions and improves data throughput. SMDS accesses the network at speeds of 1.17, 4, 10, 16, 25, 34, and 155.52 Mbps. As a result of these access speeds, SMDS providers must have high-capacity switches and lines that operate at data rates of T1, T3, and SONET's OC-3 (155.52 Mbps). Note that the difference of 374 Kbps between the access speed of 1.17 Mbps and the T1 speed of 1.544 Mbps is used for network overhead.

Clarification

In data communications, T1, T3, and OC-3 are used as indication of bandwidth, not protocols. The bandwidth of T3 is 44.736 Mbps , or 672 DS0s. We refer to it as 45 Mbps for convenience.

Asynchronous transfer mode (ATM) A network technology based on transferring data in cells or packets of a (small) fixed size for transmitting video, audio, and computer data over the same network.

Like ATM (discussed later), SMDS uses 53-byte cells. Unlike ATM's cell format, however, SMDS's cell format is based on the IEEE 802.6 specification. The differences occur in the five-octet header. For instance, SMDS employs the ITU's E.164 addressing system so addresses are analogous to the telephone numbering system. ATM, on the other hand, does not currently use E.164 addressing. Despite the differences in the headers, SMDS will run over ATM switches that will provide wide area networking for BISDN. Therefore, SMDS networks can become components of the BISDN networks that will be implemented later.

FIGURE 6.6 **Seven Fields Comprise the SMDS SIP Level 2 Cell**

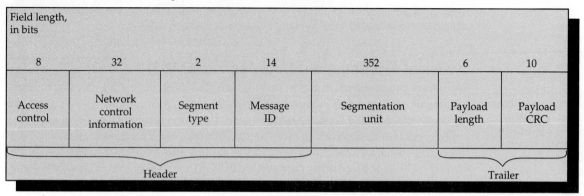

The following briefly summarize the functions of the SIP level 2 protocol data unit (PDU) fields illustrated in Figure 6.6:[4]

- **Access control:** Contains different values, depending on the direction of information flow. If the cell was sent from a switch to a CPE device, only the indication of whether the level 3 PDU contains information is important. If the cell was sent from a CPE device to a switch, and if the CPE configuration is multi-CPE, this field can carry request bits that indicate bids for cells on the bus going from the switch to the CPE device.

- **Network control information:** Contains a value indicating whether the PDU contains information.

- **Segment type:** Indicates whether the cell is the first, the last, or a middle cell from a segmented level 3 PDU. Four possible segment type values exist:
  00 Continuation of message
  01 End of message
  10 Beginning of message
  11 Single-segment message

- **Message ID:** Associates level 2 cells with a level 3 PDU. The message ID is the same for all the segments of a given level 3 PDU. In a multi-CPE configuration, level 3 PDUs originating from different CPE devices must have a different message ID. This allows the SMDS network receiving interleaved cells from different level 3 PDUs to associate each level 2 cell with the correct level 3 PDU.

- **Segmentation unit:** Contains the data portion of the cell. If the level 2 cell is empty, this field is populated with zeros.

- **Payload length:** Indicates how many bytes of a level 3 PDU actually are contained in the segmentation unit field. If the level 2 cell is empty, this field is populated with zeros.

- **Payload cyclic redundancy check (CRC):** Contains a CRC value used to detect errors in the following fields. The payload CRC value does not cover the access control or the network control information fields.
  Segment type
  Message ID
  Segmentation unit
  Payload length
  Payload CRC

---

[4] http://www.cisco.com/univercd/cc/td/doc/cisintwk/ito_doc/smds.htm.

Advantages of SMDS

Since SMDS is a public network, it provides any-to-any connectivity with no theoretical distance limitation. SMDS subscribers can send data to any endpoint on the SMDS network. This is especially attractive to businesses that need to communicate with many other organizations, such as customers, suppliers, and business partners. Because SMDS enables subscribers to achieve mesh connectivity with fewer access lines and less network equipment than point-to-point private lines, SMDS makes the connection of widely scattered LANs economical. However, the real cost advantage of SMDS may lie in usage-based pricing in which a subscriber pays a minimum fee for leasing the access line, plus a per-byte charge for data transmitted.

SMDS is based on a single set of standards defined primarily by Telcordia. This allows multivendor equipment to interoperate on the same network. Many regional Bell operating companies or their counterparts sell SMDS service. BellSouth, for instance, offers T1 SMDS service in Atlanta, Birmingham, Charlotte, and Nashville and T3 SMDS over ATM. Thus, SMDS also operates transparently with different technologies, like ATM and frame relay.

SMDS offers enhanced network services, such as call screening, call verification, and call blocking. It supports common protocol architectures such as TCP/IP, Novell, AppleTalk, and OSI (open systems interconnection). SMDS allows subscribers flexibility in selecting different access speeds for each of their sites and in defining specific user groups. It also makes adding or dropping sites as simple as adding or dropping a telephone number. Finally, SMDS provides **bandwidth-on-demand,** so a subscriber with a 10 Mbps access speed can burst to 34 Mbps for specified increments of time as long as the average bandwidth utilization is 10 Mbps.

Disadvantages of SMDS

One major disadvantage is that SMDS cannot easily handle delay-sensitive traffic such as voice and video traffic. SMDS's overhead can be described as moderate to high. Its use for delay-sensitive traffic is hindered because SMDS cannot guarantee the timing of cell arrivals. As mentioned previously, 374 Kbps of the 1.544 Mbps bandwidth on a T1 line is devoted to overhead. Another problem with SMDS is its implementation cost. These initial costs include the implementation of new SMDS switches and the modification of SMDS access equipment.

Frame Relay versus SMDS

Similarities

Frame relay and SMDS are both services that provide wide area connectivity for LANs at moderate-to-high data rates. Both are fast packet-based data services that perform packet assembly and disassembly, error checking, and flow control only at the endpoints. Since frame relay and SMDS are data services, neither is well-suited for voice, video, or other delay-sensitive traffic. For this reason, both services are designed to interface with ATM, which integrates data, voice, and video. For instance, frame relay frames can be encapsulated into the payload portion of ATM cells. This process converts frame relay packets into cells at the network entrance, moves the cells through the network, and reconverts the cells back into frames at the network exit. With this approach, a carrier can offer both frame relay services as well as cell-based services such as SMDS and ATM. (As previously mentioned, SMDS uses ATM-like 53-byte cells and operates over ATM switches.)

Differences

Frame relay is connection-oriented (point-to-point), whereas SMDS is connectionless (point-to-multipoint). Connectionless (datagram) service provides any-to-any connectivity without establishing a logical connection before the exchange of information. Thus, connectionless service is more flexible but is more difficult to manage and troubleshoot. Frame relay and SMDS also differ in transmission rates. SMDS may operate at the OC-3 (155.52Mbps) rate, while frame relay operates at less than T3 (45 Mbps). Thus, SMDS is more appropriate for applications such as large file transfers. Frame relay, however, works well for low-volume applications like electronic mail. Another difference between frame relay and SMDS is in the units of data to be transported. Frame relay uses variable-length packets or frames. SMDS, on the other hand, uses fixed-size, 53-byte cells. These differences can be important to users as they determine the service that best meets their business needs or requirements.

Synchronous Optical Network (SONET)

Synchronous optical network (SONET) was conceived at Bellcore (later Telcordia) as an optical communications interface standard in 1984. By 1988, an international version known as **synchronous digital hierarchy (SDH)** was adopted by **International Telecommunications Union (ITU).** Since then, a significant effort has been made to harmonize the difference between SONET and SDH. Thus, SONET and SDH now use the same basic structure and are designed to interoperate at 155.52 Mbps and higher transmission rates. SONET's main purpose is high-speed, highly reliable, serial digital transmission over optical fiber cable. SONET provides carrier mechanisms to flexibly supply, multiplex, and manage transmission rates beyond T1 and T3 speeds. Therefore, SONET provides a means for taking advantage of the high-speed digital transmission capability of optical fiber.

SONET transmission rates are multiples of a basic signal rate of 51.84 Mbps. This 51.84 Mbps rate is referred to as **Synchronous Transport Signal 1 (STS-1)** when an electrical signal is used or **Optical Carrier 1 (OC-1)** when an optical signal is employed. Higher rates are formed by combining multiple OC-1/STS-1 signals to form an OC-*n*/STS-*n*. This is accomplished by interleaving bytes from *n* OC-1/STS-1 signals that are mutually synchronized. As previously mentioned, SONET and SDH interoperate at SDH's lowest defined rate of 155.52 Mbps, which corresponds to OC-3/STS-3 (three OC-1/STS-1 signals equal 3 * 51.84 Mbps = 155.52 Mbps). Table 6.5 lists several SONET rates and the equivalent SDH rates.

In addition to the rates listed in Table 6.5, SONET supports transmission speeds of 13.27 Gbps (OC-256) and higher. Therefore, SONET defines both high-speed transmission over fiber and a consistent multiplexing scheme.

Advantages of SONET

SONET's advantages stem from several areas. First, SONET defines a multiplexing standard for combining lower-speed digital channels into high-speed digital transmission signals. This allows SONET equipment to support future SONET interfaces as well as T1 and T3. Thus, the amount of network equipment for each node is significantly reduced and multiplexing is simplified.

Second, SONET defines standard optical interfaces for interconnecting fiber terminals from different vendors. This allows network owners to purchase and deploy fiber optic

TABLE 6.5   SONET/SDH Data Rates
The following table lists the hierarchy of the most common SONET/SDH data rates.

Optical Level	Electrical Level	Line Rate (Mbps)	Payload Rate (Mbps)	Overhead Rate (Mbps)	SDH Equivalent
OC-1	STS-1	51.840	50.112	1.728	
OC-3	STS-3	155.520	150.336	5.184	STM-1
OC-12	STS-12	622.080	601.344	20.736	STM-4
OC-48	STS-48	2,488.320	2,405.376	82.944	STM-16
OC-192	STS-192	9,953.280	9,621.504	331.776	STM-64
OC-768	STS-768	39,813.120	38,486.016	1,327.104	STM-256

equipment from different vendors without a concern for compatibility. "Mid-Span Meet" refers to this interconnect capability of SONET, which provides increased flexibility for network designers when adding equipment to the network or when interconnecting different carrier networks.

Third, *SONET defines a set of network management protocols*. These protocols allow the network to be *monitored, reconfigured, and maintained from a central point*. Therefore, a technician located hundreds of miles away could create access links in minutes. In addition, the network management protocols can set up loopbacks, order bit-error-rate tests, or collect performance statistics.

Next, SONET is a flexible payload structure that can accommodate most types of digital signals. As a result, SONET can transport services such as bursty asynchronous data, high- and low-speed synchronous data and voice, and on-demand services (e.g., videoconferencing). This flexibility allows for broadband services such as BISDN. One function of SONET is as the transmission medium for BISDN. Notice in Table 6.5 that **Optical Carrier 3 (OC-3), Optical Carrier 12 (OC-12),** and OC-48 are speeds defined by BISDN. In BISDN, SONET performs such functions as mapping cells into transmission systems and extracting the cell. Also, even though ATM can be supported by any digital transmission hierarchy, ATM cells are optimized when transported within the SONET payload. Therefore, *BISDN employs SONET as the transmission medium and ATM as the switching technology*.

Finally, SONET equipment has been deployed since 1989, making it a more stable standard than more recent entrants. This should lead to a longer product life, more competition, and reduced equipment prices.

Disadvantages of SONET

SONET is specifically intended for local exchange and interexchange carriers. Thus, most business users only will see the impact of SONET indirectly. More importantly, these business users are at the mercy of the LECs and IXCs to supply SONET services. Additionally, SONET equipment is expensive. These costs have deterred many LECs and IXCs from investing heavily in SONET, so the entire network was not SONET-compatible.

Asynchronous Transfer Mode (ATM)

When first defined in the late 1980s, BISDN was based on a switching technology known as **asynchronous transfer mode (ATM).** In 1990, the ATM standard was formally adopted for BISDN services. ATM is a variation of packet-switching technology that transmits fixed-length units of data (called **cells**) at very high speeds. The speeds presently specified range from 155.52 Mbps to 2.488 Gbps. ATM switches are predicted to run at rates of 100 Gbps or faster. ATM standards are developed by the CCITT, ANSI, and the ATM Forum, the latter a union of over 100 telecommunications equipment makers and telecommunications carriers.

Clarification   The CCITT (Comité consultatif international téléphonique et télégraphique) was supplanted by the International Telecommunications Union (ITU) in 1992 as the international telecommunications standards body, but literature continues to have references to CCITT.

ATM, using a technique called *cell switching,* breaks all data into cells (small fixed-size packets) and transmits them from one location on the network to another over fiber and other circuits connected by switches. The 53-byte-size cell consists of a 5-byte header and a 48-byte information payload, which results in a bandwidth efficiency of 91 percent (see Figure 6.7). The use of a 53-byte cell reduces the queuing delay for high-priority cells since the wait behind a low-priority cell is decreased. Also, the fixed size means that

FIGURE 6.7
**53-Byte ATM Cell
(Bullet)**

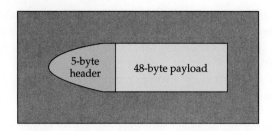

the processing can be done by a simple hardware circuit that allows faster processing speeds. Another important advantage of the 53-byte cell is that small cells meet the low delay requirements necessary for voice data. Remember that with usual packet switching, packets can be of varying sizes with each packet containing its own unique address information. Thus, more processing overhead is involved in reading the address and determining the length of each packet. ATM combines the best features of packet and circuit switching.

ATM is a connection-oriented transfer service. However, ATM is very flexible and can accommodate both connection-oriented and connectionless network services. Two types of ATM connections exist: virtual channels (VCs) and virtual paths (VPs). A virtual channel describes the connection between ATM end-user equipment, while a virtual path refers to a group of VCs that all have the same endpoint. Once the virtual channels are established, routing tables are created that route the cell to the appropriate output link. The use of VCs and VPs simplifies the network architecture, increases network performance and reliability, and reduces processing and connection setup time. Data are broken into fixed-size cells at one point, transmitted to the destination as quickly as possible, and reassembled in the original order.

ATM is a connection-oriented transfer mode. Before sending cells that carry user data, a virtual connection between source and destination has to be established. All packets of a connection follow the same path within the network. During the connection setup, each switch generates an entry in the Virtual Path Identifier (VPI)/Virtual Channel Identifier (VCI) translation table. This enables the switch to move an incoming packet from its VP/VC to corresponding outgoing VP/VC. As an advantage, this kind of routing requires a smaller header. Just a locally valid address (i.e., VPI/VCI) has to be carried in the packet.

Advantages of ATM

One of ATM's advantages lies in its scalability. First, ATM will operate on different physical media, moving from unshielded twisted-pair to optical fiber cable. Second, ATM is media independent and allows 150 Mbps, 600 Mbps, and 2.4 Gbps links on the same network. Finally, ATM switches provide bandwidth-on-demand, so users can send as much data as they have within a specific amount of time and pay accordingly. ATM's fixed cell size can be rapidly handled by routers, and there is less delay for voice and video.

Another advantage is ATM's ability to support different kinds of traffic, such as voice, video, and data, either separately or in multimedia applications. Likewise, ATM is both distance and protocol independent, while serving local and wide area requirements. Thus, ATM allows the user to mix and match channels of varying bandwidth and data types. These advantages lead to a wide variety of applications such as high-speed LAN interconnection; computer disaster recovery; supercomputer access; medical imaging; and multimedia for distance learning, collaboration, and concurrent engineering. ATM is viewed as a unifying technology for LANs and WANs as well as for multimedia applications. The scalability and flexibility of ATM mean that the technology can provide the sort of unification that was heretofore impossible. In order to increase user satisfaction and productivity and decrease network operating cost, network managers should be aware of these potential applications available through ATM. (See Figure 6.8 for the ATM process.)

FIGURE 6.8 **ATM Process**

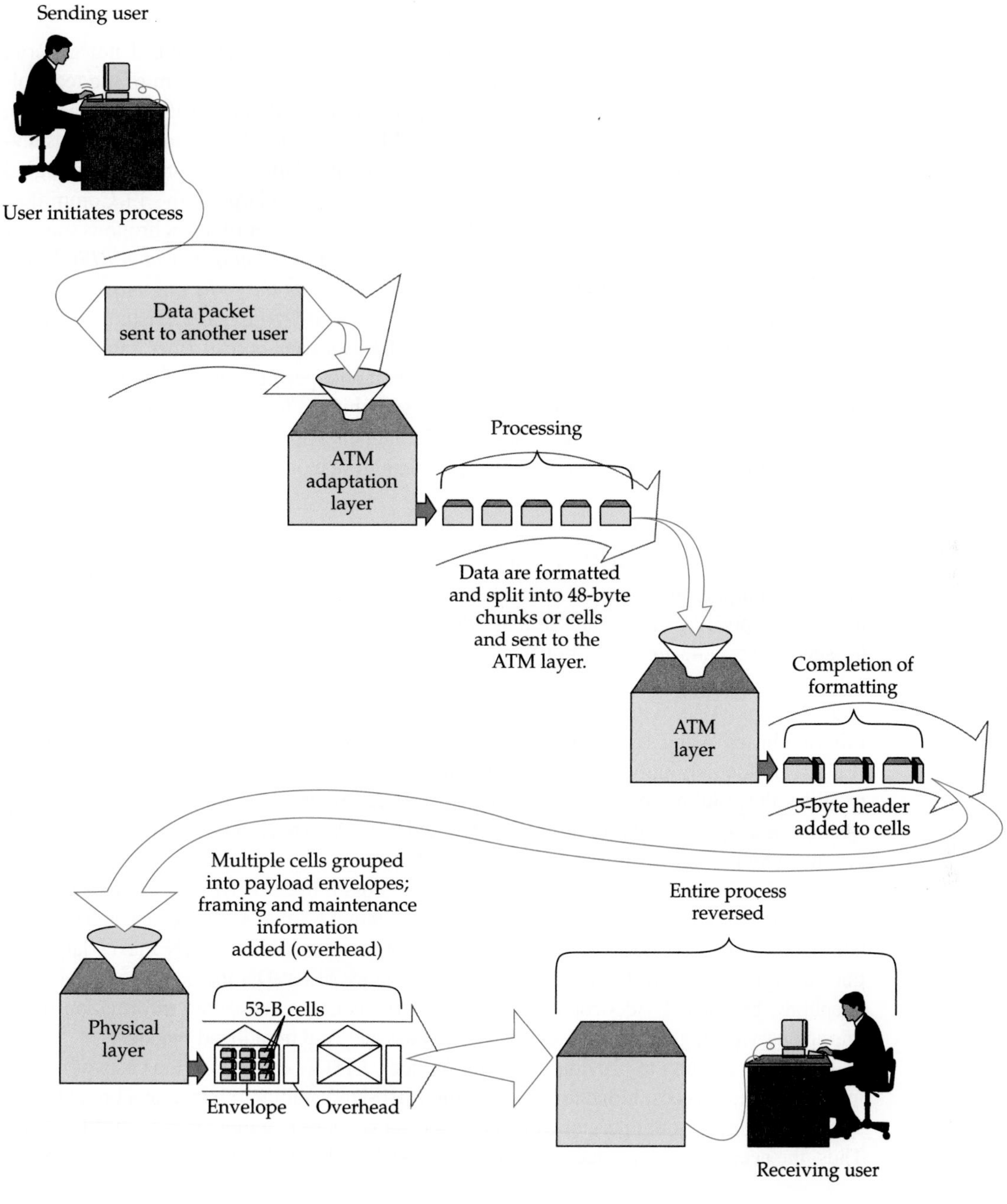

Sending user

User initiates process

Data packet sent to another user

ATM adaptation layer

Processing

Data are formatted and split into 48-byte chunks or cells and sent to the ATM layer.

ATM layer

Completion of formatting

5-byte header added to cells

Multiple cells grouped into payload envelopes; framing and maintenance information added (overhead)

Entire process reversed

Physical layer

53-B cells

Envelope Overhead

Receiving user

Limitations of ATM

One of the drawbacks of ATM technology is the expense. Additionally, ATM creates a new set of problems for network designers and planners. For example, network designers and planners must determine when ATM will meet their needs, and then find the correct combination of commercially available ATM carrier services and equipment. Network managers must examine the compatibility expense when addressing the impact on their networks of MANs and WANs.

Broadband Integrated Services Digital Network (BISDN) Technology

In the beginning of this section, we described BISDN as a generalized, if not generic, description of broadband. The term also is used to denote a specific form of the technology. In order to provide the services that require large bandwidth, BISDN is based on transmission speeds and capacities at the 155.52 Mbps, 622.08 Mbps, and 2.488 Gbps levels. These bit rates derive from ITU G Series Recommendations for synchronous digital hierarchy (SDH), synchronous optical network (SONET) standards of the T1 Committee, and the ITU I Series Recommendations that support the concept of asynchronous transfer mode (ATM)–based BISDN. Therefore, *the transmission medium over which BISDN will operate is described by SDH/SONET with ATM employed as the switching mode.*

BISDN also differs in several ways from ISDN. First, ISDN uses the existing telephone network infrastructure of copper wires, whereas BISDN uses optical fiber cable. Second, because ISDN is primarily a circuit-based network, it only performs packet switching on the D channel. BISDN, however, uses only packet switching. Finally, ISDN channel bit rates are prespecified, whereas BISDN uses virtual channels without any prespecified bit rate. The only limitation on BISDN rates is the physical bit rate of the user-to-network interface.

Advantages of BISDN

As a point of reference, ISDN has not been widely adopted due to cost and inadequate data rates for many applications of interest. For instance, a 50-megabyte file takes over five minutes to move at T1 transmission speeds ((((50 megabytes*10 bits/byte)/1.536 Mbps)/60 sec/min) = 5.43 minutes) but would take 12 times longer with ISDN (65.1 minutes using both B channels). This same file can be moved using BISDN in 3.2 seconds at 155.52 Mbps. Also, with LAN speeds in the broadband range (e.g., 100 or 1,000 Mbps Ethernet buses, 4 and 16 Mbps token rings, and 100 Mbps fiber distributed data interface [FDDI] protocols), higher speeds are needed for LAN-to-LAN interconnection. BISDN provides the solution to these problems. BISDN can be used to interconnect LANs over wide areas at speeds equal to or greater than today's LAN speeds. Thus, BISDN solves the problems by providing high data rates over a distance.

Several fields are emerging as adopters as the possibilities of BISDN are realized. For instance, teleradiology transmits medical imagery among hospitals, physicians, and patients, which requires large bandwidth due to the size of the data files. BISDN makes large file transfers quick and easy. Other commercial applications of BISDN include broadband video telephony, broadband videoconference, video surveillance, video/audio information transmission service, high-speed telefax, video mail service, and broadband videotex.

Finally, in addition to providing various services, BISDN can reduce the costs of operating a network. This cost reduction is the result of BISDN's ability to integrate a broad mix of services so the network operator can handle special services in a more standardized manner. Thus, the need to build service-specific transmission and switching systems is minimized.

BISDN's Role

BISDN is a connection-oriented packet service based on ATM cells. These cells can be transmitted in the traditional packet-switching mode or in the payload of a SONET envelope. Since ATM cells are optimized when transported using SONET, several ATM manufacturers have announced products that use SONET as the physical transport for ATM cells. Therefore, BISDN offers ATM over SONET as a common transport platform.

The goal of BISDN is one network capable of supporting multiple services, such as frame relay and SMDS, on the same platform. This would allow two frame relay users to connect to one another over a BISDN (ATM/SONET) backbone network or a frame relay user to

TABLE 6.6 **Comparison of the Broadband Technologies**

Features	X.25	Frame Relay	SMDS	SONET	ATM	BISDN
Service	X	X	X			
Switching					X	X
Connection-oriented		X			X	X
Connectionless (datagram)	X		X			
Fixed packet			X		X	X
Variable packet		X		X		
Data	X	X	X			
Data + Transmission				X		
Media				X		
Suitable for real-time processing	No	No	No	Yes	Yes	Yes
Compatibility		X				X.25
			X		X	Frame relay
		X			X	SMDS
					X	SONET
		X	X	X		ATM
			X			BISDN

connect to an SMDS user over the BISDN (ATM/SONET) backbone network. Currently, however, frame relay and SMDS are implemented on service-specific platforms, such that frame relay networks are built using frame relay switches and SMDS networks are built using SMDS switches. (Table 6.6 shows a comparison of various broadband technologies.)

Implication Gigabit Ethernet, a transmission technology based on the Ethernet frame format and protocol used in LANs, provides a data rate of 1 billion bits per second (one gigabit). Gigabit Ethernet is defined in the IEEE 802.3 standard and is currently being used as the backbone in many enterprise networks. Gigabit Ethernet is carried primarily on optical fiber (with very short distances possible on copper media). Existing Ethernet LANs with 10 and 100 Mbps cards can feed into a Gigabit Ethernet backbone. An alternative technology that competes with Gigabit Ethernet is ATM. A newer standard, 10-Gigabit Ethernet, is also becoming available.[5]

BROADBAND TECHNOLOGIES: ISSUES FOR BUSINESS ORGANIZATIONS

With LANs advancing to the 10-gigabits-per-second range, the need has arisen for LAN interconnection over wide areas and at high speeds. Services such as SMDS (data), SONET (data and voice), and BISDN (data, voice, video, image, graphics, fax, etc.) provide high-speed LAN interconnection, which improves communications and the sharing of information across an organization, between an organization and its partners (suppliers and customers). In addition, all of these services can be supported by ATM switches. Similarly, SMDS and BISDN both are designed to operate over SONET transmission facilities. Therefore, broadband technologies will provide multiple services, so network managers must select the services that match their requirements.

Network managers must closely monitor the changes and developments in broadband services, so the proper combination of services may be acquired in the most cost-effective manner. These broadband services might present some problems for network managers but, at the same time, create many opportunities. From high-speed LAN interconnection,

[5] http://searchnetworking.techtarget.com/sDefinition/0,,sid7_gci212193,00.html.

to videoconferencing, to large file transfer, to multimedia applications, broadband services are at the center of a firm's ability to obtain and maintain a competitive advantage through telecommunications. With the greater bandwidth and flexibility provided by frame relay, SMDS, BISDN, ATM, SONET, and Gigabit Ethernet, new applications exist to exploit and gain an advantage over competitors who do not realize the benefits of broadband telecommunications capabilities. Therefore, those managers who ignore the opportunities presented by broadband technologies risk placing their firms at a competitive disadvantage.

Case 6-1

Level 3 Communications

Steve Manthe

Level 3 Communications is an international communications and information services company with approximately 5,200 employees in 17 countries. Level 3's IP network spans approximately 19,000 miles in North America and 4,000 miles in Europe, connected across the Atlantic by 3,800 miles of undersea fiber optic cable. Level 3's end user computing environment is sophisticated and intensely complicated. Level 3 has centralized the majority of its operations, as well as its MIS and telecommunications infrastructure and support, within its headquarters in Broomfield, Colorado. Secondary operations centers are also located in Atlanta, Georgia, and London. These facilities provide backup, overflow, and 24*7 coverage to ease the workload on Broomfield facilities. In addition to these three major facilities, Level 3 also operates a global fiber network and, by necessity, facilities along each stretch of the network to keep the data packets moving. Level 3 operates 38 gateways (data centers similar to a PSTN's central office) for each major metropolitan market it serves as well as 578 POP (point-of-presence) sites and 656 ILA (in-line amplifier) sites that regenerate the fiber's signal along the route.

LEVEL 3'S TELECOMMUNICATIONS ENVIRONMENT

All data traffic and most of the voice traffic sent or received by Level 3 employees is carried by Level 3's own IP network. Employees making calls within their facility or to other IP-enabled Level 3 sites utilize VoIP trunking through their own (3)Tone VoIP solution. Level 3 uses VoIP-enabled handsets from Lucent and Cisco and, where their VoIP solution is not practical and/or available, Level 3 relies on local CLECs for plain old telephone service through a PBX or **Centrex** system, both antiquated solutions provided by the competition Level 3 is trying so desperately to bury. Transfer of data is also handled by Level 3's own OC-192, SONET-protected IP network and handed off to other providers if the data packets are headed for a destination not served by Level 3. Finally, in addition to highly available wired connectivity for both voice and data, Level 3 employees also have access to a variety of wireless options depending on their level in the organization. VPs and most directors each has a Blackberry device that acts as a pager, cell phone, and MS Outlook mailbox, all in one small unit. Most other employees are assigned two-way text pagers with service provided by Skytel. Blackberries, cell phones, pagers, and a remote access Citrix client make it difficult for any Level 3 employees to remove themselves from the workplace as it tends to follow them wherever they go!

Case 6-2

Organizational Networking

M.C. Kettelhut had been extremely successful with four retail sporting goods stores in Dallas, Frisco, Weatherford, and Waco, Texas. When his expansion began, he neither had the resources nor the expert knowledge to invest in data communications, so he had relied on **wide area telecommunication service (WATS)** and POTS to communicate within his organization and to his suppliers and customers. He now found an opportunity to purchase one of three similar companies located in the south (serving Virginia, North Carolina, and Tennessee), the west (serving Nevada and Arizona), and the midwest (serving Nebraska, Kansas, and Missouri). M.C. had two decisions: first, which company to purchase. All three were good opportunities, with products, customers, coverage areas, and suppliers similar to his. An added feature was that each was having financial difficulties despite their market strength, making them attractive takeover candidates. The second decision M.C. had to face was how to manage the new organization. If he chose only one opportunity, he could delegate the management of his present organization to his able assistant and concentrate on making the new acquisition profitable. However, the more he looked at the possibilities of each of the alternatives, the more he realized that each presented an excellent opportunity for someone employing the same management style and organizational methods that had made him so successful in Texas. The acquisitions of all three companies at the same time posed a complex communications as well as management situation.

After several sleepless nights and a number of conversations with his banker and legal advisor, M.C. believed that the best investment strategy was to take over all three firms. He visited each one, spending time with the owner and management team. He paid particular attention to the financials, how the stores were managed, the customer base, and relations with suppliers. M.C. found a great deal of overlap in suppliers and saw a chance to reduce the number of suppliers as well as improve his buying power. Of great importance was that dealing with a select group of suppliers meant his communications could be more direct. This probably meant better financial arrangements. M.C. was also vitally concerned with control of far-flung operations.

M.C. returned to Dallas and made arrangements with his bank and legal firm to acquire the three sporting goods companies, their resources, and their debts. Before he signed the final papers, he organized a meeting with the suppliers who held much of the debt through credit that had been extended to the companies and outlined his plan for reorganization. With the suppliers' agreement, he did not have to transfer the outstanding debt to the bank, saving interest and keeping M.C. in control. At the same time, M.C. discussed how he might create a communications system internally and externally that included the suppliers. His position was that closer coordination with them would help eliminate some of the existing overstock problems and reduce the outstanding credit debt as well as make the organization's operations smoother in the future.

QUESTION

Design the total corporate communications network infrastructure for M.C., drawing a network, naming the component parts, and describing the equipment used. Consider reliability, accessibility, performance, and security. (There are a total of 17 suppliers who are dispersed all over the United States and Canada. M.C.'s acquired companies were headquartered in Richmond, Virginia; Las Vegas, Nevada; and Topeka, Kansas.) Use netviz to design a proposed network.

Summary

Networking has become essential for organizations to operate effectively and remain competitive, resulting in requirements expanding beyond local areas to wider and wider geographic areas. Coupled with the need to expand network reach is a need to move more data at faster speeds. Driving the need for bandwidth, in addition to greater use by organizations, is the convergence of multiple media, multiple technologies, and multiple opportunities. The transportation "bandwidth" of the U.S. interstate highway system changed the nature of business; the connectivity of IP networks and greater bandwidth is changing business in a very different way. There are several competing broadband technologies and media choices to be considered in choosing or building wide area networks (WANs). Managers need to understand these choices and their associated benefits and constraints.

This chapter presented technical detail that may seem beyond the scope of the "management" level of networks. Effective management requires an understanding of the technology involved. Even high-level management will have to deal with definitions and protocols to make good decisions, decisions that will provide the cost advantage and the customer support required for a competitive advantage.

When dealing with WANs, there are some very basic terms that have importance. We have a choice of circuit switching, packet switching, and message switching. There is a choice of dedicated lines or virtual circuits as well. Managers who lack understanding of WAN technologies are less likely to choose wisely. They ought to have a fundamental knowledge of the terms and technologies, such as packet switching, frame relay, SMDS, SONET, ATM, BISDN, and Gigabit Ethernet. The technologies, offerings from vendors, bandwidth demand, and overall available infrastructure continue to evolve. This evolution means network managers must be alert to changes that offer significant advantages to their organizations.

Key Terms

asymmetric digital subscriber line (ADSL), *p. 168*

asynchronous transfer mode (ATM), *p. 177, 181*

bandwidth-on-demand, *p. 179*

basic rate interface (BRI), *p. 173*

bearer (B) channel, *p. 173*

broadband, *p. 167*

broadband integrated services digital network (BISDN), *p. 173*

cell, *p. 181*

Centrex, *p. 186*

circuit switching, *p. 170, 171*

connectionless, *p. 177*

convergence, *p. 165*

customer premises equipment (CPE), *p. 175*

customer service area (CSA), *p. 169*

Data Services One (DS1), *p. 166*

Data Services Zero (DS0), *p. 166*

datagram, *p. 171*

delta (D) channel, *p. 173*

digital subscriber line (DSL), *p. 167*

electronic mail (e-mail), *p. 165*

error detection, *p. 173*

Ethernet, *p. 167*

frame format, *p. 173*

frame relay, *p. 175*

high-data-rate digital subscriber line (HDSL), *p. 170*

integrated services digital network (ISDN), *p. 168*

interexchange carrier (IXC), *p. 177*

International Telecommunications Union (ITU), *p. 180*

Internet service provider (ISP), *p. 167*

local access and transport area (LATA), *p. 177*

local exchange carrier (LEC), *p. 177*

message switching, *p. 172*

noise, *p. 165*

Optical Carrier 1 (OC-1), *p. 180*

Optical Carrier 3 (OC-3), *p. 181*

Optical Carrier 12 (OC-12), *p. 181*

packet assembler/ disassembler (PAD), *p. 171*
packet data network (PDN), *p. 170*
packet-switched network, *p. 171*
packet switching, *p. 172*
permanent virtual channel (PVC), *p. 175*
protocol, *p. 172*
regional Bell operating companies (RBOCs), *p. 177*

signal-to-noise ratio, *p. 169*
switched multimegabit data service (SMDS), *p. 177*
synchronous, *p. 165*
synchronous digital hierarchy (SDH), *p. 180*
synchronous optical network (SONET), *p. 180*
Synchronous Transport Signal 1 (STS-1), *p. 180*

T3, *p. 167*
T-carrier system, *p. 167*
Terrestrial One (T1), *p. 166*
virtual circuit (VC), *p. 171*
wide area network (WAN), *p. 170*
wide area telecommunications service (WATS), *p. 187*
X.25, *p. 172*

Recommended Readings

Communications of the ACM (http://www.acm.org/pubs/cacm/) is the flagship publication of the ACM and one of the oft-cited magazines in the computing field.

The ATM Forum (http://www.atmforum.com) is an international nonprofit organization formed with the objective of accelerating the use of ATM (asynchronous transfer mode) products and services through a rapid convergence of interoperability specifications. In addition, the Forum promotes industry cooperation and awareness.

The DSL Forum (http://www.DSLforum.org) is a consortium of nearly 200 leading industry players covering telecommunications, equipment, computing, networking, and service provider companies. Established in 1994, the Forum continues its drive for a global mass market for DSL broadband, to deliver the benefits of this technology to end users around the world over existing copper telephone wire infrastructures.

Management Critical Thinking

M6.1 Should organizations develop their own WANs or rely on public (the Internet) or commercial networks? What are the critical decision parameters?

M6.2 A firm that operates in more than 20 nations needs a 24*7 network for high data transmission rates. What parameters should the telecommunications manager examine?

Technology Critical Thinking

T6.1 How does the IT department determine which WAN protocol to adopt for in-house networks? Is there a penalty in choosing the wrong protocol?

T6.2 What are the technical trade-offs between frame relay, SMDS, SONET, ATM, Gigabit Ethernet, and ISDN? Which cost the most initially and in terms of data rate?

Discussion Questions

6.1 Why should a packet-switched network be considered in moving data?

6.2 What are the necessary characteristics of a global network that has large CAD files to transfer? What if the files are needed in color and 3-D?

6.3 What are the differences in error checking for a circuit-switching network and a packet-switching network?

6.4 What are the broadband choices that corporate communications managers need to consider today?

6.5 Is BRI ISDN a genuine option for any but very small organizations?

6.6 What are the emerging applications that might be considered bandwidth-hungry?

6.7 If an organization wishes to have teleconferencing and multimedia communications between three domestic locations and one in Mexico, what are the bandwidth needs?

6.8 What are the major issues that the telecommunications manager needs to be concerned with in making the broadband technology choices for a firm?

6.9 What organization can satisfy their WAN requirements with DSL and the Internet?

6.10 What are the implications of free bandwidth?

Projects

6.1 Refer to current issues of *Communications Week, Network Magazine,* and so on, and bring diagrams of global networks to class. Compare the networks and provide your evaluation as to their (1) reliability, (2) cost versus expense, (3) appropriateness, (4) technology, (5) adaptability, and (6) utility.

6.2 Why should telecommunications managers build dedicated global networks? Evaluate the advantages/disadvantages of outsourcing the networks versus building your own.

6.3 Visit a hospital and determine how broadband telecommunications are employed. What applications demand the greatest bandwidth in medicine? Determine if "telemedicine" is practiced and the technologies that are applicable.

6.4 Contact both your local telephone company and long-distance provider and determine broadband alternatives that they offer and the costs.

6.5 Research the Ford Motor Company's use of broadband in its global teleconferencing system. What technologies are presently employed and what are the bandwidth requirements now and planned for the future? Determine the history of this system and the reported results obtained by its use.

6.6 Build a table listing the advantages and disadvantages of broadband networks. Include propagation, susceptibility to noise, expandability, administration, and so forth.

6.7 Projects to deploy networks involve several feasibility choices. What are the categories that telecommunications managers ought to consider when choosing between packet-switched versus circuit-switched architectures?

6.8 Contrast costs and maximum data rates of ATM and DSL.

6.9 Using the Internet and the ATM Forum, determine who is actively using ATM. Determine the use of ATM on the Internet backbone.

6.10 Determine the currently offered features and cost of X.25 and frame relay in your area.

References

Aber, Robyn. "An SMDS Glossary." *Business Communications Review,* June 1992 supplement SMDS, pp. 30–31.

Aber, Robyn. "SMDS Service One Year after Kickoff." *Business Communications Review,* June 1993, pp. 51–54.

Aber, Robyn. "SMDS Solves Users' Needs." *SMDS Today: Networks in Action,* Sept 1993, pp. 2, 5–6, 8, 10, 14.

Ali, M. Irfan. "Frame Relay in Public Networks." *IEEE Communications Magazine,* March 1992, pp. 72–78.

Bell, Trudy E. "Telecommunications." *IEEE Spectrum,* January 1991, pp. 44–47.

"Broadband Transition." *Communications News,* October 1993, p. 6.

Byrne, William R.; George Clapp; Henry J. Kafka; Gottfried W. R. Luderer; and Bruce L. Nelson. "Evolution of Metropolitan Area Networks to Broadband ISDN." *IEEE Communications Magazine,* January 1991, pp. 69–70ff.

Chen, Tai. "Frame Relay, SMDS, and HSCS: Comparing Features and Performance." *Telecommunications,* May 1992, pp. 19–20, 22.

Cheung, Nim K. "The Infrastructure for Gigabit Computer Networks." *IEEE Communications Magazine,* April 1992, pp. 60–68.

Ching, Yau-Chau, and H. Sabit Say. "SONET Implementation." *IEEE Communications Magazine,* September 1993, pp. 34–40.

Clarkson, Mark. "All-Terrain Networking." *Byte,* August 1993, pp. 111–14, 116.

Clarkson, Mark A. "Hitting Warp Speed for LANs." *Byte,* March 1993, pp. 123–24, 126, 128.

Cox, Tracy; Frances Dix; Christine Hemrick; and Josephine McRoberts. "SMDS: The Beginning of WAN Superhighways." *Data Communications,* April 1991, pp. 105–108, 110.

Crowl, Steve. "ATM for Multiservice Wide Area Networks." *Business Communications Review,* February 1993 supplement ATM, pp. 11–15.

Dagres, Todd. "Frame Relay's Day Will Dawn." *Business Communications Review,* April 1993, pp. 28–32.

Delisle, Dominique, and Lionel Pelamourgues. "B-ISDN and How It Works." *IEEE Spectrum,* August 1991, pp. 39–42.

Finneran, Michael. "The Impact of SONET on Network Planning." *Business Communications Review,* June 1992, pp. 51–55.

Finneran, Michael. "SONET: Access to the World." *Business Communications Review,* July 1993, pp. 62–63.

Frame, Mike. "Broadband Service Needs." *IEEE Communications Magazine,* April 1990, pp. 59–62.

Garciamendez-Budar, Edsel. "The Emergence of Frame Relay in Public Data Networks." *Telecommunications,* May 1992, pp. 24, 26, 28, 30, 32.

Gasman, Lawrence. "The Broadband Jigsaw Puzzle." *Business Communications Review,* February 1993, pp. 35–39.

Giancarlo, Charles. "Making the Transition from T3 to SONET." *Telecommunications,* April 1992, pp. 17–20.

Gupta, Sudhir. "Interworking: Frame Relay, SMDS, and Cell Relay." *Telecommunications,* February 1993 supplement InteNet, pp. 51–52, 57.

"Industry Outlook: Multimedia and the Future of Communications." *Telecommunications,* April 1992, p. 10.

Johnson, Johna Till. "SMDS: Out of the Lab and onto the Network." *Data Communications,* October 1992, pp. 71–72, 74, 76, 78, 80, 82.

Karpinski, Richard. "The Bell Tolls for LEC Data Strategies." *Telephony,* February 1, 1993, pp. 34, 38, 40.

Kessler, Gary C. *ISDN.* 2nd ed. New York: McGraw-Hill, 1993.

Kleinrock, Leonard. "ISDN—The Path to Broadband Networks." *Proceedings of the IEEE,* February 1991, pp. 112–17.

Lee, Byeong Gi; Minho Kang; and Jonghee Lee. *Broadband Telecommunications Technology.* Norwood, MA: Artech House, Inc., 1993.

Lowe, Sue J. "Data Communications." *IEEE Spectrum,* January 1992, pp. 39–41.

McQuillan, John. "Keeping ATM's Promise of Scalability." *Business Communications Review,* October 1993, pp. 10, 12.

McQuillan, John. "SMDS: Home Run or Strikeout?" *Business Communications Review,* July 1990, pp. 14–15.

McQuillan, John. "Why ATM?" *Business Communications Review,* February 1993 supplement ATM, pp. 1, 3.

Miller, Thomas C. "SONET and BISDN: A Marriage of Technologies." *Telephony,* May 15, 1989, pp. 32–35, 38.

Mollenauer, James F. "The Impact of ATM on Local and Wide Area Networks." *Telecommunications,* March 1993 supplement InteNet, pp. 35, 38–39, 42–43.

O'Brien, Bob, and Dolores Kazanjian. "Sprint Plans All-SONET Network." *Telecommunications,* March 1993, pp. 9, 10.

Schriftgiesser, Dave. "SMDS: A Phone Service for Computers." *Business Communications Review,* June 1992 supplement SMDS, pp. 4–9.

Schriftgiesser, Dave, and Roger Levy. "SMDS vs. Frame Relay: An Either/Or Decision?" *Business Communications Review,* September 1991, pp. 59–63.

Sinnreich, Henry, and John F. Bottomley. "Any-to-Any Networking: Getting There from Here." *Data Communications,* September 1992, pp. 69–72, 74, 76, 78, 80.

"SMDS Now, ATM Later?" *Data Communications,* June 1992, p. 18.

"SMDS, SONET and ATM." *Business Communications Review,* June 1992 supplement SMDS, p. 7.

Smith, Gail. "Planning for Migration to ATM." *Business Communications Review,* May 1993, pp. 53–58.

Stallings, William. *ISDN and Broadband ISDN.* 2nd ed. New York: Macmillan Publishing, 1992.

Stallings, William. "The Role of SONET in the Development of Broadband ISDN." *Telecommunications,* April 1992, pp. 21–24.

Strauss, Paul. "Virtual LANs Pave the Way to ATM." *Datamation,* August 15, 1993, pp. 20–22, 24.

Stuck, Bart. "Can the Carriers Deliver on SONET's Full Potential?" *Business Communications Review,* June 1993, pp. 44–48.

Super, Tom. "InfoVision: Visions of the Information Age." *Visions of the 21st Century: An RBOC Perspective,* November 20, 1992.

Toda, Iwao. "Migration to Broadband ISDN." *IEEE Communications Magazine,* April 1990, pp. 55–58.

White, Patrick E. "The Role of the Broadband Integrated Services Digital Network." *IEEE Communications Magazine,* March 1991, pp. 116–19.

Williamson, John, and Steven Titch. "Gazing Toward the Broadband Horizon." *Telephony,* October 5, 1992, pp. 34–39.

Wilson, Carol. "ATM: Hype or Happening?" *Telephony,* October 5, 1992, p. 60.

Wyatt, John C. "ATM Technology: The Emerging Opportunities." *Telecommunications,* November 1992 supplement InteNet, pp. 43–45.

Chapter **Seven**

The Internet, Intranets, and Extranets

"The Internet is like a 20-foot tidal wave coming and we are in kayaks. It's been coming across the Pacific for thousands of miles and gaining momentum and it's going to lift you and drop you." *Andy Grove*

When it comes to real WANs, nothing compares to the Internet. This network of networks has evolved to be a major factor in the way people, businesses, and governments communicate. We need to gain a basic understanding of the Internet and the World Wide Web (WWW), the technologies, languages, protocols, and infrastructure. The use of the Internet internally by creating intranets and externally with extranets involves large business applications. These uses have spawned many new and important technologies. This chapter discusses the Internet and WWW technology; the next discusses applications.

The Internet is the ultimate wide area network (a *network of networks*), giving us access to the world, its information, and its markets. The idea of the Internet is just an extension of simpler forms of connectivity, except that this **network of networks** has special tools, capabilities, and sources of information that continue to change the way we live and do business. The Internet, intranets, and extranets are important to our study and use of telecommunications because of the ways they have changed our lives. Each of them will be covered in the following material. To this point, we have discussed the value of connectivity among small groups and large populations.

Caveat

The Internet and its capabilities are changing on an almost daily basis. We have covered Internet basics, but because of the rapidly changing technology, you will likely be aware of technologies and capabilities not covered here.

THE INTERNET: THE NETWORK OF NETWORKS

network
Two or more nodes connected by one or more channels.

The Advanced Research Projects Agency (ARPA) set out to demonstrate the feasibility of a packet-switched network and subsequently created the **Advanced Research Projects Agency Network (ARPANET).** The experiment was very successful. This led to the creation of NSFNet by the National Science Foundation, for the purpose of connecting governmental and university research agencies. Connected to or connected with each of the networks mentioned have been thousands of other networks, each privately administered. The secret to the success of the **Internet** has been that each network is administered as required, funded individually, and standardized to give connectivity among them all. Using a standardized protocol (Transmission Control Protocol/Internet Protocol) and providing gateways among the networks has produced what we now know as the Internet.

Clarification

Relative to most college students, the Internet's development and evolution are ancient history, as they grew up with its capabilities. For those born before the Vietnam war, it is recent history. "The Role of Government in the Evolution of the Internet" by Robert E. Kahn, which appeared in the August 1994 issue of the *Communications of the ACM*, gives an excellent history of the evolution of the Internet and the involvement of several U.S. governmental agencies. The governmental agency experiment in packet-switched networks has grown and evolved to change our world.

The Internet was called the *information superhighway* when it was first introduced to the general populace as it was seen as a virtual replica of the U.S. Interstate Highway System or the Autobahn in Germany. As it has developed, this phenomenon provides connectivity to all industrialized places on the earth and many that are not. As a point of reference, the Internet is four things:

1. Global connectivity.
2. A network of networks.

3. A global WAN.
4. Not the World Wide Web.

Put simply, the Internet is *connectivity* and the **World Wide Web (WWW)** is the (graphical) *interface*. With the connectivity of the Internet, organizations and individuals have e-mail, news groups, and transfer files; with the WWW these same entities have an interface that draws them into use of the Internet. The combination of the connectivity of the Internet and the interface of the WWW made these technologies the Gargantua that changed the world.

Clarification

Because the Internet is connectivity, early users of the service, seeing a need for utilities that were easy to use and offered wide access to people and information, created a set of text-based software services. These capabilities have been replaced by easier-to-use Web-based tools.[1]

* *Archie* is a collection of resource discovery tools, developed and maintained by McGill University, for locating files at hundreds of anonymous FTP sites by using a file name search. Archie also is available at other designated sites around the globe. If you are looking for a specific file, Archie can help you locate an FTP site that has it.

* *Gopher* is a menu-driven system developed at the University of Minnesota. The Gopher name was derived from Minnesota's nickname as the Gopher State, as well as the concept that the Internet search will "go for" files containing information you need. Gopher combines features of electronic bulletin board services and databases with parts of FTP, Archie, WAIS, and telnet into one easy-to-use navigation tool. Gopher simplifies locating and retrieving ASCII text documents from various sources of information.

* *Veronica* (Very Easy Rodent-Oriented Net-Wise Index to Computerized Archives) locates and indexes titles of Gopher items by keyword search. A Veronica search typically searches the menus of hundreds of Gopher servers, perhaps all the Gopher servers that are attached to the Internet.

* *WhoIs* provides information on registered users and network names, including their postal addresses. The main WhoIs database runs at the Network Information Center (NIC) with administrative and technical contacts for domains automatically registered when the applications are processed.

* *WAIS* stands for Wide Area Information Server and is a client-server system developed to help users search multiple Internet sites at one time and retrieve resources by searching indexes of databases. WAIS searches are fast, and the results can either be scanned online or mailed to your network address.

Clarification

The Internet, or simply the Net, is the publicly accessible worldwide system of interconnected computer networks that transmit data by **packet switching** using a standardized **Internet protocol (IP)** and many other **protocols.** It is made up of thousands of smaller commercial, academic, domestic, and government networks. It carries various information and services, such as electronic mail, online chat, and the interlinked Web pages and other documents of the World Wide Web.[2]

Figure 7.1 shows the original T3 **backbone network.** This 45 Mbps backbone, evolving to early broadband speed (155 Mbps), radiates T3 and T1 branches, giving connectivity among the tens of thousands of smaller network Internet service provider (ISP) **points of presence** and individual **access points.**

[1] Lynda Armbruster, *Internet Essentials* (Indianapolis, IN: Que College, 1994).
[2] http://en.wikipedia.org/wiki/Internet.

FIGURE 7.1
NSFNet T3 Backbone
Service, Circa 1992

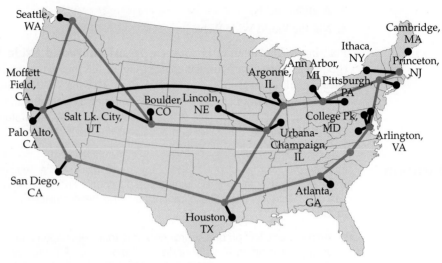

● = Core Nodal Switching Subsystem (CNSS)
● = Exterior Nodal Switching Subsystem (ENSS)

PUTTING A GRAPHICAL INTERFACE ON THE INTERNET

The WWW is the part of the Internet that propelled it from obscurity to explosive growth.

The capabilities that follow are the result of transition from text-based to multimedia and hyperlink-based interfaces, much like the movement from DOS to Windows. The basis of this movement is the software that allowed files to be retrieved and displayed with ease; for example, the World Wide Web software developed by Tim Berners-Lee, who led a team at Switzerland's European Particle Physics Laboratory (CERN for short, for the name Conseil Européen pour la Recherche Nucléaire).

Clarification

The Internet is a phenomenon that started in the early 1970s and bloomed in the early 1990s. The permit of commercial use at the beginning of 1994 sped an already rapid rate of growth and the introduction of the World Wide Web (WWW) that same year accelerated use further. The Internet became a resource that changed the way individuals and organizations communicate, operate, and share resources. By 1996, the need for a new generation of Internet-working was recognized as commercial organizations began to use a significant amount of the Internet bandwidth. The *Next Generation Internet (NGI)* was initiated, an evolution of the NREN project to connect all libraries in the United States with a 3 Gbps circuit. **Internet2** was enacted to develop the applications and technologies for the next generation in 34 U.S. universities. (See Case 7-1.)

The *World Wide Web* (WWW or the Web) retrieves resources as a powerful hypertext and hypermedia browser of databases. **Hypertext** is text with pointers to other text and media, and **hypermedia** might involve images, sound, animation, and so forth, in addition to text. This public domain software goes beyond Gopher and WAIS as a global information system with an easy-to-use graphical interface that provides access to almost all existing Internet-based information. The information appears on *home pages,* providing access to individuals and organizations with only the click of a mouse button.

The WWW is possible because of several technologies, ranging from the GET command of the http protocol to the tags that display information as wished. The following Clarification lists a few home pages created with these technologies, oriented toward college financial aid.

Clarification
WWW Sites for College Financial Aid

Matching students to scholarships – www.fastweb.com

- *The most complete source of local, national, and college-specific scholarships*
- *Search and compare detailed college profiles including tailored scholarship matches*
- *Part-time jobs around the corner, internships around the country*

> **The Federal Information Exchange** lists assistance available from federal agencies http://www.sdsc.edu/SDSCwire/v2.9/fedix.html:
>
> **Minority Online Information Service** gives information on federal scholarships and fellowships: http://www.molis.us
>
> **The Ambitious Student's Guide to Financial Aid** has details on grants and scholarships from colleges, state, and the federal government: http://www.octameron.com/productsframe.html
>
> **The Bookstore**—Sallie Mae, the nation's number one paying for college company. Plan for college, find federal and alternative student loans, apply online. . . . How to Pay for College—a practical guide for families . . . www.salliemae.com/
>
> **Student Financial Aid Information** maintains a calendar that shows you what to do and when to do it: http://www.wellsfargo.com/student/planning/index.jhtml
>
> **The Foundation Center** and **The Princeton Review** are comprehensive guides to where to look for aid: http://fdncenter.org and http://www.princetonreview.com/college/finance/default.asp

Clarification

A *home page* is the opening page of a Web site, the defined address of the URL, that is, either index.html or home.htm.

Further information on the workings of the World Wide Web, go to http://www.w3c.org.

FORMATTING LANGUAGES[3]

Standards bodies and companies have created formatting protocols for languages that define the format and presentation of text and graphics. Some languages were created before the World Wide Web for large repositories of text. These languages, such as SGML, set the stage for the ones we use today on the Internet. All such languages use tags to format or structure the document.

- **Tags** are commands inserted in a document that specify how the document, or a portion of the document, should be formatted. Tags are used by all format specifications that store documents as text files. For example, **Bolded text** adds emphasis; <u>Underline</u> does also.
- **Standard Generalized Markup Language (SGML)** is a system for organizing and tagging elements of a document. SGML was developed and standardized by the International Organization for Standards (ISO) and does not itself specify any particular formatting; rather, it specifies the rules for tagging elements. These tags can then be interpreted to format elements in different ways. The growth of the Internet, and especially the World Wide Web, created renewed interest in SGML because the Web uses HTML, which is one way of defining and interpreting tags according to SGML rules.
- **HyperText Markup Language (HTML)** is the authoring language that is used like a word processor to create documents on the World Wide Web, such as WWW home

[3] http://www.webopedia.com, http://www.lntranet.co.uk, http://www.amdahl.com.

pages. HTML is similar to SGML, although it is not a strict subset. HTML defines the structure and layout of a Web document by using a variety of tags and attributes to include multimedia and specify hypertext links. These allow Web developers to direct users to other Web pages with only a click of the mouse on either an image or word(s). Whereas simple text, even with good fonts of varying sizes, is flat to view, images make pages pull you into them.

- **Extensible Markup Language (XML)** is a specification developed by the W3C. XML is a pared-down version of SGML, designed especially for Web documents. It allows designers to create their own customized tags, enabling the definition, transmission, validation, and interpretation of data between applications and between organizations.

- **Java** is the programming language and environment designed to solve a number of problems in modern programming practice. It is a simple, object-oriented, distributed, interpreted, robust, secure, architecture-neutral, portable, high-performance, multithreaded, and dynamic language. Java applets are small programs used to create animation, sound, and graphics on Web pages. Java is well suited for applications development over the Internet because it can be run on any modern computer. **JavaScript,** in its more modern form, is an object-based scripting programming language based on the concept of prototypes. The language is best known for its use in Web sites but is also used to enable scripting access to objects embedded in other applications. It was originally developed by Brendan Eich of Netscape Communications Corporation under the name Mocha, then LiveScript, and finally renamed JavaScript. Like Java, JavaScript has a C-like syntax, but it has far more in common with the Self programming language than with Java.[4]

Clarification

Many are familiar with two formats used in files, for example, .RTF and .PDF files. These formats are used for distributing files on and off Web pages. Although these formats may compress the document, their purpose is standardization, the ability to be read across applications, or the ability to be printed as stored, including graphics.

- *Rich Text Format* (.RTF) is a standard formalized by Microsoft Corporation for specifying formatting of documents. RTF files are actually ASCII files with special commands to indicate formatting information, such as fonts and margins.

- *Portable Document Format* (.PDF) is a file format developed by Adobe Systems. PDF captures formatting information from a variety of desktop publishing applications, making it possible to send formatted documents and have them appear on the recipient's monitor or printer as they were intended.

HyperText Transfer Protocol (HTTP) is the standard used by the Web server and its clients that will send HTML files over the Internet from the server to clients requesting them. Thus, **browsers** (client software that retrieves and displays the HTML file) such as **Netscape Navigator®** and Microsoft's **Internet Explorer®** **(IE)** use the command "http://" on the WWW to access a home page; that is, a site with HTML documents. In a nutshell, the command "http" commands the browser to send the characters that follow the "//" to a network domain service, where a search is made of a database to find the characters and the attendant unique IP address. The unique device with that IP address is queried and a default file, that is, index.html or home.htm, is returned to the browser where the HTML capability displays it.

The two standards discussed previously (HTML and HTTP) provide the basis for a whole new kind of access to information. Creating multimedia files in a standard way allows client software to be built that not only can retrieve the files from an HTTP server, but also open them and display them as part of the request. And since the file can contain

---

[4] http://en.wikipedia.org/wiki/Javascript.

hyperlinks to other files (even when they reside on other computers), a user now has the ability to navigate information with a point-and-click interface from what appears to be standard textual documents. This technology takes away the complexity of accessing information on distributed computers.

Clarification

If the browser has visited a Web site recently, its IP address may be in memory, history file, or on a local DNS.

Every computer on the Internet has an IP address associated with it that uniquely identifies it. IP addresses are analogous to telephone numbers—when you want to call someone on the telephone, you must first know their telephone number. Similarly, when a computer on the Internet needs to send data to another computer, it must first know its IP address. IP addresses are typically shown as four numbers separated by decimal points, or "dots." For example, 10.24.254.3 and 192.168.62.231 are IP addresses.

If you need to make a telephone call but you only know the person's name, you can look him or her up in the telephone directory (or call directory services) to get the telephone number. On the Internet, that directory is called the **domain name system** (or **domain name server**), or **DNS** for short. If you know the name of a server, say www.cert.org, and you type this into your Web browser, your computer will then go ask its DNS server what the numeric IP address is that is associated with that name.

The DNS is a system that stores information about hostnames and **domain names** in a type of distributed database on networks, such as the Internet. Of the many types of information that can be stored, most importantly it provides a physical location (IP address) for each domain name, and lists the mail exchange servers accepting e-mail for each domain.[5]

BROWSERS AND SEARCH ENGINES

A *browser* is a software application, a client that resides on the workstation and interprets the response from the HTTP command to the server. Tim Berners-Lee of CERN developed the first browser; Mosaic, developed at the University of Illinois was the first publicly provided hyperlink-based Internet information browser and World Wide Web client that provides transparent access. A browser can recognize and manipulate a large range of data types and services, including multimedia-based resources such as sound and animation.

Netscape Navigator® is the original premier graphics and hyper-based Internet information presenter and browser that is the follow-up to Mosaic and was developed by the same people. The developers have since organized Mozilla.org and developed the *Firefox*® browser.[6]

Internet Explorer (IE)® is Microsoft Corporation's competitor to Netscape; it is included in the Windows operating system, giving it a significant advantage for adoption. One feature of these browsers is the access to **search engines.** Both include a "net search" button that takes the user to a page that includes access to several search facilities. Once the search engine is chosen, search words are included. The value of the search engine has spawned several dedicated capabilities, such as *Dogpile*® and *Google*®.

One feature of the WWW capability of the Internet is the cataloging (databasing) of Web pages and their content. Automated agents, called robots, Web crawlers, bots, and spiders, intelligently search the WWW and record the headers and other identifying data of all HTML, PDF, and other, pages for which they are allowed access. This information is then placed in a database by the owner of the robot. A user may then choose a search facility (search engine) on the Internet, put in search criteria, and then will receive a response as

[5] http://en.wikipedia.org/wiki/Dns.

[6] http://www.mozilla.org/products/firefox/.

to what home pages exist in the database listing that match those criteria. Additionally, the search engine attempts to present them in order of relevance.

Implication

People wishing to create e-mail lists to use for spam and other questionable uses also use Web crawlers to find e-mail addresses on Web pages. Cookies can be accessed in a similar manner to acquire e-mail addresses.

Implication

Search bots gather information on all Web pages they find and index all words in a database. Thus, when you "search" on a keyword(s), all Web site addresses with that word(s) are already indexed on the computer so that the response appears "instantaneous."

Ever wonder how Amazon.com knows it's you when you arrive at their Web site? Cookies are the answer. On your hard drive, go to c:\Documents and Settings*machinename*\cookies; where *machinename* is the name of your machine. Here, you will find all the cookies on your machine.

Malware is any software written to do undesirable things, for example, viruses, keylogging.

Anyone who has done a *Google* search is familiar with the two listings of the responses to a query. Those on the left side of the page, generally numbering in the millions, show all sites that meet the criteria. Those sites on the right side of the page are the result of owners participating in a program called *Adwords®*. With *Adwords*, the site owner selects a set of words that are likely to be used in a search for that site and pays *Google* when the search user "clicks through" the listing on the right side of the page, Sponsored Links. The more the person pays Google for the word, the higher in the list their site will appear.

A specific value of the browser is that it provides a universal interface for programs. Instead of a programmer writing how a new program will display information, that is, the interface, s/he writes it for the browser. Thus, any client with a browser and authentication can access a program that has the browser interface. Add to this the search window (*Firefox* provides continuous access to a *Google* search via a window on the URL line) and the interface provides the ability to query the Internet about any subject being considered.

A way for a Web site to know who has returned to it is to place a cookie on the client machine. A **cookie** is a message given to a Web browser by a Web server. The browser stores the message in a text file on the user's machine and is sent back to the server each time the browser requests a page from the server. The main purpose of cookies is to identify users and possibly prepare customized Web pages for them as they return to a site. Cookies also may hold and provide personal information to the server or other people who query it. Some servers require that cookies be enabled so that information can be stored for future usage.

Browsers are the interface to the WWW/Internet; they also are the target for intrusion. Because Microsoft programs are so widely used and written to be easy to incorporate into other applications, they are candidates for intruders and **malware.** IE is a prime candidate for such attacks; one way to lessen this threat is to use other browsers.

Implication

New kid on the block . . . "The most full-featured Internet power tool on the market, *Opera®* includes pop-up blocking, tabbed browsing, integrated searches, and advanced functions like Opera's groundbreaking E-mail program, RSS Newsfeeds and IRC chat. And because we know that our users have different needs, you can customize the look and content of your Opera browser with a few clicks of the mouse."[7]

ACCESS TO THE INTERNET

For as little as $5.47/month, PeoplePC Online offers dial-up service with virus blocker, Internet call waiting, e-mail with spam controls, and pop-up blocker.[8]
NetZero Internet Services offers three

A company that provides retail access to the Internet is called an **Internet service provider (ISP).** This is a simple concept; they provide their services either through dial-up (POTS, ISDN) or broadband (cable or DSL) wired connectivity or via wireless (802.11 or satellite) access points in retail establishments such as Starbucks®. What you

---

[7] http://www.opera.com/products/desktop/.
[8] http://home.peoplepc.com/.

levels of service: free for 10 hours/month dial-up, $9.95 for unlimited dial-up, and $14.95 for broadband access.[9] For many college students, their university is their ISP when they live on campus. If you access the Internet at Starbucks® or some other retail **wireless access point (WAP)** that is your ISP. In Starbucks's case, T-Mobile® handles the wireless capability and access to the Internet. Your DSL or cable Internet access provider is your ISP.

find at the ISP's site is not so simple. Early providers such as *Prodigy*® and *America Online (AOL)*® started life as dedicated servers/mainframes to which the subscriber connected for service. The main menu was all you got, and services ranged from e-mail to purchase of flowers.

The next type of ISP did only that: they provided access to the Internet and let you figure out the rest. In the middle is where we are today. Most ISPs provide their own information on the home page; *BellSouth.net*®, which is the ISP for one of the authors, has news, sports scores, e-mail, and advertisements. At this point, the user is not on the Internet; they have used dial-up or broadband to access the ISP's server and see what the ISP wishes them to see. The access is via a browser, so a standard URL line is available at the top for truly going to the Internet. (As the home page of the ISP is the default or home address, the uniform resource locator (URL) of the ISP is shown in the URL line.) AOL, *PeoplePC*®, *Netscape Internet Services*®, and *MSN (Microsoft Services Network)*® provide similar material on their home page (not the Internet, but a server) plus have instant messenger buttons. Wikipedia lists 27 ISPs; MSN is not included.

Implication

Probably the current major distinction of Internet access is whether it is dial-up (56 Kbps modem) or broadband (T1 or higher). This bandwidth difference, to some extent, determines the value of the Internet. Although dial-up provides the same access, the broadband bandwidth allows for much more activity in a shorter period of time. This feature most likely determines how long a person will stay on the Internet and how detailed the activities will be. Since so much of the material provided on the Internet is in graphic form, multimedia drives a demand for bandwidth, for example, broadband.

SURAnet (Southeastern Universities Research Association network) provides networking services for a variety of industries. SURAnet was one of the United States' largest regional Internet providers.

Most organizational local area networks have a gateway to the Internet. This means worldwide connectivity plus the World Wide Web graphical interface at LAN speeds, as constrained by the specific link to the Internet. At our university (see Figure 11.2 in Chapter 11), the authors' computers are nodes on the College of Business 10/100 Mbps Ethernet LAN, which connects to the rest of the university by fiber-based Gigabit Ethernet. Initially we connected to the SURAnet via two T1 channels, which combined in Birmingham, Alabama, and then connected to a T3 link in Atlanta, Georgia, via a T1 circuit. This has evolved to our having two fiber links, each over 20 Mbps, to Atlanta, Georgia, where they access T3 or greater conduits to MAE East, the connection point to the Internet backbone. The university has an OC-12 (622 Mbps) connection as a member of Internet2.

PROTOCOLS

Computers can't just throw data at each other any old way. Because so many different types of computers and operating systems connect to the Internet via modems or other connections, they have to follow communications protocols. The Internet is a very heterogeneous collection of networked computers and is full of different access protocols, including PPP, TCP/IP, and SLIP.

TCP/IP: Transmission Control Protocol/Internet Protocol

TCP/IP forms the basis of the Internet and is built into every common modern operating system (including all flavors of Unix, the Mac OS, and the latest versions of Windows). TCP

---

[9] http://www.netzero.net/.

FIGURE 7.2   Ethernet TCP/IP Packet

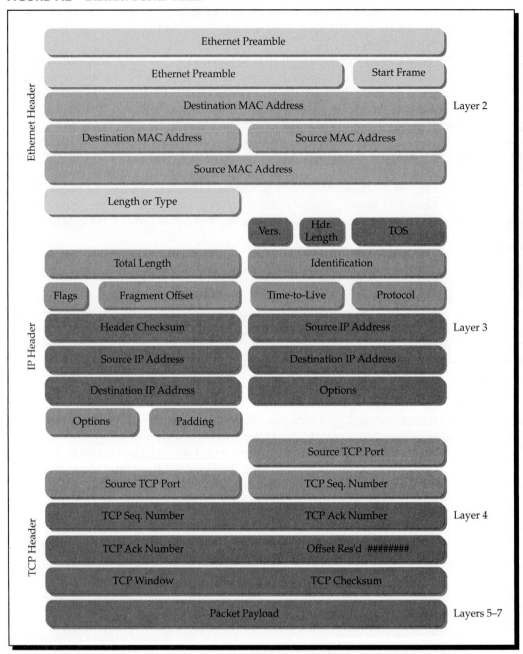

is responsible for verifying delivery from client to server. IP is responsible for moving pack-ets of data between nodes. Refer to Chapter 3 for details on TCP/IP. Figure 7.2 illustrates a TCP/IP packet. Notice the complexity of the packet, redundancy, and error checking.

Clarification   What can you do on the Internet? There are books on this subject as well as files you can access free from the Internet. One such book is *Internet Essentials* by Lynda Armbruster.[10]

[10] Lynda Armbruster, Internet Essentials (Indianapolis, IN: Que Publishers, 1994).

Serial line Internet protocol (SLIP)
A standard for connecting to the Internet with a modem over a phone line. It has serious trouble with noisy dial-up lines and other error-prone connections, so look to higher-level protocols like PPP for error correction. CSLIP is a compressed version of SLIP to enhance speed.

Point-to-point protocol (PPP)
The Internet standard for serial communications. Newer and better than its predecessor, SLIP, PPP defines how your modem connection exchanges data packets with other systems on the Internet.

As you will find there and in other books such as *Internet for Dummies*, you can sign up for ListServ to automatically send you text files on a large number of subjects, use File Transfer Protocol (FTP) to move files, and use telnet to connect and log on to other computers as a guest user. These services (collectively called the TCP/IP suite) are described below.

- *ListServ* is the program that operates discussion groups on the Internet. Subscribers can send mail to this special program to be automatically distributed to each person on the list, a convenient way of communicating without human intervention. (Note: There is a ListServ from the White House.)
- *File transfer protocol (FTP)* is a part of the Internet protocol suite of capabilities that enables a user to access files on a remote computer and move files between two computers connected to the Internet.
- *Telnet* is a program that enables you to communicate with other computers and is typically used for remote access to host computers on which they have an account or to publicly accessible catalogs and databases.
- *Usenet* started as a network on which to discuss specific topics and has expanded. Usenet newsgroups are broken down by topics, ranging from business to science to education.

ACCESS POINTS TO THE INTERNET: NAPS AND MAES

The U.S. MAE (metropolitan area Ethernet) services are Internet network traffic exchange facilities located in various U.S. cities. The Internet backbone is supported by four vendors, MCI being one. Boxes 7.1 and 7.2 describe MCI's view of MAE services. MAE East, MAE West, and MAE Central are often referred to as the primary backbone access points.

- **Metropolitan area Ethernet (MAE)**[11] is a network access point where Internet service providers (ISPs) can connect with each other. The original MAE was set up by a company called MFS and is based in Washington, D.C. Later, MFS built another one in Silicon Valley, dubbed MAE West. In addition to the MAEs from MFS, there are many other NAPs. Although MAE refers really only to the NAPs from MFS, the two terms are often used interchangeably.
- A **network access point (NAP)** is a public network exchange facility where ISPs can connect with one another in peering arrangements. The NAPs are a key component of the Internet backbone because the connections within them determine how traffic is routed. They are also the points of most Internet congestion.

Clarification
Chicago's Network Access Point (NAP)

- Tier 1 Internet traffic exchange connectivity in Chicago serving ISPs, research centers, and universities.
- Well over 140 domestic and international customers.
- Delivery of IP packets via ATM cell relay at DS3, OC-3c, and OC-12c speeds.
- World's largest ATM exchange.
- Robust SBC global network.
- Leading-edge next generation Internet (NGI) development.

SBC, now AT&T, provides a network access point (NAP) in the Midwest encapsulating IP packets in asynchronous transfer mode (ATM) cells to provide benefits of fast transfer

[11] http://www.webopedia.com.

MCI MAE Facilities

BACKGROUND

In 1993, the National Science Foundation awarded MCI the status of NAP (network access point) to the NSFnet backbone in Washington, D.C. Prior to the awarding of this NAP status, MCI and a small group of ISPs helped create an exchange "point" that eventually grew into what we currently call a MAE, or an Internet networks traffic exchange facility. Since then, MCI has set up seven MAE sites nationwide where Internet service providers can establish peering relationships for the exchange of IP traffic.

Today, MCI operates three of the major interconnect points for the Internet in the United States: MAE East, MAE West, and MAE Central. In fact, a major portion of the traffic that flows between ISP networks passes through MCI MAE facilities.

MAE East is located in the Washington, D.C., metropolitan area and connects all of the major ISPs as well as European providers. MAE West is located in California's Silicon Valley providing a second interconnection point linking major ISPs in that area. Many are connected at MCI in San Jose, while a smaller number (but individually larger in size) are connected at NASA Ames Research Center, home of the western Federal Internet Exchange (FIX West). The two sites are linked together with multiple 155 Mbps (OC-3) circuits. MAE Central is located in Dallas, Texas, and is the latest addition to our Tier-1 facilities. In addition to these Tier-1 facilities, MCI has two tier-2 regional MAE sites in operation: MAE Houston and MAE Los Angeles.

WHAT IS A MAE?

A MAE (metropolitan area Ethernet) is the MCI facility where ISPs connect to each other to exchange Internet traffic—an Internet networks traffic exchange facility. The easiest way to think of it is as a LAN switch where all the "pieces" of the Internet connect together in order to exchange traffic at high speeds. The MAE forms part of the "Inter" in Internet.

A MAE does *no* routing of data. The **routing** function is performed by the routers that are normally connected to the MAE and that are owned and managed by the ISPs. In fact, the router is the only device that connects to a MAE. ISP hosts, as long as they act as routers, may occasionally be connected to switched ports at the MAE.

A MAE and other Internet traffic exchange points (e.g., NAPs and the CIX) are not connected to each other via dedicated links. If an ISP needs to have connections to multiple MAE locations, then the ISP will build its own backbone network.

TIER-1 AND TIER-2 MCI MAE SITES

MCI's MAE facilities are grouped into two categories: Tier-1 and Tier-2 MAE facilities. Tier-1 MAE locations are regarded as "national" connection points and have a combination of the following characteristics: One or more of the major ISPs (e.g., Sprint, MCI, UUNET, ANS, Netcom, BBN, PSI) connect to the MAE. There is a high-speed FDDI switch installed offering dedicated higher-speed connections. The sites designated as Tier-1 MAE locations are MAE East, MAE Central, and MAE West. Generally, routers with very large routing tables are needed at Tier-1 MAE sites. Examples of these routers are Cisco 4700-M or Cisco 7xxx series routers.

Tier-2 MAE locations are regarded as "regional" connection points for regional ISPs and have the following characteristics: presence of regional or smaller ISPs with few or no major ISPs present and Ethernet switch and FDDI concentrator so lower speed or shared access services only. Existing sites designated as Tier-2 MAE locations are MAE Los Angeles and MAE Houston. Generally, smaller routing tables are needed at Tier-2 MAE locations and so smaller routers are satisfactory. An example of this kind of router is a Cisco 4500-M.

Please note: Effective spring 1998 MCI no longer sells access to the Tier-2 locations. These facilities remain active and in service; however, no new connections are being sold.

MAE SERVICES KEY BENEFITS

- Cost-effective way for ISPs to connect to other ISPs.

speeds and access to a vast community of other ISPs, content providers, and institutions with whom you can peer. This NAP already hosts connections for well over a hundred ISPs and other organizations.

World's largest ATM exchange point: As one of the original four NAPs in the United States, our downtown Chicago NAP was originally founded on ATM technology. In fact, it's the preeminent ATM exchange point in the world, with all the advanced scalability required of a Tier-1 NAP.

- Presence of multiple ISPs at a MAE.
- Low entry-level requirements for an ISP to connect to a MAE.
- Colocation space for housing ISP equipment.
- MCI's broad geographic coverage.

COMMON MISCONCEPTIONS RELATED TO MAE SERVICES

MAE services are not connections to the Internet. This is one of several common misconceptions related to MAE services. Following are more key points.

MAE services do not include peering agreements with any of the ISPs that are connected at a MAE. The ISPs must do the "peering" on their own. This involves negotiation and technical cooperation to enable the two ISP networks to pass traffic.

MAE services do not include transit with any of the ISPs that are connected at a MAE. Typically, a smaller ISP would become a customer of the ISP that provides the transit.

MAE services do not guarantee that there will be specific major ISPs at a MAE. While it is our objective to connect the key ISPs at a MAE, it is the decision of each ISP as to whether they interconnect at the MAE and with whom they negotiate peering.

MCI does not provide IP addresses or AS (Autonomous System) numbers. IP addresses are provided by third parties on behalf of MCI. These IP addresses only apply to physical ports at the MAE. AS numbers and IP Class addresses are provided by the InterNIC.

PEERING ARRANGEMENTS

It is a requirement for an incoming ISP to have peering or transit arrangements with the other ISPs connected to the MAE. MCI cannot arrange peering. There is no single ISP in charge of the peering. Peering agreements are bilateral agreements between individual ISPs. Each ISP must negotiate independently and needs to contact individually every ISP with which they wish to peer. MCI can and will only offer guidance (i.e., we can facilitate introductions to other ISPs). MCI cannot provide peering.

COLOCATION

The MAE colocation facilities are a secure environment where ISPs may locate routing equipment for direct connection to the MAE. If the ISP customer takes colocation space instead of having the demarcation point extended to their premises, the Ethernet or FDDI port (demarcation) appears at the ISP's rack space. In the case of HSSI, the customer gets half rack of colocation as well as an extension to their premises. Colocation space is available on a first-come, first-served basis.

MAE MANAGEMENT

Management of a MAE is done via a dedicated Sun workstation. The MAE devices are managed using SNMP. The management workstation does not share a network segment with any service providers' router.

Custom applications run on the Sun management workstation to collect performance statistics on the MAE, monitor MAE equipment, and generate alarms. It also monitors connections and generates connection lists.

It is the responsibility of the ISP connected to the MAE to configure and manage their router equipment.

SERVICE LEVELS AND LEAD TIMES

Service levels:

- Target mean time to repair = 2 hours
- Target availability = 99.99 percent, 24 hours per day, 7 days per week at all sites

Lead times:

- On-net (lit) building, no collocation: 30 calendar days
- Off-net building, no collocation: 45 calendar days
- Non-MCI city, no collocation ICB
- Collocated equipment: 30 calendar days (dependent on space availability)

Source: http://www.mae.net/doc/maedesc/maedesc1.html.

Our ATM exchange handles a peak busy-hour traffic load of over 3.0 Gbps, supporting connection speeds of DS3, OC-3c, and OC-12c. There are four ATM switches internetworked with multiple OC-48c trunks, resulting in negligible cell loss. This centralized **topology** enables you to augment facilities rapidly when needed. In the seven years of SBC NAP operation, there has never been a congestion condition resulting from switch or intermachine trunk capacity issues.[12]

[12] http://www.aads.net/pdf.html.

WHO ARE WE?

Founded in July of 1995, our mission is to provide robust Internet connectivity to Internet service providers (ISPs) at affordable prices in an effort to develop a level playing field among the Internet community.

Our current network design and equipment selection are engineered to provide our customers with a fault-tolerant connection to the Internet utilizing multiple NAP connections and peering arrangements. Through this network our customers receive robust global transit. For those already connected at a NAP, virtual transit is available. Nap.net access services get you as close to the Internet as possible . . . right to the front door . . . by directly connecting you to the gateway for its national backbone. Our service is nonrestrictive . . . you control your content without limitations, restrictions, or inflated costs!

WHERE ARE WE?

Currently we connect to the "Net" at the Chicago NAP, MAE East, and MAE West. We have a point of presence at each of these locations as well as in Dallas and Seattle. Additional connections are continually being engineered.

Our routing equipment is deployed with the latest in technology releases. NAP.net is one of the first network service providers to utilize true ATM DS3 connectivity to a NAP. Currently, our Chicago NAP connection is an OC-3. At the meet points on the East and West coasts, we connect at the FDDI level (100 Mbps). All locations utilize ATM DS3 on ramps to our fully meshed DS3 ATM backbone. NAP.net's equipment has been installed with redundant power supplies and battery backup to ensure electronic survivability.

NAP.net's Tier-1 network includes extensive peering arrangements and backup connections to provide global connectivity continuously. Our outstanding network performance standards provide our customers with rock-solid national backbone network services.

SERVICES WE PROVIDE

Our services include

- Internet connectivity: T1 (DS1).
- Ethernet: standard, measured.
- ATM DS3: full rate (FTS), rate shaped (STS), burstable (BTS), measured (MTS).
- Fast Ethernet.
- OC-3.
- Virtual transit—3 MBps and up.
- MegaPOP service, which helps ISPs to expand dial-up coverage without the high cost of hardware and line charges!
- News feeds.
- Secondary DNS.
- Multicast routing (Mbone).
- IP addresses.
- Internic registration assistance for IPs, domains, AS numbers.
- Equipment colocation space in our facilities.
- Usage reports available for each customer, real time, online.
- Quality service performance that's measurable.
- Global IP transit, Tier-1 networking.
- Network operations center monitoring, 24 hours a day, 365 days a year.
- 100 percent port assignment via direct circuit connections for your circuit(s) only.
- Facility coordination for all your telecommunications needs.
- Technical support, 24 hours a day, 7 days a week.
- 120 V/AC and 48 V/DC, providing electrical stability at all times.
- Pricing—a value in comparison to any aggravated service provider
- Nonrestrictive so you control your content and address without limitations or restrictions—fully resellable services.

We do not compete with our customers (we are not a competing ISP).

Source: http://www.nap.net/who.html.

WEB ADDRESSES

Uniform resource locators (URLs) are the global addresses of documents and other resources on the World Wide Web. The first part of the address indicates what protocol to use, describing the way the browser gets to the resource, and the second part specifies the IP address or the domain name where the resource is located. For example, the two URLs

You can find your machine's IP address at http://www.lawrence goetz.com/programs/ ipinfo/; find the IP address of a Web site by **START |RUN| cmd PING url.**

below point to two different files at the domain www.pcwebopedia.com. The first specifies an executable file that should be fetched using the FTP protocol; the second specifies a Web page that should be fetched using the HTTP protocol:

> ftp://www.pcwebopedia.com/stuff.exe
>
> http://www.pcwebopedia.com/index.html
>
> http://www.auburn.edu (looks for index.html) is the same as http://131.204.2.251.

Domain Names

The WWW is based on the use of HTTP protocol and URLs. The HTTP is the command to go to the URL address and return (i.e., GET) the default file of index.html or home.htm, process it through a browser, and display the text and graphics using HTML. URLs are addresses, ending with an extension that defines the usage. For example, the URL http://www.auburn.edu defines the home page for Auburn University in Alabama. Here, HTTP is the protocol and get-command and www.auburn.edu is the address. There is nothing magic about an address starting with "www". For example, the Web site for the authors' book site is http://telecom.auburn.edu/JEI. Thus, Web addresses follow a convention but have great variety.

Clarification

A **domain name** is a name of a computer on the Internet that distinguishes it from the other systems on the network. They are sometimes colloquially (and incorrectly) referred to by marketers as "Web addresses." Every Web site, e-mail account, and so on, on the Internet is hosted on at least one computer (server). Each server has a unique IP address that is nothing but a set of numbers, such as "207.142.131.235." To access a particular Internet service, one can specify its IP address in an appropriate application, such as an FTP client; however, because it is difficult to remember numbers, an IP address can be associated with a fully qualified host name (a domain name), such as "www.wikipedia.org." Domain names also provide a persistent address for some service when it is necessary to move to a different server, which would have a different IP address.

Each set of letters and numbers between the dots is called a *label* in parlance of the domain name service (DNS). There are some rules about the size and makeup of labels. Each must start with a letter or number, and then may be made up of letters, numbers, and hyphens, to a maximum of 63 characters. These are the rules imposed by the way names are looked up ("resolved") by the DNS. Some top level domains impose more rules, like a minimum length, on some labels.[13]

Every domain name has a suffix that indicates which top level domain it belongs to. There are only a limited number of such domains, although the number is increasing. The first six examples that follow are types of use; the last three are countries of origin.

- gov: *Gov*ernment agencies
- edu: *Edu*cational institutions
- org: *Org*anizations (nonprofit)
- mil: *Mil*itary
- com: *Com*mercial business
- net: *Net*work organizations
- ca: *Ca*nada
- th: *Th*ailand
- de: *De*utschland (Germany)

[13] http://en.wikipedia.org/wiki/Domain_name.

Because the Internet is based on IP addresses, not domain names, every Web server requires a DNS server to translate domain names into IP addresses.

Registering a Domain Name

Register your domain name, create and publish your Web site, host your Web site, and set up your professional e-mail service.[14]

The only way to register and start using a domain name is to use the services of a domain name registrar. The domain name industry is regulated and overseen by ICANN, the organization that is responsible for certifying companies as domain name registrars. At one time, there was only one domain name registrar—Network Solutions, Inc.—but today there are dozens of accredited registrars. Only a domain name registrar is permitted to access and modify the master database of domain names maintained by InterNIC. The master database contains the documentation on all of the domain names registered to date.

There are many laws that regulate the **registration of domain names,** such as registering a copyrighted name or registering a domain name for the sole purpose of blocking someone else from using it. Check your domain name registrar for the laws that affect how to register your domain name. If you are unsure if a given domain name has already been taken, check a WhoIs server that keeps track of all the registered domain names.

What Are Static and Dynamic Addressing?[15]

Every computer on the Internet has an IP address associated with it that uniquely identifies it. However, that address may change over time, especially if the computer is

- Dialing into an Internet service provider (ISP).
- Connected to a LAN and using NAT.
- Connected behind a network firewall.
- Connected to a broadband service using dynamic **IP addressing.**

Static IP addressing occurs when an ISP network manager permanently assigns one or more IP addresses for each user. These addresses do not change over time. However, if a static address is assigned but not in use, it is effectively wasted. Since ISPs have a limited number of addresses allocated to them, they usually make more efficient use of their addresses by using dynamically assigned addresses at time of logon. When a *static IP address* is assigned, the same address is used each time the device joints the network. This makes it easy for other machines to reach it as its address does not change; it also makes it a nice target for intrusion for the same reason.

Dynamic IP addressing allows the ISP to efficiently utilize their address space. Using dynamic IP addressing, the IP addresses of individual user computers may change over time. If a dynamic address is not in use, it can be automatically reassigned to another computer as needed. ISPs often assign a dynamic IP address, allowing the reusing of the address. Many home users have a dynamic IP address; if a static address is assigned, the ISP may charge for this service. The alternative is a dynamic IP address, which may change with time.

Address Resolution Protocol (ARP)

When the network layer protocol has an IP address to send a message to or get data from, it must convert the IP address into a physical address (called a *DLC address*), such as an Ethernet address. A host wishing to obtain a physical address broadcasts an ARP request onto the TCP/IP network. The host on the **IP network** that has the IP address in the request then replies with its physical hardware address. There is also Reverse ARP (RARP), which can be used by a host to discover its IP address. In this case, the host broadcasts its physical address and a RARP server replies with the host's IP address.[16]

---

[14] http://www.register.com/.

[15] http://www.cert.org/tech_tips/home_networks.html.

[16] http://www.webopedia.com/TERM/A/ARP.html.

What Is NAT?

Network Address Translation (NAT) provides a way to hide the IP addresses of a private network from the Internet while still allowing computers on that network to access the Internet. NAT can be used in many different ways, but one method frequently used by home users is called *masquerading*.

Using NAT masquerading, one or more devices on a LAN can be made to appear as a single IP address to the outside Internet. This allows for multiple computers in a home network to use a single cable modem or DSL connection without requiring the ISP to provide more than one IP address to the user. The same is true for a LAN: multiple users for one IP address. Using this method, the assigned IP address can be either static or dynamic. Most network firewalls support NAT masquerading.

Another value of NAT masquerading, or proxy, is that the outside world does not know the true IP address and its assigned machine or machines. This is a form of security.

What Are Network Ports?

TCP guarantees delivery; UDP is lower overhead and does not guarantee delivery.

TCP (transmission control protocol) and UDP (user datagram protocol) are both protocols that use IP. Whereas IP allows two computers to talk to each other across the Internet, TCP and UDP allow individual applications (also known as *services*) on those computers to talk to each other.

In the same way that a telephone number or physical mail box might be associated with more than one person, a computer might have multiple applications (e.g., e-mail, file services, Web services) running on the same IP address. *Ports allow a computer to differentiate services such as e-mail data from Web data. A port is simply a number associated with each application that uniquely identifies that service on that computer.* Both TCP and UDP use ports to identify services. Some common port numbers are 80 for Web (HTTP), 25 for e-mail (SMTP), and 53 for domain name system (DNS).

Dynamic Host Configuration Protocol (DHCP)[17]

The **Dynamic Host Configuration Protocol (DHCP)** is an Internet protocol for automating the configuration of computers that use TCP/IP. DHCP can be used to automatically assign IP addresses, to deliver TCP/IP stack configuration parameters such as the subnet mask and default router, and to provide other configuration information such as the addresses for printer, time, and news servers.

The protocol[18] assigns dynamic IP addresses to devices on a network. With dynamic addressing, a device can have a different IP address every time it connects to the network. In some systems, the device's IP address can even change while it is still connected. DHCP also supports a mix of static and dynamic IP addresses. Dynamic addressing simplifies network administration because the software keeps track of IP addresses rather than requiring an administrator to manage the task. This means that a new computer can be added to a network without the hassle of manually assigning it a unique IP address. Many ISPs use dynamic IP addressing for dial-up users.

INTERNET2

As a result of the congestion created on the Internet by commercial enterprises, a number of corporations, research groups, and universities have formed a consortium to create Internet2. Previously called the Abilene Project, the fiber-based network provides very-high-speed

---

[17] http://www.dhcp.org/.
[18] http://www.webopedia.com/TERM/D/DHCP.html.

connectivity to these members. The purpose is research as opposed to commerce. Internet2 is available to members only and not to the Internet. Case 7-2 describes this capability.

Clarification   Internet2 is a consortium being led by 207 universities working in partnership with industry and government to develop and deploy advanced network applications and technologies, accelerating the creation of tomorrow's Internet.[19]

INTRANETS AND EXTRANETS

Intranets

In simple terms, **intranet** is the descriptive term used for the private implementation of Internet technologies within a corporate organization; it is the use of Internet technologies, that is, HTML, browser, http, and so forth, on a private LAN. This implementation is performed in such a way as to transparently deliver the immense informational resources of an organization to each individual's desktop with minimal cost, time, and effort.

An *intranet* is a private network based on TCP/IP protocols (an internal Internet) belonging to an organization, accessible only the organization's members, employees, or others with authorization. An intranet's Web sites look and act just like any other Web sites, but the firewall surrounds an intranet and fends off unauthorized access from outside. Like the Internet itself, intranets are used to share information. Secure intranets are now the fastest-growing segment of the Internet because they are much less expensive to build and manage than private networks based on proprietary protocols. Examples of intranet usage are provided in the next chapter.

Clarification   A firewall is a system designed to prevent unauthorized access to a private network. It can be implemented through hardware, software, or both. Firewalls are the first line of defense from the outside world, for example, the Internet. More on this in Chapter 11, "Network Security."

Extranets

extranet
An invitation-only group
of trading partners
conducting business via
the Internet.[20]

Extranet refers to an intranet that is partially accessible to authorized outsiders. Whereas an intranet resides behind a firewall and is accessible only to people who are members of the same company or organization, an extranet provides various levels of accessibility to outsiders, often via the public Internet. You can access an extranet only if you have a valid username and password (authentication) and your identity determines which parts of the extranet you can view. To provide adequate security in this environment, virtual private networks are often used, providing an encrypted (i.e., secure) tunnel through the insecure Internet.

Extranets are becoming a very popular means for business partners or customers to exchange information, particularly in business-to-business usage. FedEx and UPS allow customers access to parts of their intranets via the extranet so that customers can check on the status of shipped parcels. Extranets are a natural evolution taking advantage of the Internet infrastructure and previous Internet investments to focus communications to exchange information and share applications with business partners, suppliers, and customers. As the business environment evolves to incorporate more interorganizational systems, the extranet is an obvious choice for establishing high-speed connectivity with suppliers, customers, outsourcers, and other partners.

[19] http://www.internet2.edu/about/.
[20] http://www.misq.com/misc/presentation/geis_big/sld012.htm.

The Extranet and EDI

EDI was described in the last chapter as a way for different computer systems to communicate about common business documents. When an organization provides extranet access to a partner, supplier, or customer, it may be replacing the EDI technology with more direct access.

THE INTERNET AND SECURITY

We leave the subject of security until Chapter 11. Much of the concern for network security is a direct result of the Internet and access from it to private networks. *As with personal security, the price of Internet freedom is eternal vigilance.*

Case 7-1

NREN and Its Present State in Relation to Internet2

Teresa Lang

WHAT IS NREN?

Today, the National Research and Education Network (NREN) is the National Aeronautics and Space Administration's (NASA) research wide area network (WAN). The network uses gigabit technology capable of moving data at 622 Mbps and is part of a national effort to develop innovative networking technologies. NASA uses the NREN to test applications that are under development. Once the new networking technology is tested and complete, NASA plans to use it to improve the design of air and space vehicles, permit Earth system modeling, and provide access to vast databases.

Most recently, the NREN has been used to prototype applications linking high-speed network users with high-end computation resources, including the ability to view images that have been created and stored at remote sites. NASA expects the emerging technologies developed and tested with the assistance of NREN will significantly benefit their overall networking applications.

The technologies currently under development include multicasting, quality of service, gigabit networking, hybrid networking, adaptive middleware, and traffic engineering. NREN evaluates the alternative approaches to implementing these technologies and prototypes solutions in context specific to NASA applications. [1]

WHAT IS INTERNET2?

Internet2 is a closed membership research network. This membership consists of universities, the U.S. government, and some private-sector businesses. Those working in academia are using the network to build applications for teaching and research and establishing digital libraries. Businesses are working with and incorporating the new networking protocols and software into their future products with plans to use them in commercial applications.

Internet2 actually consists of two interconnected networks. The first one is called Abilene, has a transfer rate of 2.4 Gbps, and was built by Cisco and Nortel. The second network was commissioned by the National Science Foundation and is called the very-high-performance backbone network system (VBNS). The VBNS has a transfer rate of 2.4 to 40 Gbps, or 64 times faster than NASA's NREN.

The two networks evolved separately and are now interconnected to one another as well as to research networks around the world. Although the two networks are interconnected with one another, neither is connected to the commercial Internet. The commercial Internet is the network referred to by the general public as "the Internet." [2]

A goal of Internet2 is to provide better and more appropriate networking capabilities to support the research and educational activities of the nation's major universities. Internet2 functions as a closed system, available only to members for experimenting with new network technology, new collaboration technology, and new instructional technology.

Internet2 is used to prototype new quality-of-service controls, multicast techniques, and network-based videoconferencing. It also is used to facilitate improvements in commercial networks by sharing newly developed technology with both the education community and the private sector. As a result, networks are expected to become more flexible and efficient due to advances made with Internet2. [3]

FROM NREN TO INTERNET2

So, both NREN and Internet2 are working toward some of the same advances in technology. Then why are there two initiatives working toward the same end?

NREN was first established in the early 1990s, and was further empowered with federal funding when Congress passed the U.S. High Performance Computing Act of 1991. (This is when Vice President Gore took credit for the Internet.) The Act was developed and passed with the intent of coordinating government, industry, and the academic community toward the development and use of emerging technologies.

The 1991 Act established the Federal Networking Council, which included the National Science Foundation (NSF), the Department of Defense (DoD), the Department of Energy (DOE), NASA, and other agencies. The Council was to develop and establish the NREN.

The NREN ended up to be a federal program composed of two parts. The first part involved research on high-speed networks, and the second part was the aggregation of the federal research networks.

The Information Infrastructure Act of 1993 clarified the use of the term "the network" referred to in the NREN section of the High Performance Computing Act of 1991. In the 1993 Act, "the network," previously considered to be the NREN, was defined to be the Internet— the same Internet that the general public uses today, but before it was turned over to the private sector for commercial use. This in effect left the NREN without a purpose or identity.

For a while after this, NREN was known as the Interim Interagency NREN, or IINREN. It included the NSFNet backbone, and all the networks were allowed to use the backbone for research and education purposes. The NREN changed from technology innovation to the furtherance of the development of the present reality. [4]

The term NREN today refers to what is now part of the technology research continuing at NASA, one of the original agencies involved in the project. It is used to connect scientists all over the country.

The NREN Testbed provides a nationwide high-performance Wide Area Network (WAN) research platform for early deployment and testing of emerging networking technologies and for prototyping of applications and collaborative processes that will be enabled by these technologies. Hence, the testbed is used both to conduct network research and to support application research. The testbed provides connectivity between selected NASA centers and peers with other high-performance network testbeds to enable NASA scientists, engineers and researchers to reach their partners within other Federal agencies and academia. The testbed infrastructure has been upgraded several times as new networking technologies are developed. NREN is currently investigating the use of optical networking to achieve a 10-fold increase in throughput. (http://www.nren.nasa.gov about/index.html)

Multicasting is also called point-to-point transmission. It is the simultaneous delivery of IP packets to multiple endpoints. This technology allows a collection of different applications to efficiently distribute large data sets to many different sites.

Quality of service is the commitment of resources to specific applications to ensure that performance limitations such as bandwidth, latency, jitter, and packet loss are maintained within acceptable ranges. This technology facilitates efficient sharing of resources among several users while giving preferential treatment to selected applications when the resources are limited.

Middleware is an adaptive technology that enables distributed multimedia applications to adapt to available CPU and network resources. [1]

Although the focus and the definition of NREN changed in the early 1990s, the initiatives to develop faster, more efficient networks continued.

The NSF transferred the responsibility for the Internet, as the general public now knows it, to the private sector in 1995. The Internet had grown so quickly that the NSF could no longer maintain it, so it was turned over to the private sector for commercial use and profit.

The Internet continued to grow and usage increased. This continued to the point the Internet became so clogged that universities and scientists were unable to effectively use the Internet to achieve their research goals. So the NSF (one of the original agencies involved in developing the first Internet) began searching for ways to improve networking capabilities to assist scientists and the academic community.

In 1997, work began on the Next Generation Internet (NGI). The federal government once again supported the effort with funding and coordinated an effort to develop advanced research networks at six agencies. These agencies included the Defense Advanced Research Projects Agency (DARPA), the Energy Department, NASA, the National Institutes of Health, the National Institutes of Standards and Technology, and NSF.

The federal government has three objectives for approving funding for the Internet2 initiative. The first is to connect universities and/or national labs at speeds 100 to 1,000 times faster than the Internet. The second is to provide a test bed for next-generation network technology. The third is to demonstrate how advanced networks will enable future scientific research, national security, distance education, environmental monitoring, and health care. [3]

A goal of NGI is to maximize the potential for fiber-optic cables. Researchers are working to create smarter vehicles to take advantage of the capacity already available and perfecting videoconferencing technology.

The NSF entered into a five-year agreement with MCI to build a national network called the very-high-performance backbone network system (VBNS). [5]

Two years after the inception of VBNS, several universities came together and formed the University Corporation for Advanced Internet Development (UCAID). This group serves as an alliance for the academic researchers, hardware and software vendors, and service providers involved in the project. UCAID first called this initiative Internet2 and the network Abilene. The initiative shares much of its focus with that of the NGI's initiative, and the combined efforts have become known as Internet2 (although UCAID claims to be the "official Internet2").

UCAID worked out a five-year deal with Qwest Communications in which they donated a national OC-48 (2.4 Gbps) network to UCAID called the Abilene network. Cisco Systems Inc. provided the routers and Nortel Networks Corp. provided the switches for the research network.

The VBNS runs an OC-12 (622 Mbps) IP over an asynchronous transfer mode (ATM) backbone. It will run dual technology with one backbone based on ATM and the other IP over synchronous optical network (SONET). [6]

These fast networks are separate from the "public or commercial Internet" and therefore allow researchers to do experiments that large ISPs aren't willing to do on their own networks. Commercial networks are expected to be reliable in a way that the research backbone doesn't have to be. Internet2 also gives vendors and service providers the chance to test and experiment with new technology before putting it on the market. [6]

INTERNET2 TODAY

October 4, 2001, Internet2 was used for the world's largest virtual Internet videoconference event. The conference was scheduled to take place in Austin, Texas, on October 1–3; however, due to travel complications related to the World Trade Center and Pentagon tragedies on September 11th, the conference was to be canceled. Instead of canceling, organizers held the event via videoconferencing from The Ohio State University using Internet2. [6]

In a September 24, 2001, press release, Auburn University announced it is now connected to "the next generation of the Internet," Internet2. In fact, Auburn has been connected for almost a year. [7]

REFERENCES

1. NASA Research and Education Network, NASA. October 9, 2001, at http://www.nren. nasa.gov/about/overview.html.

2. "A History of Internet2." *Internet World,* October 1, 1999.

3. NERO Web pages, Oregon State University. October 9, 2001, at http://www.cs. orst.edu/,pancake/internet2/internet2/html.

4. Kahin, Brian. "Whatever Happened to the NREN?" *Telecommunications* 28, no. 9 (May 1994), p. 28.

5. Dean, Joshua. "Gazing into the Internet's Future." *Government Executive* 33, no. 10 (August 2001), p. 64.

6. "Megaconference III Cancelled in Austin; Goes Virtual from Columbus, Ohio." *PR Newswire,* October 4, 2001.

7. Emmons, Mitch. "Internet2 Ties Research Community Together at High Speed." *Auburn University News,* September 24, 2001.

Clarification
*National
LambdaRail[21]*

National LambdaRail (NLR) is a major initiative of U.S. research universities and private sector technology companies to provide a national scale infrastructure for research and experimentation in networking technologies and applications. NLR aims to catalyze innovative research and development into next generation network technologies, protocols, services and applications. NLR puts the control, the power and the promise of experimental network infrastructure in the hands of our nation's scientists and researchers.

*DWDM Technology
Enables Multiple
Dedicated
Testbeds[22]*

NLR is lighting the first fiber pair with an optical dense wavelength division multiplexing (DWDM) network capable of transmitting 32 or 40 simultaneous light wavelengths ("lambdas" or "waves"). Each of these wavelengths is capable of transmitting 10 Gbps. The unprecedented

[21] http://www.nlr.net/.

[22] http://www.nlr.net/architecture.html.

richness and flexibility of this infrastructure, combined with robust technical support services, will allow multiple concurrent large-scale experiments to be conducted. Network researchers will be able to develop and control their own dedicated testbeds with full visibility and access to underlying switching and transmission fabric.

Case 7-2

Internet Protocol Version 6: *The Future of Internet Communication*

Joe Bryce

The exponential growth of the Internet has outgrown its foundations. The current protocols were not designed for the number of connected devices and technologies being used today. A new standard is being developed to provide for increased node access, as well as better security and data flow and control.

With each passing day, the use of the Internet becomes more and more a part of human activity. From data transfer between companies, such as Wal-Mart and their suppliers, to office staff working out of their home, to students researching papers, the Internet has become an invaluable resource.

The fundamental technology of the Internet was developed in the late 1960s and 1970s and was confined to military and educational use until the mid 1990s, when commercial Internet service providers (ISPs) began offering Internet access to the general public. While Internet technology advances have seemingly become commonplace, with the advent of broadband connectivity to individual residences, satellite and wireless connectivity for mobile computing, and compression technologies for streaming live audio and video, the basic communication protocol for the Internet has remained unchanged and unprepared for the explosion of computers connected to the Internet.

INTERNET ADDRESSING

In order to connect to the Internet, a computer or other device must have an address so that other devices will know where to find it and be able to exchange information with it. This is not unlike a building having a street address so that mail and packages can be sent to an individual or business residing at that location. For the Internet, this address is known as the Internet protocol (IP) address. The current standard for IP addresses was developed in the 1970s using a 32-bit addressing system known as Internet Protocol Version Four (IPv4). This provides 2^{32}, or 4,294,967,296, addresses available for use. It would seem that four billion addresses would be enough to accommodate all users who wish to access the Internet, but as we enter the 21st century, most of these individual IP addresses have been taken. Because of this shortage, the majority of home Internet users, as well as many corporate users, are given a dynamic IP address when they connect to the Internet. They are assigned the IP address by their ISP and hold that IP address for the duration of their Internet session. When the user logs off, the IP address is removed from that user's device and passed on to another user of the ISP when s/he connects to the Internet. This is an alternative to a static IP address, which would require more addresses.

As broadband Internet connections become more accessible and affordable, the number of users with a permanent connection to the Internet will grow, meaning that the use of dynamic IP addresses will not be as effective in allowing Internet access. The last major update to the IPv4 standard came in 1981, and now, some 20 years later, Internet growth and technological innovations have passed the protocol by. The move is now being made toward a new IP addressing standard, known as **Internet Protocol Version Six (IPv6).** IPv6 is an address composed of 128 bits, which allows for 2^{128}, or over 3.4 _ 10^{38}, addresses. Not only does IPv6 allow hundreds of billions of additional computers to access the Internet, the IPv6 protocol solves some existing problems, allows for greater security, and is much easier to configure when compared to IPv4.

The Dotted Quad or Dotted Decimal Notation

IPv4 is a 32-bit address system, broken down into four eight-bit information blocks known as octets. A computer recognizes these eight-bit units as an integer ranging between 0 and 255 ($2^8 = 256$), so when expressed in integer notation, an IPv4 address is represented by four numbers, each between 0 and 255, separated by a decimal, such as 192.168.116.76. This is also known as a dotted quad address because of the four numbers separated by the decimals.

The Colon Hexadecimal

IPv6 addresses will be similar to those used in IPv4; however, with the move to 128-bit addressing, the standard decimal notation used in IPv4 would move from a series of four integer octets between 0 and 255 to a series of combinations of 32 numbers! IPv6 addresses are represented by eight groups of four numerics called hexadecimals (which use a combination of the numbers 0 through 9 and the letters a through f). In this way, the addresses can be compressed into a representation of eight character blocks separated by colons. An example of an IPv6 address would be something along the lines of 0000:0000: 0000:0000:0000:0000:c0a8:0012. One advantage of hexadecimal notation in IPv6 is that a string of four zeros can be replaced by a double colon, and leading zeros can be dropped, in order to save space. Therefore, the above IPv6 address can be rewritten as ::c0a8:12. In order for IPv4 and IPv6 to work together, all IPv4 addresses can be converted to IPv6 simply by using the dotted quad ID. For example, 192.168.1.32 would become ::192.168. 1.32. [2] This will be of the utmost importance in converting from IPv4 to IPv6, as it would be next to impossible to reconfigure domain name servers (DNSs) and all other applications that have been created and configured using IPv4. While the DNSs will have to be upgraded to handle IPv6 addressing, as will the nodes and routers, the nodes and routers will be able to maintain their current address and have it converted to the IPv6 protocol.

IPV6 SECURITY

One of the most common criticisms of the Internet is the lack of security as data pass from device to device. In order for the Internet to continue to grow, it has to be as secure as the entities with which the general public is comfortable today. Many people shy away from online banking or online shopping because they are concerned with the possibility that their information could be stolen. While there are security measures in place today that are employed at different levels to provide secure transactions or encrypt communications, such as the Secure Socket Layer (SSL) and data encryption, IPv6 has security as one of its most basic tenets. IPv6 uses optional header information to provide authentication and other security features.

Security Methods

There are two levels of security built in to IPv6. The first is the *authentication header*. This information provides authentication to a downstream device that the data did indeed originate from the server location that it says it did. The second piece of security information is the *encapsulating security header*. This provides data confidentiality between the origin and the destination devices. [4] At this level, data are encrypted at the source and the encryption key is sent to the destination device in the header of the first data packet sent.

While this information is optional, all IPv6 communications equipment must be able to encode, decode, and transmit the information in the headers. While much of this could be done with IPv4, proponents of IPv6 claim that upgrading IPv4 to IPv6 security levels would cost more than upgrading to IPv6 [4].

Configuration

With the move to the 128-bit addressing system, a faster, easier way to determine the IP address of a computer is required. In many cases today, the address is determined by the network administrator and manually entered into the system. This is difficult enough using today's 32-bit addressing [4] but simple when compared to IPv6. In order to solve this problem at the next level, IPv6 features several forms of autoconfiguration. When plugged into a network, the IPv6 device is able to gather information about itself and the local network it is on and create a basic IP address. It will then send out a short message querying the rest of the network to determine if the IP address it has assigned itself is unique. If so, the device establishes the chosen IP address as its own. [4] If the query receives confirmation of a duplicate IP address on the network, it creates a different IP address, most often by adding a random number to the initial address, and sends a query through the network to determine the uniqueness of this IP address. [4] This process is repeated until the device has confirmation that it has, in fact, assigned itself a unique IP address.

This autoconfiguration will help to address one of the growing problems with the current IP address system, illegal—or rogue—IP addressing. In many instances, a network administrator will create a local area network (LAN) using the TCP/IP protocol for communication between nodes. The administrator may implement the network using IP addresses that are made up. While this is fine for an internal network, once the LAN is connected to the Internet, the results of this rogue addressing can be catastrophic. [4] If the network administrator has chosen internal IP addresses that are already in use on another network, information meant for one device could be rerouted to another device. At the least, routers at network access points (NAPs) must spend time deciphering which is the correct address to pass the information on to—a process that makes for congestion at the NAP.

DATA FLOW

Anycasting

One of the many features of the Internet is that the information is passed from computer to computer by sending data packets over many different routes. The routing of packets through various channels allows for faster exchange of information. If the direct pathway between points A and B becomes too crowded, some of the data can be sent from point A through point C and then on to point B. The information is then collected at the receiving computer and assembled into the correct order for decoding and use. Under the IPv4 protocol, there are 70,000 possible routes for traffic to take over the Internet. This can lead to data being sent on long, circuitous paths to reach its destination. Depending on how traffic is routed, some connections get bogged down, while others remain relatively congestion free. IPv6 offers increased control over the data flowing from device to device. IPv6 offers an advantage in routing as it can determine the best pathway and instruct all data to be passed through those transmission points. This feature is known as *anycasting* and uses what is termed an *anycast address*. [5] Anycasting also allows data to be passed along a specific set of nodes or networks, such as those that belong to one specific ISP. [5] With the anycast method, data can be guaranteed to stay within a network or certain networks for security and/or bandwidth considerations.

Drop Priority

IPv6's *drop priority* feature determines the type of information that is being passed from device to device and assigns it a priority. The highest priority is give to the communications that negotiate the connection between devices and keep the connection active. Priority levels then get lower as bandwidth needs get higher. Low-bandwidth applications such as voice are given higher priority than bandwidth-intensive applications such as streaming audio and video. [4] If the traffic on a channel reaches a point where bandwidth becomes an issue, information is stripped from the traffic with the lowest priority, and therefore the most bandwidth intensive. This allows all users to function with as little interruption as possible.

IMPLEMENTATION AND OUTLOOK

IPv6 is already experiencing sporadic adoption and implementation. The newest releases of open source operating systems support IPv6, and it is available in *Microsoft Windows XP;* however, the Microsoft OS lists it as being prerelease code. Microsoft states that the company cannot be held responsible for issues that occur when using IPv6 in the XP environment, due to the fact that the "software is made available for research, development, and testing only and must never be used in a production environment." [3]

APPLICATION SUPPORT

At the application level, the move to IPv6 is appearing mostly in the open source domain. Currently, there are patches for the most popular open source server applications, such as the Apache Web server and Sendmail e-mail server. The newest versions of the Mozilla Web browser are supposed to include native support for IPv6, as does Sendmail 8.10. Microsoft also claims that Internet Explorer, along with the Windows XP versions of telnet and FTP will support access to IPv6-enabled hosts. [1]

IPV6 IN THE FUTURE

Ipv6 has been under development for almost a decade now; initial proposals were finalized in July 1995. Although the IPv4 network is still functioning well, the need for movement to IPv6 is at hand. The rollout process needs to begin so that there is time for network administrators to become accustomed to the use of IPv6 and the differences between it and IPv4. There will be some level of resistance to the implementation of the new standards, as it is human nature to resist a change in something that we have reached a comfort level with. However, once users become familiar with the added benefits of IPv6, the increased security measures will help usher more people into the world of e-commerce. Its ease of use in configuration, especially for those individuals that do not have the technical knowledge or skills to configure the Internet connectivity for a device, will allow more users and more devices to access the Internet; and the sheer number of available addresses using the 128-bit system will allow for those devices, and many more, to become part of the global information network.

As with the rollout of all new technologies, there will be some bumps on the road, and it is best that these bumps be handled while the current IPv4 network is still functioning well. At some point in the not-too-distant future, the proliferation of Internet-connected devices—computers, cell phones, PDAs, home security systems, and even home appliances—will overwhelm the limits of IPv4. It is best to have the replacement online and functioning before that breaking point is reached, or else the instantaneous data flow that the world's population has become accustomed to may just come to a screeching halt.

REFERENCES

1. IPv6 Information Page, "IPv6 Enabled Applications," http://www.ipv6.org.
2. Jang, Michael. *Mastering Red Hat Linux 9*. Alameda, CA: Sybex, 2003, pp. 616–19.
3. Microsoft, Help and Support Center, Microsoft Windows XP Professional, 2001.

4. Morton, David. "Understanding IPv6." *PC Network Advisor* 83 (May 1997), pp. 17–22. Available at http://www.pcsupportadvisor.com/nasample/c0655.pdf.

5. Starobinski, David, and Ari Trachtenberg. *IPv6*. Boston: Laboratory of Networking and Information Systems, Boston University. Available at http://nislab.bu.edu/sc546/sc441 Spring2003/ipv6/index.html.

Summary

One of the most profound developments in networking has been the Internet. This network of networks has been a major force for change in a variety of ways. It is essential that the basics of the Internet be understood and the essential components be described so that future evolution may be comprehended adequately. We have provided some of the facts such as the WWW, languages, protocols, access points, infrastructure, and the related concepts of intranets and extranets in this chapter.

As more and more uses of the Internet are implemented, we should be cognizant of the underlying technologies and rules for use. When new applications are discovered and deployed, there will be great implications for the future evolution of the Internet. We have already seen the NGI initiative and the Internet2 deployment. The greatest impact may be on how business is conducted. We cover this in Chapter 8, "Internet Applications."

Key Terms

access point, *p. 195*
ARPANET, *p. 194*
backbone network, *p. 195*
browser, *p. 198*
cookie, *p. 200*
domain name, *p. 199, 207*
domain name server (DNS), *p. 199*
domain name system (DNS), *p. 199*
Dynamic Host Configuration Protocol (DHCP), *p. 209*
Extensible Markup Language (XML), *p. 198*
extranet, *p. 210*
hypermedia, *p. 196*
Hypertext, *p. 196*
HyperText Markup Language (HTML), *p. 197*
HyperText Transfer Protocol (HTTP), *p. 198*
Internet, *p. 194*

Internet Explorer® (IE), *p. 198*
Internet protocol (IP), *p. 195*
Internet Protocol Version 6 (IPv6), *p. 216*
Internet service provider (ISP), *p. 200*
Internet2, *p. 196, 209*
intranet, *p. 210*
IP addressing, *p. 208*
IP network, *p. 208*
Java, *p. 198*
JavaScript, *p. 198*
malware, *p. 200*
metropolitan area Ethernet (MAE), *p. 203*
Netscape Navigator®, *p. 198*
network access point (NAP), *p. 203*
Network Address Translation (NAT), *p. 209*
network of networks, *p. 194*

packet switching, *p. 195*
point of presence (POP), *p. 195*
point-to-point protocol (PPP), *p. 203*
protocol, *p. 195*
registration of a domain name, *p. 208*
routing, *p. 204*
search engines, *p. 199*
serial line Internet protocol (SLIP), *p. 203*
Standard Generalized Markup Language (SGML), *p. 197*
tags, *p. 197*
topology, *p. 205*
uniform resource locator (URL), *p. 206*
wireless access point (WAP), *p. 201*
World Wide Web (WWW), *p. 195*

Recommended Readings

Business 2.0 (http://www.business2.com/b2/).

Fortune (http://www.fortune.com/fortune/).

Computerworld (http://www.computerworld.com/), America's number 1 publication for IT leaders.

Management Critical Thinking

M7.1 How should management evaluate the choice among use of a VAN, a private network, or the Internet?

M7.2 What are the major considerations that managers need to assess the cost effectiveness of establishing an Internet presence?

M7.3 What are the management considerations of allowing outside access to the organization via an extranet?

M7.4 What types of organizations can operate without an Internet presence?

Technology Critical Thinking

T7.1 What will be the major advantages of IPv6 over version IPv4?

T7.2 If your firm wants to begin EDI exchange with its major customer, how can you recommend use of a VAN or the Internet from a technological view?

T7.3 What are the technical considerations of allowing outside access to the organization via an extranet?

Discussion Questions

7.1 What technology is required to create an intranet within an organization?

7.2 Who is Marc Andreesen?

7.3 What are the ways you can access the Internet from your community?

7.4 What is a listserv? Do you belong to one?

7.5 What is the likely impact of Internet2 on business?

7.6 What are the major features that the Internet of the future should incorporate?

7.7 With the rapid deployment of broadband in the United States, are SLIP and PPP still a topic of interest?

7.8 In an age of terrorism, should the MAEs have greater redundancy?

7.9 Explain the differences between the Internet and the WWW. What has been the major impact of each?

7.10 Do you have a personal Web page? Where is it hosted? How did you get your domain name? What did it cost?

Projects

7.1 Access the Internet's World Wide Web and search for a topic of interest, such as health care; sports cars; *Sports Illustrated*®; CAD/CAM; travel; museums such as the Louvre; skiing; lodging in Boone, North Carolina; stock prices. Compare the level of information and ease of access to what you might find in your library.

7.2 Get a free trial to one of the online services and search for the same topic and compare with the services of a simple ISP.

7.3 Find an organization with an intranet and determine the problems or issues involved with creating one.

7.4 Look in the preface of this text and send the author(s) e-mail, commenting on the approach of the book.

7.5 Identify and classify the components of five IP addresses.

7.6 Find an organization with an intranet and determine its resources. Describe how this is being well used and what is lacking.

7.7 Divide the class into teams and have each develop an intranet. Compare the results.

7.8 Install *Firefox*® browser (http://www.mozilla.org) and install the applets of your choice. Make a comparison with *Internet Explorer*®.

7.9 Locate and identify the cookies on your hard drive. Can you determine their source? What are the implications?

7.10 (a) Design a Web page on paper; (b) create the Web page from (a) using the tool of your choice. Attempt to create the page within one hour and evaluate your results.

Chapter **Eight**

Internet Applications

In this chapter, we continue the examination of the Internet, intranets, and extranets. The concepts of eBusiness and eCommerce and common applications are outlined. The use of portals has become widespread and so have the various versions. We cover the major concerns of doing Web-based business as well as some current issues surrounding Internet conduct.

The previous chapter investigated the technology of the Internet and its companions. This chapter addresses how this technology is used in applications. We begin by exploring intranets, Internet technology on an organizational LAN. Just as browsers and HTTP protocol can be used on the public Internet, they have great value in adding a standardized graphical interface to LAN access, supporting the distribution of organizational documents.

THE INTRANET–USING INTERNET CAPABILITIES WITHIN THE ORGANIZATION

The Demands of Business Today

Competition has reached a new level of intensity in virtually all industries. Mere survival, let alone success, requires that a business perform at unprecedented levels of effectiveness. The new pressures on business include the following (we discuss this in detail later but provide only a preliminary glance at this point):

- **Shortened product lifecycles.** Time to market is becoming an ever-more-significant factor on the ability to achieve market share, profitability, and even survival as product lifecycles continue to become shorter.

- **Increased cost pressures.** The need to control costs, with the corresponding desire to improve productivity, continues unabated with renewed emphasis on the productivity of the knowledge workers.

- **Increased demand for quality and customer service.** As competition builds, the increase in customers' expectations for responsiveness and personalized support is beginning to change the culture and operation of many industries.

- **Changing markets.** The only constant for business is that things will change. The need and ability to respond to ever-changing market forces continue to push the need to adopt and implement technology to be able to rapidly react.

- **New business models.** Constant change is now pushing into the very core of many corporations with corresponding new business models emerging for the way in which organizations and people work together. These include teleworking, virtual corporations, collaborative product development, and integrated supply chain management.

While each issue requires multifaceted strategies, the common link is the need to enable and expand communications within the organization, between partners, and out into the marketplace. The internal adoption of Internet technology to create the corporate intranet can make significant contributions to each of these critical areas.

In simple terms, *intranet* is the descriptive term used for the private implementation of Internet technologies within a corporate organization; it is the use of Internet technologies, that is, HTML, browser, HTTP, and so forth, on a private LAN. Intranets affect an organization's operation, efficiency, development, and even its culture. To fully understand what is meant by an intranet, we need to look at several areas, namely

- Today's demands on business.
- The Internet and its technologies.

- The Internet versus an intranet.
- The intranet revolution.

Before investigating each area in detail, let's explore a single, simple example that puts the impact of an intranet in context.

Imagine this scenario. Your company has 20 sites and 1,000 people who need timely access to company news, corporate policy changes, human resource procedures—even simple, but crucial, documents such as phone books, product specifications, and pricing information. Normally, you use printed matter such as employee handbooks, price lists, sales guides, and so on. This printed material is both expensive and time-consuming to produce, as well as not contributing directly to the bottom line.

Once created, there is the question of distribution and dissemination. How can you guarantee that all your people have received exactly what they need? How can you be sure they have the latest and correct versions? How can you ensure that they even know that important policy details or other information has changed or is now available? The simple answer is, with printed technology, you can't.

Add to this the problem that, due to the changing nature of any organization in today's frenetic business world, the shelf life of any internal printed matter is reducing so rapidly that, in many cases, it is out-of-date before it reaches the people that need it. Many corporate hours are lost just confirming and verifying the validity of information.

Next, consider the direct cost of preparation, typesetting, production, distribution, and mailing. Add labor costs and overhead and the fact that during any financial year most documents require reprint in ever-increasing frequencies.

For example, a standard price book may cost approximately US$15 each to produce. Add the distribution cost, and multiply this by the number of people who need it, and then by the number of times per year it is produced. We can very easily see the substantial cost that is required to deliver just a single, accurate document to one of our employees to allow him/her to perform the job. But if you also add the hidden cost of the people verifying accuracy and quality of the information, the cost becomes even more astronomical. And this is just one document!

Today's cost-cutting environment demands that you do more for less. But you cannot eliminate these internal communications tools. In fact, we know that increased communication is absolutely essential within companies. Also, we know these increased demands on our busy staff mean they do not have the time to waste chasing down the correct price or product description. In today's competitive business arena, timely access to accurate information is crucial.

The above example assumed 20 sites and 1,000 employees, but in reality this problem is equally important to a single site with 20 people. Accurate, timely communication and information flow are essential in today's world.

Is the Internet actually competitive to the intranet? The answer is obviously a resounding *no!* To put it all in context, the Internet continues to define the technologies available for external communication, whereas an intranet is the application of these technologies within your organization and centered around the corporate LAN.

The problem described above is not new. Attempts to exploit existing computer technologies were implemented with different degrees of success. Implementations usually had built-in gross inefficiencies and expenses—for example, mail that resulted in the unnecessary stuffing of employee mailboxes or client-server databases that put an inordinate and expensive load on the MIS teams who end up being responsible for the maintenance and update of the information.

The solution to the problem requires technology that

- Can deliver information on demand-as needed.
- Can guarantee the information is the latest and most accurate available.
- Ensures information can be held at a single source (although there is no need for that source to be the source of all information).

- Allows information to be maintained by the people who would normally maintain and prepare the original information.

The solution to this problem is provided by just one of the technologies available under the generic heading of the Internet. Different problems require different solutions, and the use of the full spectrum of Internet technologies within an organization will generate one of the biggest corporate IT revolutions.

The Internet and Intranets

In creating an intranet, there are several main reasons why the Internet technologies have such a dramatic impact on the scope of business networking applications. These include

- **Universal communication.** Any individual and/or department on an intranet can interact with any other individual/department and beyond to partners and markets.
- **Performance.** On an inherently high-bandwidth network, the ability to handle audio clips and visual images increases the level and effectiveness of communication.
- **Reliability.** Internet technology is proven, highly robust, and reliable.
- **Cost.** Compared with proprietary networking environments, Internet technology costs are surprisingly low.
- **Standards.** The adoption of standard protocols and APIs such as MIME, Windows Sockets, TCP/IP, FTP, and HTML delivers a fast-track series of tools that allow infrastructures to be built, restructured, and enhanced to meet changing business needs as well as allowing standards-based intercommunication between external partners, agencies, and potential customers.

Multipurpose Internet Mail Extensions (MIME) is an Internet Standard for the format of e-mail.

The Intranet Revolution

Internet technologies are actually extremely well suited for developing internal corporate information systems: intranets. Use of the corporate intranet can contribute to gaining and maintaining a competitive advantage. All employees may have updated and current information on a near-real-time basis. Some of the applications of this technology follow:

- **Publication of corporate documents.** Along with oft-mentioned human resource guides, these documents can include newsletters, annual reports, maps, company facilities, price lists, product information literature, and any document that is of value within the corporate entity. This is one area where significant cost control can be achieved as well as much more efficient, timely, and accurate communication across the entire corporate organization.
- **Access into searchable directories.** Rapid access can be provided to corporate phone books and the like. These data can be mirrored at a Web site or, via scripts, the Web server can serve as a **gateway** to back-end preexisting or new applications. This means that, using the same standard access mechanisms, information can be made more widely available in a simpler manner. Electronic documents tend to be living documents with continuous updating whereas printed documents are updated only periodically, and thus become out-of-date.
- **Corporate/department/individual pages.** Internet technology provides the ideal medium to communicate current information to the department or individual. Powerful search engines provide the means for people to find the group or individual who has the answers to the continuous questions that arise in the normal course of doing business.
- **Simple groupware applications.** With HTML forms support, sites can provide sign-up sheets, surveys, and simple scheduling. As intranet technologies continue to evolve, the press has been treating the technologies as alternatives to major **groupware** applications

(e.g., Lotus Notes®) to the point that confusion has been caused as to the appropriateness of each area of technology. The intranet technology can be used to complement or serve as an alternative to groupware products. It is a matter of scale, cost, time scale, openness, and taste.

- **Software distribution.** Internal administrators can use the intranet to deliver software and updates on demand to users at a far lower cost than using physical media.
- **Mail/e-mail.** This was, perhaps, the killer application for the Internet. With the move to intranet mail products, with standard and simple methods for attachment of documents, sound, video, and other multimedia between individuals, e-mail has become the communications method. Mail is essentially individual-to-individual, or individual-to-small-group, communication. With the emergence of Web technology, there are now better and more appropriate tools for one-to-many communication, which historically is where mail systems have been overburdened and overburdening to the point of reducing their effectiveness.
- **User interface.** With HTML, an end user comfortable interface can be built, only limited by the creator's imagination. The beauty of using intranet technologies for this is that it is so simple.

The Paper Reduction Act minimizes the paperwork burden for individuals; small businesses; educational and nonprofit institutions; federal contractors; state, local, and tribal governments; and other persons resulting from the collection of information by or for the federal government Intranets do this nicely.

Intranet technologies provide the tools, standards, and new approaches for meeting the problems of today's business world. The beauty of most of these technologies is that they are simple and, in their simple elegance, phenomenal power can be unleashed. Communication is the key to business success. Exploitation of intranet is the key to effective and efficient communications.

E-BUSINESS AND E-COMMERCE

Electronic commerce (e-commerce) is a dynamic set of technologies, applications, and business processes that link enterprises, consumers, and communities through electronic transactions and the electronic and physical exchange of goods, services, information, and capital. It is the exploitation of IT to deliver services and conduct business. E-commerce improves commerce through the use of many core technology tools: the Web, electronic data interchange, electronic mail, electronic funds transfer, electronic benefits transfer, electronic catalogs, credit cards, smart cards, and other techniques. Therefore, e-commerce is not just about using the Web as a storefront. It involves shortening the supply chain, streamlining distribution processes, improving product delivery, reducing inventory-carrying costs, and many other measurable activities. E-commerce strategies allow businesses to leverage electronic alliances to speed the delivery of products and services to market.

IBM's practice has evolved to where e-commerce is different from **electronic business (e-business).** E-commerce involves buying and selling goods and services **online,** usually in the form of **business-to-consumer (B2C)** or **business-to-business (B2B).** This is opposed to the much larger concept of e-business, which is conducting business electronically, everything from sending e-mail to advertising on the Web or creating an intranet so that your HR department can post online policies and procedures manuals. So, to IBM, e-commerce is a subset of e-business.

Business-to-Consumer (B2C)

The part of electronic commerce that was at first perceived to be the focus of action was business-to-consumer, the retail point of contact. Here was a chance for even small firms to establish a virtual storefront equal to those of the largest corporations. Another great advantage was the idea of *disintermediation* that could occur as the dot-coms often had no inventory, no warehouses, no wholesalers, but used drop-shipment from the manufacturer directly to customers. Thus, the B2C firm could operate with a fraction of the costs of a traditional store. Benefits also include increased sales and revenues, expansion to global markets, as well as finding new sources of revenues.

B2C has become a "normal" part of marketing and selling. Consider any offer advertised on television, from vacuum cleaners to stock trading, and you will find an 800 number *and* a Web site to make the purchase. The value of the B2C Web site is that the potential buyer can get additional information without revealing his/her presence; talking on the telephone means you are visible to someone *and* they will likely try to sell you additional products.

A significant part of B2C is for the consumer to find the product of choice at the price of choice. Search engines and comparative shopping Web sites (pricewatch.com, priceline.com) assist here. Discount sales outlets such as www.overstock.com and discount sites such as geeks.com bring shoppers for the bargains. Industries consolidate at Web sites such as Travelocity.com and Orbitz.com to allow the consumer to shop across companies. Some comparative shopping sites send out e-mails with the buy of the day (http://www.reseller-ratings.com/), further bringing the shoppers to the sites. With such services, consumers soon learn that first they should search online and then shop online.

http://www.thegeeknext door.com was started by two graduate students while in school. They shared one-half million dollars in net income their second year.

Business-to-Business (B2B)

Frequently, businesses are able to reengineer their supply chains and establish stronger partnerships through B2B applications. B2B applications range over a wide variety of areas. For example, there are uses that connect to all parts of a firm's value system such as suppliers, transportation, sales and marketing, service, and so forth. Cisco Systems is often cited as an example of success using B2B. Cisco uses its extranet for online sales and electronic purchasing, resulting in several economies. Their customers find that they can get access to prices, configuration suggestions, order status, invoice checking, and technical support. Cisco saves in marketing expenses, administrative costs, and lower customer service costs. In addition, they achieve improved customer service and happier customers, all contributing to increased sales.

The frenzy of enthusiasm for B2B has caused vendors to recast ERP as "extended ERP," which helps the firm transform to an e-business by providing enterprisewide planning for B2B. This effort is aimed at exploiting the Internet for core business functions. Most firms attempt to measure the benefits of e-business. The areas that they track are customer service, knowledge of customer preferences, marketplace presence, brand recognition, supply-chain efficiency, and cycle times with their supply-chain partners.

Clarification

Enterprise resource planning systems (ERPs) are management information systems that integrate and automate many of the business practices associated with the operations or production aspects of a company. ERP systems typically handle the manufacturing, logistics, distribution, inventory, shipping, invoicing, and accounting for a company. Enterprise resource planning or ERP software can aid in the control of many business activities, such as sales, delivery, billing, production, inventory management, and human resources management.[1]

B2B would seem to mean company A placing an order with company B via the Internet. A variation of this is reverse auctions. Organizations wish to purchase goods at the lowest price. Consider a large company that uses tens of millions of pounds of starch, a commodity, each year; saving a penny a pound means a lot of money. Reverse auctions place the specification on the Web site, as would have been done with a request-for-quote. Instead of sealed bids, the bids are displayed as the auction progresses, allowing all players to reconsider their last bid. As a manager in such an auction company said, "in this forum, there is no money left on the table."

[1] http://en.wikipedia.org/wiki/Erp.

Consumer-to-Consumer (C2C)

The global connectivity of the Internet and the interface of the WWW support transactions directly between consumers (**consumer-to-consumer** or **C2C**). For example, the online auction *eBay*® is a way for the owner of an item to advertise it directly to prospective buyers, without a retail or commercial intermediary. eBay[2] is the world's largest trading community where millions of people buy and sell millions of items every day. For buyers, this means finding great deals on all kinds of items; for sellers, it is the ability to market your product to millions of daily visitors. This makes using eBay one of the most efficient ways to sell just about anything.

Implication   Before you buy anything, search eBay.com to see what the going price is.

Government-to-Customers (G2C or E-government)

> E-government reduces cost and speeds up services; reduces travel and pollution; employees can work off-shift; could be done off-shore; should increase citizen satisfaction.[3]

The public expects service from the government and that it uses the Internet more than ever before. Polling data from the Pew Foundation, for example, show that over 40 million Americans went online to look at federal, state, and local government policies, and over 20 million used the Internet to send their views to governments about those policies. This and similar data show that if the U.S. government can harness the power of technology, it will be meeting expectations of an increasingly wired citizenry.

In his February 2002 budget submission to Congress, U.S. President Bush outlined a management agenda for making government more focused on citizens and results, which includes expanding **electronic government,** or **e-government.** E-government uses improved Internet-based technology to make it easy for citizens and businesses to interact with the government, save taxpayer dollars, and streamline citizen-to-government communications. E-government does not mean putting scores of government forms on the Internet. It is about using technology to its fullest to provide services and information that are centered around citizen groups.

> Our success depends on agencies working as a team across traditional boundaries to better serve the American people, focusing on citizens rather than individual agency needs . . . I thank agencies who have actively engaged in cross-agency teamwork, using e-government to create more cost-effective and efficient ways to serve citizens, and I urge others to follow their lead.

The president's e-government strategy identified several high-payoff, governmentwide initiatives to integrate agency operations and information technology investments. The goal of these initiatives is to eliminate redundant systems and significantly improve the government's quality of customer service for citizens and businesses.

Implication   E-government goes hand-in-hand with the U.S. federal Paperwork Reduction Act; what better way to reduce paper than to move to electronic access, electronic storage, and electronic processing?

An example of how one state approached e-government is shown in Box 8.1.

Mobile Commerce

Mobile commerce (m-commerce) is often referred to as the next generation of e-commerce. It includes the selling of goods and services through wireless devices such as

---

[2] http://www.ebay.com.
[3] http://www.whitehouse.gov/omb/egov/about_backgrnd.htm.

Alabama e-government services are divided into functional categories:

- Citizen—Find Citizen Services including Jobs, Vital Records and Certificates, Taxes, Motor Vehicles and more.
- Business—Find Services and Resources for doing business in and with the state.
- Education—Find Services and Educational Resources for parents, teachers and administrators.
- State Employee—Find Services and Resources of interest to State Employees.
- Online Services—Find Services you can use online, from downloads to database searches, to e-gov transactions.

Source: http://www.state.al.us/.

personal digital assistants (PDAs), laptop computers, cellular phones, and any other hand-held wireless device. The concept has taken hold in Europe where cellular phones and wireless PDAs have gained widespread popularity among Europeans of all ages. (See http://www.bluetooth.com.)

Retail establishment such as *Starbucks* and *Panera Bread* provide free wireless (802.11b) hotspots in their stores. This draws people in because (a) they need access and (b) they can be productive while they eat or have coffee. These are prime candidates for mobile commerce; they are on the move, they have time to surf, and they have discretionary money.

Implication

From a base of almost 900 million at the end of 2001, worldwide cellular subscriber numbers are forecast to rise to 1.9 billion by the end of 2006.[4] Despite a challenging environment in the handset market caused by low replacement rates, global subscriber growth prospects remain buoyant. Meanwhile, the personal digital assistant market continues to expand and these two technologies are seen as natural complements, occupying the same space. The wireless "appliance" of the future, resembling the crew's brooch on the Federation Starship *Enterprise*, will mean that billions will be wirelessly connected and, therefore, potential m-commerce users.

A futuristic feature of mobile commerce is on sale now. The consumer has a wireless device, for example, cell phone, PDA, or *Star Trek* brooch, and has registered with a number of stores online. As s/he walks through the mall, one of the stores detects the consumer's presence, looks up his/her preferences, sees that s/he is a good customer, checks out his/her wish list, and immediately notifies the customer that the shirt s/he has on the list is on 10 percent sale, even though it is not so advertised. This is catering to your best customers with mobile commerce; this is customer relationship management.

Clarification

Customer relationship management (CRM) is a philosophy that says that the only form of competitive advantage is to place the customer at the center of all systems. It subscribes to the philosophy that a customer will remain loyal by service, not by price alone.

Business on the Internet (Web Services)

The explosive growth of the Internet has been fueled, in part, by a rush by companies to develop a presence there. Although many articles have been written about business on the World Wide Web, it was difficult to assess in the early days. Within a year of being allowed

[4] http://www.3gnewsroom.com/3g_news/jan_02/news_1796.shtml.

to add commerce sites on the Internet, many thousands of companies had been making money online. There appear to be four distinct revenue models for doing business on the Web.[5] Examples of each follow:

1. Direct selling or marketing a firm's existing products or services.
2. Selling advertising space.
3. Charging fees for the actual content accessible on a WWW site.
4. Charging fees for online transactions or links, for example, service.

Implication

Commercial use of the Internet was first allowed in 1994.

In its simplest definition, the Internet is connectivity among millions of nodes. This does not, however, define how it is used; software defines how it is used. An example of this is, of course, the graphical interface of the World Wide Web. Another equally intriguing use is for telephone communications via Internet Phone® and MSNetworking®. These applications, and others, use the multimedia sound card on computers to carry on long-distance phone conversations, at the price of Internet connectivity alone. Some foresee this as the pending doom of the long-distance providers. This is unlikely as the applications now stand. However, it is interesting to see that the addition of software on each end to the sound cards in the machines makes it a telephone circuit.

Direct Selling

Before the Internet, in general, and the Web, in particular, telemarketing and e-sales meant use of the telephone. With the installation of a Web site (perhaps a Web storefront), anyone can have exposure to the global community. Direct selling on the Web means the buyer is dealing directly with the seller for the purchase of goods or services and the seller could be a standard retail outlet, a wholesaler, or an individual. In the early days of Web marketing, small companies appear to have had more impact than large firms with comprehensive channels of distribution. Several small firms found that the Web catapulted them into global distributors overnight while, in contrast, larger firms found that the Web provided them with new niche market channels. Some large corporations such as *Holiday Inn* found the Internet to be valuable in increasing reservations. Firms and services like Holiday Inn and Orbitz were offering both free information as to services and the ability to reserve (i.e., buy) services at the same site.

Implication

All sales transactions were, in days gone by, person-to-person. Then the postal system was used in combination with catalogs, starting the mail-order business. Some stores instituted use of the telephone to call in catalog orders; this was the beginning of electronic commerce. The next step, the one that has made e-commerce business so possible, was the creation of the information-based financial system, that is, widely accepted credit (cards). Still, with catalogs and telephones (TV was next), consumers could call in an order, pay with credit (information), and have the item shipped. The final step has been made possible by the Internet; Web sites supplement the telephone, credit is used, business increases. No company exemplifies this better than *Dell Computer*. Without storefronts, Dell has become the leader in computer sales due to telephone and Web-based e-commerce.

Part of the Internet sales scheme involves informing the buyer of the transaction progress via e-mail and using e-mail to inform him/her of the buy of the week. One of the authors receives advertising e-mails each week from a shoe company, a book company, and three computer sales companies, to name just a few. These e-mails are generally in full HTML graphic form, giving the look and functionality of a Web page inside the e-mail client. They are to inform and to sell.

[5] Kate Maddox, Mitch Wagner, and Clinton Wilder, "Making Money on the Web," *Information Week*, September 4, 1995, pp. 30–40.

In markets where wholesalers and brokers are common, direct sales on the Internet have drastically changed the marketing chain. Additionally, the store is open 24 hours per day, accessible from anywhere on earth. Thus, electronic commerce is more than a different form of marketing and selling; it is a change in the players and the roles played.

Direct sales complement in-store sales. Circuit City sends e-mails to customers advertising merchandise. The items can be ordered online for shipping, ordered online for in-store pickup, or bought in-store. The Web site and stores don't complete; they complement and supplement.

Selling Advertising Space

Consumers of the "television generation" are accustomed to receiving entertainment that is funded by advertisement. Public TV is the exception. With advertisement-funded programming, most of what we pay to the cable or satellite companies is for transport. When we wish to be free of advertisement, premium channels or pay-per-view are options. In general, we accept the fact that advertisements fund much of our entertainment.

Many Web sites fund their activities by selling advertising space. Go to http://moneycentral.msn.com/ for financial information and you will see several advertisements. The information is "free," funded by the number of "eyeballs" on the page. The extension of this is the inclusion of pop-up ads on Web sites, those little windows that open up when we enter or exit the main site. Some use the pop-up for funding the basic services; some press this funding mode to an extreme as a dozen ads appear before the user can get off the site.

Charging for Content

Organizations that historically sold content on printed media are opting to use the Internet to extend this same service. A Web page is a natural mode for getting subscriptions to printed or electronic media. Internet access is also a natural for delivering content earlier than the printed copy would arrive, even organized in a personal format. These same sources may not only deliver images of the printed matter, for example, the *New York Times,* but send e-mails of breaking news. Television news channels also provide access to material and even send e-mails for breaking news, but they rely on advertising to fund the environment. Newspapers and magazines rely on charging for content when the content is of significant substance.

A different form of charging for content is the adult material industry. Historically, a few publishers provided printed material; now hundreds of entrepreneurs provide electronic delivery of this material as Web sites, for a fee. Such Web sites cost significantly less than the publication of a magazine and can reach a much greater, global market.

Charging for Service

Many services on the Internet are "free" because there is a hidden revenue stream that supports them. Pages where advertising appears was noted previously. *Yahoo* offers free accounts, e-mail, and disk space, all of which is paid for by advertising. If, however, additional disk space is desired, there may be a charge. On *Google,* the *Adwords* system is a great way to be selected in the search engine but at a cost. Although you may receive Web hosting free at your ISP, within the limits of the supplied disk space, the ISP service itself is a cost to you and, obviously, a revenue stream for them. Some financial institutions charge for Internet access to your accounts; some don't. Another form of charging for service would be the interactive adult sites where the live material is provided on a pay-as-you-go basis. Similarity, some companies use chat rooms on a fee-basis for their help desk. Because of the large number of potential customers, charging for content or service, even at very low rates, has a very high financial potential.

Telephone Service on the Internet

Because the Internet is "free" (which generally means it has a fixed cost for access and few variable costs for use) and because many organizations have their own IP networks, an obvious use is for voice telephone, that is, Voice over IP (VoIP). The Cisco Systems corporation, consisting of some 30,000 people with many offices around the globe, utilizes VoIP for all internal voice needs. This gives greater control and avoids significant long-distance charges. Companies such as *Vonage®, Verizon®,* and *SunRocket®* offer VoIP over your broadband connection, for a fee. *Skype* (http://www.skype.com) does the same for free if you are only connecting to an Internet-connected computer; connecting to other wired telephones requires a subscription fee.

Clarification

Vonage is a revolutionary new phone service that uses your broadband Internet connection. If you have a broadband Internet connection, Vonage can relay your voice over the Internet using a regular phone and a phone adapter. This is a technology known as VoIP. All you need is a broadband connection (such as DSL or cable modem) and a regular corded or cordless touch-tone phone. We provide a free phone adapter when you sign up for service through Vonage.[6]

Web Sites in Support of Customer Service

An aspect of enhanced onsite service has been employed by firms such as *Federal Express, UPS,* and even the *U.S. Postal Service.* They, for example, allow customers to track packages directly through their Web pages (http://www.fedex.com, http://www.ups.com, http://www.usps.com). Firms that use the delivery services provide a link from your e-mailed confirmation of shipping to the carrier, so you only have to click on the link or go to the shipping company via your browser and input the provided number. In a different vein, Sun Microsystems Inc. claims to have saved several millions of dollars by providing downloadable software program corrections (patches) and product literature via the Net. Microsoft will even install a program on your computer that either reminds you to go looking for updates or does it for you automatically. Many, if not most, mature technology providers post answers to frequently asked questions (FAQs) to give instant service while saving the company the cost of a help-desk person. These services are extremely valuable for customers who have nontraditional schedules and customers across time zones.

There are several expansions of electronic commerce that are underway that promise a means of significant financial infrastructure via the Web. Many banks, for example, support not only access to your account, but information on and assistance with other services. Many other services are being offered that seem to be only limited by the imagination of providers, as many have found by accessing the wrong site. In Denver, Colorado, Bronco fans can find details about schedules, players, and items for purchase, and even can converse with individual players via e-mail. Fans of the *Dilbert®, BC®,* and *Frank and Ernest®* cartoons can receive a cartoon each day and perhaps a newsletter each month.

Electronic Payment Makes Electronic Commerce Work

The fast, electronic transfer of funds has increased the speed of commerce. It started with the **electronic funds transfer (EFT)** capability of the banking industry. This is a very secure information system between banks and the Federal Clearing House to move funds instead of money. The result is immediate transfer with no financial float. Next, **electronic data interchange (EDI)** supported the transfer of standard business documents between computers as data files. One portion of this development was instructions to banks that resulted in EFT funds movement. With the advent of retail e-commerce, a form of EFT/EDI was required. The result was PayPal and digital cash.

---

[6] http:www.vonage.com.

PayPal

PayPal® and its competitors use the established credit card or bank check systems as their infrastructure. When the payee has established an account, all that is required to pay for a transaction, such as an auction purchase on *eBay®*, is to go to PayPal's Web site and provide the e-mail address of the payee and payer and the amount. PayPal then moves the money from account to account and sends both parties an e-mail of the transaction. This transfer takes place within an hour and requires no postage or physical instrument. There is a cost to the recipient, though sometimes significant.

Implication   Not all consider PayPal a wonderful idea. Go to http://www.chat11.com/Problems_With_Paypal to see some comments, which seem to center on PayPal's limited liability and control of funds in their control.

E-cash or Digital Cash

A facilitator of the mobile and electronic commerce revolution is **electronic cash (e-cash).** E-cash allows individuals to make payment for purchases online through their checking account. E-cash differs from credit card and check payments online by allowing users to make small purchases, typically five dollars or less. E-cash or **digital cash** will enable users to make micro payments for goods such as colas or snacks from vending machines or merchants using their PDAs or cellular phones.

Security of Transactions[7]

Consumers and businesses are concerned about doing business on the Internet; they fear for confidentiality and security of information that is collected and transmitted during an online transaction. If the data transferred are not secure, the user could suffer financial losses, even identity theft. To counter this threat, companies provide assurance though *web seal programs* that they can trust the site being used.

Web Seals

Trust is a critical component of the virtual transaction. Various web site seal programs have evolved to address such issues as privacy, security, and business practices. These programs are designed to increase Web user trust and confidence in the virtual vendor and in the virtual transaction. While the subject of security is presented in Chapter 11, the following are some of the companies providing security in transactions on the Internet. *Verisign* may be the most recognized **web seal,** but they are not the only one.

- **BBBonline:** A Better Business Bureau program (http://www.bbbonline.org/), BBBOn-Line's mission is to promote trust and confidence on the Internet through the BBBOn-Line Reliability and Privacy Seal Programs. BBBOnLine's Web site seal programs allow companies with Web sites to display the seals once they have been evaluated and confirmed to meet the program requirements.
- **BizRate.com**
- **CPA WebTrust**
- **AICPA program**
- **TruSecure**
- **TRUSTe®** (http://www.truste.org/) is an independent nonprofit that enables trust based on privacy for personal information on the Internet. They certify and monitor Web site privacy and e-mail policies, monitor practices, and resolve thousands of consumer privacy problems every year.

---

[7] http://ecommerce.etsu.edu/Web_seals.htm.

- **Verisign** (http://www.verisign.com/)
- **Payment Service Programs**
- **Escrow.com**
- **TradeSafe**

Do We Need Secure E-mail?

With three trillion e-mail messages per year, the question arises as to the need to secure these messages. Organizations wonder if they need to use encryption for this medium as they increase their use of it for official communications and even product ordering. Many believe that the medium is inherently secure due to the large volume and, thus, low probability of interception. Yahoo believes secure e-mail has value and provides the service; Netscape, meanwhile, developed a standard of its own.

Cost of Being on the Internet/Web

For low-cost Web hosting, see www.godaddy.com . . . as low as $3.95/month.

What does it cost to get Web presence (a home page)? In financial terms, it may be free, as many ISPs include space for hosting, but the owner must create the site. An advertisement on TV shows that you can get a turnkey system for as little as $59.95 in the form of a kit offered by *Internet Treasure Chest*®. Business may pay as high as $100,000 when graphic designers and online databases are involved. In business terms, the cost of not having a Web presence may be much higher. IBM's general manager for Internet Application Services has said, "Not being present on the World Wide Web will soon be the equivalent to not having a fax machine. In the not-too-distant future, not doing business on the World Wide Web will be equivalent to not doing business at all."

One of the hurdles that organizations face is that there is too much information for the searcher. If looking for generalized information, then a search engine may suffice. If, however, the user is desiring information about a specific organization and its services, the placement of that information in an easy-to-access form is a challenge. One way to achieve this is via a *portal,* an access point to all of the information of an organization, in an organized manner.

Weblogs

A *weblog,* or *blog,* is a Web-based location to tell the world what you think or place publications consisting primarily of periodic articles. Blogs range in scope from individual diaries to arms of political campaigns, media programs, and corporations. They range in scale from the writings of one occasional author to the collaboration of a large community of writers. Many weblogs enable visitors to leave public comments, which can lead to a community of readers centered around the blog; others are noninteractive. For the history of blogs, see http://en.wikipedia.org/wiki/Blogs.

PORTALS[8]

While the term *portal* itself is not new, it is not clearly defined within the marketplace. TandemSeven defines a **portal** as a Web application that uses a common interface and technology to tie together multiple related, but independent, content pieces or transactional applications through **portlets.**

1. **Portlet.** The means to access content or transactional applications though a user interface (UI). Different portal packages call these different things: portlets, gadgets, iViews, and so forth.

[8] http://www.tandemseven.com/SolutionFocus_BusPor.html.

2. **Common interface.** This is made of two components: the UI standards that go into the design and development of the portlets and the technology used to display and organize portlets within a Web application.
3. **Related.** The applications are related from an end-user's perspective. When users go to the site, they are able to do everything they need to do in order to accomplish the business goals for which they are responsible.
4. **Independent.** The portlets cross internal and external organizational boundaries.

One may encounter labels such as "customer portals, B2B portals, enterprise application portals, knowledge portals," based on their primary orientation. Regardless of label, portals serve as a way to consolidate information access from a single Web site. The labels provide a primary orientation indication.

Business Portals

Business portals are a cost-effective means of communicating information and transacting business with customers, suppliers, partners, and employees. Through a single point of entry, portals enable users to access the critical business systems and information that reside inside and outside of the company, fostering a closer, stronger relationship.

Customer Portals

A company's internal business processes should be transparent to customers. Portals are a good way to make it easier for customers to do business, without having to drastically change the way the company does business. Portals have the ability to hide complex internal processes by creating a single point of entry and a consistent user interface for your customers.

Implication   *Compass Bank of Alabama* has a service called *MyCompass.com*. This is a portal, or single point of contact, for all of your bank(s), credit card(s), and investment(s) information for which you have online access. The account is free; you create it and then enter the name of the institution in which you have funds, your userid, and password. When you have entered all the institutions, for example, several banks, some credit cards, and multiple stock accounts, the system queries each one when you go to Dashboard. This happens very quickly. It displays all of your stocks on the left side, summed from multiple accounts, along with credit card summaries. On the right side, it shows bank summaries and then more detailed information from the sources. After you have set up the system, you can see *all* of your financial assets in one place, summed to show your net worth. Of significant value is that the due date for credit card payments in listed by the amount due, which is the updated amount of charges. Also shown is the activity for each account you have, for example, checks that have cleared, stock dividends, and so on. To look at the system to sign up for it, go to http:www.compassweb.com and click on mycompass on the right side, third block down. Again, it's free to anyone; a Compass Bank account is not required.

Business-to-Business Portals

When customers are companies that have unique and complicated preexisting relationships, it is necessary to offer them richer functionality than is typically provided in a customer portal. Much of the functionality that would appear within a customer portal also would appear in a business-to-business portal, for example, order tracking, account history, inventory information, maintenance information, scheduling, and so forth. However, these more complex customers need to see this information in their own companies' terms—using their accounting codes, billing information, routing information, and corporate lingo. A portal that could accomplish that for them, rather than forcing them to translate or interpret data, would be of great benefit to them and provide a competitive advantage.

A business-to-business portal that cuts across two organizations' business processes provides increased value. For those companies that share data feeds, a portal can provide portlets that perform related tasks across the companies (for example, a portlet might triple-check receiving to make sure that the items ordered were the same as the items delivered and the items billed).

Enterprise Application Portals

Many companies have probably spent the last decade implementing dozens of applications, groupware applications, and Internet applications. While enterprise applications have increased productivity, they also have probably led to information overload. Many employees have difficulties wading through, or even finding, all of the applications that have been developed for them. An enterprise application portal can play a central role in helping employees (or suppliers or partners) access a wide range of applications.

The enterprise portals can dramatically improve the usability and users' experience of enterprise applications. A common, familiar user interface can be developed at the portal level and applied across enterprise applications. User roles can be developed and implemented to personalize access to information across many applications. By enabling users to access information from enterprise applications, groupware applications, and Internet applications, employees, suppliers, and partners can be propelled to a new level of collaboration.

Several uses of business applications employing portals have been reported. Dell has been highly successful in use of its portal for customers, suppliers, and sales coordination. Cisco Systems has employed a knowledge portal for its service and support advocacy group in order to provide a single point of contact for large clients. The portal provides service and support managers with immediate access to needed resources and with orientation and reference information. The portal was well received and now Cisco is moving to incorporate the concept to other areas of the organization.

The portal concept also is employed by governments and for children. FirstGov.gov is the U.S. government's official Web portal for all government transactions, services, and information. ALFY.com is a Web portal for children, providing for free a fun, entertaining, and educational mega-site for children worldwide, with lots of interesting games, free e-mail, music, stories, and more.

Implication

Chick-Fil-A (CFA) has a portal that all franchisees and corporate store managers can use. This provides a standard interface for everyone at CFA. This portal provides e-mail and a common inventory ordering system for each store. The owner orders the 285 items of inventory him/herself via the portal; a process that used to take hours each week now can be accomplished in two 10-minute sessions per week. This corporate site has access to the corporate information and documents necessary to successfully run a CFA store and has forums for franchisees to offer helpful tips to each other.

One of the most valuable features of the CFA portal is the human resources area. This area provides guidelines for screening applicants and hiring employees. Instead of having to educate each new employee on all company policies, employees can go through browser-based tutorials, saving managerial time when a new employee is hired. The HR area allows employees to make changes to benefits without management intervention.

SPECIAL CONSIDERATION OF THE INTERNET AND INTRANETS

Security will be covered in Chapter 11; much of it relates to the Internet or Internet-related processes. Governments and users alike have concerns about Internet access, or lack thereof. Taxation and voting are two significant topics, ones that will be receiving high exposure in the very near future.

Digital Divide

A concern of the U.S. government is how to close the gap in the **digital divide,** a common phrase to describe the separation between those with access to the Internet and those without. Within the United States, the digital divide is not as serious as it appears at first glance. It appears that the digital divide is primarily income based. As the cost of computers and Internet access continues to decrease, this gap will get smaller. Efforts already have been successful in placing computers with Internet access in schools and libraries so that the young may learn essential computer skills. It is reported that 90 percent of the U.S. population have access to the Internet through these resources. Because of the advantages of Internet access, this will continue to be an issue, an issue of have and have-not. In light of this, the U.S. **Federal Communications Commission (FCC)** is considering including broadband under the universal access program. Until recently, universal access supported same-pricing for wired telephone service regardless of location of the customer. Adding broadband access to the program would mean high-speed Internet access would be available to more people who do not live within an access area now.

The Impact of the Internet in Underdeveloped Countries

During the time since the introduction of the WWW and its browsers, the world has seen not only industrialized nations take advantage of this capability but underdeveloped and developing countries gain global access and presence. Just as a small 250-watt AM radio station can have a global presence on the Internet, developing countries can advertise to and communicate with the world. A negative aspect of this global visibility is that it may bring commerce that depends on nonexistent transportation services or may introduce ideas and practices not accepted by that country. Internet access by developing nations is a way to speed up the process only if they choose to provide the infrastructure. At the time of the September 11, 2001, attack on the World Trade Center in the United States, there was only one ISP in Afghanistan, a country that had less than one television set per 1,000 citizens.

The Internet is proving to be a boon for nations that have large land areas but relatively underdeveloped infrastructures. Thus, we expect to see large impacts in countries such as China and Brazil. The impact in Brazil is indicated by Figure 8.1 and the fact that the country is expending so much to provide connectivity.

Tax on Internet Sales

The consideration and effect of state sales taxes in general and a federally imposed Internet tax in particular are of great concern in the United States. The concern is the value of the revenue generation from the tax as opposed to the inhibiting total sales effect of that tax. Historically, one state in the United States cannot tax the goods sold in another state. This was to inhibit the creation of barriers to trade but was based on physical transactions. When mail order became the first remote purchasing, followed by the first electronic sales via telephone orders, the tax assessors and collectors saw that they were losing tax revenue. Considering that some states in the United States of America face significant deficits due to shortfalls on income or overextending services, this is an issue that will not disappear.

Voting via the Internet

The U.S. presidential election of 2000 raised the question as to whether voting over the Internet was feasible. The answer is that it is technically feasible, but it is not (yet) politically feasible. What this means is that we have the technology to receive the vote of an individual from anywhere in the world over the Internet, but the validity of that vote cannot yet be guaranteed. So, what does it take to cast the vote and what does it take to guarantee its validity?

FIGURE 8.1 **Brazil's National Research Network**
With more computers on the Internet than in all the rest of Latin America, Brazil is adding to the
infrastructure to woo users, despite high access cost.

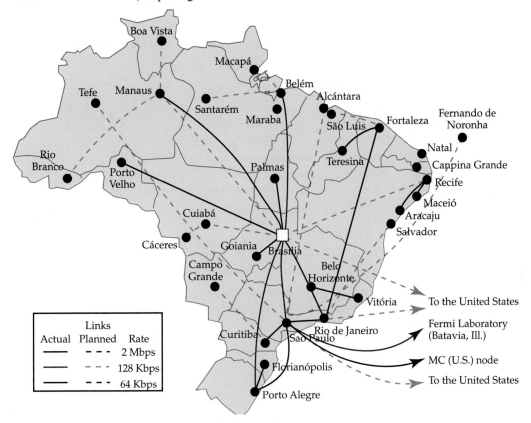

Internet voting has the same requirement as any proper secure system. It must have several elements. The first element is authentication. **Authentication** is the process of determining whether someone or something is, in fact, who or what it is declared to be. It is the ability to identify a system or network user through the validation of a set of assigned credentials, or tokens, that cannot be shared, copied, or altered. The second element is **authorization,** the ability to determine if the person accessing the application is authorized to do so and access is appropriate. The next element is **clarity of the interface.** This means that the user not only understands the layout of the computer screen (the interface) but is comfortable with it and the interface is nonambiguous. Next is **security,** that feature that ensures that neither the user nor anyone else can inappropriately access, view, copy, change, or destroy data in the system. If the data are about an individual, we must also add **privacy,** as security is not enough. Security is required first at the time of request for access; this is accomplished by authentication. Then it is enabled by making sure that no one can electronically eavesdrop on the input device or data communications lines and intercept the information as it is entered or is in transport. Security of transport connectivity, for example, the network, is an ongoing problem as hackers and viruses abound. Finally, we must secure the data in storage, for example, the database or file. This ensures that the data cannot be lost due to a catastrophic event to the physical media, nor can it be accessed by an unauthorized person.

The IT industry believes that it can implement all of the above features that require Internet technology. However, the one feature that is not now present is **validation,** a nonambiguous, nonsharable, noncopyable, unique identifier for an individual. At this point in history, the closest we have is a biometric identifier. In the presence of a human, a surrogate of biometrics can be used; for example, one person determines the identification of another via a photograph. However, photographs don't work on the Internet, so there must be an electronic, nonsharable, noncopyable element (token) similar to a fingerprint or retina or lens scan. Until this token is available, the problem of identification will block Internet voting in most cases.

Accommodation on the Internet

The Internet can provide significant accommodation for the mobility impaired; however, the sight impaired have a real problem as there is so much graphical and textual information. Federal and state laws exist that cover instances where Web sites must be accessible by those with a disability such as vision. For example, providing audible information when the mouse moves over an icon can help those who cannot see. When working in the public sector, care must be given to the rules that govern accommodation.

Challenges to Academic Honesty

One of the authors received a letter from the superintendent of The Virginia Military Institute, his alma mater. It contained the following comment:

> Incidentally, among the new challenges to our Honor Code, and to codes at other military academies, is a new species of what we might call Internet Crime—in which academic materials are downloaded, and then offered as original unattributed work.–*Josiah Bunting, Superintendent, VMI*

The Internet, uniquely, allows the student or professional writer to easily search and copy material. It can then be presented as his/her own. It denies the originator his/her fair due. This is dishonest and a blatant form of plagiarism.

Pornography and Hate Groups on the Internet

New technology is often adopted for uses not acceptable to the general public. The existence of **pornography** and **hate group** material on Web sites is of concern to citizens, parents, and lawmakers. While states and governments generally have the power to legislate within their domains, the Internet is global and is outside the reach of any specific state, government, civic, or religious organization. Regardless of the style (or form) of government, when the technology exists for reaching across borders, the influx of ideas cannot be suppressed. The U.S. government tried to impose its standards of morality on the content of the Internet in the Telecommunications Act of 1996; the act (Internet Decency Act) was found unconstitutional within three months. Nations, states, cultures, and other groups may wish to protect themselves from various influences but find themselves limited by the very global nature of the Internet that provides the material.

One way individuals may protect themselves or their children from materials or actions they deem offensive or dangerous is to place a software barrier between the home or office computer and the Internet. On a LAN, this means software on the server that inspects the destination and/or content of the URL or e-mail and applies rules of acceptable behavior. In the home, parents can place programs such as *Net Nanny*®, *CyberSitter,* or *Cyber Patrol*® on each machine and create a file of restrictions, such as the child including his/her name on a Web site or in e-mail. U.S. laws require parental consent for information gathering from children under the age of 13 and require reasonable use.

Implication
- CyberSitter provides over 30 categories of filtering, making it the most complete Internet filter available. Now records all Instant Messenger chat conversations for AOL (AIM) and Yahoo Messengers.[9]
- CyberPatrol's Internet safety software lets you take control of your Internet access . . . even when you can't be there![10]
- Net Nanny stops illicit material from invading your child's computer by filtering and blocking Web content while they surf. Time limits give parents the tools they need to restrict the times of day their kids use the Internet and the amount of time they spend online. Net Nanny stops many different kinds of applications from communicating on the Internet. That way, parents can block "file sharing" or "chat," without knowing about all of the different kinds of programs available. Your private information is valuable to those who would use it to exploit your family. Net Nanny protects your information by filtering it out of the data that leaves your computer in e-mail, chat rooms, and on the Web.[11]

Compression on the Internet

The technique of compression is discussed elsewhere. Because home pages on the World Wide Web often contain significant graphic images and video clips, and users move around large files, compression is important to the conduct of the Internet. A specific solution to this matter on Web sites is to use small graphic files with large files available upon request. For example, www.tigerdirect.com and www.bestbuy.com generally have photographs of items on the specification page, with a link to create a larger photo for investigation of downloading.

IP Television

Is the idea of entertainment programming over IP networks, that is, the Internet, a farfetched idea? Broadcast satellite and digital cable already convert the analog video to continuous stream digital for conduct. Packetizing it for conduct over an IP network is just an additional small technical step. Then, small stations have a global reach and programming can be stored on servers for viewing at any time from anywhere, for a fee.

IP-TV has two features of interest to us. Delivering TV entertainment over the IP-based Internet will make programs universally available and any advertising globally effective. It also would use significant amount of public bandwidth. Both authors receive entertainment via DirecTV, not on the Internet. One receives XM satellite radio via air and via the Internet (32 Kbps streaming). Using the Internet for continuous broadcast of commercial and noncommercial video entertainment will strain it beyond our expectations.

Summary

Organizations have seen no greater impact than the application of the Internet and its technology. Business uses of the Internet and its technologies continue to evolve. This trend is likely to continue into the foreseeable future and limited only by the creativity of those adapting to the new environment. An example of this business adaptation is the trend towards deploying systems employing wireless devices with Internet connectivity. With this technology and ubiquitous WiFi access, "Road Warriors" have continuous and mobile connectivity. Over time, the distinction between an intranet and the Internet will be blurred, separated only by the organization's firewalls. Portals of all forms are becoming essential to the organization's natural outreach. Extranets will become more important as organizations create partnerships with all types of participants. In addition, as more employees and outsiders are provided means of access, security concerns must be considered apace with the

---

[9] http://www.cybersitter.com/.

[10] http://www.cyberpatrol.com/.

[11] http://www.netnanny.com/p/page?sb=product.

application of the technology. While companies can not afford to ignore provision of Internet, extranet, and intranet services, they can ill afford to ignore the increased security required because of the threats this sort of access brings.

The frenzied adoption of applications in the business environment tend to reinforce the Internet's capability as a business resource that can't be downplayed. Those businesses that truly understand the potential of Internet connectivity have the capacity of creating Internet applications that can provide sustainable competitive advantage and potential for growth. Those who fail in this arena are likely to hasten the demise of the organizations.

Key Terms

authentication, *p. 237*
authorization, *p. 237*
business-to-business (B2B), *p. 225*
business-to-consumer (B2C), *p. 225*
clarity of the interface, *p. 237*
consumer-to-consumer (C2C), *p. 227*
digital cash, *p. 232*
digital divide, *p. 236*
electronic business (e-business), *p. 225*

electronic cash (e-cash), *p. 232*
electronic commerce (e-commerce), *p. 225*
electronic data interchange (EDI), *p. 231*
electronic funds transfer (EFT), *p. 231*
electronic government (e-government), *p. 227*
Federal Communications Commission (FCC), *p. 236*
gateway, *p. 224*

groupware, *p. 224*
hate group, *p. 238*
mobile commerce (m-commerce), *p. 227*
online, *p. 225*
PayPal, *p. 232*
pornography, *p. 238*
portal, *p. 233*
portlet, *p. 233*
privacy, *p. 237*
security, *p. 237*
validation, *p. 237*
web seal, *p. 232*

Recommended Readings

Business 2.0 (http://www.business2.com/b2/).

Fortune (http://www.fortune.com/fortune/).

Computerworld (http://www.computerworld.com/), America's number 1 publication for IT leaders.

Management Critical Thinking

M8.1 What are the most important Internet capabilities for a modern business? Can you locate a business firm that could significantly benefit from expanding or adopting Internet-based applications?

M8.2 What is the impact of the loss of Internet service on an e-business engaged in both B2C and B2B? What actions may be taken to limit negative impact?

Technology Critical Thinking

T8.1 What would the technology requirements be to implement e-government? Would this expense be justified on a cost-benefit or political basis?

T8.2 What is the difference of a city or state government installing simple Web access to information as opposed to a portal?

Discussion Questions

8.1 To what extent does the digital divide exist in your community? Your school?

8.2 What technology for use on the Internet is applicable for accommodation of those with special needs?

8.3 Consider the next state or federal elections. Is Internet voting a reality? What is inhibiting it?

8.4 What have you purchased on QVC or the Home Shopping Network on TV, and what have you purchased via the Internet? How were the purchases different?

8.5 In what ways can the Internet be used for an organization's network planning, implementation, and operation strategies?

8.6 Does spam have a measurable impact on the network's capacity? What laws are being enacted to help?

8.7 Most members of the class download music via the Internet, right? Are any streaming movies or TV programs? Why or why not?

8.8 What are some customer service applications that a firm can use the Internet for?

8.9 What sort of review process should be developed by the firm for its Web site(s)?

8.10 Should EDI be considered a part of e-commerce? Why?

Projects

8.1 Find two companies' Web sites and compare them in terms of usability and content.

8.2 Determine the use of the Internet by three major airlines.

8.3 Find an organizational portal and provide an analysis and critique of its linkages.

8.4 Find a company who has a significant customer care capability. Are they using the Internet to support this? Why or why not?

8.5 Divide the class into groups of three persons each. Each group should choose an item they wish to purchase. Use a search engine on the Internet to find (a) a comparison shopping site, (b) a discount outlet, (c) a normal outlet, and (d) prices on eBay. How do the prices compare? How would this exercise affect your buying of this item?

8.6 Contact local companies and determine who has intranets and extranets. Do they believe they work?

8.7 Locate a firm that uses a portal and determine the benefits derived.

8.8 Find an example of a global business using the Internet. What are the examples of the technology deployed? List alternative examples and the advantages/disadvantages of each (e.g., cost, level of service, bandwidth constraints, etc.).

8.9 Determine the "hot" jobs connected to the Internet. Make a rank-ordered list by demand and salary.

8.10 Determine how a local bank transmits electronic payments between financial institutions. To what extent does it use the Internet for any service?

8.11 Determine how many of the class have younger brothers and sisters under the age of 13. In how many of these families do they use software to protect the child during Internet access? What software do they use? Why do the families have the software or why do they not have the software?

Part **Four**

Wireless Networks

People using wired nodes are, by definition, tethered to some part of the system. Use of wireless appliances frees the users from this constraint while maintaining connectivity. This untethered connectivity comes at a security price. This section addresses the basic technologies of the wireless domain as well as management and security issues.

Chapter Nine

Wireless Networks: The Basics

Chapter Outline

There are a myriad of wireless formats. We review the traditional categories of fixed, portable, and nomadic forms and look at evolving versions. Along with the formats, we need to understand basic equipment and the connectivity needed in the relevant area of coverage. Some of the regulatory constraints and bodies also are covered. The rapidly evolving deployment of wireless broadband and related technologies are discussed. Management and security issues will be discussed in a later chapter.

We discussed the basic concepts of wireless (radiated) technology in Chapter 3. It's now time for a more in-depth investigation and even a comparison with wired (conducted) technology. As previously noted, all wireless applications rely on a **radio** or infrared light carrier. Since there are no wires, wireless is untethered. Most people have experienced wireless with AM/FM radio and television, although they may have been tethered for power. Many have become untethered with battery power. The TV/stereo remote control is a true wireless device, as is the garage door opener. A relatively new wireless device is a GPS receiver on a boat, car, or even handheld.

All of the above devices are indeed wireless, but they are simplex, for example, receive-only for AM/FM/TV/GPS and transmit-only for the various remotes. Rather than just these one-way devices, consumers want at least half-duplex interactive capabilities. The cell phone meets consumer expectations in voice and for limited data communications. More advanced cell phones can access the Internet and send e-mail.

Wireless will likely be the connectivity of choice because it allows the user to be untethered and mobile.

The purpose of networks is their ability to provide connectivity. Wired networks are designed to connect fixed or stationary nodes. Because people tend to be mobile, the tethering to stationary wired nodes becomes a hindrance. The rapid proliferation of cellular telephones kindled demand for other wireless devices that can provide connectivity to the traditional wired networks. For example, PDAs' wireless connectivity enables their mobile users to stay connected. It's not uncommon for mobile workers to use a wide variety of wireless appliances, ranging from notebook computers to hybrid PDAs (pocket PCs, smart phones, camera phones) in order to have wireless connectivity. The present trends toward wireless network routers that can connect entire buildings without a cable infrastructure and increased **bandwidth** may portend wireless networks as alternatives to wired ones.

PDAs and other portable devices have far more value when they can communicate and interact with other instruments. This is all leading to a world of wireless connectivity, everywhere. A former CEO of Hewlett-Packard remarked, "Every process in the world will be transformed from physical, static, and analog, to digital, mobile, virtual, and ultimately personal."

EQUIPMENT

Recall that *a network is two or more nodes connected by one or more channels*. Wired, or conducted, networks have a physical **medium** as the channel; **wireless networks** use the space medium. Each requires a complement of equipment to create the network.

Wired

From Chapter 5, we know that a wired network starts with a **network interface card (NIC)** as the entree or insertion point of the **node.** This is connected to some sort of channel, such as UTP, coax, or optical *connecting cable*. If there are only two nodes, there is a single cable between the NICs; if there are more than two nodes, there must be a *connecting node*, or *hub* or *switch*, to provide a common point for the connecting cables. And there is always a **network operating system (NOS).**

TABLE 9.1
Types of Wireless Technology

	Fixed	Portable	Nomadic
Point-to-point	Microwave	Laptop	
Point-to-multipoint	Satellite	Satellite, LMDS/MMDS	Cellular, Wi-Fi, packet radio
Mesh-enabled architecture	Access point	Repeaters	802.11

Wireless

Wireless networks have the same functionality in equipment as wired. A node has a *wireless NIC*, which is the intermediary connectivity to the nodes' processor plus a radio transmitter and **receiver.** The *channel is the space medium.* In the hub-and-spoke network, an **access point** performs the function of the wired hub and may provide connectivity to a wired network. If a **mesh-enabled architecture (MEA)** is used, the NICs contain routers to convey packets to other nodes as if they were access points.

TRADITIONAL WIRELESS FORMATS[1]

The domain of wireless technologies covers several facets, which we have grouped into fixed, moveable, and nomadic (moving). Additionally, they may be simplex, dual-simplex, and full-duplex in form. A review of these categories will begin our discussion of emerging technologies. (See Table 9.1.)

Fixed, Portable, and Nomadic[2]

In Chapter 2, we noted that wireless installations can have any combination of fixed, movable, or moving nodes. (Please refer to Table 9.1.) The idea of fixed wireless nodes is well understood as they are replacements for wired circuits. The idea of portability is where wireless technology starts to show its value and the concept of having nomadic nodes, totally untethered and unbound within the limits of the access point, creates networks of significantly greater value. Humans are seldom stationary, so networks that force them to remain at fixed nodes are unnatural, artificial, and arbitrarily limiting.

Point-to-Point Links

Point-to-point wireless systems are typified by the parabolic antennas of a microwave link, that is, one unit communicates to one, and only one, other unit. *It's a form of wireless cable.* The nodes can communicate reliably as long as the two end points are close enough to one another to "see each other" and escape the effects of **RF interference (RFI)** and attenuation. If unable to achieve a reliable connection initially, sometimes relocating the antennas or boosting the transmit power can achieve the desired **reliability.** When the end points are farther apart than a single hop (i.e, one transmitter and one receiver) can support, for example, greater than about 30 miles,[3] or have obstacles in the way, repeating units can make the links work. Thus, if the need is for an analog or digital circuit or channels between two points 200 miles apart, such as in the jungles of Guyana for a mining operation, towers would be constructed every 30 or fewer miles so that the

[1] http://www.commsdesign.com, http://www.pcmag.com, http://www.sensorsmag.com, http://www.infoworld.com, http://www.technewsworld.com, http://www.surfability.com, http://www.faulkner.com, http://www.intel.com.

[2] "Nomadic" (adj): (of groups of people) tending to travel and change settlements frequently; "a restless mobile society." http://www.dictionary.com.

[3] The 30-mile limit is the line-of-site limit due to the curvature of the earth for terrestrial stations. Aircraft have a limit of about 400 miles line-of-sight.

TABLE 9.2
Wireless Bandwidth and Range

Technology	Data Speed	Range
Cell phone	19.2 Kbps	5 mile cell
PDA	19.2 Kbps	3 miles
Wi-Fi	1–100 Mbps	50–200 feet
WiMax	100 Mbps	30 miles

antennas on the towers have a good line-of-sight to their partners. This may require changing directions along the way to work around a mountain or avoid setting a tower in the middle of a lake. The signal is sent, received, amplified, and resent; that is, the intermediate towers are acting as simple repeaters. This type of circuit may be multiplexed into multiple channels. This use of wireless reduces the number of structure placements and avoids digging trenches or the legal considerations involving **right-of-way.** In truly inaccessible areas, such as jungles or rugged mountains, the tower may have to be delivered via helicopter due to a lack of access roads. Although this example may seem out of the ordinary, a common use of this same technology is to deliver channels to cellular towers from a central switching office.

Fixed wireless can operate in a regulated part of the radio **frequency** spectrum or it may use unregulated **spectrum** and/or **spread-spectrum** technology. Additionally, it can use above-radio-frequency optical spectrum. While the standard point-to-point microwave systems have a **range** of up to 30 miles, **wireless T1** systems generally reach under two miles in distance, and then only in reasonably fair weather. Either of these technologies can be used as primary circuits or backup circuits. A bank in Atlanta has three wired T1 circuits from the headquarters to the operations center and three wireless microwave T1 circuits as backup. (See Table 9.2.)

Point-to-Multipoint Links

Often there are multiple receiving points for a single transmitter; that is **point-to-multipoint links.** An example of this is **satellite** communications, where the ground transmitter and orbiting transponder act as a single unit but are being received by many dispersed units. Originally, satellite data communications were simplex transmissions but are now half-duplex, or, better stated, dual-simplex as different frequencies are used, creating dual channels for the to-node and the from-node transmissions. Uses range from car dealership inventory accounting and management and parts manual distribution to *DirecTV*® entertainment. (See Figure 9.1.) Reliability of these satellite installations is good, although heavy rain or snow can degrade or interrupt transmission. While the examples here are fixed receivers, *XM*® and *Sirius*® satellite radio exemplify nomadic receivers.

Mesh-Enabled Architecture (MEA)

Unlike traditional hub-and-spoke wireless networks, this **architecture** produces a point-to-point-to-point, or peer-to-peer, system called an **ad hoc,** *multi-hop network.* (See Figure 9.2.) Initially created like an 802.11 network where a node can send and receive messages, in a **mesh network,** a node also functions as a router and may relay messages for its neighbors. Through the relaying process, a packet of wireless data will find its way to its destination, passing through intermediate nodes with reliable communication links. Like the Internet and other peer-to-peer router-based networks, a mesh network offers multiple redundant communications paths throughout the network, providing reliability in case a node is disabled or destroyed. Additionally, wireless mesh can be put up or torn down quickly to assist in many specialized situations, such as security or emergency services.

FIGURE 9.1 **Point-to-Multipoint Links**

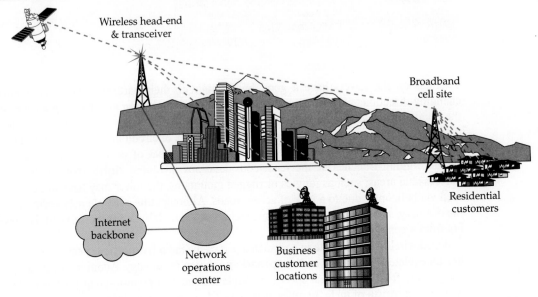

FIGURE 9.2
Wireless Mesh-Enabled Architecture (MEA) Network

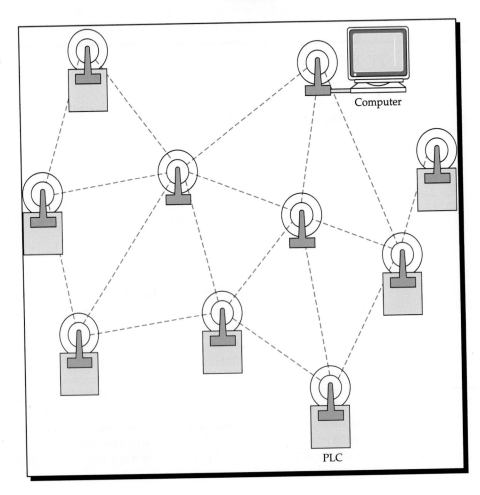

TABLE 9.3
Comparison of
Networks

	Suitability in Industrial Applications		
Topology	Reliability	Adaptability	Scalability
Point-to-point	High	Low	None (two end points)
Point-to-multipoint	Low	Low	Moderate (7–30 end points)
Mesh	High	High	Yes (thousands of end points)

Implication   MEA can significantly extend the range of a wireless network, at the expense of processing on individual nodes.

WIRELESS LOCAL AREA NETWORKS (WLANs)

Wireless local area networks (WLANs) are rapidly replacing wired networks in many instances. They are the wireless equivalent to an Ethernet bus network, with the *access point* acting as the hub, the radio as the medium, and CSMA/CA as the NOS. WLANs use high-frequency radio or optical waves in free space rather than on a physical medium to communicate between nodes.[4] It is simply an evolution of the standard LAN to establish an untethered environment. (See Table 9.3.)

Implication   Many pieces of network equipment started life as a simple capability, with other devices providing greater capability. What started as a hub evolved into a switchable hub and then a bridging hub. Access points have done the same, with most of them now being included as part of a wireless router.

Wireless communications via radio requires a center-frequency carrier onto which the information of interest is modulated. It can take two forms: continuous or packet-based. *Continuous wave* radio communications establishes the carrier between source and destination, solely uses the total channel, and continuously transmits information, although it may do so in an asynchronous manner. It can be **unidirectional,** that is, point-to-point, providing a replacement for a dedicated wired circuit, or it can be **omnidirectional,** giving a one-to-many environment.

Wireless LANs have evolved from early incompatible systems with low security, to systems with 54 and 108+ Mbps speeds and "reasonable" security.

The second variant of radio communication is **packet radio,** sharing the environment with other radiating sources. The packet mode breaks the file or continuous stream into **packets,** adds destination and error-checking information, and broadcasts into the **reception area,** that is, the geographic area in which signal strength is adequate for a node to receive. One packet-radio protocol is **IEEE 802.11,** or **Wireless Fidelity (Wi-Fi),** using the public spectrum with coverage over a few hundred feet in normal environments.

IEEE 802.11 uses (analog) radio, with a **carrier frequency** in the 2.4–2.5 GHz and 5.1–5.8 GHz ranges. The specific frequency is both standard and country dependent. Free space is used as the common channel, meaning that all 802.11 devices within range of the transmitter will receive the signal if it is in its *collision domain* (hearing range). The packets of 802.11, like Ethernet, have a header and a data portion; 802.11 also has a preamble before the header. All receivers within the collision domain receive the packet; the preamble "wakes" them up, then they read the header and discard the packet unless that device is designated as the destination. The destination device, of course, reads the packet.

Implication   Some say that 802.11 will carry up to 1,800 feet. Experience indicated a 100-foot range is more common. Actual range is dependent on output power of the devices, obstacles, and line-of-sight distance.

[4] http://www.webopedia.com/TERM/W/WLAN.html.

FIGURE 9.3
Wireless LAN

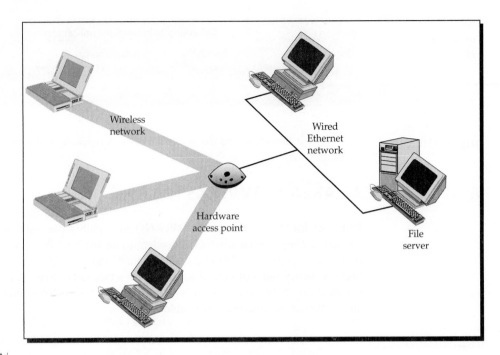

802.11a/b/g and CSMA/
CA are to wireless
networks as Ethernet is
to wired networks.

Implication

Packet radio can potentially make better use of limited spectrum than continuous broadcast by creating "more channels."

Clarification

802.11 refers to a family of specifications developed by the IEEE for wireless LAN technology. 802.11 specifies an over-the-air interface between a wireless client and a base station or between two wireless clients.[5]

There are several specifications in the 802.11 family:

Where 802.11a/b/g
(Wi-Fi) supports a
wireless LAN, 802.16
(WiMax) supports a
wireless MAN.

- *802.11* applies to wireless LANs and provides 1 or 2 Mbps transmission in the *2.4 GHz band* using either frequency-hopping spread spectrum (FHSS) or direct-sequence spread spectrum (DSSS).
- *802.11a* is an extension to 802.11 that applies to wireless LANs and provides up to 54 Mbps in the *5 GHz band*. 802.11a uses an orthogonal frequency division multiplexing encoding scheme rather than FHSS or DSSS.
- *802.11b* (also referred to as 802.11 High Rate or Wi-Fi) is an extension to 802.11 that applies to wireless LANS and provides 11 Mbps transmission (with a fallback to 5.5, 2 and 1 Mbps) in the *2.4 GHz band*. 802.11b uses only DSSS. 802.11b was a 1999 ratification to the original 802.11 standard, allowing wireless functionality comparable to Ethernet.
- *802.11g* applies to wireless LANs and provides 20+ Mbps in the *2.4 GHz band*.

Implication

802.11a has many more nonoverlapping channels than either 802.11b or 802.11g so more access points can be placed closer together, giving much more throughput in a smaller area.

Simple IEEE 802.11 Wi-Fi networks have one base station, or access point, that controls communications with all of the other wireless nodes connected with that network. (See Figure 9.3.) Often referred to as a hub-and-spoke or star topology, this architecture is similar to wired "home-run" systems in which all of the signals converge on one terminal

[5] http://www.webopedia.com/TERM/8/802_11.html.

block. The reliability of these networks is set by the quality of the RF link between the central access point and the end points. In industrial settings, it can be hard to find a location for an access point that provides dependable communications with each end point. Moving an access point to improve communications with one end point will often degrade communications with other end points.

Wireless Access Points

The equipment mentioned previously connects portions of a network or connected networks together. To extend the network to the wireless domain, a node is required that can communicate to the wired and unwired environments. This is the job of a **network access point (NAP).** To connect wireless nodes to a wired network, a wireless access point is made a part of the wired network. The access point is wired on one side and a radio on the other. The access point communicates with the wireless network interface cards of the wireless devices and acts as a bridge or router between wired and wireless environments.

Wireless Routers

Firewall
a device that protects a
private network from
intrusion.

A wireless router and access point, as shown in Figure 9.4, connects to a wired network and operates as a radio transmitter-receiver to the wireless world, providing the function of a router and firewall. This *D-Link*® is one of a variety of wireless routers and access points with firewall on the market; this one uses the 802.11g protocol and operates at 108 Mbps. The firewall provides features of NAT with VPN pass-through, MAC filtering, IP filtering, URL filtering, domain blocking, scheduling, and 64/128-bit **encryption.**

FIGURE 9.4   Wireless Router and Access Point

Source: http://www.dlink.com/products/?pid=6.

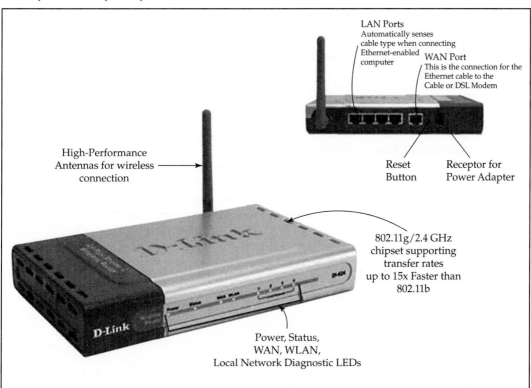

LAN Ports
Automatically senses
cable type when connecting
Ethernet-enabled   WAN Port
computer   This is the connection for the
Ethernet cable to the
Cable or DSL Modem

High-Performance
Antennas for wireless
connection

Reset   Receptor for
Button   Power Adapter

802.11g/2.4 GHz
chipset supporting
transfer rates
up to 15x Faster than
802.11b

Power, Status,
WAN, WLAN,
Local Network Diagnostic LEDs

FIGURE 9.5
MAC Layer of
OSI Model

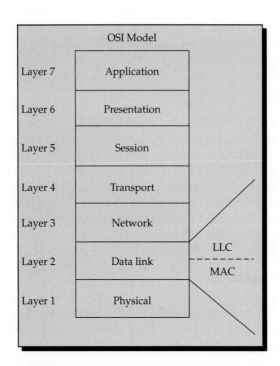

802.11 Media Access Control (MAC) Layer

The **media access control (MAC)** layer is a subset of the link layer (data link of OSI model), which is adjacent to the physical layer in an IP-based network. The data link layer in an 802.11 network performs several functions. The MAC layer is the one that handles the mobility issues of an 802.11 network. (See Figure 9.5.)

Filtering
whether MAC, IP, or URL, restricts access to those on a list.

In the 802.11 MAC layer standard, the flow of information is performed on a *best-effort* basis, which is also called *connectionless*. **Connectionless** links are those in which the receiving end of the link does not verify the receipt of data by the transmitting link. This technique uses CSMA/CA in a *best-effort* architecture; there are no guarantees that the data sent will be received successfully. One of the things an 802.11 system does to help ensure the successful receipt of information is to send the information repeatedly, which is called *chipping*. The MAC layer also provides *security,* typically handled at the *presentation layer* of the OSI model. The security measure compliant with this standard is **Wired Equivalent Privacy (WEP).**

Most common modulation schedules used for radios are binary frequency shift keying (BFSK or FSK) or phase shift keying (BPSK), QPSK, and QAM. BFSK sends a "one" with one frequency and a "zero" with another frequency. BPSK will send two states, a "one" with one phase and a "zero" with another phase. QPSK has four states to represent 00, 01, 11, 10 with four phase states. QAM modulates the carrier frequency in both phase and amplitude, getting four, five, six, seven, or eight bits per sine wave. As the modulation complexity increases, so does the probability of an error in transmission. The 802.11 standard has an *auto-rate negotiation* where the radios will automatically downshift to less complex modulation and spreading techniques in order to maintain higher levels of robustness.

Implication

IEEE 802.11g uses OFDM, which is where its superior speed compared to 802.11b comes from, yet it uses the same frequency bands and therefore the same antennas as 802.11b. Also, unfortunately, if there is any 802.11b in use around an 802.11g network, the speed of the 802.11g network drops significantly to near 802.11b levels.

Spread Spectrum

Spreading techniques distribute the information over a number of channels, while a modulation technique modulates the information over each of the channels. Direct-sequence spread spectrum (DSSS), frequency-hopping spread spectrum (FHSS), **code division multiplex access (CDMA),** and **orthogonal frequency division multiplexing (OFDM)** are spreading techniques. Coded *Orthogonal Frequency Division Multiplexing (COFDM)* is the spreading technique used in 802.11a, 802.11g, and *802.16 (WiMAX).*

In order to provide security and reliable transmission, WLANs use one of two types of spread spectrum radio: direct-sequence and frequency-hopping spread spectrum. **Direct-sequence spread spectrum (DSSS)**[6] transmissions multiply the data being transmitted by a "noise" signal. This **noise** signal is a pseudorandom sequence of 1 and –1 values, at a frequency much higher than that of the original signal, thereby spreading the energy of the original signal into a much wider band. The resulting signal resembles white noise, like an audio recording of "static," except that this noise can be filtered out at the receiving end to recover the original data, by again multiplying the same pseudorandom sequence to the received signal.

FHSS—Catch me if you can!

Frequency-hopping spread spectrum (FHSS) has transmissions where the data signal is modulated with a narrowband carrier signal that "hops" in a random but predictable sequence from frequency to frequency as a function of time over a wide band of frequencies. The signal energy is spread in the time domain rather than chopping each bit into small pieces in the frequency domain. This technique reduces **interference** because a signal from a narrowband system will only affect the spread spectrum signal if both are transmitting at the same frequency at the same time. If synchronized properly, a single logical channel is maintained. The transmission frequencies are determined by a spreading, or hopping, code. The receiver must be set to the same hopping code and must listen to the incoming signal at the right time and correct frequency in order to properly receive the signal. Current FCC **regulations** require manufacturers to use 75 or more frequencies per transmission channel with a maximum dwell time (the time spent at a particular frequency during any single hop) of 400 ms.

Infrared (Optical)

Beyond the radio frequency range of 30 MHz to 300 GHz lies the optical spectrum. Although analog, it is used in either an analog or digital mode. Infrared can be unidirectional or omnidirectional. Many PDAs, laptop computers, and cell phones have IR ports to provide moveable, point-to-point connectivity. Infrared technologies are good for short distances (<50 feet), won't pass through walls and other opaque objects, and have a bandwidth in the low 100 Kbps.

CONNECTIVITY—AREA OF COVERAGE

Wired networks are known and named by their area of coverage. (See Figure 9.6.) These include *personal area networks,* which are confined to the individual; *local area networks,* which may be a floor of a building, the building, or a campus; *backbone networks,* which connect LANs, thus extending the reach of each LAN; and, finally, **wide area networks,** which give a reach that may include the globe. With wireless networks, we consider the environment or reach, ultimately dictating the method of achieving the connectivity.

[6] http://en.wikipedia.org/wiki/DSSS.

FIGURE 9.6   Geographic or Layer Network Concept

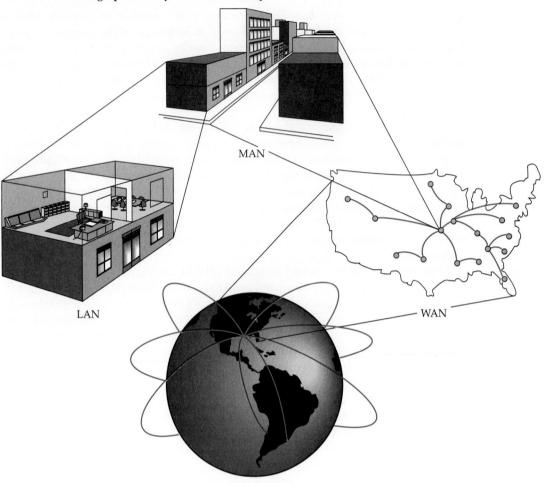

MAN

LAN

WAN

Global network

Room(s)

Although one's first thought for wireless connectivity might be the cell phone, which now has local and global reach, we start at a very local range of departure. The room might be a conference room, an office, or a warehouse with few to many people. The point is to have access in the room and even confine the access and communications to that room. If there are no physical obstacles, *infrared (IR)* connectivity offers easy, low-speed access. For example, many laptop computers come with an IR port, which was originally designed to be used with a matching port on a printer, negating the need for a physical circuit. PDAs send information from one to another or to a printer via the same IR port and the PDA can synchronize with a laptop the same way.

802.11 Wi-Fi is a natural solution for communications ranging from a room to a city.

For most applications, *802.11 packet radio* with a single access point could provide connectivity to all nodes in the room and extend the connectivity to a wired network. The low power of the 802.11 devices isolates them naturally, an advantage and a disadvantage. A retail example of this application is the use of **Wi-Fi hotspots** in coffee shops, airline lounges, hotels, and fast food restaurants, giving customers temporary access to a wireless network and the Internet.

Clarification

9 Steps to Set Up Your Home Wi-Fi Network. News flash! You can turn your home into a wireless network complete with blazing fast Internet connections from each PC in the house and even to PCs outside—without the encumbrance of cords. Transforming your home into a wireless hotspot is relatively inexpensive and not nearly as complicated as you might think. Our editors have compiled the following article to guide you in your quest to create a home Wi-Fi network of your very own.[7]

Building/Warehouse/Campus

When the area of coverage extends to several rooms, a floor of a building, an office building, or even a campus, *802.11 with multiple access points* provides good coverage. The 802.11 protocol is designed to assume multiple signals, so receiving packets from several access points or even packets from the same access points that bounce off objects are not a problem. As the area of coverage increases, the number of access points increases unless a *mesh-enabled architecture* is used. An alternative is to simply place repeaters in the environment to extend the range of the network.

City

Wireless connectivity in a region as large as a city or metropolitan area is presently provided by *cellular telephones*. The data channels of these nodes and their cell access points can transfer data packets while the voice channels provide continuous audible communications. Voice communications on some of these same devices are converted to packet radio in the form of *Voice-over-IP (VoIP)*. The range of the nodes is constrained to the cell in which they reside, but the reach is global as the analog and digital signals can be carried either wired or wirelessly as far as network connectivity exists.

Implication

Some cities see the value of 802.11 connectivity citywide. Singapore was one of the first to institute Wi-Fi hotspots around the city. Following suit, Philadelphia, Pennsylvania, was next to announce plans to make the city a wireless hotspot. Should the city support of 802.11 hotspots become common, Wi-Fi could become a connectivity of choice, even a voice connectivity of choice with Wi-Fi and VoIP.

Extending wireless connectivity to cities uses *WiMAX*. **WiMAX** is a standards-based wireless technology that provides high-throughput **broadband** connections over long distances to fixed, portable, and nomadic users. WiMAX can be used for a number of applications, including "last mile" broadband connections, hotspot and cellular backhaul, and high-speed enterprise connectivity for businesses. An implementation of the **IEEE 802.16** standard, WiMAX provides **wireless metropolitan area network (WMAN)** connectivity at speeds of up to 75 Mbps, and as far as 30 miles radius. However, on the average, a WiMAX base-station installation will likely cover between three and five miles.[8] The ultimate complement to Wi-Fi, the WiMAX wireless MAN standard can be used to backhaul 802.11 hotspots and WLANs to the Internet, provide campus connectivity, and enable a wireless alternative to cable and DSL for last-mile broadband access. It provides up to 50 kilometers of service area range, allows users to get broadband connectivity without needing direct line-of-sight with the base station, and provides total data rates of up to 280 Mbps per base station—a sufficient amount of bandwidth to simultaneously support hundreds of businesses with T1/E1-type connectivity and thousands of homes with DSL-type connectivity with a single base station.[9] Initial installations of WiMAX will be fixed at both ends, with mobile clients being adopted later.

Clarification

WiMax is to the MAN as Wi-Fi is to the LAN.

[7] http://www.tigerdirect.com.
[8] http://www.intel.com/netcomms/technologies/wimax/index.htm.
[9] http://www.wimaxforum.org/about/.

A mobile complement to WiMAX is **wireless broadband (WiBro).** WiBro is part of the IEEE 802.16 family of wireless Internet specifications and is expected to offer speeds up to 1 Mbps for devices traveling at up to 60 kilometers per hour.

Implication

The capabilities and uses of 802.16 WiMAX are continuing to expand. To keep up with this technology, go to http://www.wimaxtrends.com/feature.htm.

State (PSC regulated)

Wired and wireless services offered to the public generally come under the regulation of the state **public service commission (PSC),** also called the *public utilities commission* in some states. This comes in the form of a *tariff,* which is a list of telecommunications services to be offered and the associated charges. For example, *T-Mobile®* offers hotspot access points for 802.11 access to the Internet with hotspots in many convenient locations. Since 802.11 uses a portion of the public spectrum, there is no regulation there; since the hotspot is offered to the public, a tariff may have to be filed with the state PSC and possibly with the **Federal Communications Commission (FCC).** The FCC has regulatory authority over radio spectrum nationally and services offered across state lines. Thus, if the service spectrum is outside of the spectrum set aside for public use, for example, 900 MHz, 2.4 GHz, and 5.3 GHz, FCC approval is required. An example would be point-to-point microwave unless the technology used is spread spectrum.

Nationwide (FCC Regulated in the United States)

Extending a service to cover more than one U.S. state or all U.S. states, as noted previously, requires permission from the FCC. Cellular service was auctioned off by the FCC, which exercised its control at that time, involving service and spectrum. Once the cellular network is in place, other considerations take place. For example, use of the cellular data channel for **short message service (SMS)** would be an example of multistate service that would not require FCC additional approval. *Nextel®* offers **push-to-talk** communications among its customers nationwide. Push-to-talk is provided by voice-over-IP and the IP networks in question would be private networks once the communications reached the cell tower. Private networks are not regulated.

National (PTT Regulated outside of the United States)

PSC for state regulation, FCC for U.S. national, and PTT for non-U.S. national.

Crossing a national boundary is not difficult with wireless technology. It is, however, a different matter when it comes to governmental permission. Nations originally took the stance that data were goods and goods crossing national boundaries are subject to taxes. But what about voice and data communications—are they goods? Crossing national boundaries via satellite communications is commonplace; the sticky part probably arrives when the service offered is easier to "see." If cellular traffic is regulated in the United States and in Canada by different governments, then the movement of that traffic would likely be co-regulated. The FCC equivalent agencies in countries outside of the United States are called **postal, telephone, and telegraph (PTT)** agencies.

RADIO FREQUENCY IDENTIFICATION (RFID)

The **Radio frequency identification (RFID)**[10] system has two components: tags, or chips, which are radio transponders attached to physical items; and readers that energize and query the tags, receive, translate, and forward data to networks. The tags, or chips, contain unique

---

[10] Katherine Albrecht, "RFID: Tracking Everything, Everywhere," http://www.boycottgillette.com/RFID_overview.html).

FIGURE 9.7
Exxon Mobil
Speedpass
RFID in action for
over five years.

Source:
http://www.speedpass.com/how/
index.jsp.

Speedpass® is easier than using a credit or check card. No more waiting for authorization and then signing receipts. No more searching for cash and waiting for change. To use it, simply point the Speedpass "wand" at the area of the pump or register that says "Place Speedpass Here." Your Speedpass automatically—and immediately—communicates your payment preferences. It knows what credit or check card you wish to use. Speedpass even knows whether or not you want a receipt. Speedpass operates on a radio frequency that transmits an identification and security code. No personal or financial information is ever stored on your Speedpass device. The catch? There isn't one. Speedpass is FREE and it's easy to use.

information. When energized by the reader, the chip transmits an identifying signal that communicates with the reader. The chips are often smaller than a postage stamp, providing incorporation into packaging and individual items, such humans, animals, food, money, credit cards, and virtually anything imaginable as a means of verification. RFID tags have built-in antennas that allow them to receive and respond to radio-frequency queries from the RFID system.

RFID products[11] read or write data to RF tags that are present in a radio frequency field projected from RF reading/writing equipment. Data provide identification and other information relevant to the object to which the tag is attached. It incorporates the use of **electromagnetic** or electrostatic coupling in the radio-frequency portion of the spectrum to communicate to or from a tag through a variety of modulation and encoding schemes.

Implication

The promise of RFID is the automatic collection of data. As a carton of goods with tags attached passes a reader installed at a loading dock, the reader logs the tag IDs into a database, so both the shipper and receiver know exactly when the delivery was made.

Clarification
The size of an RFID
chip can be as large as a
postage stamp or as
small as a grain of rice,
to be easily hidden
within or on a product.

RFID uses an active transmitter to query a passive data storage device. Remote car door entry and garage door devices are not RFID in that they both transmit a signal that activates a passive device, but the device does not reply with a signal.

Two types of tags are offered by RFID tag manufacturers: passive and active. Because passive tags cost less to manufacture, most are of the passive variety. While they do not have their own power supply, the incoming radio-frequency scan from the RFID system provides enough power for the RFID chip in the tag to send a response. Because they have no power supply, passive RFID tags usually provide brief responses, such as an ID number. Passive RFID tags can be extremely small and can be interrogated by a RFID reader at ranges that vary from less than an inch to five yards or so. (See Figure 9.7.) Active RFID tags have a power source (a battery with a life of up to several years) and often have longer read ranges (in some cases, tens of yards). They have larger memories than passive tags, as well as the capacity to store additional information sent by the transceiver.

[11] http://rfid.globalspec.com/.

RFID: How Can It Be Used?[12]

RFID is meant to replace the current inventory marking system, the barcode-based Universal Product Code (UPC), with the **Electronic Product Code (EPC)** system. EPC was designed to track items through the supply chain whereas UPC was originally a point-of-sale identifier. While the original intent of EPC was a concern for backroom supply management, some businesses want to use it for item-level tagging.

A major push for RFID at an item level comes from an unlikely place—Bentonville, Arkansas, the headquarters of the world's largest retailer, Wal-Mart. The company is pressuring its suppliers to adopt RFID or else lose the products' shelf space. Other companies including Gillette, Exxon Mobil, Pratt-Whitney, Boeing, Airbus Industries, Delta Airlines, Kimberly-Clark, and other Fortune 500 companies are implementing RFID in some form or another. The U.S. military has found it useful in supply management in their military operations. No matter how it is implemented, the RFID system enables vital information to be matched with a database.

Implication

RFID is a technology with significant potential to control the supply chain and reduce cost but is facing social resistance because of privacy concerns. A major concern for the use of RFID on consumer products is that they remain active after the sale. The question remains as to how to disable the device when it passes the checkout counter. If it is imbedded in the product, removal will be difficult; its presence may be forgotten. Having the antenna destroy all data is a possibility, but destroying the ability of the chip to transmit is far better. Where possible, the chip should be removed and physically destroyed.

Implication

ITSPA surveyed members of its advisory board and its technology committee about technology trends affecting small-to-medium businesses (SMBs). The company came up with a total of six trends that will affect SMBs next year, most of which should come as no surprise. They include Voice over IP (VoIP), mobility and wireless networking, radio frequency identification (RFID), security, utility computing, and servers.[13]

The business-level benefits of EPC over UPC include the tags' unique identification and user friendliness at the reading stage (point-of-sale): they do not require line-of-sight to be read nor do they have to be stationary. If stores implement the system, it is likely they will also be equipped with embedded tracking systems to monitor movement of products. In addition to these benefits, businesses believe that the system will reduce costs by reducing theft, misplaced merchandise, and other inventory burdens. Proponents feel the technology will drive global competition and will likely have strategic utility in providing a competitive advantage.

Implication

Several pharmaceutical firms have initiated use of RFID labels to better track their products. The labels can contain date of manufacture, sales destination, and so forth. This can help reduce counterfeiting of some popular drugs and enable police to identify the location from which stolen drugs were taken, thus helping to positively identify that they were stolen.

RFID Case: Use in the Mining Industry

Should parents RFID tag their children as well as their pets?

Radio frequency identification (RFID) has received a lot of press as Wal-Mart and the U.S. Department of Defense announced the requirement for their major vendors to use this technology. Other industries have used RFID for nearly a decade. Specifically, the mining industry has used this technology to manage resources, control productivity, and enhance safety.

[12] Assistance in this material was provided by Ryan D. Taunton.
[13] http://www.integratedmar.com/ecl-usa/story.cfm?item=18806.

Should government RFID criminals? Will they want to RFID all of us?

Consider an open pit mining operation. There are several sources of ore of differing grade; a number of loading machines; servicing of hundreds of haulers, over miles of roads, to several storage locations for ore-by-grade. Using RFID, this environment can be better managed, enhancing the flow of materials and utilization of equipment and personnel and bolstering the safety environment.

Starting at an ore deposit, shovels will be removing the waste material (overburden), placing it in a hauler to be moved to a waste site. RFID sensors at the deposit and along the route note the movement of haulers, which have RFID nodes onboard; register the building queue of waiting haulers; note the speed of movement of the trucks; and ensure that the grade of ore, or overburden in this case, is taken to the correct dumping location. As ore is removed, the information from RFID sensors on the hauler can note this in a database, indicating reserve remaining. When the ore is placed in storage, the database will note the size of the storage by grade. Should the queue of haulers at a shovel become excessive, the computer can alert management that there is either a breakdown of equipment or that the haulers need to be rerouted.

Safety is enhanced as RFID allows computer-based systems to know which equipment and even which personnel are in specific areas. This is especially important in an underground mine, where personnel and moving equipment may not safely co-exist. Personnel and equipment can be identified by the RFID sensors to alert management in the case of unsafe conditions. RFID-tagged individuals can be located when they are out of voice or radio range. This can extend to saving lives in the event of an accident, such as a cave-in, when the sensors continue to work.

OTHER WIRELESS STANDARDS

Bluetooth (802.15.1)

Bluetooth[14] is a cable-replacement technology using the **IEEE 802.15.1** standard. Conceived initially by Ericsson, before being adopted by a myriad of other companies, **Bluetooth** is a standard for a small, cheap radio chip to be plugged into computers, printers, mobile phones, and so on. A Bluetooth chip is designed to replace cables by taking the information normally carried by the cable and transmitting it at a special frequency to a receiver Bluetooth chip. The projected low cost of a Bluetooth chip (about $5) and its low power consumption mean they could be placed anywhere. The interest in Bluetooth is expanding as Bluetooth chips are being considered in freight containers to identify cargo when a truck drives into a storage depot, or a headset that communicates with a mobile phone in your pocket, or even in the other room.

Clarification

Bluetooth is a standard developed by a group of electronics manufacturers that allows any sort of electronic equipment—from computers and cell phones to keyboards and headphones—to make its own connections, without wires, cables, or any direct action from a user. Bluetooth is intended to be a standard that works at two levels:

- It provides agreement at the physical level—Bluetooth is a radio-frequency standard.
- It also provides agreement at the next level up, where products have to agree on when bits are sent, how many will be sent at a time, and how the parties in a conversation can be sure that the message received is the same as the message sent.[15]

Bluetooth[16] was intended to unify different technologies like computers and mobile phones. It provides a way to connect and exchange information between devices such as

14 http://www.palowireless.com/infotooth/whatis.asp.
15 http://electronics.howstuffworks.com/bluetooth2.htm.
16 http://en.wikipedia.org/wiki/Bluetooth.

TABLE 9.4
**Competitive Wireless
Technologies**

Protocols	ZigBee 802.15.4	Bluetooth 802.15.1	Wi-Fi 802.11b/a/g	GPRS/GSM 1XRTT/CDMA
Application Focus	Monitoring and control	Cable replacement	Web, video, e-mail	WAN, voice/data
System Resources	4 KB–32 KB	250 KB+	1 MB+	16 MB+
Battery Life (days)	100–1000+	1–7	0.1–5	1–7
Nodes per Network	255/65K+	7	30	1,000
Bandwidth	20–250 Kbps	720 Kbps	11–108 Mbps	64–128 Kbps
Range (meters)	1–75+	1–10+	1–100	1,000
Key Attributes	Reliable, low power, cost effective	Cost, convenience	Speed, flexibility	Reach, quality

personal digital assistants (PDAs), mobile phones, laptops, PCs, printers, and digital cameras via a secure, low-cost, globally available, short-range radio frequency. Bluetooth lets these devices talk to each other when they come in range, even if they're not in the same room, as long as they are within 10 meters (32 feet) of each other. It is a wireless radio standard primarily designed for low-power consumption, with a short range and with a low-cost transceiver microchip in each device.

Cell phones with integrated Bluetooth technology also have been sold in large numbers and are able to connect to computers, PDAs, and, specifically, hands-free devices. BMW was the first motor vehicle manufacturer to install hands-free Bluetooth technology in its cars, adding it as an option on its 3 Series, 5 Series and X5 vehicles. Since then, other manufacturers have followed suit, with many vehicles, including the Ford F-150 truck, the Toyota Prius, and the Lexus LS 430. The Bluetooth car kits allow users with Bluetooth-equipped cell phones to make use of some of the phone's features, such as making calls, while the phone itself can be left in a suitcase or in the boot/trunk, for instance.

The Bluetooth standard also includes support for more powerful longer-range devices suitable for constructing a wireless LAN. Every Bluetooth device can simultaneously maintain up to seven connections, but only one active connection at a time. These groups (maximum of eight devices: one host and seven slaves) are called **piconets.** The specification also enables the possibility of connecting two piconetworks, with one master device acting as a bridge, resulting in a **scatternet.** Every device can be configured to constantly announce its presence to nearby devices in order to establish a connection. It is also possible to password protect a connection between two devices so that no one can listen in.

The protocol operates in the license-free ISM band at 2.45 GHz. It reaches speeds of 723.1 Kbps. In order to avoid interfering with other protocols that use the 2.45 GHz band, the Bluetooth protocol divides the band into 79 channels and changes channels up to 1,600 times per second.

Bluetooth should not be compared to Wi-Fi, which is a faster protocol requiring more expensive hardware that covers greater distances and uses the same frequency range. While Bluetooth is a cable replacement creating personal area networking between different devices, Wi-Fi is a cable replacement for local area network access. (See Table 9.4.)

ZigBee (802.15.4)[17]

ZigBee is a proprietary set of high-level communication protocols designed to use small, low-power digital radios based on the **IEEE 802.15.4** standard for wireless personal area networking (WPAN). The technology is designed to be simpler and cheaper than other WPANs such as Bluetooth. The most capable ZigBee node type is said to require only

[17] http://en.wikipedia.org/wiki/ZigBee.

about 10 percent of the software of a typical Bluetooth or wireless Internet node, while the simplest nodes (RFIDs) are about 2 percent. ZigBee is aimed at applications with low data rates and low power consumption. ZigBee's current focus is to define a general-purpose, inexpensive, self-organizing mesh network that can be shared by industrial controls, medical devices, smoke and intruder alarms, building automation, and home automation. The network is designed to use very small amounts of power so that individual devices might run for a year or two with a single alkaline battery.

ZigBee operates in the unlicensed 2.4 GHz, 915 MHz, and 868 MHz ISM bands. The radio uses DSSS, which is managed by the digital stream into the modulator. The data rate is 20 Kbps per channel, and transmission range is between 10 and 75 meters (33~250 feet). The software is designed to be easy to code for small, cheap microprocessors. The radio design utilized by ZigBee has few analog stages and uses digital circuits wherever possible. Most vendors plan to put the radio on a single chip.

ZigBee wireless networking technology is to be included in professional installation kits for light switches, thermostats, security system controls, and other relatively simple electronic controls. ZigBee sends small packets of data over distances of up to 20 feet between nodes. The data piggybacks on intermediary nodes but is only acted upon by the specified recipient node. Thus, the control instruction data can be transmitted over an entire network, given available pathways. ZigBee can be used in mesh networks by up to 65,000 nodes, promising to allow managers to control entire buildings. The battery life is great because the devices only turn on for the moment they are needed. (See Table 9.4.)

Ultra Wideband (UWB) (802.15.3)[18]

Ultra wideband (UWB), short-range radio technology using the **IEEE 802.15.3** standard, complements other longer range radio technologies such as Wi-Fi, WiMAX, and cellular wide area communications. It is designed to relay data from a host device to other devices in the immediate area (up to 10 meters or 30 feet).

A traditional UWB transmitter works by sending billions of pulses across a very wide spectrum of frequencies several GHz in bandwidth. The corresponding receiver then translates the pulses into data by listening for a familiar pulse sequence sent by the transmitter. Specifically, UWB is defined as any radio technology having a spectrum that occupies a bandwidth greater than 20 percent of the center frequency, or a bandwidth of at least 500 MHz.

Modern UWB systems use other modulation techniques, such as orthogonal frequency division multiplexing (OFDM), to occupy these extremely wide bandwidths. In addition, the use of multiple bands in combination with OFDM modulation can provide significant advantages to traditional UWB systems. UWB's combination of broader spectrum and lower power improves speed and reduces interference with other wireless spectra. In the United States, the Federal Communications Commission (FCC) has mandated that UWB radio transmissions can legally operate in the range from 3.1 GHz up to 10.6 GHz, at a limited transmit power of -41 dBm/MHz. Consequently, UWB provides dramatic channel capacity at short range that limits interference.

UWB[19] signaling promises a potentially revolutionary approach to radio communication. By using pulses or waveforms compressed in time, frequency energy may be spread over a very wide bandwidth to very low levels (even under the thermal noise floor). This may allow UWB radios to share spectrum with existing narrowband broadcasters without causing undue interference, thus creating many interesting and novel application opportunities.

---

[18] http://www.intel.com/technology/ultrawideband/.

[19] http://bwrc.eecs.berkeley.edu/Research/UWB/default.htm.

Clarification
The origin of ultra wideband (UWB) technology stems from work in time-domain electromagnetics begun in 1962 to fully describe the transient behavior of a certain class of microwave networks through their characteristic impulse response.[20] . . . The concept was indeed quite simple. Instead of characterizing a linear, time-invariant (LTI) system by the more conventional means of a swept frequency response (i.e., amplitude and phase measurements versus frequency), an LTI system could alternatively be fully characterized by its response to an impulsive excitation—the so-called impulse response h(t) . . .[21]

WIRELESS BROADBAND TECHNOLOGIES

The area of wireless broadband has made significant advances in the past few years and will continue to do so. The advantages are (a) much faster installation, (b) building architecture independence, (c) movability, and (d) far simpler right-of-way access. Wireless technologies fall into two basic categories: fixed installations at each end and moveable or moving destination devices. For the most part, the two forms serve different purposes.

Fixed Wireless

Wireless devices, for the most part, are line of sight in that they will not work if an obstruction is in the way. These obstructions may be trees, buildings, and even heavy rain or snow for specific frequencies. Fixed wireless falls into two realms as far as government regulations: using regulated frequency spectrum and operating in the public domain.

Short Distance; Moderate Bandwidth

The first example of fixed wireless can operate in either the regulated or public arena: the technology of *wireless T1* services. This technology substitutes wireless T1 bandwidth and services for wired. A point-to-point installation can be purchased and installed for about US$20,000 and operates over a distance of four miles in light rain. Some systems use FCC-regulated spectrum and others use spread-spectrum technologies in the 2.4 and 5.3 GHz unregulated spectrum. Both forms can quickly create a wireless channel on rooftops or walls and neither requires right-of-way permission. Each operates at reasonable power levels and with adequate security.

Terrestrial Microwave

Terrestrial microwave is a tried-and-true technology that has a proven track record. While the radio-based spectrum does require FCC registration for frequency spectrum allocation, the installation is simple and the protocols are stable. Thus, two facilities that can see each other can install microwave wireless broadband capability. Either analog or digital services may be employed over a distance of up to 30 miles.

Alternatives to radio-frequency systems are those that operate at the highest radio-frequency spectrum (in the near-light region). General Electric offers *Gem-Link*, which works in the unregulated domain. It is designed to provide up to 4 MHz of bandwidth over a four-mile range, packaged for self-installation and alignment.

Clarification
Terrestrial microwave lost favor when fiber cables matured. Microwave has seen a resurgence in that, in many cases, especially data, it is less expensive to install, it requires less right-of-way, and delays are tolerable.

[20] G. F. Ross, "The Transient Analysis of Multiple Beam Feed Networks for Array Systems," Ph.D. dissertation, Polytechnic Institute of Brooklyn, Brooklyn, NY (1963); G. F. Ross, "The Transient Analysis of Certain TEM Mode Four-Port Networks," *IEEE Transactions on Microwave Theory and Technology* 14, no. 11 (1966), pp. 528-47.

[21] http://www.multispectral.com/history.html.

MMDS/LMDS

These two sister technologies—**Multipoint Microwave Distribution System (MMDS)** and **Local Multipoint Distribution System (LMDS)**—were developed for public television and education. When not used for the original purpose, channels can be used for **cable television (CATV)** delivery of up to 25 channels to a distance of 25 miles from the centralized omnidirectional antenna to the premise-based antenna. The more recent use is in the digital domain, where it competes well for Internet access, if line-of-sight topography permits. Although FCC regulated and with a limited number of channels available, it provides a large service footprint with no cables and no right-of-way required.

Satellite and VSAT

VSAT
a combination of a satellite antenna and the transceiver, packaged together.

Both of these forms of satellite delivery rely on a geostationary orbiting transponder and fixed receiving antenna. Rare would be the U.S. car dealership that does not have a VSAT (very small aperture terminal) antenna on its roof. Satellite service is common for cable TV companies for delivery of content that is then placed on their coaxial cables, with both analog and digital channels offered. Direct broadcast service television and data are provided by *DirecTV®*, *DirectWav®*, and the *Dish Network®*. Some of these services are simplex, receive-only at the destination while others are half-duplex for data.

This technology has been noted as a fixed service. The capability for it to be used in a moveable service is established as football fans take a receiver and antenna to game day tailgating and watch other teams play before their game starts. Several airlines have installed *DirecTV* service on their airplanes, offering passengers full *DirecTV* service while moving in flight. One company is offering a service for *DirecTV* to the moving automobile.

On a more commercial basis, time and bandwidth on existing satellite systems can be reserved and rented on an hourly basis. Watching TV news, you can detect whether the newsperson is reporting via satellite (significant delay) or fiber optic cable (no delay). This service, often used to televise sports events, relies on mobile, but not moving, ground antennas and can provide HDTV bandwidth.

Implication

Simplex satellite service works well. Half- or full-duplex service has (propagation) latency delays due to the distances traveled.

The implication of satellite television is so great that satellites are being placed in orbit to create an estimated 1,500 TV channels, most of them HDTV. Consider the bandwidth this will create.

Fixed and Movable Wireless

Cellular

Cellular telephones operate from omnidirectional radio towers, located in cells of two to five miles in diameter. The existing infrastructure covers most cities of any size and all major highways, giving voice and some data coverage. Although designed for moving receptors, cellular companies are eyeing their technology for broadband, fixed wireless, Internet access. Present handheld cell phones use this system for data through short message service and Internet access, demonstrating the data capability, even though it is presently operating at less than 40 Kbps. Although higher speeds of and above 384 Kbps will not be available until generation three of most cellular protocols, the infrastructure exists for fixed-wireless lower-broadband transport. A building that requires connectivity installs one or more cellular attachments and the cell systems do the rest. A significant advantage of this technology is that it can be moved without interference with the signal.

A specific market for cellular-based broadband service, especially Internet access, is being developed for hotels. We are seeing data ports appearing for wired access and wireless access will take its place. Thus, a businessperson can enter a hotel, Starbucks® coffeehouse,

FIGURE 9.8
Wi-Fi Telephone

Source: ZyXEL

or some McDonald's® restaurants and have wireless fee-based access to a laptop or palm device. Many airlines provide Wi-Fi access (hot spots) in their terminal waiting areas and most airlines offer this service in their hosted waiting areas. A second use of this fixed form for cellular systems is light Internet access to the home, competing with DSL-light and shared cable modem service.

Implication

Broadband cellular has a significant future because the infrastructure is in place, client phones are pervasive, and people are becoming used to relying on this "telephone" for many services, both audio and video.

Wireless Fidelity (802-11a/b/g)

Using nonregulated spectrum, Wi-Fi offers mobile, moving, and non-line-of-sight voice and data. With repeaters, Wi-Fi can "see around" obstacles and with the use of wireless ad hoc mesh networks, networks can be expanded with just the addition of router-based repeaters, generally requiring no human operation. The 802.11 realm offers reasonable security for data, giving good mobility in the warehouse or on the battle field.

Portable VoIP = Wi-Fi as Cellular[22]

The Prestige 2000W VoIP Wi-Fi phone [see Figure 9.8], compatible with IEEE 802.11b wireless standard, is a perfect solution for Voice over IP applications. It allows users to make or receive phone calls as long as they are in the coverage of IEEE 802.11b or 11g wireless Access Points. By using the Prestige 2000W, users no longer have to pay expensive communication fees and can enjoy the convenience of wireless mobility. The brand new application is developed to support open standard SIP (Session Initiation Protocol), which interoperates with major SIP-based call servers, IP-PBXs and various VoIP client devices. It is not only an ideal alternative for ITSPs (IP Telephony Service

[22] http://us.zyxel.com/products/model.php?indexcate=1109113163&indexcate1=&indexFlagvalue=1079378556.

Providers) to deploy their VoIP services; it can also be the wireless handset, which is applied in corporate IP-PBX centric VoIP environment. The Prestige 2000W is very easy to use and configure. It allows users to configure with LCD screen menu or Web browser. Meanwhile, with the smart auto-provisioning mechanism, ITSPs can easily deploy and manage the VoIP services. Easy-to-use and convenient, the Prestige 2000W delivers high-quality voice functionality in a cost-effective way.

- IEEE 802.11b support
- Frequency band: 2.400 ~ 2.497 GHz
- Channel: FCC Ch1~11, ETSI Ch1~13, Japan Ch1~14
- Data Rate: 11 / 5.5 / 2 / 1 Mbps
- Output Power: 14 + 1dBm
- Sensitivity: −82 dBm@11Mbps
- Operating range: Out-door up to 300m, In-door up to 75m
- 64/128 bit WEP encryption
- Site Survey: Scan available APs in hand set's environment
- Support infrastructure (public) mode and Ad-hoc mode (option)

$400 @ Amazon.com; $199 @ http://wlanparts.com/.

WIRELESS METROPOLITAN AREA NETWORKS

The IEEE 802.16 *WiMAX* standard is a wireless metropolitan area network technology that will provide a wireless alternative to cable, DSL, and T1/E1 and is transforming the world of wireless broadband. WiMAX is a wireless networking standard that offers greater range and bandwidth than the Wi-Fi family of standards. When deployed, WiMAX will transfer data at about 70 Mbps over a distance of 30 miles to thousands of users from a base station. WiMAX will enable "next generation" wireless broadband applications and will help to fundamentally transform many businesses and enterprises.

Implication   WiMAX and Wi-Fi will soon be so pervasive that they may supplant cellular service.

WIRELESS WIDE AREA NETWORKING

The earlier discussion about WANs has been exclusively conducted on wired networks. As wireless (radiated) connectivity becomes more and more routine, the wireless portions of the total end-to-end connection may be considered a portion of the WAN. That is, the user may be communicating between the wireless node and an access point that is directly connected to a PDN PAD. This idea extends the reach and use of the wired WAN and will likely push for greater bandwidth in WANs to keep up with the high bandwidths of wireless links, such as 802.11g running at 54 Mbps.

When using push-to-talk walkie-talkie devices, the intent is to use WAN connectivity, in that the users may be remote and the communications are not homogeneous. Even though cellular providers have nationwide push-to-talk service using UTP and fiber, the users see this as a wireless WAN, just as the cellular system itself. The real impact of wireless WANs will become apparent when the data portion of cell phones increases and the user is unaware of whether s/he is using cellular or Wi-Fi technology. This will likely be the killer application that pushes broadband technologies into the wireless domain, whereas they seem primarily in the wired domain at this time. As instant messenger (IM) is presently wired and short message service (SMS) is wireless, these two separate services have the same objective and will likely merge into a wireless realm. Then, as IM started as text and then added **convergence** in the form of graphics and images, so will SMS, resulting in the need for greater bandwidth. This is part of the push for greater data bandwidth to connected devices, such as cell phones and PDAs.

Meanwhile, wireless technologies, whether near or remote, have greater security issues than their wired counterparts. The nature of wireless is to radiate information into the unprotected airwaves. Thus, wireless WANs, just like wireless local connectivity, must concentrate on security in the form of encryption in addition to the spread-spectrum mode of transmission. (See Table 9.5.)

Because wireless devices are untethered, they often become the end node of choice, whether the network is a LAN, MAN, or WAN. For the interim, wired broadband may be the transport of choice, but connection to wireless devices is vital.

Implication

A businessperson without wireless connectivity is not really in business.

TABLE 9.5   **Characteristics of Wireless Devices**

Wireless Device, P-to-P, P-to-MP, Nomadic	Voice, Data	Analog, Digital, Technology	Use	Bandwidth	Range	Frequency (MHz)	Specification
Cordless telephone P-to-P, limited nomadic	Voice	A, DFM radio, spread spectrum	Portability, wire replacement	3 KHz	100 ft 200 ft 500 ft	34, 900 2,400 5,300	
TV/stereo/DVD remote P-to-P	Data	D (pulse) IR	Control	100 bps	50 ft	IR	
Cellular telephone P-to-MP, nomadic, push-to-talk	Voice D/SMS Voice	D, CDMA, TDMA, GSM D = VoIP	Nomadic voice, Internet access, short message service Walkie-talkie	3 KHz	5-mile cell	900 1,900	CDMA TDMA AMPS
Personal digital assistant Nomadic	Data	D	Internet access	19.3 Kbps			
Laptop IR port P-to-P	Data	IR	Connectivity	115 Kbps	50 ft	IR	
Bluetooth P-to-MP	Data	D	Connectivity		10 meters		802.15.1
Wireless Fidelity (Wi-Fi) P-to-P, P-to-MP, nomadic	Data	Digital, packet radio, spread spectrum	Network connectivity	11 Mbps 54 Mbps 108 Mbps	150 ft 150 ft 300 ft	2,400 5,300 2,400	IEEE 802.11b IEEE 802.11a IEEE 802.11g
Mesh-enabled architecture P-to-MP, nomadic	Data	Same as 802.11 by routing	Network extension through retransmission	Same as 802.11			
WiMAX P-to-MP	Data	D	Metropolitan area network (Wireless-MAN)	75 Mbps	30 miles		IEEE 802.16
Microwave P-to-P					30 miles		
Direct broadcast satellite, VSAT P-to-MP	TV, radio, data	D	Video-audio entertainment Internet access		22,000 miles		
LMDS/MMDS P-to-MP	TV, data	A, D	Broadcast TV, Internet access		25 miles		
GPS P-to-MP, nomadic	Data	A	Positioning		Global		
Satellite telephone	Voice						
Ultra wideband	Data	D	Messaging, data				802.15.3
ZigBee							802.15.4

Case 9-1

Can the Technology of RFID Ensure Security?

Randy Conley

In World War II with German forces just across the English Channel, England needed a technology for identifying incoming aircraft as either "friend" or "foe." The first radio frequency identification (RFID) system was created, called IFF, or identification, friend or foe. Although the newer technology to identify units through wireless transmission was created in 1969 and then patented in 1973, the commercial success of the technology has had a slow start. Recently however, science fiction has become science reality as a seemingly endless stream of applications for the technology is developed. RFID is now being used for tracking inventory, livestock, and employees. Other applications being developed include implanting a tag in important documents (including cash in Europe) or implanting chips underneath the skin of people so that Alzheimer's patients and children can be tracked. This burst in usage also has attracted attention from the technology's critics, who have focused on the social, ethical, and legal risks associated with implementing the technology. In addition, the security of the devices themselves and the information they contain has become a hot topic for the technology to overcome.

WHAT IS RFID?

> Radio frequency identification, a technology similar to bar code identification. With RFID, the electromagnetic or electrostatic coupling in the RF portion of the electromagnetic spectrum is used to transmit signals. An RFID system consists of an antenna and a transceiver, which read the radio frequency and transfer the information to a processing device, and a transponder, or tag, which is an integrated circuit containing the RF circuitry and information to be transmitted.[23]

Stated differently, RFID has two components. The first is a tag that is placed on the item to be tracked. This solid state device, that is, chip, is passive until activated; that is, it has no battery but can store information in a nonvolatile format. The second part of RFID is the transceiver/antenna unit, which queries the tag, energizes it, and either places information on the chip or reads information from it. Thus, the chip's onboard information can be updated as it passes specific points or it can simply reply to multiple queries with who it is.

WHO IS USING THE TECHNOLOGY?

To fully grasp the drastic changes that RFID could and most likely will bring about, we must consider the many companies and groups currently interested in the technology. First, Wal-Mart, the largest retailer and employer in the United States, has planned and, in some areas of the country, begun tagging its entire inventory with RFID tags. The company began testing individually tagged products. They remain dedicated to RFID and have continued to develop RFID tagging by pallet. [8] In order to maximize the profitability of their estimated US$3 billion investment, Wal-Mart must have corresponding RFID systems with its suppliers. According to multiple sources, Wal-Mart, who also perpetuated the widespread usage of the bar coding system, [3] has required its 100 top suppliers to tag all of their products by 2005. It is their intention to know, as the trucks are being unloaded at their distribution centers, the content of every box and then be able to allocate the appropriate products in correct

[23] http://www.webopedia.com.

proportions to Wal-Mart stores whose RFID readings show a low stock of those products. Complying with Wal-Mart's demands, Gillette has acquired 500 million tags from Alien Technology, a company that focuses on cutting-edge technology. [5] Others will definitely follow as Procter and Gamble, Sun Microsystems, Exxon Mobil, Visa, Michelin, Gap, Prada, Home Depot, and many more have begun using, servicing, or creating the technology. The federal government also has begun using RFID to track and direct shipments of military weapons and equipment. Fire departments and other rescue and criminal defense teams (police and S.W.A.T.) are researching the possibility of personal identifier tags to locate individuals in emergency situations. The European Central Bank has even investigated the possibility of embedding RFID tags in the euro note as an anticounterfeit procedure. [3]

APPLICATIONS AND BENEFITS OF RFID TAGS AND BADGES

RFID is a technology that will replace and augment barcoding. It can carry more information, the information can be changed during the process, and visible, line-of-sight access is not required. It is these features that make the technology's potential so great, for example, the collection, storage, and continuous tracking of items. The use of RFID for retail ventures expands the benefits of UPC and barcodes in such areas as quick and constant inventory systems and antitheft technologies. These inventory systems can save the companies millions of dollars in labor and loss due to theft and can save less tangible resources such as time. Delta Airlines has begun testing RFID for customer bags to reduce baggage loss and ease the changes to routing when customers change flight plans. [3] Also, the use of tags can allow instantaneous collection of product purchasing data to the company. Wal-Mart already compiles data every time an item's barcode is scanned, but the RFID system would make it easier for other companies to build systems to collect this data upon checkout. A customer walks out of the store with items with tags, instantly the product types, brands, and purchase times are recorded. If items are purchased using a personal identification badge (discussed below), a target market can be established using the purchaser's information.

Often discussed in the RFID field is the movie *Minority Report*. The movie contains scenes where individuals enter a shopping center and the advertisements on billboards instantly change to reflect his previous purchase information. Such a system could be made possible by RFID technology.

Other applications include a chip containing an individual's credit card information that could be instantly accessed as the customer leaves the store. These technologies have been available and used in the marketplace for quite some time. Exxon Mobil's *Speedpass*® was made available nationally in 1997. The Texas Instrument technology allows users, with the touch of an RFID token attached to his/her keychain, the ability to purchase gas at the pump and hurry through the checkout line inside the store. The RFID device responds to the RF query and transmits the credit card and personal information of the user to the pump, even including the answer to the age-old question "would you like a receipt?" Similarly, the Seattle Seahawks have used TI RFID systems to create the PowerPay token. Using Secure Marketing Architecture for Retail Transactions (SMART), payment is wirelessly transmitted as the transaction takes place. Lines at the stadium have all but disappeared. [1]

The result of using these technologies is an improvement in the bottom line for the companies and a greater satisfaction and convenience for the customers. Wal-Mart, for instance, is predicted to save $6.7 billion by eliminating the labor needed to scan the numerous barcodes encountered during shipping. Another $600 million in revenue is expected for the reduction of empty shelves in Wal-Mart stores and $575 million in loss could be avoided by making it more difficult for employees of the company to steal products. Finally $300 million will be saved by simply having a better idea of the location of products through the supply chain and $180 million in storage costs could be avoided. [8] Customers will benefit by saving both time and money. It may be possible for customers to walk out of the store without waiting at a register. Also, a trickle-down effect would place a portion of corporate savings in the pockets of employees or in discounted prices on products purchased every day. With all the prospective benefits for these technologies, the question could be asked, what is all the fuss about?

THE ISSUE OF SECURITY

The issue of the security of RFID tags has been founded on some serious allegations. The first is that the signal transmitted by the device, in response to a query, is easily disrupted. Mark Baard writes in "Is RFID Easy to Foil?" that the signal emitted from an RFID tag can be easily limited by many common materials including aluminum foil, most liquids, and even human flesh. This fact indicates that most tags can be rendered useless as a security device if held in a closed fist or placed in a bag lined with aluminum foil. These concerns were confirmed by Matt Reynolds, a principal at an RFID systems development company, ThingMagic. He stated at a Massachusetts Institute of Technology RFID privacy workshop held in November of 2003, "Any conductive material can shield the radio signals. There are all kinds of ways to render the tags inoperable." [2] If the tags are rendered inoperable, a thief could easily walk out of the store with valuable objects hidden from RFID readers.

A second security concern for the RFID industry is the securing of data contained on the RFID tag or badge. RFID badges could contain information such as credit card numbers, account numbers, PIN, name, addresses, and phone numbers. This information enables the user to utilize the badge for quick identification and purchase. However, what if this information is stolen by a rogue transmitter querying the RFID device? Much like a credit card, the information on the RFID badge would give a thief the ability to make purchases under the original owner's account, that is, identity theft. The difference between theft of credit card and theft of RFID information is that for RFID theft, no contact need occur between the thief and the victim. RFID can be stolen from across the room. In addition, the use of badges for employees would give a thief the ability to steal company information from an employee carrying his badge. Dan Kaminsky describes such a scenario in "RFID Security":

> Passive RFID systems are powered by the outside world—the evil demon of Cartesian yore is handing over the battery. Given a cooperative RF field, the chip spews the same bits, over and over and over again.
>
> When an employee is standing in front of the legitimate badge reader, this is a good thing. When an employee is sitting on the subway on his way to work and some guy walks by with a power source and 13.56Mhz sniffer in his briefcase . . . well, I guarantee you that briefcase ain't going to beep "Thank you for your access credentials, I'll be you now." All the attacker needs to do is [to] forge a standard plastic badge and covertly trigger a transmitter when approaching the door—there's no way for anyone to know the badge wasn't the source of the RFID transmissions! [5]

The ability to easily steal information or product from a company using RFID tags may cause more problems for companies and outweigh the benefits. In a time of highly intelligent electronic thieves, placing valuable information on a vulnerable data chip could result in devastating consequences.

In the instance of European Central Bank, where they are discussing embedding euros with RFID technology, a thief does not need to steal data, only ping it. A thief could stalk the streets of London, pinging passersby, to see who has enough money to make a mugging worthwhile.

CAN THE TECHNOLOGY OF RFID ENSURE SECURITY?

In responding to the security issues of RFID, it is easiest to divide the technology into its applications: inventory management and RFID security systems, employee identification, and customer and credit card instant identification.

Inventory Management and RFID Security Systems

For inventory management, RFID is incredibly beneficial. Instant and constant inventories, easy importation and exportation of items, and data information collection make the investment in RFID a must for companies in the next decade.

Though RFID as a security management system is troublesome, it will not be harmful for an organization to switch to the new technology. Current electromagnetic devices are equally troublesome and their signals are just as easily disrupted. If RFID will have any effect, it will be an improvement. By taking a nightly inventory rather than a monthly or quarterly inventory, a retail store can more quickly determine which items were stolen during the course of the day and identify employees or customers who had access to those items. The RFID system is capable of monitoring inventory even more often, such as hourly, which enables companies to identify the hour during which inventory disappeared. Then they can review security tapes and would be more efficient at determining who stole missing items. Finally, sections of the store that have more consistent theft can be identified and actions can be taken accordingly.

Employee Identification

The use of RFID in employee identification is a more complicated matter. Information security is of utmost importance for corporations today, especially in the field of sensitive, such as militaristic, research. With the ease of intercepting data available in RFID badges, technology must be updated in order to protect valuable company information. Kaminsky describes several methods to combat this threat. [5] A read/write system could be designed for the badges. This option could be used to maintain unique but constantly changing IDs for employees of an organization. Basically, when an individual uses his ID with the company reader, it verifies the identification number and writes a new random number to the ID. This would reduce the window of opportunity for an attacker to obtain and use an employee's ID to access company data. The attacker would have to steal an identification code and enter the system before the originally intended user. An automatic alert could be developed to identify the use of repeat codes. If the attacker attempts to access the system after the code has been used, he could be immediately identified. Even if the intended user was the second individual to use the code, the organization would be able to detect the break-in.

A second addition for RFID security could drastically improve the effectiveness of the first and keep the badges cheap and simple. A squeeze sensor could be created so that two contacts must be touching for the ID to be transmitted. This would prohibit the attacker from publicly intercepting the transmission. A third suggestion is to create an encryption code. With the power limitation of a badge, Kaminsky was skeptical of the possibilities of encryption. A random responder to his article suggests a fairly low-tech but efficient way to slow the attacker's download of the badge's information and simultaneously contain "dummy" information to randomize the code. The complicated process is best explained by him:

> . . . [I]t seems like without using too much power (and I believe power is the only signifi-
> cant limiting factor for some applications), you could throw a shift register (for decoding
> serial data) and a ROM onto the RFID chip. You fill the ROM with random data, and you
> have the scanner transmit an address. Then look up the address and transmit the data you

find there. Basically, it's a single-use password. As I understand it, you can build a ROM without any transistors at all, and thus presumably with very low power consumption. (Power consumption not proportional to the capacity of the ROM, that is.)

Obviously, it would be possible to eventually uncover the contents of the ROM just by pinging it wirelessly, but that would take quite a while if the ROM contains a few megabits of random data. Especially if [you] build in a delay that only allows one scan response every minute. (You could pull that off not with a clock but with some kind of hardware-based delay, like a capacitor.)

Still, you might say, someone might spend 20 minutes next to you on the train and discover 10% of the ROM's contents, and then they'd have an 0.1 (10%) probability of being able to use your ID to gain access to something. If this is not good enough, then have the scanner transmit 15 random addresses in a row and require 15 correct responses from the ROM. If you know 10% of the ROM's contents, your chances of getting all 20 right are $0.1 \wedge 15$, or about 0.0000000000001%. Seems safe enough to me. Even if you have 50% of the contents of the ROM, your chances are still only about 0.003%. For extra added protection, it might be feasible to have the device track which random data has/hasn't been broadcast so that the device is eventually "used up."

Using these three technologies, the ability of attackers to steal employees' identification information from RFID badges would be nearly neutralized.

Customer and Credit Card Instant Identification

The same three technologies suggested for preventing theft of employee identification could also be effective for protecting customers. However the concern of imposing on privacy is always an issue when new information technology is developed. With millions of credit cards to be replaced by RFID tags, there will be ample opportunity for criminals to steal valuable information. The systems should be thoroughly and repeatedly investigated to ensure the security of personal information available on RFID badges. Horror stories of identity theft will no doubt surface in the early years of this technology and will, in all likelihood, always occur in some minuscule portion of the consumer market. However, these occurrences are the cost of convenience; it will be up to society to decide if their existence is enough to disband the technology for more primitive technologies. I do remind everyone that current technologies of credit cards and online shopping and banking are frequently plagued with similar attacks.

FINAL THOUGHTS

RFID technology is here. There is no denying that within 10 years nearly everything we purchase will be equipped with tags. Millions of people and billions of objects will carry personal identification badges containing credit card information and account numbers. Driving the adoption of this technology is the potential for cost savings, inventory management, and retail scanning, along with personal convenience. The security risks are real but so are the solutions. There will always be a risk of theft in a capitalistic society, but, through experience and testing, technology will be developed to improve safety and security.

REFERENCES

1. Texas Instruments.11-Nov-2003. "PowerPay RFID Payment and Marketing Solution Speeds Concession Purchases at Seahawks Stadium with Technology from Texas Instruments." November 11, 2003. At http://www.ti.com/tiris/docs/news/news_releases/2003/rel11-11-03.shtml.
2. Baard, Mark. "Is RFID Easy to Foil?" *Wired News,* November 18, 2003. At http://www.wired.com/news/privacy/0,1848,61264,00.html.

3. Granneman, Scott. "RFID Chips Are Here." June 27, 2003. At http://www.theregister. co.uk/content/55/31461.html.

4. Haley, Colin. "Are You Ready for RFID?" internetnews.com, http://www.internet-news.com/wireless/article.php/3109501, November 14, 2003.

5. Kaminsky, Dan. "RFID Security." Doxpara Research, http://www.doxpara.com/read. php/security/rfid.html, November 17, 2002.

6. McCullagh, Declan. "RFID Tags: Big Brother in Small Packages." News.Com, http://news.com.com/2010-1069-980325.html, January 13, 2003.

7. Meta Group. "RFID Security Scares Ignore Facts." ITworld.com, http://www.itworld. com/nl/it_insights/12102003/.

8. Roberti, Mark. "Analysis: RFID—Wal-Mart's Network Effect." Ziff Davis Media Inc., http://www.cioinsight.com/print_article/0,1406,a=61672,00.asp, September 15, 2003.

Case 9-2

Building a Home Network

Brandon Billingsley

(Continuation of project from Case 4-1)

Wireless networking is replacing wired networking because of ease of installation. Wireless networking is a very good solution if a person does not have a way to physically run wire to every room in a home and it also allows connectivity while moving around the house. Wireless networking is inherently insecure and steps must be taken to ensure the integrity of the network and data. The steps that you should follow in setting up a wireless network are

1. Buy a wireless- capable router and broadband connectivity from a cable or phone service provider.

2. Connect the router to a power source and connect any computers that are not going to be connected wirelessly with category 5 networking cable or connect wirelessly to the open network with a wireless-capable computer, for example, a laptop. If your laptop does not come with wireless capability built in, purchase an 802.11b or g wireless PCI card or a USB wireless adapter.

3. Log in to the software on the router and change the default password to something you can remember, change the SSID to something other than the default, and turn SSID broadcasting off Figures 9.9 and 9.10 show the admin screens where you can do all of this.

4. While logged in to the router software, go to the Security tab and enable WPA security if all of your networking cards support it. If your wireless network cards do not support WPA, enable WEP encryption but be aware that this is just a deterrent to anyone trying to get into your network and can be broken rather easily. Figure 9.9 shows the screen where you can enable your wireless security.

5. Go to each machine that will be connecting to the network wirelessly and enter the key for the type of encryption you used into each one. Figure 9.11 shows the screen that would come up when you try to connect to a protected wireless network. Enter the key into each computer and you will connect.

6. After you have completed the previous steps to secure your network, then connect your router to the DSL or cable modem with category 5 network cable.

FIGURE 9.9
Changing Wireless Access Point Default

Source: Linksys

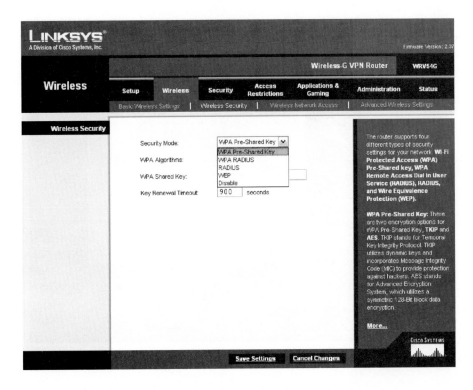

These steps will allow you to connect to the Internet and to any other computers that are connected to the router with a moderate level of security. Additionally, you could set up MAC address filtering to ensure that only the computers you have physically touched and know the MAC address of can connect to your network to increase the level of security.

FIGURE 9.10
Changing SID and Security

Source: Linksys

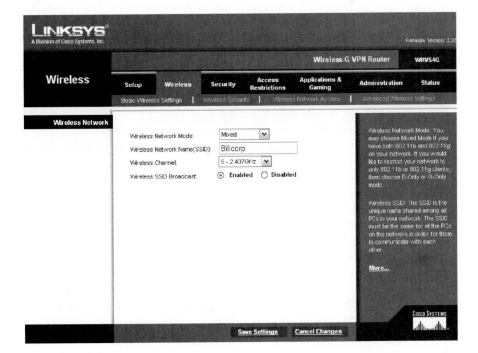

FIGURE 9.11
Protected Wireless
Network Log-on

Source: Linksys

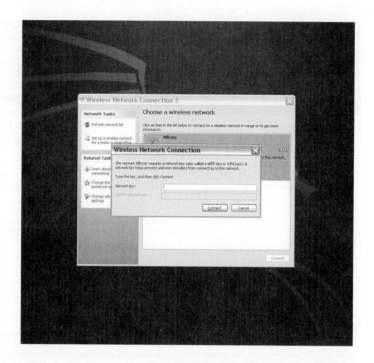

Remember that the bandwidth shared is on a random basis. That means that each node may be receiving or sending data at different times and, therefore, have access to the total bandwidth of the DSL or cable modem, to the limits that the router may restrict any node. The ISP service will generally severely cap the upload bandwidth to present users from installing servers.

Summary

The nature of wireless devices provides important capabilities in both the untethering of network nodes and the extension of bandwidth. While data communications connectivity began with wired networks, the movement to wireless capabilities is logical and cost-effective. Individuals are mobile and global and wireless capabilities support their needs. Sending a company's representative into the field without a cell phone is unthinkable, so why should we not have equipped machines and nomadic people inside the company with similar supporting capabilities.

We have noted previously that convergence drives bandwidth. This is evident in wireless devices as the pager is integrated into the cell phone along with voice recording, a camera, a viewing screen for e-mail, and Internet access. All of these capabilities make the user more productive; all require more bandwidth than the original voice cellular capability was designed for. Thus, the increase in wireless capabilities is driving the need for greater bandwidth, which is represented in the movement from 11 Mbps 802.11b to 108 Mbps 802.11g and its metropolitan connectivity in WiMAX. Cities are investigating making their entire footprint a Wi-Fi hotspot. The idea that began in coffee shops and moved to fast food and pizza parlors is expanding to encompass whole cities, giving these metropolitan areas a competitive advantage in attracting business.

As all organizations strive to reduce cost to remain competitive, the wireless capability of RFID takes center stage. Although not without opponents as to security and privacy issues, RFID is a wireless capability that seems to be able to change the way business runs, from the way inventory is managed to checkout of retail counters.

The brooch that the crew of the Federation starship *Enterprise* wear is only an extension of the capabilities we discussed in this chapter. The technology will be developed. Will organizations understand their ability to seize upon them and change the competitive landscape?

Key Terms

access points, *p. 246*
ad hoc, *p. 247*
architecture, *p. 247*
bandwidth, *p. 245*
Bluetooth, *p. 259*
broadband, *p. 255*
cable television (CATV), *p. 263*
carrier frequency, *p. 249*
code division multiplex access (CDMA), *p. 253*
connectionless, *p. 252*
convergence, *p. 266*
direct-sequence spread spectrum (DSSS), *p. 253*
electromagnetic, *p. 257*
Electronic Product Code (EPC), *p. 258*
encryption, *p. 251*
Federal Communications Commission (FCC), *p. 256*
frequency, *p. 247*
frequency-hopping spread spectrum (FHSS), *p. 253*
IEEE 802.11, *p. 249*
IEEE 802.15.1, *p. 259*
IEEE 802.15.3, *p. 261*
IEEE 802.15.4, *p. 260*
IEEE 802.16, *p. 255*
interference, *p. 253*
Local Multipoint Distribution System (LMDS), *p. 263*
media access control (MAC), *p. 252*

medium, *p. 245*
mesh network, *p. 247*
mesh-enabled architecture (MEA), *p. 246*
Multipoint Microwave Distribution System (MMDS), *p. 263*
network access point (NAP), *p. 251*
network interface card (NIC), *p. 245*
network operating system (NOS), *p. 245*
node, *p. 245*
noise, *p. 253*
omnidirectional, *p. 249*
orthogonal frequency division multiplexing (OFDM), *p. 253*
packet, *p. 249*
packet radio, *p. 249*
piconets, *p. 260*
point-to-multipoint link, *p. 247*
point-to-point link, *p. 246*
postal, telephone, and telegraph (PTT), *p. 256*
public service commission (PSC), *p. 256*
push-to-talk, *p. 256*
radio, *p. 245*
radio frequency identification (RFID), *p. 256*
range, *p. 247*
receiver, *p. 246*

reception area, *p. 249*
regulation, *p. 253*
reliability, *p. 246*
RF interference (RFI), *p. 246*
right-of-way, *p. 247*
satellite, *p. 247*
scatternet, *p. 260*
short message service (SMS), *p. 256*
spectrum, *p. 247*
spread spectrum, *p. 247*
terrestrial, *p. 262*
ultra wideband (UWB), *p. 261*
unidirectional, *p. 249*
wide area networks, *p. 253*
Wi-Fi hotspot, *p. 254*
WiMAX, *p. 255*
Wired Equivalent Privacy (WEP), *p. 252*
wireless broadband (WiBro), *p. 256*
Wireless Fidelity (Wi-Fi), *p. 249*
wireless local area network (WLAN), *p. 249*
wireless metropolitan area network (WMAN), *p. 255*
wireless network, *p. 245*
wireless T1, *p. 247*
ZigBee, *p. 260*

Recommended Readings

Mobile Enterprise (http://www.mobileenterprisemag.com/) and www.Mobile EnterpriseMag.com are targeted at decision makers responsible for selecting and purchasing mobile and wireless solutions and implementing them into the business processes of the enterprise.

Mobile Computing (http://www.mobilecomputing.com/) is the premier magazine for mobile computing and communications, especially end-user devices.

Information on Mobile Computing (http://www.mobileinfo.com/Education/magazines.htm).

Messaging News (http://www.messingnews.com) is the one-stop Web site for mobile computing and wireless information.

Network Magazine (http://networkmagazine.com).

Network World (http://www.networkworld.com/) understands the critical role technology plays to ensure the success of any business today and can connect you with today's technology decision makers, key influencers, and buyers.

Management Critical Thinking

M9.1. What are the organizational implications of moving all voice and data communications over private or public wireless networks?

M9.2. Many firms have increasingly incorporated communications technology into their products; for example, Ford's MobileEase allows drivers to place and receive hands-free phone calls through the vehicle's audio system with *Bluetooth*® technology. How can a manufacturer of other consumer goods employ data communications technology to differentiate its products?

Technology Critical Thinking

T9.1 What are the technical implications of an organization replacing all of its data communications with wireless technology?

T9.2 What are the implications of broadband packet radio?

Discussion Questions

9.1 What are the potential benefits that a sales organization might find deploying wireless applications to its sales force?

9.2 What applications of wireless would seem to favor infrared over radio-based systems?

9.3 What would you recommend as ways to mitigate primary concerns surrounding RFID deployments?

9.4 Discuss the differences between Wi-Fi, Bluetooth, and ZigBee from a consumer standpoint.

9.5 What are the major issues facing business if ubiquitous mesh wireless networks emerge to connect huge numbers of the world's population.

9.6 Bluetooth technology allows personal area networks without wires. What uses would you find for this?

9.7 Discuss the differences between Wi-Fi, Bluetooth, and ZigBee from a organizational standpoint.

9.8 Discuss how you and your classmates are using 802.11.

9.9 What has been the environmental impact of wireless systems? Include microwave, cellular, radio, and so on.

9.10 Does the pager still have a place in wireless communications? Where do you find it being used?

Projects

9.1 Evaluate an RFID application for (a) the automobile industry, (b) the medical delivery system, and (c) law enforcement.

9.3 Find a business that relies upon wireless applications and describe them in terms of technology, benefits, and costs.

9.3 Determine how two firms use wireless today and what technologies they expect to deploy over the next five years.

9.4 Investigate the present status of WiMAX. Is it being used? How?

9.5 Contact a company with a sizeable warehouse. How are they using wireless technologies?

9.6 Perform an Internet search for additional information on Wi-Fi, Bluetooth, and ZigBee. Are they complements or competitors?

9.7 UWB seems like an outer-space technology. Is it being practically used? What are some uses you can think of?

9.8 Evaluate the use of a wireless network versus a wired network in terms of bandwidth alternatives.

9.9 Determine the present social status of RFID. Search the Internet for alternate views of its use.

9.10 Draw a home WLAN. Access http://www.Tigerdirect.com or http://www.pcconnections.com and price it.

9.11 Call large companies and ask how they are actually using RFID. Does use match projections?

Chapter **Ten**

Wireless Networks: Issues and Management

As wireless networks are developed and deployed, several significant issues arise. Such management issues as access points, points of failure, total costs of ownership, maintenance, and support are discussed in this chapter. Protocol choices and regulatory issues abound as we seek intensive use of the limited spectrum. New management issues arise and require attention. Training and certification of network professions is a topic of interest. Wireless security is an issue of such magnitude that it is discussed in detail in Chapter 12.

This chapter continues the discussion of wireless technology, focusing on concerns and issues. Wired network technology is relatively mature, even while it evolves; wireless technology, in general, and packet-radio technology, in particular, are just now evolving to a point of stability. There are features of nomadic wireless systems that are specifically different from any fixed wired systems.

Wireless devices range from fixed microwave to moving satellite receivers to nomadic cell phones and PDAs. The movement is towards consolidation of all of our personal wireless devices. As a single device matures that can support voice and data communications plus location-based services, our dependence on this technology will increase even more.

Wireless networks have specific value due to ease of installation and extension; they have problems because of range variations due to obstacles and environment conditions. When settling on a wireless network, right-of-way concerns may be lessened but logistic issues may be increased. Fixed nodes tend to be the easiest to support, although remote devices may require exceptional considerations. Nomadic clients are generally trivial to add to a network but a much more serious consideration for support.

WIRELESS ISSUES

Voice communications began as wired, analog telephone technology; data communications began as wired, digital telegraph technology. In the early 1960s, Bell Labs developed the pulse code modulation technology that made it possible to convert analog signals to a digital form, providing for compression, **encryption,** and greater noise control. Optic fiber provided significantly greater **bandwidth** for these digital data in a, basically, noise-free environment. **Broadband** technology supports many new capabilities. A significant problem is that wired is all tethered. Wireless technologies promised to "cut the cord" and allow users to overcome constraints on connected mobility. We should recall that wireless technologies remain extensions of the wired infrastructure for broad connectivity.

Wireless capabilities have advantages and disadvantages. As cell or cordless phone users will readily attest, and all things being equal, being untethered (wireless) is the only way for voice communications. With the advent of short message service on cell phones and **personal digital assistant (PDA)** access to e-mail and the Internet, untethered personal digital communications had arrived. Wired-equivalent security, however, has not; all things are not equal.

Number of Access Points Required

Where a wired network required a central circuit long enough to reach all of its users, a **wireless network** requires that the reception area encompass all wireless users. This generally means one or more **access points (APs).** Consider a single access point wireless network, as would be found in a conference room, small office, coffee shop, restaurant, or unobstructed area of up to 1,000 feet radius. (Such would be the case for a technology exhibition in Las Vegas, allowing attendees **hotspot** access for e-mail and the

Internet). However, as *obstructions increase,* the *number of access points required increases.* As the number of users increases, the number of access points required increases because the bandwidth is used up. (Back to the technology show; a sniffer found that there were dozens of access points operating in the same frequency spectrum with hundreds of users. This caused unbelievable conflict and packet collisions, resulting in such a reduced bandwidth and interference that many users could simply not log on.) This illustrates that as the need for bandwidth or reach increases, the number of access points required increases. Also, if the building has metal studs in the walls or the terrain has obstructions, the number of access points required increases to overcome these problems. (When we installed 802.11b access in our College of Business building, the number of access points required was 15, where only four were planned. The culprit was metal studs in the walls.)

Implication
How Much Band-width Do You Get?

Wireless LANs have limited transmission **capacity.** Networks based on 802.11b have a bit rate of 11Mbps while networks based on 802.11a have a bit rate of 54 Mbps. *Media access control overhead alone consumes roughly half of the normal bit rate.*

Implication

The D-Link AirPlus XtremeG DWL-G800AP is a High Speed 802.11g Wireless Range Extender that can be configured to perform in two operation modes: Wireless Access Point or Wireless Repeater. Adding the DWL-G800AP to your existing wireless network enables you to extend the wireless signal, particularly useful when walls, ceilings, or other architectural obstacles inhibit the wireless signal from reaching its desired destination. When used as an access point, the DWL-G800AP is fully compatible with 802.11b-compliant devices. Security features include WPA and 128-bit WEP encryption for stronger network security.[1]

Implication

Capacity is shared between all the users associated with an access point, and since load balancing doesn't exist on access points, network performance can be improved dramatically if the appropriate number of access points are available to users.[2]

Complexity

The definition of **complexity** states that the problems in a network will increase as the square of the number of devices or the number of vendors. Thus, a network with 15 APs will experience 14 times greater potential for problems than the same network with four APs ($15^2 = 225$; $4^2 = 16$; $225/16 = 14$).

More Points of Failure

Wireless networks would seemingly be simpler to construct and, therefore, be less prone to failure. Each end node of a wired or wireless network will have a NIC; wired networks will have a physical circuit from the NIC to the hub; wireless uses space. A wired network can support dozens of end nodes with a single hub or switch; wireless networks may have to have multiple access points due to distance and obstacles. (Remember that our College of Business required *15* access points instead of four). As the number of nodes increases, the complexity increases and the probability of a failure increases-that is, there are more **points of failure.**

Packet versus Cell

Cellular voice systems operate on continuous radio technology, using hub-and-spoke, that is, star, topology within a **cell.** The (shared) data channel passes information as packets between the tower and the indented receiving node. The information may be SMS data, data downloads from the host or Internet, e-mail, or requests for service to the tower.

---

[1] http://shopping.howstuffworks.com/xPF-Dlink_D_Link_DWL_G800AP_802_11g_Wireless_Range_Extender_IEEE_802_11g_54Mbps_2_4GHz_wireless_range.
[2] http://www.computerworld.com/mobiletopics/mobile/story/0,10801,86951,00.html.

When requested by the data channel, a continuous, dedicated voice channel is established between the tower and the cellular telephone. The (cell) channel remains until the call is terminated or the phone leaves the cell. It is this dedication of the channel that causes blocking during times of high voice traffic, such as on game day at many southeastern university stadiums. Should a tower of a provider fail, all traffic to that tower, that is, all traffic in that cell, would cease as all but the nodes at the edge of the cell would be too far from another tower to establish service.

Packet-radio service, that is, 802.11 protocol, also may use a hub-and-spoke topology in a small reception area, such as a localized Wi-Fi hotspot, because only one access point would be used. For all but the small, localized domains, multiple access points would be used, in somewhat of an overlapping cell mode. The 802.11 protocol, using packets, does not establish dedicated channels, thereby making better utilization of the spectrum and bandwidth. However, with the failure of an access point, all receiving nodes in that reception area would cease to operate unless they were within range of another access point. Where cellular technology tries to define specific cells for each tower, 802.11 may be created with redundancy, that is, more than one access point serving any receiving node. Because the nodes expect duplicate packets, redundancy works well. In the case of redundant coverage of a reception area, the loss of one access point will not stop service, but it will reduce bandwidth.

Implication

Wireless networks require adapters in the computers (built right into some models, such as notebooks built around Intel's Centrino®). They communicate with a wireless base station. Range varies with a number of factors including the location of the base station, the arrangement of its antennas, and the type of construction of the building. (Steel and concrete reduce the range, while wood-frame buildings are relatively invisible to the radio waves.) Notebooks with a built-in antenna generally get better wireless performance than models that use a smaller antenna built into a plug-in card. Even the material of a notebook computer's case can affect performance. Apple's high-end G4 Titanium Powerbook, for instance, has worse wireless range than the same company's much lower-priced plastic-shelled iBooks.

Officially, 802.11g base stations are supposed to have a range of about 300 feet (100 meters) out of doors; in the real world, range varies. But that implies standard antennas. (*It's also extremely optimistic; try 50–90 feet being usual.*) Using specialized antennas on the transmitting base stations and on the receiving computers, range can be improved dramatically.[3]

If a notebook or desktop computer does not have 802.11 built in, it may be added easily by a PCI card or USB plug-in that resembles a thumb drive.

As mesh-enabled architecture is deployed, it will provide inherent redundancy and extension of area services to any access point within its reception area. The MEA device may be an unsuspecting user or overt repeaters. In any case, this topology and protocol will overcome the problem of node failure.

Range

The wireless remote control unit for your TV/stereo is nomadic and uses IR signaling. Its reach is from the user to the TV/stereo set. Just about all other wireless **nomadic nodes** use radio communications. The reach of a network is the range between the access point and the receiving node. Reach is dependent on the output power of the transmitting node or access point (or tower), the presence of obstacles, and environmental factors that cause path loss. Most devices operating in the public domain spectrum are limited in their power output, thus limiting range. Examples of obstructions range from metal studs in walls, to lead paint in older buildings, hills, trees with wet leaves, and the metal of houses in trailer parks. Even with a clear view, the range may be limited as pollution and rain cause path loss.

[3] http://www.zisman.ca/Security/networks_printer.htm.

Advertisements tout that 802.11 can have ranges up to 1,800 feet. Realistically, bandwidth drops off significantly after 50 feet, so some consider the range of 802.11 to be less than 150 feet. This and the reception area's environment will determine the number of access points required.

Implication

A *Pringles*® can Wi-Fi antenna is an inexpensive way to get greater distance for Wi-Fi. The point is that it is using a very directional antenna; thus, gain up to 9 db over an omnidirectional antenna is possible. For those wishing only to eat *Pringles* and not play with the can, Cantennas are available for purchase.[4]

Cables and cable connectors induce signal loss, while antennas are designed to introduce signal gain. These products cite a loss or gain rating in product specs. A loss of −3 dB cuts power (mW) in half, while a gain of +3 dB doubles power. A −10 dB loss reduces power to one-tenth, while a +10 dB gain increase power 10 times. The actual formula is (Pdbm = 10log-PnW), but using 3's and 10's can approximate signal gain/loss without a logarithmic calculator.[5]

Clarification

Because omnidirectional antennae radiate in all directions, the transmitted energy dissipates rapidly with distance.

Inverse-square law: The physical law states that irradiance, that is, the power per unit area in the direction of propagation, of a spherical wavefront varies inversely as the square of the distance from the source, assuming there are no losses caused by absorption or scattering. For example, the power radiated from a point source, for example, an omnidirectional isotropic antenna, or from any source at very large distances from the source compared to the size of the source, must spread itself over larger and larger spherical surfaces as the distance from the source increases.[6]

Interference

Wireless networks generally require line-of-sight (LOS) communications, although signals will pass through walls that are not opaque to RF energy. **Interference** to wireless signals can occur in several ways. First, physical obstacles block the "view." Examples of such objects are hills, walls with internal metal, and even trees, especially trees with leaves and particularly wet leaves. This necessitates either moving the transmitter or receiver or placing a repeater that provides communications around the obstacle. Next, the environment itself may cause interference. For example heavy rain, fog, snow, or significant airborne pollution may interfere with the signal. Finally, microwave ovens, cordless phones, and other appliances that use the same 2.4 GHz frequency as 802.11b/g devices will cause interference, thus limiting effective range. When interference exists, actions must be taken or the wireless network will not meet its goals.

Implication

What is the residential location with the least chance of getting any distance for wireless, especially 802.11? Try mobile home parks (trailer parks). Consider that almost all of them are made of metal and metal is not only opaque to RF energy, it reflects any that hits it. Thus, to get out of the trailer, you have to have the AP at the window and aim it to where you wish to receive, hoping there are no other homes in the way. One of our students tried this with dismal results.

Battery Life

Wired equipment has historically required A/C or DC power to operate. This necessitated backup power in case of power outages and now line conditioners, surge protectors, and uninterruptible power supplies are needed for continuous, regulated power. **Fixed wireless** nodes require the same considerations. Mobile wireless devices require a mobile power

---

[4] http://www.turnpoint.net/wireless/has.html.

[5] http://expertanswercenter.techtarget.com/eac/knowledgebaseAnswer/0,295199,sid63_gci976404,00.html.

[6] http://www.atis.org/tg2k/_inverse-square_law.html.

Wi-Fi Friendly Cordless Phones

Ken Colburn

Is there a practical solution (other than switching to a 5.8 GHz phone or turning off the cordless phone) to eliminating interference between a 2.4 GHz cordless phone and a wireless network?

Wi-Fi or wireless networks are wildly popular in part due to the very low cost of the equipment. The low cost of the equipment is due in part to the use of an unregulated and unlicensed radio frequency for making the connections. This unregulated frequency (2.4 GHz) is becoming congested with lots of consumer devices such as portable phones, wireless computing devices, and even some microwave ovens that can cause debilitating interference for Wi-Fi networks.

Since the 2.4 GHz spectrum is the most popular for cordless phones and wireless networks, not only do you have to worry about your own household's use of each technology, your neighbor's choices could also impact you (especially if you live in an apartment or condo or have close proximity to many neighbors). Your cheapest option is probably to get a new cordless phone system but not necessarily a more expensive 5.8 GHz system. There are actually two kinds of 2.4 GHz cordless phones available: one that is *wireless network friendly* and one that will wreak havoc. The difference is in the transmission method of the cordless phones. The most common (that will wreak havoc) uses a modulation process called *FHSS (Frequency Hopping Spread Spectrum)* while the wireless friendly units incorporate DSSS (Direct Sequence Spread Spectrum).

FHSS hops around bull-in-a-china-shop style with no regard for any other devices, searching for the best channel in the 2.4 GHz spectrum to use while DSSS uses a more polite way of choosing the best channel. DSSS is specifically designed to co-exist with other devices (which is why Wi-Fi networks use it) and phones that incorporate it will have a manual channel selector switch which keeps it in a more confined area of the spectrum.

As for dealing with intrusions from a neighbor's wireless network, you may want to change your broadcast channel. By default, most Wi-Fi access points are set from the factory to use channel 6, which means it is right in the middle of the available spectrum. If your neighbor decides to install a wireless network and leaves the access point set at channel 6 as well, it is competing with your access point for the same channel. Try setting yours to either channel 1 or 11 so that you can be as far away as possible from the common channel 6 and then see if everyone is willing to turn off their broadcast so that your computer does not know that their access point exists and vice versa. Windows XP is designed to constantly monitor the radio waves for a stronger Wi-Fi signal and this "wandering eye" can often cause disconnections as well.

Source: http://channels.lockergnome.com/windows/archives/20050812_wifi_friendly_cordless_phones.phtml.

source, usually batteries. For example, choosing a new cell phone will entail a consideration of **battery life** in addition to the features and size of the device; the smaller the size, the smaller the battery. (U.S. soldiers now go into battle with 7 to 12 batteries each to support on-person equipment and wireless network access.[7])

Using a battery to power a mobile wireless device solves the problem of power conditioning and momentary spikes and brownouts. The problem that is introduced is the requirement for backup batteries and/or recharging facilities. Battery life is a critical constraint to the use of nomadic wireless nodes. With that in mind, vendors are already selling "power packs" and solar charging panels and investigating fuel cells to provide a **portable,** replaceable/renewable/rechargeable power source for nomadic devices. (See Table 10.1.)

Implication
Power over Ethernet (PoE)

A number of access point manufacturers (Lucent, Symbol) are now offering **Power over Ethernet** add-on's for their access points. A PoE module inserts DC voltage into the unused wires in a standard Ethernet cable (pairs 7–8 and 4–5). The idea is to supply the AP's power and UTP ethernet connectivity requirements via a single ethernet cable. This works great in areas where you may not have power and/or ethernet easily accessible, like a roof. This also allows you to more easily place the AP closer to the antenna, thus reducing signal loss over antenna cabling.[8]

[7] John Matlock, Major, U.S. Army.

[8] http://www.nycwireless.net/poe/.

It has been reported that there are currently over 1.5 billion users of portable electronic devices world wide and this number is expected to reach 2 billion by 2007, with reported annual sales of approximately 650 million devices per year, representing new and replacement sets. In this market, device manufacturers are continuing to add more and more entertainment, communication and other features on their handsets, including digital cameras, internet access, video games, messaging, PDA applications, MP3 players, FM radios and even television broadcasts. This trend is consistent with the strategies of mobile operators (service providers) world wide who are requiring that products have greater capabilities in order to increase air time usage.

This huge and growing market needs one very important factor—long-lasting, uninterrupted, portable power to run these highly functional power thirsty new devices, which batteries cannot supply having reached close to their maximum capabilities. Our disposable Power Pack is a portable auxiliary power source that delivers far longer operating time than any of today's batteries. It provides power to operate and charge even the most modern cell phones (including the most advanced "3G" cell phones with a full range of functionality), digital cameras, PDAs, MP3 players, hand-held video games and other devices with similar power requirements. When a device's battery is running low or is discharged, the Power Pack allows the continued use of the device while at the same time charging the battery. And, it can repeat this process a number of times.

A specific reason for choosing PoE over directly wired power is that wired power has significant additional installation requirements, such as conduit, whereas an existing Ethernet cable is already installed according to standard. Placing power over Ethernet occurs at no additional installation cost.

Right-of-Way

While wireless service may be subject to regulatory oversight and tariffs, it overcomes a major problem of right-of-way. **Right-of-way** is the legal requirement to obtain the permission of the property holder before access or construction on his/her property. It may involve a fixed fee and/or a recurring fee. For example, wired last-mile connectivity generally involves not only significant construction costs, but gaining right-of-way permissions. Long-haul communications may have fewer property owners with whom to deal, but the property owners must give permission before construction may proceed, even if it means only placement of a tower for lines. Digging a ditch means disrupting an area up to five feet wide for the entire length of the installation. Placing poles means placements every 50–100 feet; larger towers may be 200–500 feet apart. This all involves agreements as to the use of the ground and easements under the wires for the entire distance of installation and access permission to maintain facilities.

Implication   Companies on the east coast of South America chose to run fiber cables into the Atlantic Ocean to avoid right-of-way problems and constraints, including swamps and estuaries of rivers.

Wireless installations avoid much of this right-of-way problem. For the last-mile consideration, there often will be no consideration at all, especially if the transmitting antenna is omnidirectional and the receiving antennas are all within line of sight and installed on the receiver's property. Satellite communications is the best of all cases when it comes to avoiding right-of-way considerations; LMDS/MMDS is a close second; and cellular and microwave, a not-too-distant third. For long-haul communications, microwave towers can be placed up to 30 miles apart, requiring only one tower every 30 miles as opposed to poles every 50–100 feet or line towers every 200–500 feet. And the only easement required is access to the microwave tower installation point.

It's Not all Free

All wired (conducted) and all wireless (radiated) nodes require electrical power at the transmitting and receiving sites. UTP and coaxial circuits have repeaters about every half mile;

TABLE 10.2
TCO for 48-User
Wired Bus Network

Item	Qty	Cost Each	Total Cost	
NICs	48	$20.00	$960.00	
Hub/switch	2	$95.00	$190.00	
NIC cables	48	$3.50	$168.00	
Network operating system	1	$1,000.00	$1,000.00	
Hub-to-hub cable	1	$200.00	$200.00	
Shipping	1	$40.00	$40.00	
Equipment subtotal				$2,558/00
Design and engineer hours	30	$40.00	$1,200.00	
Cable installation	50	$40.00	$2,000.00	
NIC/hub installation	50	$20.00	$1,000.00	
System test hours	2	$40.00	$80.00	
Installation subtotal				$4,280.00
Maintenance contract/year	3	$500.00	$1,500.00	
Repair parts (5 NICs, 1 hub)	1	$215.00	$215.00	
Cost of moving nodes	10	$50.00	$500.00	
Recurring subtotal				$2,215.00
Total cost of ownership			**$9,053.00**	

How do you refuel a microwave tower generator on an isolated mountain top?

fiber can run for 10–40 miles before regeneration. While satellite and LMDS/MMDS are single-point transmitters (satellite in space with its own power source and LMDS/MMDS terrestrial), each receiver requires power. Microwave systems are an even greater concern as the towers may be in remote areas that are not served by commercial power companies, thus requiring local generators and fuel. (Solar systems are only being used for lower-power systems.) Therefore, if the right-of-way and construction problems mentioned earlier for a circuit in the jungles of Guyana were solved by use of microwave towers, each tower would require a generator, fuel replenishment, and security.

Expense: Total Cost of Ownership

The concept of **total cost of ownership (TCO)** takes into account the cost of equipment, installation, and testing plus all costs incurred over the life of the system. For example, a simple portable telephone in a home may be a US$15.00 (Sears) purchase, ignoring the cost of the trip to get it or assuming that shipping is free. It requires no recurring expenses other than charging power and is discarded if it fails. Assuming the charger runs full out for the three-year life of the phone, the electrical cost is US$8.50. Thus, the TCO of this device would be US$23.50.

As Table 10.2 illustrates, the equipment for a 48-user wired network may cost only US$2,558 to obtain, but it costs another $4,280 to have it installed if the building was not already wired and $2,215 to maintain, adding up to US$9,053 over its lifetime. The installation estimate is based on a new or easy-access building. Older buildings or those with solid walls, such as concrete, will be significantly more difficult and costly to wire. As an allied cost example, the installation of a single electrical outlet in one author's cubicle in industry took a jackhammer to drill through the reinforced concrete floor.

A wireless network, assuming that four access points will be required, could cost as illustrated in Table 10.3. While this illustration does not favor wireless from the equipment purchase point, the wireless could be installed significantly more quickly and cheaply and expanded with ease and minimal cost and the TCO is lower. Wireless bandwidth could be added through the addition of access points plus faster NICs, whereas the wired network would require replacement of NICs, cables,[9] and possibly hubs/switches.

[9] If the wired network was built using CAT 3 wiring, the wiring would have to be replaced with CAT 5 wiring as bandwidth approached and exceeded 100 Mbps.

TABLE 10.3
TCO for 48-User
Wireless Network

Item	Qty	Cost Each	Total Cost	
NICs	48	$50.00	$2,400.00	
Access point	4	$350.00	$1,400.00	
NIC cables	0	$3.50	$0.00	
Network operating system	0	$1,000.00	$0.00	
Access point cables	3	$20.00	$60.00	
Shipping	1	$40.00	$40.00	
Equipment subtotal				$3,900.00
Design and engineer hours	10	$40.00	$400.00	
Cable iInstallation	3	$40.00	$120.00	
NIC/hub installation	4	$20.00	$80.00	
System test hours	1	$40.00	$40.00	
Installation subtotal				$640.00
Maintenance contract/year	3	$500.00	$1,500.00	
Repair parts (5 NICs, 1 AP)	1	$215.00	$215.00	
Cost of moving nodes	0	$50.00	$0.00	
Recurring subtotal				$1,715.00
Total cost of ownership			**$6,255.00**	

Implication

One author worked in an aircraft factory that covered over 100 acres of floor space. When the mainframe terminals were initially placed in the factory and offices, over 1,000 miles of coaxial cable were run. Later, it was realized that the terminals did not require coax bandwidth and UTP, often extra POTS pairs, was used, at a significantly lower cost. This was a time when wireless was not an option; it continues to not be an option in cases such as this due to DoD security.

Time to Deploy (Implementation Issues)

As noted in the previous paragraph, time to deployment of wireless networks can be significantly less than that of a wired network. This is, obviously, due to the lack of a need to install most connecting cables. This case continues when it comes to expansion and upgrade. All of these considerations are enhanced if there are right-of-way considerations, such as may occur in the case of multiple buildings that are not adjacently located.

What about Nomadic Nodes?

Movable nodes can be accommodated with wired networks but at a significant cost relative to wireless networks. Consider the cost, planning, and time delay involved with the movement of a wired desk telephone. One of the features of Rohm digital telephones of the 1990s was that the instrument could be moved to a new insertion point and the switch would recognize the instrument. This was an obvious enhancement over the standard analog instrument, as no technical intervention was required. What happens, however, when either an instrument is moved or added to a place where no cables exist? For a wireless network, movable nodes can be relocated anywhere in the reception area with no planning, cost, or time-delay considerations; the node is simply moved.

If the node in question is nomadic, it is not possible to serve it with a wired channel; wireless is required. The node was created to assume that its location is constantly changing, which necessitates a larger reception area, which can be achieved with multiple access points.

Capabilities

The idea of complexity was mentioned earlier; the more components or the greater the number of relationships, the great the complexity of a device or system. Consider that a nomadic device is often a hybrid set of technologies, housing analog voice, digitally encoded voice (VoIP), digital e-mail and Web browsing, and onboard games. This entails multiple technologies and various forms of input, ranging from keyboard to stylus, video capture, address book database with images, audible tones, GPS-enabled technology, and various size displays. Once in the field, the multitude of features greatly increases the complexity of use and support. The nomadic wireless device now includes the technology for connectivity *and* the applications for use, which are indistinguishable by the user. Earlier personal computer users called the help desk first for aid with the machine, then for aid with the applications. Wireless users will have a more difficult time differentiating the two technologies, thus the complexity of maintenance and support will increase.

Maintenance and Customer Support

Maintenance is a subset of customer support; maintenance implies support of the equipment whereas customer support implies aid to the user. In reality, they both are designed to help the user achieve his/her communications needs. When the consideration is the user asking the customer support personnel a question, there would be no difference between a wired or wireless environment. This, however, is where the lack of difference ceases, as wired environments are fixed in location and wireless environments are remote and changing. In either case, there must be a way to maintain the network and user nodes to support the user. The question we address is how to provide the support.

The first assumption for equipment and user support is to use **in-house** capabilities, either existing or newly created. The second option is to *contract with the vendors* of the equipment to supply hardware, software, and user support. A third option is to contract with a third-party organization to provide all services. An advantage is often to chose to outsource the support. This brings resources to bear that may not exist internal to the organization, quickly, and with a minimum of cost. While it may appear cheaper to use or create the expertise in-house, the TCO concept will favor the **outsourcing** of the task. Equipment support may be outsourced to organizations that have facilities near each concentration of users, or else the users will need to be able to mail or ship wireless nodes for replacement or repair. As for user aid and support, the choice would likely be an organization with groups that can cover all of the time zones in which the users are likely to need support. In-house support groups would tend to be concentrated in a single location, giving little support to remote users. Multiple-shift support is expensive, so it is advantageous to use organizations that are created to provide this support, for example, centers in the United States, England, and India.

Caveat

Outsourcing support offshore, for example, a foreign country, is not a simple matter. It may provide a cost and time coverage benefit, but making the contractual arrangement takes time and political savvy. An organization would be well advised to never outsource core competencies; the management of information and data communications systems should be considered a core competency.

Software Differences from Back Office

The technology for wireless networks is more than the wireless protocols and radio-based access points; it includes different user applications. This is a direct result of the user interface on the wireless node. Applications on desktop computers changed when the operating systems changed from DOS to GUI. This was, however, due to greater interface capabilities on the same, though evolving hardware. Wireless nodes, at least the nomadic ones, have significant constraints of on-board processing power, storage, and, most of all, interface input and output viewing accommodation. Consider that most nomadic nodes have very small if any input features and the screens are also miniature to make the nodes small and convenient. All of these parameters sum to mean that the applications that would be used on a desktop or laptop computer are not equally usable on a nomadic wireless node. Thus, new software must be created and supported, keeping reasonable comparability with the originating applications. A spreadsheet on a PDA is very different than one on a desktop. Although wireless nomadic devices offer significant value, the user software and interface are far different than customary back office applications. Increasingly, vendors offer means to interoperate with "normal" systems or at least be able to transfer data seamlessly so that the PDA can be more easily integrated into the organization's architecture.

Signaling Speeds: Bandwidth

Wired (conducted) media generally have greater bandwidth than wireless (radiated). There is no question that optic fiber can out perform any other technology, wired or wireless. This makes it most useful for long-haul, high-bandwidth applications. A key question is what bandwidth is required for the user and how will the network supply this total multi-user bandwidth? The answers tend to evolve as technology evolves.

Wired local area networks have seen the migration from 2 to 10 to 100 to 1,000 megabit service, with a resulting requirement to replace all node and channel equipment. This has occurred over as short a time frame as 25 years. The increase in bandwidth has been fueled by adding nodes to the network and increasing the requirements of each node by **convergence.** Just as network users moved from text-based to graphics-based material, resulting in the requirements for greater bandwidth, so has the wireless world. Because wireless users were accustomed to the touch and feel (interface and material) of the matured wired network, they expected the same level of material on their wireless devices.

The IEEE 802.11g specification calls for 54 Mbps; proprietary engineering creates double that bandwidth. Some manufacturers call this the turbo mode, producing 108 Mbps.

Wireless networks will continue to experience a demand for greater bandwidth. Where cell phones began operating at a data rate of 19.2 Kbps, 802.11g channels offer up to 108 Mbps, an increase of over 500,000 percent. This is driven by the increase in the number of users and types of applications. As with wired networks, wireless requirements have moved from text to streaming video. A specific feature of wireless networks is that increased demands for bandwidth can be accommodated by the addition of access points, unlike that for wired systems. Wired networks ultimately have a finite pipeline/bandwidth; wireless networks can increase the bandwidth to users with additional access points as long as the total bandwidth serviced by the access points is adequate.

Proprietary Signals Protocols

Wireless systems have different protocols, depending on the capability. As noted in Chapter 2, cellular telephone relies on two primary protocols: **code division multiple access (CDMA)** and **time division multiple access (TDMA).** Not only do the two protocols not **interoperate,** the cell systems don't interoperate because of different carrier frequencies assigned to different systems. When observing a cell tower, the number of cell antenna sets, for example, the number of three-antenna layers, reveals vendor differences. In one part of our city, there are four towers within a square mile; three have two antenna sets each and the other has six. The logic of this separation is discrete separation by vendor. This means that when a user goes outside of his/her vendor's tower system, s/he may enter the **roaming** domain. Roaming can be good and bad. The good is that a vendor's network is extended through reciprocal arrangements; the bad is that there may be a roaming fee charged.

For Wi-Fi packet radio, 802.11b and 802.11g, to interoperate, they use the same frequency and basic technology, just the bandwidths are different. Unfortunately, 802.11a does not interoperate with either b or g due to a different frequency assignment.

Regulatory Issues[10]

Because radio-packet-based services in particular, and radio frequency carrier-based services, in general, may broadcast into the *regulated spectrum,* state and federal/national agencies have a vested interest. The U.S. Congress has given regulatory authority to the **Federal**

---

[10] http://www.wrf.com/publications/publication.asp?id=153952232004.

Communications Commission (FCC); a primary charge is to specify classes and *uses of spectrum,* setting aside some for public, unlicensed usage. Assigned with the bands of spectrum use are allowed *power levels.* If transmitters wander out of their assigned spectrum channels or exceed power limits, interference results to those who receive transmissions that were not intended for them. In the days of citizen band (CB) radio, some truckers would install output amplifiers along with heavy-duty antennas. The result was the increase of broadcast and reception range along with significant interference as the signals bled into adjacent channels. A similar problem exists in the home as 2.4 GHz telephones, and even microwave ovens, interfere with 2.4 GHz-based **wireless access points.** The FCC is especially sensitive to interference in the 800 MHz public safety band.

Interference of great concern occurs in environments where the user devices may be transmitting in spectrum used by *critical equipment,* such as in health care facilities and on commercial aircraft. While the Intel Corporation put wireless *Centrino*® technology into laptops to seek a competitive advantage, its use is not allowed on many commercial flights.

The newest technology on the consumer market is **radio frequency identification RFID.** Since it is designed to be activated by passing through an RF field, any environment of RF energy at the appropriate frequency will cause interference. Examples found are bug zappers, cellular towers, walkie-talkie radios, and forklifts.

The FCC is continuously being requested to allocate additional spectrum for a specific user or use category. As the total spectrum is finite, regulatory agencies can only *reallocate spectrum.* At conflict are military, general civilian, safety, and public use of existing spectrum and requests for spectrum for new uses. One way that vendors are addressing these spectrum problems is via new technology. For example, CDMA provides 8 to 15 times the capacity of the analog cellular it replaced through spread spectrum. TDMA/GSM makes efficient use of cellular telephone spectrum by using time division multiplexing. With a large installed base of TDMA/GSM, moving to CDMA would be difficult, even if proved to provide more capability, for example, more channels for the same amount of spectrum. Meanwhile, however, categories of "smart" radios are being developed to increase the *spectrum efficiency* beyond these technologies.

Ultra wideband (UWB) technology design calls for overlapping several spectrum bands (interference), but it uses a technology that makes the signals appear as low-level **noise** to other than the intended receiver. This is a way to use existing allocated spectrum without interfering with present users.

Counter to the efficient use of spectrum is the consideration for **quality of service (QoS).** Users demand not only increasing bandwidth but good-quality voice with good wireless coverage. This adds more pressure on bandwidth usage and demands for additional spectrum.

Cost oftentimes seems to allocate scarce resources; the FCC is considering greater use of spectrum leasing as a way of meeting the competing demands. This means that a wireless application may confront not only regulatory issues but significant cost issues. *Spectrum leasing,* that is, auction, bankrupted some providers in Europe as the customer base did not support the significant up-front cost.

802.11 *Wi-Fi* and 802.16 **WiMAX** broadcast in the public domain spectrum. As wireless becomes even more prevalent, the traffic in the 900 MHz, 2.4 GHz, and 5.8 GHz bands will continue to increase, using up the bandwidth. Portable phones have moved from 900 MHz for just that reason. As new technology is developed to use the **unlicensed spectrum,** the condition will only get worse. For example, when a city provides citywide Wi-Fi access, there will be a great tendency to use Wi-Fi in place of cellular, greatly increasing the demand for 802.11 spectrum and lowering it in the cellular bands.

Voice over IP is thought of as a wired technology. VoIP is important in our discussion of wireless issues because many wireless conversations travel a portion of their trip over wired IP networks. Additionally, cellular technologies of push-to-talk and instant messaging use VoIP technology. As users start to view wireless devices as just another client with processing power with richer displays, use of gaming, video streaming, and other forms of entertainment, they will demand more and more VoIP wireless bandwidth.

Emergency **E-911** services started with wired telephones and quickly moved to cellular phones as they increased in popularity. As voice communications become a commodity and technology of simple phones moves to multitasking wireless nodes, such as PDAs and handheld computers, E-911 must be considered for each of these devices. Originally this meant radio triangulation to locate a wireless device, for example, cell phone, requesting help. With GPS-enabled devices, the instrument broadcasts its location at the time of the E-911 call. As the number of E-911-compliant devices increases, so does the pressure on the safety band.

Health Hazard Potential

Do you think cell phones pose a health hazard? How many hours a day do you hold one within four inches from your retina and brain?

Whereas wired data communications pose no **health hazard,** wireless devices potentially may. One concern is for wireless access points that broadcast RF energy with people nearby. A more specific health concern has been for wireless devices held close to the user's head, such as the case of cell phones. Specifically, cell phones are held within four inches of the *user's brain and retina,* both of which are sensitive elements of the body. Although research has not indicated a specific threat, many users still have a grave concern for this hazard.

A second, and more likely, health hazard is the concern for the *disposal of wireless devices,* especially their batteries. There were more than 153 million cell phone users in the United States by 2005, and because phones have a relatively short life—averaging 18 months to two years—there are many retired phones not being put to good use, said Travis Larson, spokesman for the Cellular Telecommunications & Internet Association.[11] Nokia notes that the cell phone demand for China alone in 2006 will be 200 million new devices with their batteries.

The U.S. EPA estimates that more than 350 million rechargeable batteries of all types are purchased annually in the United States.[12] Batteries are a unique product comprised of heavy metals and other elements that make things "portable," such nickel cadmium, alkaline, mercury, nickel metal hydride, and lead acid. It is these elements that can threaten our environment if not properly discarded. Batteries may produce the following potential problems or hazards:

- Pollute the lakes and streams as the metals vaporize into the air when burned.
- Contribute to heavy metals that potentially may leach from solid-waste landfills.
- Expose the environment and water to lead and acid.
- Contain strong corrosive acids.
- May cause burns or danger to eyes and skin.

The solution? Search the Internet for "recycle cell phones" or "recycle batteries" and see the number of agencies that offer to help.

SECURITY

The considerations of security are of such importance that we devote two full chapters to them. Please refer to Chapters 11 and 12 for this coverage. Meanwhile, realize that wireless communications are always in the open and, thus, always susceptible to interception. To intercept messages on a wired network, physical attachment is required; this is not true

---

[11] http://www.foxnews.com/story/0,2933,107794,00.html.
[12] http://www.earth911.org/master.asp?s=lib&a=electronics/bat_env.asp.

for wireless networks. The worse case is that others listen to 802.11 traffic totally in a passive mode, that is, their presence is undetectable.

MANAGEMENT

Issues and management tend to be different sides of the same coin. That is to say that issues must be addressed by management procedures. The issues discussed previously were concerns or problems that needed to be addressed. The following are concerns, but of a more positive nature.

Network Administration for the Wireless World

Network administration is the process of designing, installing, maintaining, and upgrading networks and the attached equipment. This involves the physical circuits, client machines, servers, and connecting devices such as switches and routers. Wireless networks beyond their wired connectivity are primarily access points and (nomadic) clients.

If the future is truly based on a mobile wireless environment, then network administration will have to be remote from most of the supported nodes. Support of a wired LAN means that the administrators are within walking or driving distance of the nodes and circuits requiring maintenance and repair. For a network of constantly moving nomadic nodes, administration must take place remotely, without knowing where the nodes are located. This implies a need to access remote, wireless nodes as well as an ability for upgrading them without touching them.

The process of remote network administration began with wired LANs and **point-to-point** access with capabilities such as *PcAnywhere®* and *Carbon Copy®*. With each of these, the administrator is able to take over the node in question and see what the user saw, oftentimes solving problems by making changes on the fly. Consider how this problem intensifies with a wireless network. Access points are the simple part, as they are stationary. As will be discussed in Chapter 12, the installation of (unauthorized) rogue access points adds a new sort of problem. Meanwhile, the number of clients will tend to change as will their capabilities. All of them will require support. We discuss this matter of support to a greater extent in Chapter 14.

Training and Certification

Training

As technology changes increasingly from wired to wireless, specialized training will be required. First, engineers and designers will use different standards for implementation and different protocols, along with radio towers and their complement of hardware, power, and physical security. Technicians must learn to troubleshoot and maintain new types of equipment with the nodes fixed, movable, and nomadic. This means trouble and repair at a distance, requiring a new skill set. Finally, users will have to learn to use new equipment in different ways. As always, these users will find innovative ways to implement and use the equipment, resulting in the need for changes in design, installation, administration, and maintenance.

Certification

In the IT industry, **certifications** can lead to better opportunities and better compensation. Certifications can be viewed as a surrogate for experience for the new employee and differentiate him/her from other individuals.

Certification proposes to address two issues: knowledge/training and recognition. Network administrators must understand how to install, maintain, repair, and upgrade equipment and channels. Wired networks have well-established certification programs, organized around vendor equipment such as *Microsoft®, Cisco®, Linux®, and Novell®.* The wireless world of network administration begins with the *Certified Wireless Network Professional (CWNP)* program.

The CWNP Program[13] is the industry standard for wireless LAN training and certification. Obtaining any of the CWNP certifications brings valuable, measurable rewards to network professionals, their managers, and the organizations that employ them. The CWNP Program has three levels of knowledge and certification covering all aspects of wireless LANs. These levels allow students to pursue as much knowledge about wireless networking as they need, including the critical area of wireless security.[14]

- **Foundation:** *CWNA*™ *(Certified Wireless Network Administrator)* is the foundation of the CWNP program. Covering a broad range of wireless networking topics, CWNA brings technical people new to wireless networking up to speed quickly. For those already familiar with wireless LANs, earning the CWNA certification fills in any gaps in their knowledge, and officially proves expertise to help provide a competitive edge.
- **Advanced:** *CWSP*™ *(Certified Wireless Security Professional*™*)* ensures that the student understands how to secure a wireless LAN from hackers and protect valuable information on the network. CWSP offers the most thorough information available on how attacks occur and how to secure your wireless network from them. Topics include advanced processes and techniques for keeping enterprise wireless network data secure.
- *CWAP*™ *(Certified Wireless Analysis Professional)* certification is an advanced wireless LAN certification, focusing entirely on the analysis and troubleshooting of wireless LAN systems. The CWAP-certified individual will be able to confidently analyze and troubleshoot any wireless LAN system using any of the market-leading software and hardware analysis tools.
- **Expert:** *CWNE*™ *(Certified Wireless Networking Expert)* credential is the final step in the CWNP program. By successfully completing the CWNE practical examination, network engineers and administrators will have demonstrated that they have the most advanced skills available in today's wireless LAN market.

Convergence Will Drive Wireless Capability and Bandwidth

Convergence has impacted the desire for bandwidth in something as simple as *e-mail*. Originally, people sent only text messages and a 300-baud modem was sufficient in most cases, with a 9.6 Kbps modem in event of more extreme cases. When there was demand to add graphics and images, then voice, and video, the bandwidth requirements soared.

For a wireless convergence case, consider the simple **pager.** Initially, the system simply sent a tone to the receiving unit and the user made a phone call to a unique number. The next advance was to send voice, not a very successful venture. Next, digits of a variable telephone number or name were sent. This was followed by digital data being sent, allowing the user to receive many lines of information to be stored on the pager and reviewed at will. Now, advanced pagers have reply capabilities. As these changes occurred, the technology changed, necessitating new equipment and investment.

Personal digital assistants (PDAs) started life as stand-alone text-only memory devices. Then users wanted to synchronize them with their laptop and desktop computers, necessitating wired and then wireless connectivity and specialized protocols. Next, users wanted to be able to receive messages, as on cell phones, which required simplex communications at first and then bidirectional communications as with pagers. Now, PDAs are Internet-enabled and can send and receive e-mail. This has changed the hardware, software, and bandwidth requirements as the data sent may include images from a digital camera as well.

---

[13] http://www.cwnp.com/cwnp/certifications.html.
[14] http://www.cwnp.com/about/industry_standard.html.

Case 10-1

Wireless Application Protocol

Todd Dugo

ABSTRACT

Wireless Application Protocol (WAP) is a communication protocol based on an open-standards-based architecture that was developed for digital mobile communications. Using a microbrowser embedded inside a cell phone, a mobile user can access WAP content (similar to HTML) such as stock, news, and weather information over the wireless network with the help of a WAP gateway and/or support servers. Like most information technologies, WAP has some security issues that have been identified and are being addressed. The future for WAP looks fairly promising as the de facto standard for the mobile World Wide Web for the foreseeable future.

INTRODUCTION

Most (if not all) of today's Internet users are undoubtedly familiar with the World Wide Web (WWW) and have probably spent hundreds (if not thousands) of hours "surfing" the Web from the comfort of their offices, home desktop computers, and laptops. However, many mobile users also want to access the same type of information available via the Internet from their mobile digital wireless devices such as digital cell phones. The technology that enables mobile users to do this is Wireless Application Protocol (WAP). This paper describes the background of the development of WAP, its architecture and major network components, some of the security issues, the future direction of WAP, and finally some conclusions.

BACKGROUND

WAP is a communications protocol. A *protocol* is defined as a standard or set of rules or guidelines that govern the interaction between people, between people and machines, or between machines and are developed and implemented according to a specified standard or architecture. [10] WAP release 2.0 is the current version of the standard, based on an open architecture. The WAP Forum (part of the Open Mobile Alliance) serves as the focal point for development of WAP architecture and standards. [3]

One might ask why we have to have another standard or protocol for accessing the WWW when we already have well-known WWW technology standards such as Hyper-Text Transport Protocol (HTTP), HyperText Markup Language (HTML), Web browsers such as Netscape Navigator, and JavaScript. The answer is that many of these standard WWW components we use when surfing the Web from our homes and offices are not suitable for the current relatively low bandwidth and multitransmission standards (e.g., GSM, TDMA, CDMA) of the wireless networks and the small user devices found in the mobile wireless environment. WAP, on the other hand, is optimized for the mobile wireless environment and is designed to overcome mobile issues such as latency, low bandwidth, and limited display sizes and to take advantage of hand-controlled navigation features found on wireless devices. WAP content is also scalable from a two-line text display on simple mobile devices to full graphic content on the latest mobile devices that have larger and/or color screens. [2] The WAP architecture is discussed in the next section.

WAP ARCHITECTURE AND NETWORK COMPONENTS

Protocol

As mentioned earlier, WAP is based on an open architecture. An open architecture is an equipment and software design that lets the hosts interoperate based on standard protocols. [10] This means you can purchase a digital WAP-enabled cell phone and use it to access WAP content over a wireless connection without regard to the manufacturer of the cell phone as long as the cell phone was designed to WAP standards and specifications. The WAP protocol architecture is illustrated in Figure 10.1. As one can see, it is a modular architecture consisting of several layers similar to the TCP/IP protocol. A discussion of each of the WAP protocol layers is beyond the scope of this paper.

WAP Network

A simplified view of the typical WAP network elements is shown in Figure 10.2. In this simplified WAP network, a mobile user initiates a session from the WAP microbrowser, contained in the cell phone, via a URL request (much like using a Web browser on a home/office computer). The request is then routed through the wireless network and usually to a WAP gateway or proxy inside the Internet. The WAP gateway or proxy then performs any necessary conversions and/or translations and then routes the request to the appropriate content server. [2,11] The content server, in turn, may query a support

FIGURE 10.1   **WAP Protocol Architecture**

Source: WAP Architecture Version 12, July 2001, p. 18.

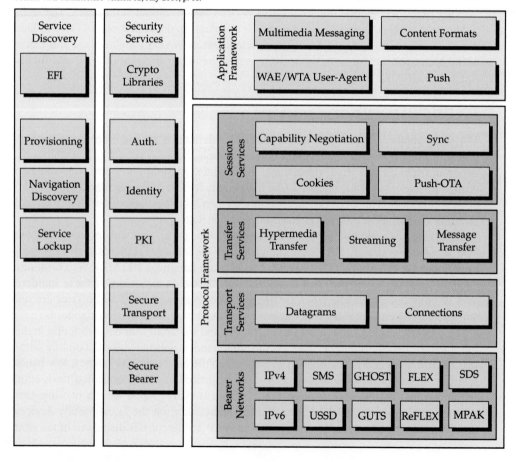

FIGURE 10.2 Simplified WAP Network

Source: http://www.iec.org/online/tutorials/wap/; WAP Architecture Version 12, July 2001.

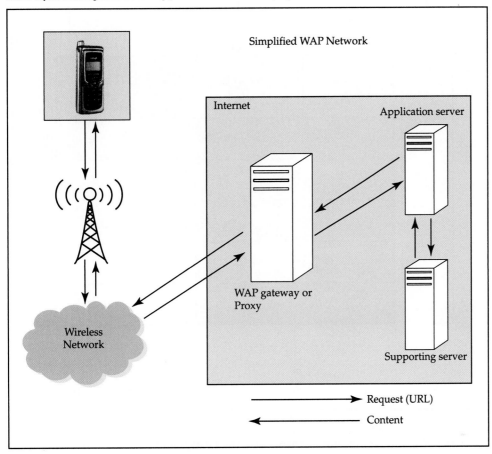

server to provide additional support such as PKI, user profile, or provisioning information. [11] The response information or content is then transmitted back to the mobile user in byte code, where it is interpreted and then displayed on the mobile device's screen. [2,11]

Microbrowser

The microbrowser is a simple Web-based client (sort of like a super-scaled-down Netscape Navigator) embedded in the user's mobile device. Many major Internet sites such as Yahoo! Mobile have Web content that is specifically designed for viewing using a microbrowser. [5] Many wireless providers such as Alltel Corp. provide WAP applications such as games, stock quotes, sports scores, news, and weather to wireless customers (usually for a service fee). The information from these applications is displayed on the mobile device's screen. [1]

WAP Gateway

The architecture of the WAP gateway is shown in Figure 10.3. Most of the intelligence needed for browsing WAP content is placed inside the WAP gateways and the WAP applications and content are located on servers. This allows developers to place most of the resources demanded on the gateways and servers instead of in tiny mobile devices that have limited memory and processing power. This makes WAP a client-server based architecture. [4]

FIGURE 10.3
WAP Gateway Architecture

Source: http://www.iec.org/
online/tutorials/wap/topic04.
html.

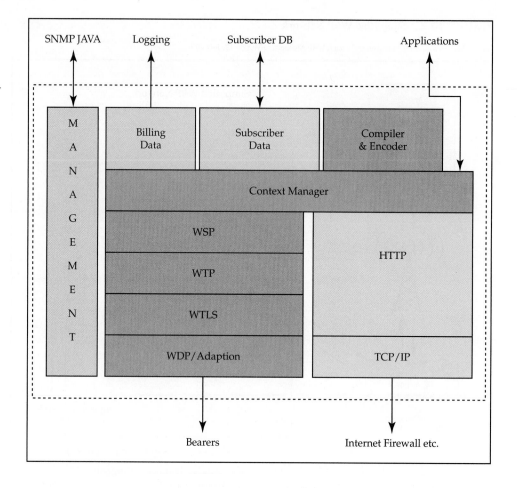

The WAP gateway or proxy consists of several datagram-based protocols:

- WAP datagram protocol (WDP). This layer is the transport layer that sends and receives messages via any available bearer network.
- Wireless transport layer security (WTLS). This layer is an optional security layer that has encryption facilities to provide the secure transport service required by many applications, such as e-commerce.
- WAP transaction protocol (WTP). This layer provides transaction support, adding reliability to the datagram service provided by WDP.
- WAP session protocol (WSP). This layer provides a lightweight session layer to allow efficient exchange of data between applications.
- HTTP interface. This interface retrieves WAP content from the Internet when requested by the mobile device. [2]

In general, the WAP gateway or proxy converts data between the above datagram-based protocols and the Internet-based protocols (i.e., TCP/IP and HTTP). The gateway also performs domain name service lookups of the servers named by the client in the request URLs. [2,11]

WAP Content

WAP content is created using WAP markup language (WML), which is based on Extensible Markup Language (XML), and WML script, which is similar to a scaled-down version of

JavaScript. A WML page is known as a deck, and a deck is composed of a set of cards. The two are related together by links. When a user accesses a WML page from a mobile phone, all the cards contained in the page are downloaded from the WAP server to the phone. The user can then navigate between the cards using the phone's microbrowser without having to access the WAP server again. [6] A WAP gateway or proxy also can create WML data from HTML-based data as required. [4] These technologies create content that is suitable for display on the small screens of today's generation of digital mobile devices. [2]

WAP Security Issues

Many Internet users have personally dealt with a security issue such as a computer virus or worm or know of people or companies that have. Is WAP also vulnerable to security issues like those found on the Internet? Unfortunately the answer is yes. In 2004, the Cabir virus was the first reported virus capable of attacking and spreading by cell phones. It currently infects phones running the Symbian operating system that is used by many mobile phone manufacturers. The virus (really a worm) appears to be a benign "proof of concept" design and only displays the message "Caribe" on the infected cell phone's screen. Most phones are safe from this virus as the virus appears to only spread using **Bluetooth** connections to other Bluetooth-enabled devices; thus, you have to have a Bluetooth-enabled phone in order to become infected with this particular virus. [7]

Another WAP security issue that has recently been identified is known as WAPjacking. In WAPjacking, the malicious code (a type of Trojan horse) manipulates the WAP settings on your cell phone. It then changes your cell phone's WAP home page, and then switches the call to a premium rate number. [8]

Experts working WAP security issues are currently focusing on securing the WAP gateways, running virus detection software on the WAP gateways, or getting rid of the WAP gateways where possible since they appear to be the most vulnerable to attack. Other items of importance are improving the use of user authentication and data encryption. WML may itself be vulnerable to exploitation as well as the WML scripting language. [9].

In sum, not many cell phones have been trashed by a WAP virus, worm, or Trojan. This is not to say there aren't any malicious people out there who are not interested in wreaking havoc on mobile users; it is probably just a matter of time until these mobile devices are exploited in a more harmful way.

FUTURE DIRECTION

The collaboration by numerous parties in developing and standardizing the WAP protocol via the WAP Forum has been accomplished on an unprecedented scale. WAP has the potential to do for mobile telecommunications what Netscape Navigator did for the Internet. [2]

As discussed earlier, WAP was primarily developed to work over low-bandwidth wireless connections and on small mobile devices with limited processing power and small screen displays. [2] Many cell phone manufacturers have incorporated WAP technology into their latest digital mobile products. As these mobile devices evolve and become more powerful, and new higher-speed wireless Internet technologies become available in the mobile domain, will WAP evolve to take advantage of the increased capabilities, or will it become obsolete and supplanted by more standard/traditional Internet technologies that we are more familiar with such as HTTP, HTML, and JavaScript? Only time will tell.

As WAP technology matures and proliferates, hackers will undoubtedly increase their efforts to target mobile users and exploit the weaknesses in the technology. The cell phone manufacturers and the software vendors (who supply the operating systems for those phones, WAP gateways, and supporting servers) will undoubtedly continue to strive to secure their products and protect their customers from malicious hackers.

CONCLUSION

WAP appears to have been developed with unprecedented speed and collaboration to enable multiple cell phone manufacturers, multiple wireless communications providers, and multiple content providers to offer mobile users access to WWW-like content from the Internet. Thus, WAP appears to be the standard for mobile WWW access for the foreseeable future. As WAP technology matures, security will undoubtedly become more of a focus. To hackers, the focus will be to exploit the vulnerabilities of WAP; to developers, the focus will be to eliminate those vulnerabilities; to users, the focus should be to be aware of the vulnerabilities inherent in WAP and take the necessary measures to minimize the potential for harm.

REFERENCES

1. http://www.alltell.com/axcess/apps.html.
2. http://www.iec.org/online/tutorials/wap/.
3. http://www.openmobilealliance.org/about_OMA/index.html.
4. http://www.gsmworld.com/technology/wap/intro.shtml.
5. http://mobile.yahoo.com/.
6. http://www.w3schools.com/wap/wap_basic.asp.
7. http://enterprise-security-today.newsfactor.com/story.xhtml?story_title=Cabir_Virus_First_To_Attack_Cell_Phones&stor y_id=25403.
8. http://www.thefeature.com/article?articleid=100702&threshold=-1&ref=3407111.
9. http://www.techsupportalert.com/pdf/b1423.pdf.
10. H. Carr and C. Snyder, *Management of Telecommunications,* 2nd ed. (New York: McGraw-Hill, 2003).
11. Wireless Application Protocol Architecture Specification ver. 12, Wireless Application Protocol Forum, Ltd., July 2001.

Summary

With the rapid proliferation of wireless technologies, organizations may be faced with some new issues and management problems. Transitioning a large part of the networks from wired to wireless means deriving new parameters for evaluating relative costs and performing administration and maintenance. As with the introduction of any new technologies and ways of doing business, it is vitally important to address the skill sets required for operation. This usually means training and certification of specialists who can keep the systems running to support the organization. Managers should become familiar with these requirements if they are to gain the true advantages that can accrue from implementing wireless solutions to a wide variety of problems and opportunities. Wireless coverage continues to advance into public places, homes, cars, and aircraft. Some envision the end results as transparent, seamless access to services regardless of location—the network—convergence revolution concluded.

There is ample evidence that the widespread proliferation of wireless networks has brought forth many issues because of the turbulence accompanying the deployment. Along with the promise of untethering have come the problems associated with an unstable market and evolving standards. If one simply looks at the standards of CDMA, CDMA2000 1X, CDMA2000 2X, W-CDMA, EDGE, GSM, TDMA, GPRS, WiFi a/b/g and WiMAX, it is easy to see the dilemma facing managers in their choice for voice and high-speed wireless data services. These standards are only a small part of the wireless data puzzle as firms seek to adopt the services that approach their wire-line broadband bit rates.

Many firms have not yet addressed several of the relevant facets of wireless network adoption. Among these issues are infrastructure, both internal and carrier. There is a significant cost difference and an expense differential over the life of the systems chosen. Managers need to examine factors such as TCO, time to deploy, carrier support, bandwidth, architecture, and trends that might make a choice one that won't last. There also should be an implementation plan that seeks logical integration with the organization's wired architecture. There is a real need to examine the skills inventory required to administer wireless networks that is different from the existing skills so that the deficiencies may be made up.

Issues facing managers include security, technological usage, and productivity gains. The increased vulnerability of wireless means that the enthusiasm with their adoption needs to be tempered by the realities of protocol shakeout, costs, impact on productivity, and integration with the company's technical plans, goals, and objectives. This must be done with an awareness of the evolutionary trends with the Internet, standards, and security of the firm's data.

Key Terms			
	access points (APs), *p. 279*	hotspot, *p. 279*	radio frequency identification (RFID), *p. 289*
	bandwidth, *p. 279*	in-house, *p. 287*	
	battery life, *p. 283*	interference, *p. 282*	
	Bluetooth, *p. 297*	interoperate, *p. 288*	right-of-way, *p. 284*
	broadband, *p. 279*	maintenance, *p. 287*	roaming, *p. 288*
	capacity, *p. 280*	network administration, *p. 291*	time division multiple access (TDMA), *p. 288*
	cell, *p. 280*	noise, *p. 289*	total cost of ownership (TCO), *p. 285*
	cellular, *p. 280*	nomadic nodes, *p. 281*	
	certification, *p. 291*	outsourcing, *p. 287*	ultra wideband (UWB), *p. 289*
	code division multiple access (CDMA), *p. 288*	pager, *p. 292*	unlicensed spectrum, *p. 289*
	complexity, *p. 280*	personal digital assistant (PDA), *p. 279*	Voice over IP, *p. 289*
	convergence, *p. 288*	point-to-point, *p. 291*	WiMAX, *p. 289*
	E-911, *p. 290*	points of failure, *p. 280*	wireless access point, *p. 289*
	encryption, *p. 279*	portable, *p. 283*	
	Federal Communications Commission (FCC), *p. 288*	Power over Ethernet, *p. 283*	Wireless Application Protocol (WAP), *p. 293*
	fixed wireless, *p. 282*	quality of service (QoS), *p. 289*	wireless network, *p. 279*
	health hazard, *p. 290*		

Recommended Readings	
	Mobile Enterprise (http://www.mobileenterprisemag.com/)
	Mobile Computing
	Messaging News (www.messagingnews.com)
	Network Magazine (networkmagazine.com)
	Network World
	Information Week

Management Critical Thinking	
	M10.1 What are the major factors influencing the decisions for an organization to adopt wireless networks versus wired networks?
	M10.2 Using the criteria of cost, speed (bandwidth), scalability, expandability, distance, and security, evaluate a 12-node network covering 10 states in the United States for wired versus wireless implementation. Assume distance of 100 to 1,250 miles between nodes.

Technology
Critical
Thinking

T10.1 If an organization's departments are geographically dispersed and located globally, what wireless technology constraints must be addressed to achieve maximum interoperability.

T10.2 What are the support requirements for a wireless-based sales force?

Discussion
Questions

10.1 Determine the most important characteristics of nomadic devices for mobile sales in the pharmaceutical industry.

10.2 What are the considerations when choosing a new or replacement cell phone?

10.3 Examine the degree of maturity of the various technologies supporting wireless networks. What is the shift in the technologies; that is, has 802.11b disappeared in favor of 802.11g and is 802.11n active?

10.4 How can physicians employ wireless to improve patient care and save lives?

10.5 What advantages does wireless offer in a manufacturing environment where mission-critical components can halt productions if they break down?

10.6 What are the "hot" privacy issues surrounding RFID for consumer products? What technologies have been developed to enhance RFID privacy?

10.7 Is the use of 802.11 extending antennas feasible? How does their basic design create a constraint?

10.8 How can a wireless network deal with congestion?

10.9 How does E-911 work for VoIP over broadband, for example, Vonage, when a person in Texas chooses a telephone number in the Vermont area code?

10.10 When establishing a new e-business and adding remote offices, is it less expensive to deploy local wireless networks or wired networks?

Projects

10.1 Find a business that could significantly benefit from adopting wireless networks. Determine the devices, number of nodes (fixed and mobile), support infrastructure, and so forth.

10.2 Build a table or matrix for the business in project 10.1 above in order to compare alternative costs.

10.3 Search the WWW to locate case histories that show how managers use wireless effectively to remotely manage global enterprises.

10.4 Determine the organizations offering training/certification in the wireless arena. What are the relevant costs?

10.5 Determine (1) the vendors/carriers in the present wireless marketplace and (2) the wireless networking vendors.

10.6 Have teams choose emerging wireless technologies or standards and list them in these categories: (1) those with productive use in the next two years; (2) those with use within the two- to five-year time frame; (3) those that should be successful in five to 10 years; and (4) those that will take more than 10 years to be productive for the firm.

10.7 Graph the adoption numbers for satellite radio from 2000 through the latest year. List the possible business implications.

10.8 What are the alternatives to IEEE 802.15.1 (Bluetooth) devices? Compare *Bluetooth* and *Wi-Fi* in terms of advantages and limitations.

10.9 Consider a wireless enterprise with all 250 sales force personnel equipped with wireless PDAs. What are the major management issues? Include standards, security, capacity, access, and so forth, and list in a matrix.

10.10 Develop a listing of currently deployed wireless applications in the automobile industry. Include those incorporated in the products.

Part Five

Security

Security is the denial of unauthorized access (intrusion) and the protection of assets. Organizations, like people, must be able to provide a reasonable level of security for their facilities, systems, networks, and data. The concerns for the security of organization networks range from physical protection of the hardware to intrusion protection from virtual attacks from the Internet. Wired networks are inherently more secure than wireless as physical attachment is required for direct access where wireless intrusion does not require the same. Both, however, are open to attack from virtual attack once an intruder has gained access somewhere. Table V-1 illustrates the facets of security.

TABLE V-1
Tactics of Network
Security

Security Tactics	Prevention	Detection	Correction
Architecture			
Authentication			
Authorization			
Auditing			

Chapter **Eleven**

Network Security

With the advent of the Internet and broadband connectivity, over one-half of the homes in the United States are connected to the outside world, often on a continuous basis. This access involves risk to the user from a variety of sources. Although we have introduced security earlier, the topic is so important that we devote this and the following chapter to this vital topic. In this chapter, we focus on security in general; features unique to wireless security are discussed in the next chapter. Also included are diverse threats and responses as well as prevention measures required to protect networks and their data.

Clarification

Security is the capability to defend against intrusion and to, ultimately, protect assets (from access and disclosure, change, or destruction). The assets can be physical, virtual, or data. **Privacy** is the added security provided for assets, especially information, of a personal nature. Thus, we provide security to servers and privacy to databases that contain information about persons. **Proprietary** is information of particular importance to an organization such that disclosure could harm competitive advantage or divulge trade secrets. *Control* means to (1) exercise authoritative or dominating influence over; direct; and (2) hold in restraint; check. Thus, we provide security and privacy to control assets.

Computer security[1] is the process of preventing and detecting unauthorized use of a computer. **Prevention** measures help to stop unauthorized users (also known as "intruders") from accessing any part of the computer system. *Detection* helps to determine whether or not someone attempted to break into a system, if they were successful, and what they may have done. *Correction* responds to detection and other events in the environment to make changes to prevention. Adequate prevention eliminates the need for detection and correction.

SECURITY STARTS WITH RISK ASSESSMENT

Security is not a technical matter; it is not a management matter; it is both.

There are books on security in general, and network security, and wireless network security, in particular. We recommend them to you. Therefore, our tack will be to address this topic as a part of *risk assessment, risk management,* and *disaster planning.*

In the use of any **technology,** process, or methodology, someone should determine where things are likely to go bad. Managers must think about objectives, the systems, and procedures they have installed to achieve the objectives, and the weak points in the system's equipment, software, staffing, and procedures.

The assessment of **risk** and the preparation for **threats** and **disasters** may seem like **management** functions. While it is important for management to recognize the vital nature of risk management, designers of data communications systems and the personnel who keep them operational on a day-to-day basis must be acutely aware of threats, protection, and recovery in the event of a disaster. When the topic of hackers and denial-of-service attacks arises, the concepts of threats and recovery are a natural result. There are many other threats to data communications systems and the data they access and transmit that must be protected. Before one can adequately protect a resource, the nature of the threat must be determined. Once protection measures are in place, there must be plans to deal with the occurrence of unexpected events, such as a fire, flood, earthquake, or storm.

Clarification

Hacker is a term used to describe people *proficient in computers,* who employ a tactical, rather than strategic, approach to computer programming, administration, or security, as well as their culture (hacker culture). Popular media and the general population use *hacker* to mean a black hat hacker, that is, a network security hacker lacking in ethics. **Cracker** (also called *black-hat hacker*) is a hacker who commits the act of compromising the security of a system without permission from an authorized party.[2]

[1] http://www.cert.org/tech_tips/home_networks.html.
[2] http://en.wikipedia.org/wiki.

TABLE 11.1
Ranking of Threats to Computer Systems, circa 1990

Source: Karen Lock, Houston Carr, and Merrill Warkentin, "Threats to Information Systems: Today's Reality, Yesterday's Understanding," *MIS Quarterly*, June 1992, pp. 173–86.

Threats to Information Systems Weighted (circa 1990)	Rank	Votes
Natural disasters	1	324
Accidental entry of bad data by employees	2	270
Accidental destruction of data by employees	3	252
Weak or ineffective controls	4	149
Entry of computer viruses	5	128
Access to system by hackers	6	123
Inadequate control over media	7	96
Unauthorized access by employees	8	93
Poor control of I/O	9	67
Intentional destruction of data by employees	10	58
Intentional entry of bad data by employees	11	36
Access to system by competitors	12	31

We will use the title *hacker*, for simplicity, to mean either, with the tone of the context differentiating good or bad intent.

Risk involves threats from *internal* and *external* sources.

We tend to equate risk with something *external*, such as natural disasters that present the risk of power disruption or worse. Less obvious are internal events, such as the risk inherent in the adoption of a new computer-based system or the distribution of data processing and data storage across a country or the world via telecommunications networks. The implementation of IT involves significant risk from external sources and from the technology and process of implementation.

Table 11.1 shows the ranking of 12 threats to computers and networks in 1990. Paramount, at that time, were acts of Mother Nature and acts of employees. As we will show, the nature of threats changed significantly after the year 1994. Table 11.2 shows how the importance of threats changed in the early 2000s.

The Nature of Risk[3]

According to *Webster's Dictionary, risk* is "the possibility of loss or injury; also, the degree of the probability of such loss." The four components of risk are threats, resources, modifying

[3] Ronald E. McGaughey Jr., Charles A. Snyder, and Houston H. Carr, "Implementing Information Technology for Competitive Advantage: Risk Management Issues," in *Management of Telecommunications*, Houston H. Carr and Charles A. Snyder (New York: McGraw-Hill/Irwin, 2003), p. 435.

TABLE 11.2
Overall Mean Importance/ Ranking of Threats, Spring 2004

Source: Houston H. Carr and Jason Deegan, unpublished research, Auburn University.

Rank	Mean	Threat	1990 Rank
1	5.6	Virus	5
2	5.2	Unauthorized access by hackers	6
3	4.8	Network failure	
4	4.4	Unauthorized access by employees	8
5	4.3	Utility failure	
6	4.2	Unauthorized access by competitors	12
7	4.1	Spam	
8	4.1	Intentional data destruction by employee	10
9	4.0	Denial of service	
10	4.0	Accidental data destruction by employee	3
11	3.9	Terrorist attack (virtual)	
12	3.8	Improper system use by employee	
13	3.7	Inadequate control over hardware	4
14	3.5	Inadequate control over media	7
15	3.3	Natural disaster	1
16	3.1	Terrorist attack (physical)	

FIGURE 11.1
The Components of Risk

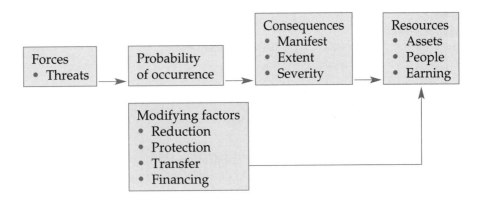

factors, and consequences. (See Figure 11.1) *Threats* are the broad **range** of forces capable of producing adverse consequences. *Resources* consist of the assets, people, or earnings potentially affected by threats. **Modifying factors** are the internal and external factors that influence the probability of a threat becoming a reality, or the severity of consequences when the threat materializes. *Consequences* have to do with the way the threat manifests its effects upon the resources and the extent of those effects.

Clarification   Risk is the expected value of the consequences of an unexpected event times the cost.

Risk assessment and **risk analysis** involve a methodological investigation of the organization's resources, personnel, procedures, and objectives to determine points of weakness. Finding such points, organizations overtly manage the risk by strengthening the weak points or passing the risk to someone else (insurance or **outsourcing** the task).

Risk management is the science and art of recognizing the existence of threats, determining their consequences to resources, and applying modifying factors in a cost-effective manner to keep adverse consequences within bounds.

The Changing View of Threats

Research done in 1990 (see Table 11.1) noted that the major threats to computer and network systems prior to major acceptance of the Internet and development of the WWW were Mother Nature and the organization's own employees. The research was repeated 14 years later, 10 years after commercial companies were allowed on the Internet and the introduction of the WWW. The view of threats has changed dramatically, to one of greatest concern for viruses, intrusion, and network failure; Mother Nature has dropped from 1st place to 15th place. As previously noted, threats come from various categories: internal versus external, intentional versus unintentional, human versus technological. Table 11.3 shows the NIST categories of threats. The other three dimensions of this table would be intent, perpetrator, and media (see Table 11.4).

TABLE 11.3
Categories of Threats/Risk

Source: NIST Special Publication 800-53.

NIST Categories	Source	
	Internal	External
Management/policy		
Operational		
Technical		

TABLE 11.4
Threats to Network
Security

Media	Source Perpetrator	Intent	
		Passive	Active
Wired	Internal		
	External		
Wireless	Internal		
	External		

Clarification

The National Institute of Standards and Technology (or NIST), formerly known as The National Bureau of Standards, is a nonregulatory agency of the U.S. Department of Commerce's Technology Administration. The institute's mission is to develop and promote measurement, standards, and technology to enhance productivity, facilitate trade, and improve the quality of life.[4]

The third listed threat in Table 11.2, network failure, has a variety of occurrences. This can be the result of an intentional event, such as a denial-of-service attack, but it also can occur due to inadequate planning. *Equipment failure* can result from poor planning, use beyond its lifetime, or inadequate maintenance. *Electrical power problems* can be caused by acts of nature, poor drivers, or inadequate planning. Organizations recognize the overt threats because of media exposure, but they may be overshadowed by the unintentional or passive threats. The following are some problems caused by unexpected threats.

Hurricanes Hugo and Andrew on the East Coast of the United States; the San Francisco earthquake; and the Hinsdale, Illinois, central office fire were very-well-publicized, significant acts of nature or accidents. Just as significant but somewhat less expected were the results of the major snowstorm in the city of Birmingham, Alabama, in early March of 1993. Over 13 inches of snow came to that *southern city,* and business halted. The city planners had not ever considered the possibility of a blizzard. The city had absolutely no snow removal equipment. If risk assessment had been complete and in place, it would have considered which telecommunications systems needed to function in spite of the snow and, of equal importance, how to protect telecommunications equipment vulnerable to water damage from the run-off as the snow melted.

Implication

The problems caused by the September 11, 2001, terrorist attacks on the World Trade Center in New York and the Pentagon in Washington, D.C., are well known because of media coverage. The destruction to network communications was severe. Considering just the technological impact of the attack, organizations recovered fairly quickly, but at great expense. Events without widespread media exposure may be as severe in the long run.

Unexpected events have caused some planners real headaches. A college in Texas placed its academic mainframe computer in the basement of a low-lying building, just above the sanitary sewer level, and the rains came. Talking about stinking grades at exam time! On the other hand, a commercial timeshare firm that knew the risk of low-lying areas for its mainframe in Chicago placed it on the fifth floor of a 10-story building. Snow fell, crushed the roof, and flooded the computer despite its lofty positioning.

A large company in south Georgia that relies heavily on computer and telecommunications technology had a well-thought-out **disaster recovery** plan when Hurricane Opal hit. The plan to call the appropriate personnel to work depended on telephone communication, which was disrupted by the storm. The only way to call in the required personnel was to drive to a radio station and make an announcement.

A less obvious problem to assess and manage is what to do when someone in an office goes on vacation, is sick, or goes on medical leave. Hopefully, there are provisions for a

[4] http://en.wikipedia.org/wiki/NIST.

trained replacement with like skills, and adequate documentation to do the job. What about a labor strike? These possible problems are less consequential than acts of nature, but more likely to happen.

What about everyday network and computer operations? We have seen mainframe computers brought to their knees by a 100-millisecond flicker of the power because there was no surge protection and no uninterrupted power supply (UPS). Does the file server on your organization's LAN have redundant components, or does it have a backup server for mission-critical functions? Are there alternate, redundant circuits for wide area network lines in case of an attack by a marauding backhoe? One telecommunications-dependent firm has buried the telecommunications trunks on their premises in deep trenches and then poured concrete on top to protect against such digging.

Implication   Organizations cannot continue without voice and data communications. Therefore, the threat of an event such as a flood, fire, hurricane, earthquake, DoS attack, flu, power failure, and so forth, is a threat against the continuance of the organization and the jobs of its employees.

Risk management must include the analysis and choice of subsequent actions to be taken to ensure that organizations can continue to operate under any foreseeable conditions, such as illness, wars, accidents, labor strikes, hurricanes, earthquakes, fire, power outages, heavy rains, oppressive heat, terrorism, or flu epidemics. Telecommunications capabilities support all facets of the company; therefore, risk analysis and management of these technologies are always in order.

Disaster Planning

Recall that the beginning of risk management is assessment, which should lead to management on a continuous basis. A specific requirement is the creation of a disaster plan in case of a severe, even catastrophic occurrence.[5] If organizations depend on telecommunications-intensive, computer-based information systems, like banks, stock exchanges, and airlines, what do they do in case of a total power outage, data-line cuts, viruses, hacker intrusions, DoS attacks, and fires? **Disaster planning** requires the documentation of procedures to allow recovery after a disaster and permit the organization to continue operations. The disaster plan provides the place, procedures, and equipment to make use of the organization's data that are secure and available for continued operations. Disaster plans are often referred to as "business continuity plans" for good reason.

Implication   The primary planning point for data in disaster preparedness is making regular backup copies of all data and safely transporting them to and protecting them at an offsite location. Whether the data will be used back at the company or at a contracted **hot** or **cold site,** continuance is not possible in the event of a severe disruption, a.k.a. disaster, without organizational data.

It is easier to attack a system than to defend it.

The firm's disaster plan will generally be monitored at a high level in the organization, ensuring that a myriad of solutions will be available to be implemented. The IT department will have its own plan, of which voice and data communications are a critical portion. In general, organizations cannot function without voice and data communications, so the IT function must know what it takes to ensure continuity. The infrastructure should be resilient against a wide variety of threats, including those of Mother Nature. More practically, the disaster plan must contain provisions to recover from a short-term outage

---

[5] The definition of severe or catastrophic event is organization dependent. An hour's outage for the worldwide SWIFT financial network may mean the loss of millions of dollars; the loss of power for two hours for a fast-food restaurant may mean closing doors for the rest of the day. A 100-millisecond flicker of power on a mainframe without protection or UPS unit may send hundreds of employees home.

or continue operations for an indefinite period. The decisions contained in the plan must be made before a disaster strikes and should cover the gamut of possibilities, from water in cable trays to fuel for backup generators.

Clarification
Advertisement[6]

Of course, we provide protection for your network and systems. SunGard is known for its disaster-proven, world-class infrastructure that includes a 25,000-mile, platform-independent global network backbone and over 3.0 million square feet of conditioned operations space with advanced HVAC controls, security systems and more. As well as advanced hardware and software, reliable tape and disk storage, and sophisticated networking services. But our Recovery solutions extend well beyond those capabilities and include:

- Systems Recovery
- Trading Services
- End-User Recovery
- Voice Recovery
- Mobile Recovery

Helping Hands

One way an organization can supplement the management of, and help reduce, risk is to outsource parts of data storage and networks. For example, *Rackspace.com* will provide highly secure servers that are maintained and available **24*7** with excellent access via broadband. This can be considered as a primary way to manage the data resource and its security and can be utilized as a response to a disaster. Additionally, *www.connected.com* and *www.Dataprotection.com* provide scheduled backup services, replacing onsite backup and, more importantly, storing the backups offsite.

An alternate way to prepare for continuation of information services after a disaster is to have reciprocal agreements with companies to share services in time of need. Many banks have such arrangement as their IT needs would be similar among like organizations. While data recovery might be provided by a company such as *Sungard,* processing could well be provided by a neighboring organization with sufficient IT capacity.

NETWORK SECURITY

No single piece of technology can be used to secure all information. A solid information security infrastructure will consist of many technologies. This typically includes firewalls to accept or deny traffic into and out of networks, encryption to ensure data integrity and confidentiality for transport mechanisms like e-mail and virtual private networks, passwords and biometrics to authenticate network users, and data backup and storage systems to ensure system continuity and recovery.[7]

Table 11.2 lists the order of severity of network and computer threats in mid-2004. It is for these reasons that the concerns for security and privacy exist.

Security is the protection of assets from access, change, copy, theft, or destruction. Security is a part of risk assessment, risk management, and disaster planning. It takes three forms: physical security, virtual security, and data security.

Physical security involves not only the locks on the server or mainframe room, but the condition and continuation of power; environmental control, for example, cooling; and protection of wiring from **traffic** and water. (As late as August 2004, a southeastern university suffered a flood in its server/mainframe room due to Hurricane Charlie.) Physical security

---

[6] http://www.availability.sungard.com/Products+And+Services/Business+Continuity/Recovery+Services/?ga.
[7] Contributing to Sarbanes-Oxley compliance with IronMail, http://www.CipherTrust.com.

also considers the nodes of the wired network, such as the environment and condition of these nodes, such as power and cooling.

We must remember that the purpose of physical security is, in addition to denying entry, disclosure, and destruction, to prevent *smash-and-grab*. It may be easier to break into a server room that has poor physical security and steal the actual hard drive than hack in. Stealing information contained on a laptop is simply as easy as stealing the laptop.

Implication

Servers or network equipment located in low-lying areas are subject to flooding; wires that are not protected from physical damage, that is, being stepped on or cut with a backhoe, may disrupt service.

Spam
any unwanted e-mail.

Some controls such as physical controls are concerned with protection of the physical property. We often visualize this as placing critical assets in locked vaults, perhaps with armed guards. With even the most secure physical environment, there must be some human access. Consequently, access controls allow access only by authorized people. In order for access controls to be effective, there must be specific authentication measures in place and enforced.[8] These measures are part of the procedural controls.

Virus
any software intended to perform malicious acts (malware).

Hackers are intruders into networks, intent on harm.

Virtual security deals with the external world to which an organization connects but over which there may be limited control, that is, the environment. It is described as virtual because it exists, not physically, and not in our data repositories, but via our connectivity to the external world. There are two specific parts of virtual security: wired and wireless.

Wired virtual security is concerned with how outsiders might access the organization's data via wired ports, for example, the Internet. The threat is manifested by spam, viruses, and hackers. We must consider threats when using VPNs, encryption, **compression,** and authorization.

Wireless virtual security encompasses all of wired virtual security but has additional aspects. These are called out in greater detail in the next chapter. What must be taken into account is that the very nature of wireless networks is to **broadcast** our data into the public domain, for example, airwaves. Wireless technology allows authorized and unauthorized receptors alike to intercept the data and/or use the same mechanism to access our networks.

Data security is the objective we seek with physical and virtual security. The point of hardware security is protection of physical assets that may contain or allow access to data. Virtual security considers an environment in which our data are attached but do not own; that is, where data resides. Data are often accessed over a network, the largest being the Internet. Data security is our ultimate aim as data are the underpinning of the organization.

Privacy is security of information dealing with personal data. While access to data is a concern for security, there must be further protection, generally in the form of authorization, when dealing with information about individuals. For example, guarding the server and database that hold student grades is a matter of security; access to those grades is a matter of privacy.

Proprietary information is of particular importance to an organization because disclosure could harm competitive advantage or divulge trade secrets. The term *proprietary* is to organizations as private is to individuals.

Caveat

Social engineering means using nontechnological means to gain access to your objective; to get people to do what you wish them to do. Social engineers are con artists.

Physical and virtual security provide protection for specific threats to data, nodes, and networks. Locks and firewalls provide specific, technology-based protections. What technology does not protect, however, is attacks that utilize **social engineering.** Why should an attacker spend hours trying to break into a network or server when s/he can do it

---

[8] Identification and authentication that can be shared, copied, or altered is a threat. With a driver's license; all three threats can be applied to this form of identification. To make identification foolproof, it must beat these three critical threats. Organizations employ some form of biometric identification, such as finger print scan, eye retina scan, iris scan, or DNA test. Even these can, with some effort, be falsified.

instead with a simple phone call? In addition to the installation of the technology is the vital need to educate and train the personnel of the organization as to the methods of social engineering.

SECURITY ISSUE: THREATS AND RESPONSES

Organizations must be aware that security is a major responsibility for the total telecommunications infrastructure of their business. This includes the security of their links to suppliers, customers, and other partners. In addition, many organizations have employees who **telecommute,** who travel or take computers to their homes or on business trips and connect to the firm from just about anywhere. Since information and its transport media are everywhere in the organization and almost everywhere the employees go, the security issue is pervasive and very broad in scope.

The manager must examine the threats to security in order to determine appropriate safeguards. Each category of threat has safeguards that may be considered as alternatives. In the final analysis, there must be a reliance on trust. Total security would mean that no one could have access. The manager is responsible for ensuring that procedures exist that put proper controls in place so that trust remains the final and not the total factor relied upon. What are some security threats? Table 11.5 provides a start.

TABLE 11.5 Security Threats

Threat	Source or Target	Consequences	Primary Defense
Users	Internal, mobile	Majority of security a control problem	Controls for prevention and deterrence; training
Programmers	Internal	Bypass, disability of security mechanisms	Properly designed control and supervision audits
Hardware	Internal	Failed protection mechanisms lead to failure	Control, detection, limitation, and recovery procedures
Databases	Internal	Unauthorized access, copying, theft	Passwords, intranet, VPN
Systems software	Internal	Failure of protection, information leakage	Controls, audits
Operators	Internal	Loss of confidential information, theft, insecure	Proper access controls, partitioning of data
Radiation (interception)	External, remote	Interception of confidential data	Shielding, access control Authentication
Spoofing	External	Fraud	Firewall, passwords
Hacking	External	Intrusion, destruction of resources	
Denial of service	External	Stoppage of real work Leakage of confidential data	Firewall, honey pot, VPN
Crosstalk	External	Loss of data	Shielding, separation
Wiretaps (eavesdropping)	External	Disruption of service, loss of resources	Procedural controls, audits Precaution/management controls, BC plans
Environmental hazards	External		
Criminal attacks	External	Disruption of service, loss of business/reputation	UPS, generators
Power outages	Internal	Disruption of service, loss of resources	Firewall, procedures
Viruses, worms, Trojan horses Access	External External	Loss of data, sabotage to equipment	Proper authentication and control procedures, physical barriers, biometric controls

Threats

Threats can take two forms: passive and active. **Passive threats** are those that occur without malicious intent, without the active participation of people, or through unintentional consequences. **Active threats** are overt attacks or the release of hostile applications for the purpose of harm. Table 11.4 indicates that these threats may be via wired or wireless media, from internal or eternal sources. The following section discusses examples of passive threats.

Passive Threats

Passive threats can come from technology, users, or Mother Nature; they are not intentional. Nonetheless, these threats can bring a system to a halt.

User Threats

A major threat, historically, has come from human error or inadvertent acts on the part of users. These threats may be the result of improper training, poor system design, inadequate procedures, programming flaws, and/or carelessness.

Systems Software

The two facets of systems software as a threat are policy and operational. Policies must exist to ensure that system updates and **patches** are installed as soon as they are available. Failure of operators to do so is both an operational and management problem. This is the first line of defense for many threats. Additionally, procedures should be devised to prevent implementation of insecure systems, unauthorized copying, and theft.

Clarification

A computer program **patch** is the process of replacing or adding a portion of a program, for example, a file, with a new file that has added features or corrects a problem, that is, fixes a bug. Patches correct weaknesses in programs that are often exploited by hackers to gain entry.

Environmental Hazards

Environmental hazards are the threats due to natural occurrences (Mother Nature) such as floods, earthquakes, tornadoes, hurricanes, avalanches, volcanic activity, tsunamis, severe snow, rain, and sand storms. They also can be events such as fires, climate control system failure, explosions, act of war, and so on. In the wireless domain, some threats such as rain simply degrade service.

Business continuity means that resources continue to be reliable, be accessible, perform as expected, and be secure. During environmental hazards, either the services provided by these assets must be moved to an offsite location or operation must continue. Protection entails backup power, fire protection and suppression, and prevention from water damage. As a result of the terrorist attack on the World Trade Center towers, several entire telephone switches were destroyed, necessitating rerouting of this traffic. Much of the wired infrastructure in the immediate area were destroyed.

The Reception of Radiated Signals

Every electronic, electro-optical, or electromechanical device gives off some type of electromagnetic signals, whether or not the device was designed to be a transmitter. This is why the use of cellular phones is not permitted on airplanes—their unintentional signals can interfere with navigational equipment. The electromagnetic **radiation** (EMR) that "leaks" from devices can be intercepted and, using the proper equipment, reconstructed on a different device. The EMR that is emitted by devices contains the information that the device is displaying, storing, or transmitting. With equipment designed to intercept and reconstruct the data, it is possible to steal information from unsuspecting users by capturing the EMR signals.

Tempest (*Telecommunications Electronics Material Protected from Emanating Spurious Transmissions*)[9] was the name of a classified U.S. government project to deal with the susceptibility of computer and telecommunications devices to emit EMR in a manner that can be used to reconstruct intelligible data.

Crosstalk

In general, **crosstalk** is the reception of signals from another circuit or channel and is generally evidenced as noise. From a security standpoint, it may be the loss of data/information because of radiation between wires. Unlike collisions, which are simultaneous packets on a single channel, crosstalk is one channel or circuit interfering with another. The primary method of defense is to twist the UTP, use coaxial cable, or use fiber strands. A simple fix if the problem is one cable interfering with another is to separate the cables via rerouting.

Power Outages

All data communications systems require electrical power.

Power outage has been covered in some detail in Chapter 5. It includes any electrical power–related event that prevents normal operations. This can include a shortage of batteries.

Active Threats

The passive threats discussed previously are the results of acts of nature or people doing unintentional acts. Active threats, in contract, are the result of overt actions taken by people who wish to do harm or gain an advantage. Examples include willful entry of erroneous data, fraudulent identification, bypass of proper authorization and authentication, entry of a virus, and other intentional security breaches. Oftentimes, the underlying force of the attack relies on the nature of humans, for example, social engineering.

Societies wish to protect themselves by enacting laws and **regulations.** During its short lifetime, the United States has seen rapid advances in technology. It has been difficult for the legal system to keep pace with the threats brought on by use of this technology. The Computer Fraud and Abuse Act of 1986 is an early attempt to recognize the vulnerability of computers and network systems and apply legal sanctions.

Implication
Computer Fraud and Abuse Act of 1986 (United States), 18 USC § 1030(a)[10]

1030. Fraud and related activity in connection with computers

(a) Whoever—(1) knowingly accesses a computer without authorization or exceeds authorized access . . . (4) knowingly and with intent to defraud . . . (5) intentionally accesses a Federal interest computer without authorization and by means of one or more instances of such conduct alters, damages, or destroys information in any such Federal interest computer, or prevents authorized use of any such computer or information . . . (6) knowingly and with intent to defraud traffics in any password or similar information through which a computer may be accessed without authorization . . . shall be punished as provided in subsection (c) of this section . . .

The USA PATRIOT Act increased the scope and penalties of this act by

1. raising the maximum penalty for violations to 10 years for a first offense and 20 years for a second offense;
2. ensuring that violators only need to intend to cause damage generally, not intend to cause damage or other specified harm over the $5,000 statutory damage threshold;
3. allowing aggregation of damages to different computers over a year to reach the $5,000 threshold;

---

[9] http://searchsecurity.techtarget.com, http://www.webopedia.com.
[10] http://www.usdoj.gov/criminal/cybercrime/1030_new.html.

4. enhancing punishment for violations involving any (not just $5,000) damage to a government computer involved in criminal justice or the military;

5. including damage to foreign computers involved in US interstate commerce;

6. including state law offenses as priors for sentencing; and

7. expanding the definition of loss to expressly include time spent investigating and responding for damage assessment and for restoration.[11]

Information Warfare—Military and Industrial

Information warfare is the process of protecting your information and network resources while, potentially, denying the adversary access to his/hers. In a military setting, it is a strategy for undermining an enemy's data and information systems while defending and leveraging one's own information edge. In an organizational setting, it primarily means protecting your information assets from a variety of adversaries. The most obvious of these, although not necessarily the most dangerous, is intrusion.

Who Are the Attackers?

When it comes to active threats, it is a case of bad people doing bad things, most of the time. Although the problem may be due to honest people bypassing rules and policies, the result may be the same to the organization.

Criminal Attacks

Computer crime allows a crime to be committed without being at the scene of the event. It involves deliberate attacks on computer and telecommunications systems that range from data tampering to fraud and extortion. We might think of the criminal attack as coming from organized crime. A criminal attack is one where the intent is financial gain for the attacker. It would be difficult to differentiate the components of the tool set used in information warfare as it contains all forms of attack.

Terrorist Attack

An act of terrorism is one designed to specifically cause harm to people, systems, infrastructures, or nations. From the standpoint of the network administrator, there are two types of terrorism: physical and virtual. While a terrorist attack can be seen as just another form of intrusion, the intent is more severe.

Protection from *physical terrorism* is a function of the organization's overall security. The network security group must be aware that organizations run on networks and an attack on network equipment is an attack on the organization. This form of attack requires a physical presence and is guarded against with physical security. The second form is *virtual terrorism,* which is attempting an intrusion or disruption of the network infrastructure. Because this does not require a physical presence, it is the most dangerous form of intrusion as it can be launched from *anywhere on earth.* Virtual attacks will likely become a major weapon for groups that wish to cause disruption of the technological infrastructure of any industrialized nation. While intrusion is the most overt act, a virtual attack may take the form of a virus launched onto the Internet, which is a threat as long as any *one* system contains a copy and there are other computers that have not updated their protection.

Hackers

The term *hacker* (n.) has two uses. The *positive use* of the term describes someone who is an expert in his/her ability to use a resource, such as a computer or network. In the *negative sense,* it describes a person wishing to break into another's computer or network. This is overt intrusion: a form of attack that poses threats of data destruction, fraud, spread of

---

[11] http://en.wikipedia.org/wiki/Computer_Fraud_and_Abuse_Act.

viruses, and so forth. This threat requires a significant amount of effort and resources to protect against, especially in the wireless domain. Hackers are a primary adversary in industrial information warfare.

Unfortunately, it does not take a (positive) hacker to perform (negative) **hacking.** Web sites[12] exist that provide free software to (a) educate novices in the technology of hacking and (b) give the tools to break into a computer. Teenagers have been known to use such software, thinking it was just a prank—thus, the *script kiddie.*

Script Kiddies

This is the category that exemplifies the saying, *a little knowledge can be dangerous.* **Script kiddy** is the derogatory term given to would-be hackers who do not possess the knowledge or skill to write their own programs but rely on ready-to-use kits from the Internet or programs written by others. They are nuisances on a good day and dangerous on a bad day because they don't understand the consequences of their actions, which may be as deadly as altering a virus and releasing it to the Internet.

Programmers

Programmers sometimes bypass or disable security mechanisms, install insecure systems, install default passwords and backdoor[13] provisions, and circumvent established procedures. Hardware operating systems (written by programmers) can have flaws that make data vulnerable to loss or interception. Hardware and its software must work together to ensure they do not contribute to lack of security.

Rogue Users

Rogue users are dishonest or unethical people, or perhaps users getting around some of the rules; they do things to be mischievous or damaging and may range from intruders to disgruntled or dismissed employees. Rogue users are generally considered a wireless problem and are extensively covered in the next chapter. However, if a machine is hacked, one common trick a hacker will use is to access a default account, such as guest, to add a rogue user account, possibly with administrator rights. Whether the hacking occurred via wired or wireless access, the addition of the rogue as a legitimate user will make further unauthorized wired or wireless accesses very easy. The first solution is to delete or rename all default accounts and apply normal password policies.

What Are the Threats?

Active threats not only come from a variety of sources, as noted previously, but take an assortment of forms. These range from overt intrusion to the more subtle approach of **electronic mail (e-mail).** All of the applications discussed in the following paragraphs are created to perform unwanted acts—software attacks that can destroy or damage data or software by attaching to other computer programs. These applications are often transported on spam, often spoofed. They are one of the tools of information warfare.

Intrusion[14]

An intrusion is the process of someone (a.k.a. hacker or cracker) attempting to break into or misuse your system. The word *misuse* is broad and can reflect something as severe as

To would-be Script Kiddies: Remember, the laws do not differentiate the intent of the offender. Intrusion is intrusion; viruses are equal; spam is not a joke.

Using default accounts and passwords is the simplest and easiest form of intrusion.

[12] For a tutorial in hacking, see http://www.hetland.org/python/instant-hacking.php.

[13] Backdoors are entry points left in applications, often by the creator, so that access can be gained later, generally for maintenance. Backdoors can be exploited by hackers by guessing the password.

[14] http://www.robertgraham.com/pubs/network-intrusion-detection.html.

stealing confidential data to something as minor as misusing your e-mail system for spam. The purpose of intrusion is change, corruption, destruction, or theft. For our study, we mean intrusion is the active work of a human at the time of the action, such as the scanning of network ports to find a vulnerability to exploit.

Many would consider intrusion the primary active threat to our systems. It is a continuing problem but probably not the most often-used method of gaining information. As you continue with the discussion of threats, you will realize that many do not require the exposure of intrusion; some don't even require technical knowledge.

Cuckoo's Egg Revisited[15]

Cliff Stol chronicled his experience with Internet-based intrusion in his 1988 book *The Cuckoo's Egg*. Cliff was given the task of reconciling a 75-cent accounting discrepancy in the computer records. He assumed it was a rounding error but found an intruder by reviewing the audit logs. This appears now as ancient history, but the events and methods are very relevant today. As with Cliff, perusing of logs may seem mundane, but you may find nuggets of gold, for example, the tracks left by the intruder.

Denial-of-Service (DoS/DDoS) Attacks[16]

A **denial-of-service (DoS/DDoS) attack** is characterized by an explicit attempt by attackers to prevent legitimate users of a service from using that service. These attacks can disable a computer or network. Depending on the nature of the enterprise, this can effectively disable the organization. Denial-of-service attacks come in a variety of forms and aim at a variety of services. There are three basic types of attack:

1. Consumption of scarce, limited, or nonrenewable resources.
2. Destruction or alteration of configuration information.
3. Physical destruction or alteration of network components.

Examples of DoS include attempts to

- *Flood a network* with packets, thereby preventing legitimate network traffic.
- *Disrupt connections* between two machines, thereby preventing access to a service.
- *Prevent* a particular individual from accessing a service.
- *Disrupt service* to a specific system or person.

Clarification

DoS attacks are very common but they are not a joking matter. In the United States, they can be a serious federal crime under the *National Information Infrastructure Protection Act of 1996* with penalties that include years of imprisonment, and many countries have similar laws. At the very least, offenders routinely lose their Internet service provider (ISP) accounts, get suspended if school resources are involved, and so forth. These attacks are sometimes also known as "nukes," "hacking," or "cyber-attacks," but we use the technically correct term of *DoS attacks*.[17]

Anonymous FTP
an account like a guest account; FTP is the TCP/IP *file transfer protocol* capability.

Not all service outages, even those that result from malicious activity, are necessarily DoS attacks. Other types of attacks may include DoS as a component; therefore, the DoS may be part of a larger attack. Illegitimate use of resources also may result in DoS. For example, an intruder may use an anonymous FTP area as a place to store illegal copies of commercial software, consuming disk space and generating network traffic.

[15] http://www.inforingpress.com/articles/cuckoos-egg.htm.
[16] http://www.cert.org/tech_tips/denial_of_service.html.
[17] http://www.irchelp.org/irchelp/nuke/.

Implication

One of our students found that an intruder had loaded 20 gigabytes of pornography on his computer. He didn't discover it until he ran out of disk space.

Not only must assets be in place to thwart such an attack, organization assets must not be used in such an attack. Such attacks generally use a number of computers, thus the term *distributed DoS*. These computers may be owned by the attacker, be part of the attacker's organizational assets, or be unwary owners of machines that have been hijacked for the attack. *Where spoofed spam may be the rifles of information warfare, distributed DoS attacks are the machine guns.*

There are two types of DoS:[18]

1. DoS vulnerability due to victim error/*inadequate care.* Many Internet sites have made an error in their security perimeter, or are running e-mail, Web, or other software with known security bugs that make them vulnerable to a denial-of-service attack. Typically, a server will crash under an attack that exploits the particular problem. Often just a small amount of traffic is used in the attack, just enough to tickle the flaw. Such attacks can be avoided 99 percent of the time with proper security practices, keeping up to date with patches, and so on. It is vital to have regular (quarterly or monthly) active security tests as part of this ongoing effort of maintaining the security bar.

2. DoS due to *brute force.* This flavor of attack does not depend on the victim having a security problem. The attack works by simply overloading systems at the victim site. For example, one such DoS may involve bringing a Web server down to a crawl, and stopping legitimate users getting pages from it, by the brute force technique of requesting the same page so fast that it cannot keep up. This attack would simply need a normal Web server at the victim site. Grabbing Web pages creates high volumes of outbound traffic (a small Web page request may cause 50K or more pages to be delivered) and a high server load. Grabbing SSL pages puts a higher "load per page grabbed" on the server due to the extra load of processing the encryption elements of the SSL process. The Yahoo attack was a brute force attack, but instead involved injecting high volumes of traffic into the site by having enough attack agents in parallel.

Implication

Nothing more than the whim of a 13-year-old hacker is required to knock any user, site, or server right off the Internet.[19]

Within a minute of the start of the first attack it was clear that we were experiencing a "packet flooding" attack of some sort. A quick query of our Cisco router showed that both of our two T1 trunk interfaces to the Internet were receiving some sort of traffic at their maximum 1.54 megabit rate (1.544 Mbps), while our outbound traffic had fallen to nearly zero, presumably because valid inbound traffic was no longer able to reach our server. We found ourselves in the situation that coined the term: Our site's users were being denied our services.

I had two priorities: I wanted to learn about the attack and I wanted to get us back online. I immediately reconfigured our network to capture the packet traffic in real time and began logging the attack. Dipping a thimble into the flood, I analyzed a tiny sample and saw that huge UDP packets—aimed at the bogus port "666" of grc.com—had been fragmented during their travel across the Internet, resulting in a blizzard of millions of 1500-byte IP packets. Mixed into this appeared to be ICMP debris from large-packet ping commands. We were drowning in a flood of malicious traffic and valid traffic was unable to compete with the torrent.

[18] http://www.nta-monitor.com/news/press-releases/00-DoS.htm.
[19] http://grc.com/dos/grcdos.htm.

The total story can be read and/or downloaded in .PDF format at http://grc.com/dos/ grcdos.htm.

Intrusion of Governmental Agencies

A number of U.S. laws allow duly authorized law officers to eavesdrop on suspected felons. This began with the wiretap law of the early 1900s and has progressed to the USA PATRIOT Act of 2001. Originally, the agency had to have the authorization of a federal judge; now some cases are allowed with just federal agency oversight. Of significance is the use of *Carnivore,* for sniffing e-mail at an ISP's port, and *Magic Lantern,* an application that is deposited on a computer to capture keystrokes (keylogging). Either may be used in the personal or organizational environment. While organizations are aware that e-mail is fair game when it leaves the protection of the organizational network, keyloggers can be placed on organizational computers without warning, with serious consequences. Software is available to detect keylogging applications. Anti-virus software generally will not detect it.

Spoofing[20]

Spoofing is the creation of TCP/IP packets using bogus header information, such as somebody else's IP address. Routers use the "destination IP" address in order to forward packets through the Internet but ignore the "source IP" address. That address is only used by the destination machine when it responds back to the source.

Implication

Cost to businesses for spam was about $23 billion per year by 2004 based on PC repairs and lost work.

IP **spoofing** is an integral part of many network attacks.

A common misconception is that IP spoofing can be used to hide your IP address while surfing the Internet, chatting online, sending e-mail, and so forth. This is generally not true. Forging the source IP address causes the responses to be misdirected, meaning you cannot create a normal network connection.

Clarification

The MAC address of the network card is a unique identifier assigned to each Ethernet card. Network administrators can locally find the MAC address of a machine by either sniffing traffic from the wire or downloading ARP tables from routers. Therefore, hackers on internal networks (such as corporations or universities) often will try to hide their MAC address. Spoofed MAC addresses also can be used locally to redirect traffic from its intended host to the hacker's machine.

There are several ways of spoofing the MAC address:

* Some Ethernet adapters can be simply configured at boot up to use a different MAC address (soft configuration).
* Most adapters use EPROMs to store the address and can be reprogrammed (hard configuration).
* It is really the TCP/IP stack that copies the MAC address into a frame; a reprogrammed stack can usually bypass the configured MAC address.

E-mail Spoofing[21]

E-mail spoofing may occur in different forms, but all have a similar result: a user receives e-mail that appears to have originated from one source when it actually was sent

[20] http://www.iss.net/security_center/advice/Underground/Hacking/Methods/Technical/Spoofing/default.htm.
[21] http://www.cert.org/tech_tips/email_spoofing.html.

from another source. E-mail spoofing is often an attempt to trick the user into making a damaging statement or releasing sensitive information (such as passwords).

Implication

All warfare is based on deception.–Sun Tzu

Examples of spoofed e-mail that could affect the security include

- E-mail claiming to be from a system administrator requesting users to change their passwords to a specified string and threatening to suspend their account if they do not do this.
- E-mail claiming to be from a person in authority requesting users to send them a copy of a password file or other sensitive information.

E-mail Spamming and E-mail Spoofing[22]

E-mail spamming refers to sending e-mail to thousands and thousands of users—similar to a chain letter. E-mail spamming may be combined with *e-mail spoofing,* so that it is very difficult to determine the actual originating e-mail address of the sender. Some e-mail systems, including *Microsoft Exchange,* have the ability to block incoming mail from a specific address. However, because these individuals change their e-mail address frequently, or it is spoofed, it is difficult to prevent some spam from reaching your e-mail inbox.

Implication

By 2002, Merrill Lynch was receiving over two million bogus and junk e-mails per day.

"When it comes to the e-mail threats, we're well protected," said Adam Nunn, security and corporate compliance manager for Province Healthcare in Brentwood, Tennessee. "E-mail security has gotten pretty good. It's the other attack vectors that worry me, like instant messaging." He cited the Fatso and Kelvir worms as examples. Both have exploited MSN Messenger in the past week.[23]

Meanwhile, IBM Lotus has a product *Sametime* Instant Messenger and Web Conferencing that includes significant security. Thus, you get what you pay for.

Phishing

Phishing,[24] short for *password harvesting fishing,* is the luring of sensitive information, such as passwords and other personal information, from a victim by masquerading as someone trustworthy with a real need for such information. It is a form of social engineering attack, a form of online identity theft. Phishing uses spoofed e-mails, possibly designed to lure recipients to fraudulent Web sites, that attempt to trick them into divulging personal financial data such as credit card numbers and account usernames and passwords.

Today, online criminals put phishing to more directly profitable uses.[25] Popular targets are users of online banking services and auction sites such as eBay. Phishers usually work by sending out spam e-mail to large numbers of potential victims. These direct the recipient to a Web page that appears to belong to their online bank, for instance, but in fact captures their account information for the phisher's use. The following Implication note is a phishing e-mail, received by one of the authors, that is very typical. Most HTML code has been removed.

---

[22] http://www.lse.ac.uk/itservices/help/spamming&spoofing.htm.

[23] *Security Wire Perspectives* 7, no. 20 (March 14, 2005).

[24] http://en.wikipedia.org/wiki/Phishing.

[25] http://www.cnn.com/2003/TECH/internet/07/21/phishing.scam/.

Implication

DO NOT provide personal information to any request that comes via e-mail.

Reply-To: <no-reply@ebay.com>
From: "help_fraud@eBay.com" <fraud@ebay.com>
Subject: eBay Warning—Your user was suspended for fraud

You have received this email because you or someone had used your account to make fake bids at eBay. For security purposes, we are required to open an investigation into this matter. THE FRAUD ALERT ID CODE CONTAINED IN THIS MESSAGE WILL BE ATTACHED IN OUR FRAUD MEDIATION REQUEST FORM, IN ORDER TO VERIFY YOUR EBAY ACCOUNT REGISTRA-TION INFORMATIONS.

Fraud Alert ID CODE: 00937614

(Please save this Fraud Alert ID Codefor your reference.

To help speed up this process, please access the following form to complete the verification of your eBay account registration informations:
    <a target=3D"_blank" href=3D"http://216.170.99.243/.eBay/index.htm"
    >http://scgi.ebay.com/verify_id=3Debay&fraud alert id code=3D00937614</a>

If we do not receive the appropriate eBay account verification within 48 hours, then we will assume this eBay account is fraudulent and will be suspended. The purpose of this verification is to ensure that your eBay account has not been fraudulently used and to combat the fraud from our community.

We appreciate your support and understanding, as we work together to keep eBay a safe place to trade. Thank you for your patience in this matter.

Regards,

Safeharbor Department (Trust and Safety Department) eBay Inc.

Please do not reply to this e-mail as this is only a notification. Mail sent to this address can not be answered.

Copyright 2005 eBay Inc. All Rights Reserved. Designated trademarks and brands are the property of their respective owners. eBay and the eBay logo are trademarks of eBay Inc. eBay is located at 2145 Hamilton Avenue, San Jose, CA 95125.

Checking the URL in the address bar of the browser may not be sufficient, as, in some browsers, that can be faked as well. However, the file properties feature of several popular browsers may disclose the real URL of the fake Web page.

Implication

"Phishing attacks represent a collaboration of the world's most skilled hackers and organized crime—instead of breaking into the bank to take money, phishers are tricking users into handing over their account information, or rather the electronic keys to the vault," said Paul Judge, chief technology officer at *CipherTrust*.[26]

Clarification

Multipurpose Internet Mail Extensions, or MIME, is an Internet standard specifying message formats for transmission of different types of data by electronic mail.

We use Novell's GroupWise for e-mail at our university. If the e-mail originates in GroupWise (as most on-campus e-mail would), no MIME encoding takes place. It's not needed within GroupWise. Only if the e-mail passes through the SMTP gateway out of GroupWise does MIME encoding take place. This is important because I can look at a suspicious e-mail safely by performing a SAVE AS process and saving the MIME copy, which is pure text. This is how I captured the phishing e-mail previously displayed.

Spam[27]

Spam is flooding the Internet with many copies of the same message, in an attempt to force the message on people who would not otherwise choose to receive it. Most spam is commercial advertising, often for dubious products, get-rich-quick schemes, or quasi-legal services. Spam costs the sender little or nothing to send outside of the fixed-cost ISP

[26] http://www.CipherTrust.Com.
[27] http://spam.abuse.net/overview/whatisspam.shtml.

monthly fee—most of the costs suffered by the carriers in the form of resource usage. As noted in the following Implication note, millions of e-mail addresses are easy to purchase.

Implication
Advertisement
12 Sep 2001—Spammers solicited donations for relatives of the victims of the 9/11/2001 terrorist attacks on the USA as a scam.

This is the one you have been hearing about.

The Limited Edition Y2K Targeted Email Address Cd

Includes 19,000,000 Highly Targeted Email addresses In Over 20 Categories

This Is Volume 3 And There Are A Limited Number Available

Visit Our Site Today For Details. . . And Receive $100 off If You Order Today[28]

Clarification
Advertisement

McAfee® *SpamKiller*® features advanced filtering technologies to stop junk e-mail, including foreign-language character-based spam. New features catch spam that rely on spelling variation matching, as well as image-only content in attempts to bypass text-based filtering rules—a practice used extensively by pornographic spammers. Scam filters help protect against "phishing," frauds, hoaxes, and other identity theft e-mails. The inclusion of "One-Click Block" functionality automatically eliminates those e-mails you label as spam all at once. An automatic import of designated friend lists ensures you never miss e-mail from your friends, family and co-workers.

New Offering: *McAfee's Secure Messaging Service* offerings are designed to reduce the infrastructure required to process messages by stopping spam, viruses and directory harvest attacks at the perimeter. The services do this by quarantining spam and virus infected e-mails before they enter the network, in some cases, reducing mail volumes by 70 to 80 per cent. Customizable spam settings and white/blacklisting functionality also give end-users personal control to manage their own definitions of what is legitimate e-mail, and what is unwanted e-mail.[29]

DO NOT provide personal information to any request that comes via e-mail.

In simpler terms, *spam is the reception of _any_ unwanted e-mail*. Although not considered an overt threat, its existence is of significant consequence because of the wasting of bandwidth and productivity. **Unsolicited commercial e-mail (UCE)[30],** a.k.a. *junk mail,* is the leading complaint of Internet users. But junk e-mail is more than just annoying; it costs Internet users and Internet-based businesses millions, even billions, per year. Junk e-mail is "postage due" marketing; it's like a telemarketer calling you collect. The economics of junk e-mail encourages massive abuse and, because junk e-mailers can get into the business very cheaply, the volume of junk e-mail is increasing every day.

Clarification

If you get spam e-mail that you think is deceptive, forward it to spam@uce.gov. The FTC uses the spam stored in this database to pursue law enforcement actions against people who send deceptive e-mail.

Pharming[31]

Similar in nature to e-mail phishing, **pharming** seeks to obtain personal or private (usually financial related) information through domain spoofing. Rather than being spammed with malicious and mischievous e-mail requests for you to visit spoof Web sites that appear legitimate, pharming "poisons" a DNS server by infusing false information into the DNS server, resulting in a user's request being redirected elsewhere. Your browser, however will show you are at the correct Web site, which makes pharming a bit more serious and more difficult to detect. Phishing attempts to scam people one at a time with an e-mail while pharming allows the scammers to target large groups of people at one time through domain spoofing.

[28] http://lists.debian.org/lsb-impl/2000/03/msg00001.html.

[29] http://www.integratedmar.com/ecl-usa/story.cfm?item=19893.

[30] http://www.cauce.org/.

[31] http://www.webopedia.com.

DNS Cache Poisoning[32]

DNS cache poisoning is a technique that tricks your DNS server into believing it has received authentic information when, in reality, it has been lied to. Why would an attacker corrupt your DNS server's cache? So that your DNS server will give out incorrect answers that provide IP addresses of the attacker's choice, instead of the real addresses. Imagine that someone decides to use the Microsoft Update Web site to get the latest Internet Explorer patch. But the attacker has inserted phony addresses for update.microsoft.com in your DNS server, so instead of being taken to Microsoft's download site, the victim's browser arrives at the attacker's site and downloads a worm.

Adware[33]

Adware is any software application in which advertising banners are displayed while the program is running. The authors of these applications include additional code that delivers the ads, which can be viewed through pop-up windows or through a bar that appears on a computer screen. Some sites will bombard the viewer with adware to the extent that disconnect is difficult. The justification for adware is that it helps recover programming development cost and helps to hold down the cost for the user.

Adware has been criticized because it usually includes code that tracks a user's personal information and passes it on to third parties, without the user's authorization or knowledge. This practice has been dubbed *spyware* and has prompted an outcry from computer security and privacy advocates, including the Electronic Privacy Information Center.

Malware[34]

Malware (for malicious software) is any program or file that is harmful to a computer user; it is created to exploit user machines. Thus, malware includes computer viruses, worms, Trojan horses, logic bombs, and spyware.

Many of the following definitions come from http://www.webopedia.com.

A virus is a program or piece of code that is loaded onto your computer without your knowledge and runs against your wishes. All computer viruses are manmade. Viruses also can replicate themselves. A simple virus that can make a copy of itself over and over again is relatively easy to produce. Even such a simple virus is dangerous because it will quickly use all available memory and bring the system to a halt. An even more dangerous type of virus is one capable of transmitting itself across networks and bypassing security systems.

A **worm** is program or algorithm that replicates itself over a computer network and usually performs malicious actions, such as using up the computer's resources and possibly shutting the system down.

Clarification

A computer worm is a self-replicating computer program, similar to a computer virus. A virus attaches itself to, and becomes part of, another executable program; however, a worm is self-contained and does not need to be part of another program to propagate itself.[35]

The purpose of most viruses and worms is to cause annoyance or harm to the recipient's computer and spread this same problem to all computers to which it is sent in replication. An additional effect is to use up significant bandwidth in the process.

A **Trojan horse** is a destructive program that masquerades as a benign application. Unlike viruses, Trojan horses do not replicate themselves, but they can be just as destructive. One of the most insidious types of Trojan horse is a program that claims to rid your computer of viruses but instead introduces viruses onto your computer.

[32] http://en.wikipedia.org.

[33] http://whatis.techtarget.com/.

[34] http://searchsecurity.techtarget.com/.

[35] http://en.wikipedia.org/wiki/Computer_worm.

Logic bombs[36] are applications that lie dormant until one or more logical conditions are met to trigger it. An early example is the Columbus Day,[37] or Friday the 13th, virus. The virus was triggered by a system date of October 13 or later. (Note that October 13, 1989, is a Friday.) Its effect was to perform a low-level format of cylinder zero of the hard disk on the target machine, thereby destroying the boot sector and File Allocation Table (FAT) information.

Logic bombs may be a favorite ploy of disgruntled employees who are forced out of, a.k.a. terminated from, an organization; it is an alternative to a backdoor. The logic bomb could be designed to destroy data or it could be used for extortion; for example, if payment is not received, the bomb will activate. The backdoor program alternative gives the former employee access to organizational assets.

Spyware[38] is any technology that aids in gathering information about a person or organization without their knowledge. On the Internet (where it is sometimes called a spybot or tracking software), spyware is programming that is put in someone's computer to secretly gather information about the user and relay it to advertisers or other interested parties. These programs can range from *Gator,* which records information about the computer and the use of the Internet, to *Magic Lantern,* which is a keylogger created to capture key strokes of felons, circumventing encryption capabilities. Spyware can get in a computer as a software virus, as the result of installing a new program or visiting a Web site, or attached to (spam) e-mail. Spyware is part of an overall public concern about privacy on the Internet.

Some include cookies in the category of spyware. The **cookie** is a well-known mechanism for storing information about an Internet user on his/her own computer. However, the existence of cookies and their use are generally not concealed from users, who also can disallow access to cookie information. Nevertheless, to the extent that a Web site stores information about you in a cookie that you don't know about, the cookie mechanism could be considered a form of spyware.

Browser Hijacking[39]

The term **browser hijacking,** according to PCStats.Com, covers a range of malicious software. The most generally accepted description for browser-hijacking software is external code that changes a user's Internet Explorer settings. Generally, the home page will be changed and new favorites will be added that point to sites of dubious content. In most cases, the hijacker will have made registry changes to the system, causing the home page to revert back to the unwanted destination even if the user changes it manually. A browser hijacker also may disallow access to certain Web pages, for example, the site of an anti-spyware software manufacturer like *Lavasoft®*. These programs also have been known to disable antivirus and anti-spyware software.

Most browser hijackers take advantage of Internet Explorer's ability to run ActiveX scripts straight from a Web page. Generally, these programs will request permission to install themselves via a popup that loads when a user visits a certain site. If the user accidentally gives them permission to install, IE will execute the program on his/her computer, changing the settings. Others may use security holes within Internet Explorer to install themselves automatically without any user interaction at all. Worse, these can be launched from popup ad windows that the user has not even intended to view.

As well as making changes to a home page and other Internet Explorer settings, a hijacker also may make entries to the HOSTS file on the system. This special file directly maps DNS addresses (Web URLs) to IP addresses, so every time you type "www.pcstats.com" (as an example), you might be redirected to the IP address of a sponsored search or porn site instead.

[36] *IT Security+ Comprehensive Study Guide,* http://www.cramsession.com.

[37] http://www.totse.com/en/viruses/virus_information/nist02.html.

[38] http://searchcrm.techtarget.com/.

[39] http://www.pcstats.com/articleview.cfm?articleID=1579.

System Hijacking and Zombies[40]

Hijacking your **system,** as opposed to hijacking only your browser, turns it into a zombie. A **zombie** is a computer that has been implanted with a daemon (a process that runs in the background and performs a specified operation at predefined times or in response to certain events). The daemon puts the host computer under the control of a malicious hacker without the knowledge of the computer owner. Zombies are used by malicious hackers to launch DoS attacks. Compared to programs such as viruses or worms that can eradicate or steal information, zombies are relatively benign as they temporarily cripple Web sites by flooding them with information and do not compromise the site's data.

Internet relay chat (IRC)
a form of instant communication over the Internet.

Zombies are used by botnets. A *botnet* is a collection of zombie machines with applications that lie dormant until brought to life by the controller. Often using the IRC capability for communications, the controller issues commands to the zombies, frequently to begin a distributed DoS. It is estimated that over 30,000 botnets exists on the Internet.

Implication
IP Spoofing versus System Hijacking[41]

IP spoofing is simply forging the IP addresses in an IP packet. This is used in many types of attacks, including session hijacking. It is also often used to fake the e-mail headers of spam so they cannot be properly traced. Session hijacking occurs at the TCP level. According to Internet Security Systems, "TCP session hijacking is when a hacker takes over a TCP session between two machines. Since most authentication only occurs at the start of a TCP session, this allows the hacker to gain access to a machine."

Social Engineering[42]

Social engineering is not a major at MIT. It's the enemy of security professionals and end users alike.

Social engineering is the art and science of getting people to comply to your wishes.

Because many aspects of information security involve technology, it's too easy for employees to think that the problem is being handled by firewalls and other security technologies.

In computer security, *social engineering* is a term that describes a nontechnical kind of intrusion that relies heavily on human interaction and often involves tricking other people to break normal security procedures. A social engineer runs what used to be called a "con game." For example, a person using social engineering to break into a computer network would try to gain the confidence of someone who is authorized to access the network in order to get him/her to reveal information that compromises the network's security. Social engineers might call the authorized employee with some kind of urgent problem as they often rely on the natural helpfulness of people as well as on their weaknesses. Appeal to vanity, appeal to authority, and old-fashioned eavesdropping are typical social engineering techniques.

Another aspect of social engineering relies on people's inability to keep up with a culture that relies heavily on information technology. Social engineers rely on the fact that people are not aware of the value of the information they possess and are careless about protecting it. Frequently, social engineers will search dumpsters and other trash receptacles for valuable information, memorize access codes by looking over someone's shoulder (shoulder surfing), or take advantage of people's natural inclination to choose passwords that are meaningful to them but can be easily guessed. Security experts propose that as our culture becomes more dependent on information, social engineering will remain the greatest threat to any security system. Prevention includes educating people about the value of information, training them to protect it, and increasing people's awareness of how social engineers operate.

Implication

The basic goals of social engineering are the same as hacking in general: to gain unauthorized access to systems or information in order to commit fraud, network intrusion, industrial espionage, or identity theft, or simply to disrupt the system or network.[43]

[40] http://www.webopedia.com/.

[41] http://searchsecurity.techtarget.com.

[42] http://searchsecurity.techtarget.com/sDefinition/0,,sid14_gci531120,00.html.

[43] Sarah Granger, *Social Engineering Fundamentals, Part 1: Hacker Tactics,* http://www.securityfocus.com/infoucus/1527.

TABLE 11.6
Security Tactics and Methods

Security Tactic	Method
Prevention	Architecture: Design, VPN Authentication: Passwords Authorization
Detection	Firewalls, IDS, Accounting, Auditing
Correction	Limit damage; modify Prevention Methods

PREVENTION MEASURES

With knowledge of the threat environment, network administrators must establish preventative measures. Prevention seeks to block a threat before it can be known to the system; thus, the best prevention sits on the perimeter of the network. As noted earlier, if the preventative measures are adequate, detection and correction may be unnecessary. Table 11.6 shows the relations between the three security tactics of prevention, detection, and correction, and methods of achieving these.

Architecture

From Chapter 2, we know that an **architecture** is an overall system plan that is implemented in a set of hardware, software, and communications products. It should be expressed in terms of logical or functional and physical configuration and design. The architecture specifies components and interfaces that make up the systems, to include protocols, formats, and standards to which all hardware and software in the network must conform. Network architecture[44] includes the design principles, physical configuration, functional organization, operational procedures, and data formats used as the basis for the design, construction, modification, and operation of a communications network.

Network design and administration are intrinsically linked to network security. The organization must achieve a judicious balance of network architecture, properly installed technology, and management policies and control. Figure 11.2 shows the overall architecture of the campus network here at Auburn University. It is this design and the resultant installation of technology and management controls that provide our level of security.

The Architecture of a Virtual LAN

VLANs were discussed in Chapter 5 as a way to segment LANs and to create a virtual community of users who are on different physical networks but appear to be on a single private network. This is a primary form of architecture that addresses security; the amount of traffic on the VLAN is reduced and nodes not on the VLAN do not see their traffic. A specific use of the VLAN protocol noted in Figure 11.2 is to place an offending node from ResNet into quarantine by changing the IP address from one VLAN to another.

Implication

The development of VLAN tagging technology has truly changed the way LANs and campus networks are managed. The flexibility added by the use of VLANs has made things possible that would never have been practical just six or eight years ago.

Virtual Private Network (VPN)[45]

When an organization uses an inherently insecure WAN like the Internet for private communications, they must do so in an architecture that provides security. Such is the case of a virtual private network. A **virtual private network (VPN)** is a way to use a public

---

[44] http://en.wikipedia.org/wiki/Network_architecture.

[45] http://searchnetworking.techtarget.com/.

FIGURE 11.2
Network Security Is a Function of Design and Installation

Source: Provided by Bliss Bailey, Director of Networks, Auburn University, AL.

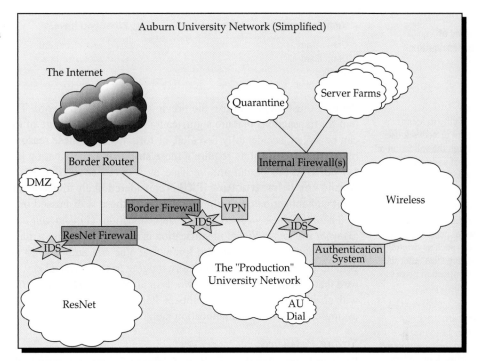

telecommunications infrastructure, such as the Internet, to provide remote offices or individual users with secure access to their organization's network. A VPN works by using the shared public infrastructure while maintaining privacy through security procedures and **tunneling** protocols such as the **Layer Two Tunneling Protocol (L2TP).** In effect, the protocol encrypts the data at the sending end and decrypts it at the receiving end, sending the data in a hidden form through a "tunnel" that cannot be "entered" by data that are not properly encrypted. An additional level of security involves encrypting not only the data, but also the originating and receiving **network addresses.** The tunnel creates an invisible channel in a very public space.

Clarification

Internet Protocol Security (IPsec) is a standard for securing Internet protocol (IP) communications by encrypting and/or authenticating all IP packets. IPsec provides security at the network layer; earlier security approaches have inserted security at the application layer of the communications model. IPsec is a set of cryptographic protocols for (1) securing packet flows and (2) key exchange. It was intended to provide either (1) tunnel mode, that is, portal-to-portal communications security in which security of packet traffic is provided to several machines (even to whole LANs) by a single node, or (2) transport mode, that is, end-to-end security of packet traffic in which the end-point computers do the security processing. It can be used to construct virtual private networks in either mode, which is the dominant use.[46]

Authentication[47]

The **four A's of security** are architecture, authentication, authorization, and accounting.

Authentication is the process of determining whether someone or something is, in fact, who or what it is declared to be. In private and public computer networks, authentication is commonly done through the use of logon **passwords.** Knowledge of the password is assumed to guarantee that the user is authentic. Each user registers initially (or is registered by someone else) using an assigned or self-declared password. On each subsequent use,

Architecture
an overall system plan that is implemented in a set of hardware, software, and communications products.

[46] http://en.wikipedia.org/wiki/Ipsec.

[47] http://searchsecurity.techtarget.com/sDefinition/0,,sid14_gci211621,00.html.

TABLE 11.7
Forms of
Authentication

Item/Security	Who you are	What you have	What you know
Credit Card	Photo	Card w/ccip and date	
Debit Card		Card w/ccip and date	Pin
Email			Logon & Password

Authentication
the ability to identify a system or network user through the validation of a set of assigned credentials.

Authorization
the ability of a specific user to perform certain tasks, such as deleting or creating files, after the authentication process has taken place.

Accounting
allows us to measure and record the consumption of network or system resources.

Authentication, authorization, and accounting are referred to as the **AAA framework**.

the user must know and use the previously declared password. The weakness in this system for transactions that are significant (such as the exchange of money) is that passwords can be stolen, accidentally revealed, or forgotten. For these reasons, Internet business and many other transactions require a more stringent authentication process. The use of **digital certificates** issued and verified by a **certificate of authority (CA)** as part of a **public key infrastructure (PKI)** is considered likely to become the standard way to perform authentication on the Internet. Each of these is discussed in the following.

There are three ways to provide authentication: (1) *who you are,* (2) *what you have,* or (3) *what you know.* Visual identification is used for item one; a major form of identification is to recognize the person and know that s/he has authority to do something. Item two generally means a token of some sort, ranging from a physical key, to a USB device or smart card that contains encrypted passwords or an *RSA SecurID Authentication Manager*®[48] token with changing six-digit logon numbers. Item three is where passwords fall. (Table 11.7 shows examples of the three authentication methods.)

Strong Passwords

Passwords are often the first line of defense for computers and other nodes on a network. From the network administrator's standpoint, passwords create a significant problem. Everyone who has an asset to protect must create a strong password. This means users and network support staff. The problem arises because (a) there is a lack of strength and (b) passwords must be changed periodically to be effective.

Although not a password, per se, making account identifiers difficult is a way to strengthen security. With difficult-to-guess accounts, particularly administrator accounts, intruders must determine at least two access words, not just the password.

Implication

Most equipment ships with publicly known default accounts and passwords. These must be included in the password management program. A list of default passwords for common vendor's equipment can be found at http://www.phenoelit.de/dpl/dpl.html. For example, D-Link DI-704 is "admin"; DI-614+ has no password.

Single passwords should be difficult to guess or devise from a dictionary. One recommendation for a password is at least eight characters in length with at least one capital and one special character. A way to do this is to make it a pass-phrase (Enterthe2dragons) or the first letters and numbers of a sentence (My girl's name is Margie; she is 23 years old = MgniMsi23y$o). A further strengthening is to have an account that requires two passwords. (Yahoo requires a password for simple account access and a second (and different) password for financial access.) The following are some suggestions for passwords.

- *Microsoft*[49] offers the following advice that should be a starting point for all users:

    You probably already know not to create passwords using any combination of consecutive numbers or letters such as "12345678," "lmnopqrs," or adjacent letters on your keyboard such as "qwerty." And you've probably heard that using your login name, your spouse's name, or your birthday as your password are also big no-nos. But did you know that you

[48] http://www.rsasecurity.com/node.asp?id=1156.
[49] http://www.microsoft.com/athome/security/privacy/password.mspx.

should never use a word that can be found in the dictionary, in any language? That's right. Hackers use sophisticated tools that can rapidly guess passwords based on words in the dictionary in different languages, even common words spelled backwards. Additionally, if you use a common word as your password, you might think you're protected if you replace letters of that word with numbers or symbols that look like the letters such as Microsoft or Password. Unfortunately, hackers know these tricks too.

- *The National Institutes of Health*[50] offers the following suggestions for passwords:

What Not to Use

- Don't use your login name in any form (as-is, reversed, capitalized, doubled, etc.).
- Don't use your first or last name in any form.
- Don't use your spouse's or child's name.
- Don't use other information easily obtained about you. This includes license plate numbers, telephone numbers, social security numbers, the brand of your automobile, the name of the street you live on, etc.
- Don't use a password of all digits, or all the same letter. This significantly decreases the search time for a hacker.
- Don't use a word contained in (English or foreign language) dictionaries, spelling lists, or other lists of words.
- Don't use a password shorter than six characters.

What to Use

- Do use a password with mixed-case alphabetic characters.
- Do use a password with non-alphabetic characters, e.g., digits or punctuation.
- Do use a password that is easy to remember, so you don't have to write it down.
- Do use a password that you can type quickly, without having to look at the keyboard. This makes it harder for someone to steal your password by watching over your shoulder.

Part of password management is ensuring that user accounts are disabled when a user is terminated for any reason or comes under disciplinary review. Disgruntled employees are prone to take actions before they leave that opens the organization to threats, such as copying and removing sensitive data, deleting data, leaving a backdoor to his/her accounts, or leaving a software bomb that will take some undesirable action at a later date.

Biometric Identification

Passwords are assumed to be something you know; they also can be something you have. **Biometric identification** is the use of a human body part for unique authorization. Biometrics may be based on a fingerprint, the eye's retina or iris, or a hand or face shape. Although biometric identification is not periodically changed, it is unique, is always with you, cannot be lost, is difficult to copy, and is difficult to steal.

Caveat   Read Dan Brown's *Angels and Demons* to see that even biometrics are not 100 percent secure.

Digital Certificate[51]

A digital certificate is an electronic "credit card" that establishes your credentials when doing business or other transactions on the Web. It is issued by a *certification authority (CA)*. It contains your name, a serial number, expiration dates, a copy of the certificate holder's public key (used for encrypting messages and digital signatures), and the digital signature of the certificate-issuing authority so that a recipient can verify that the certificate is real. Some digital certificates conform to a standard, X.509. Digital certificates can be kept in registries so that authenticating users can look up other users' public keys.

---

[50] http://www.alw.nih.gov/Security/Docs/passwd.html.

[51] http://www.robertgraham.com/pubs/network-intrusion-detection.html.

Digital certificates are useful for companies but not so much for individuals. Individuals will see them as evidence of secure Web sites, but the individual generally does not interact with them. For example, an individual receives an e-mail requesting that s/he use the included link to go to a site to update personal information. At the (bogus) Web site, the URL may appear to be secure (https://), but when the lock in the tray is clicked, it will show that there is no certificate.

Public Key Infrastructure (PKI)[52]

A PKI (public key infrastructure) enables users of a basically insecure public network such as the Internet to securely and privately exchange data and money through the use of a public and a private cryptographic key pair that is obtained and shared through a trusted authority. The PKI provides for a digital certificate that can identify an individual or an organization and directory services that can store and, when necessary, revoke the certificates. Although the components of a PKI are generally understood, a number of different vendor approaches and services are emerging.

The public key infrastructure assumes the use of public key **cryptography,** which is the most common method on the Internet for authenticating a message sender or encrypting a message. Traditional cryptography has usually involved the creation and sharing of a secret key for the **encryption** and decryption of messages. This secret or private key system has the significant flaw that if the key is discovered or intercepted by someone else, messages can easily be decrypted. For this reason, public key cryptography and the public key infrastructure are the preferred approach on the Internet.

Clarification

PGP® or *Pretty Good Privacy*® is a powerful cryptographic product family that enables people to securely exchange messages and to secure files, disk volumes, and so forth.

A public key infrastructure consists of

- A certificate authority that issues and verifies the digital certificate. A certificate includes the public key or information about the public key.
- A registration authority (RA) that acts as the verifier for the certificate authority before a digital certificate is issued to a requestor.
- One or more directories where the certificates (with their public keys) are held.
- A certificate management system.

Certificate of Authority[53]

A certificate of authority is an authority in a network that issues and manages security credentials and public keys for message encryption. As part of a public key infrastructure, a CA checks with a registration authority to verify information provided by the requestor of a digital certificate. If the RA verifies the requestor's information, the CA can then issue a certificate.

VeriSign® is the leading certificate authority, providing over 125,000 Web sites with Secure Sockets Layer (SSL) server certificates, mainly for use in e-commerce.

Secure Socket Layer[54]

The **Secure Sockets Layer (SSL)** is a commonly used protocol for managing the security of a message transmission on the Internet. SSL has recently been succeeded by **Transport Layer Security (TLS),** which is based on SSL. SSL uses a program layer located between the Internet's HyperText Transfer Protocol and Transport Control Protocol

---

[52] Ibid.

[53] Ibid.

[54] Ibid.

layers. SSL is included as part of both the Microsoft and Netscape browsers and most Web server products. Developed by *Netscape®*, SSL also gained the support of *Microsoft®* and other Internet client-server developers as well and became the de facto standard until evolving into Transport Layer Security. The "sockets" part of the term refers to the sockets method of passing data back and forth between a client and a server program in a network or between program layers in the same computer. SSL uses the public-and-private-key encryption system from RSA, which also includes the use of a digital certificate.

TLS and SSL are an integral part of most Web browsers (clients) and Web servers. If a Web site is on a server that supports SSL, SSL can be enabled and specific Web pages can be identified as requiring SSL access. Any Web server can be enabled by using Netscape's® SSLRef program library, which can be downloaded for noncommercial use or licensed for commercial use. TLS and SSL are not interoperable. However, a message sent with TLS can be handled by a client that handles SSL but not TLS.

Transport Layer Security[55]

Transport Layer Security (TLS) is a protocol that ensures privacy between communicating applications and their users on the Internet. When a server and client communicate, TLS ensures that no third party may eavesdrop or tamper with any message. TLS is the successor to the Secure Sockets Layer.

TLS is composed of two layers: the TLS Record Protocol and the TLS Handshake Protocol. The *TLS Record Protocol* provides connection security with some encryption method such as the Data Encryption Standard (DES). The TLS Record Protocol also can be used without encryption. The *TLS Handshake Protocol* allows the server and client to authenticate each other and to negotiate an encryption algorithm and cryptographic keys before data are exchanged.

Kerberos[56]

Kerberos is a network authentication protocol. It is designed to provide strong authentication for client-server applications by using secret-key cryptography. A free implementation of this protocol is available from the Massachusetts Institute of Technology. Kerberos is available in many commercial products as well.

The Internet is an insecure place. Many of the protocols used in the Internet do not provide any security. Tools to "sniff" passwords off of the network are in common use. Thus, applications that send an unencrypted password over the network are extremely vulnerable. Worse yet, other client-server applications rely on the client program to be "honest" about the identity of the user who is using it. Other applications rely on the client to restrict its activities to those that it is allowed to do, with no other enforcement by the server.

Some sites attempt to use firewalls to solve their network security problems. Unfortunately, firewalls assume that "the bad guys" are on the outside, which is often a very bad assumption. Most of the really damaging incidents of computer crime are carried out by insiders. Firewalls also have a significant disadvantage in that they restrict how your users can use the Internet.

The Kerberos protocol uses strong cryptography so that a client can prove its identity to a server (and vice versa) across an insecure network connection. After a client and server have used Kerberos to prove their identity, they also can encrypt all of their communications to assure privacy and data integrity as they go about their business.

---

[55] Ibid.

[56] http://web.mit.edu/kerberos/www/.

TABLE 11.8   **Users' Security Tasks and Responsibilities**

Basic Tasks to Be Performed on End User's Computer Environment	Category	Nonprogramming End Users (Novice) 1	Command-Level Users (Sufficient) 2	End User Programmers (Literate) 3	Functional Support Personnel (Super Users) 4	Support Group	Automated
1 Perform **virus scan** of personal machine	Security						X
2 Keep **virus definition** updated	Security					X	X
3 Keep area **clean;** vents clear	Hardware	X	X	X	X		
4 Organize/**manage** files and data		X	X	X	X		
5 Install software **updates***					X	X	
6 Install software **patches**			X	X	X	X	
7 **Defragment** computer*	Hardware/data				X	X	X
8 Perform regular **backups** *on personal space*	Data	X	X	X	X		X
8a Perform regular **backups** *on shared space*	Data					X	X
9 Install hardware/software **firewalls**	Security					X	
10 Perform basic **peripheral** upkeep	Hardware	X	X	X	X		
11 Try to resolve **basic CPU problems** (rebooting) before contacting help desk		X	X	X	X		
12 **Install** new hardware						X	
13 Perform**authentication** upkeep (passwords)	Security						X

* Contingent upon admin rights.

Authorization[57]

Authorization is the process of giving someone permission to do or have something. Logically, authentication precedes authorization. In multi-user computer systems, a system administrator defines for the system which users are allowed access to the system and what privileges of use (such as access to which file directories, hours of access, amount of allocated storage space, and so forth). Assuming that someone has logged in to a computer operating system or application, the system or application may want to identify what resources the user can be given during this session. Thus, authorization is sometimes seen as both the preliminary setting up of permissions by a system administrator and the actual checking of the permission values that have been set up when a user is getting access.

Implication

Give people minimum rights; turn off all unused services.

Training May Be the Best Prevention

Many of the threats discussed can be defended against by technology; many, however, will get through because of actions of users who don't know better. Just as one in the field of

---

[57] http://www.robertgraham.com/pubs/network-intrusion-detection.html.

psychology might say the first step in curing a problem is admitting it, in network security the first step is admitting the fact of various risks. We don't install a firewall or VPN just because they make us feel secure; we install them as a defense against the risk of intrusion and disclosure.

Any node with access to the Internet is at risk of attack or passing such an attack to other nodes. Workstation operating systems are so easy to use that users can do so with little training. Thwarting attacks, however, is not simple and not intuitive to most people. It is important that security training be a primary method of network security. This involves everyone in the organization: the technical staff and the user community. Users need to know what measures are necessary to secure their resources, such as updating virus definitions if it is not done automatically by installed systems. They need to understand the danger in opening e-mails from anyone they don't know and they must understand that social engineering is alive and well inside and outside of the department.

Table 11.8 illustrates various categories of tasks necessary to keep users safe. It shows the categories of users, by skill, and indicates whether the user, the support staff, or automation should be employed.

DETECTION MEASURES

Firewalls: Protecting the Network

In a previous discussion, equipment whose purpose was to send/allow traffic from one network to another was discussed. We have not addressed the question of how to keep out unwanted traffic. A *firewall* is a system designed to *prevent unauthorized access* to or from a private network; they can be implemented in both hardware and software, or a combination of both.

A gateway server, like the firewall that it may house, sits at the perimeter of the network, acting as, or protecting, the gateway to the world.

A **firewall**[58] is a set of related programs, located at a network gateway server, that protects the resources of a private network from users from other networks. (The term also implies the security policy that is used with the programs.) An enterprise with an intranet that allows its workers access to the wider Internet installs a firewall to prevent outsiders from accessing its own private data resources and for controlling what outside resources its own users have access to. All messages entering or leaving the intranet pass through the firewall, which examines each message and blocks those that do not meet the specified security criteria.

Basically, a firewall, working closely with a router program, examines each network packet to determine whether to forward it toward its destination. A firewall also includes or works with a proxy server that makes network requests on behalf of workstation users. A firewall often is installed in a specially designated computer separate from the rest of the network so that no incoming request can get directly at private network resources. A number of companies make firewall products. Features include logging and reporting, automatic alarms at given thresholds of attack, and a graphical user interface for controlling the firewall.

Several types of firewall techniques exist:

- A **packet filter** looks at each packet entering or leaving the network and accepts or rejects it based on user-defined rules. Packet filtering is fairly effective and transparent to users, but it is difficult to configure. In addition, it is susceptible to IP spoofing.

- An *application gateway* applies security mechanisms to specific applications, such as FTP and Telnet servers. This is very effective but can impose performance degradation.

- A *circuit-level gateway* applies security mechanisms when a TCP or UDP connection is established. Once the connection has been made, packets can flow between the hosts without further checking.

- A **proxy server** intercepts all messages entering and leaving the network. The proxy server effectively hides the true network addresses.

[58] http://searchsecurity.techtarget.com/sDefinition/0,,sid14_gci212125,00.html.

The purpose of a firewall is to keep the network protected from unwanted intrusion; it does not protect data in transport. Network protection comes in several forms, ranging from disallowing hacker intrusion to the thwarting of a *denial-of-service (DoS) attack*. A DoS is designed to jam the input queues of the target network/server and disable it. For example, DoS attacks have attempted to disable the Yahoo.com system by sending in thousands of requests per second from hundreds of known and pirated computers. The firewall would use a very fast hardware solution to inspect the incoming packets, recognize the attack, and divert the packets to an empty network (to fool the attacker) or simply discard them.

Implication   Software and hardware firewalls perform different tasks. Hardware shields from DoS entry; software may prevent DoS from exiting.

Personal Firewalls

Organizational members often use their computer(s) at home to perform work and access the firm's network resources. Therefore, a personal firewall and other protective devices are necessary. The following are admonitions from the *Firewall Guide*.[59]

If a personal firewall is the sheriff, a posse is needed to help the sheriff capture the pests sent out by Internet outlaws like spyware, browser hijackers, viruses, Trojan horses, worms, phishing, spam and hybrids thereof. A layered approach is best to protect your security and privacy:

* First line of defense—Choose an Internet service provider (ISP), an email service and/or a Web site hosting service that offers online virus, spam and content filters.
* Second line of defense—Install a hardware router with a built in firewall between your modem and your computer or network.
* Third line of defense—Use personal firewall, anti-virus, anti-Trojan, anti-spyware, anti-spam and privacy software on your desktop computer and every computer on your network.
* Important Tips—. . .
  * After installing any security software, immediately check for updates at the vendor's Web site.
  * After installing a firewall, use an online testing service to make sure that it is working correctly.

Firewalls Are Customizable[60]

Customization means that you can add or remove **filters** based on several conditions. Some of these are

* IP addresses
* Domain names
* Protocols
* Ports
* Specific words and phrases

Hardware firewalls are incredibly secure and not very expensive. Home versions that include a router, firewall, and Ethernet hub for broadband connections can be found for well under $100.

Demilitarized Zone (DMZ)[61]

The purpose of a **demilitarized zone (DMZ)** is to provide a place for systems on your network that need to have less protection than the rest of your systems. Examples of such systems include those that must be able to be seen by the rest of the Internet, such as Web and e-mail servers. The DMZ segment of your network must use public IP addressing,

[59] http://www.firewallguide.com/.
[60] http://computer.howstuffworks.com/firewall1.htm.
[61] http://searchsecurity.techtarget.com.

whereas the rest of your network can use private IP addresses using **Network Address Translation (NAT)** in the firewall to allow communications.

The DMZ is placed in conjunction with your firewall. If you have a dual-bastion-type firewall, the DMZ is between the bastion hosts that make up the firewall. If you have a single firewall machine, such as illustrated in Figure 11.3, the DMZ is on an interface of the firewall that is separate from the rest of the network that it is protecting.

Of further note in Figure 11.3 is the use of VoIP via the Internet for voice communications. This is placed separately from the organization's firewall/DMZ.

Protection from Spam

Applications exist that will filter e-mail, marking or blocking those that are suspect. Our university uses *IronMail®* at the perimeter and our *GroupWise®* e-mail clients have rules

FIGURE 11.3 **Network with Firewall and DMZ**

Source: Joseph Morris, 10th Street Solutions, Atlanta, GA.

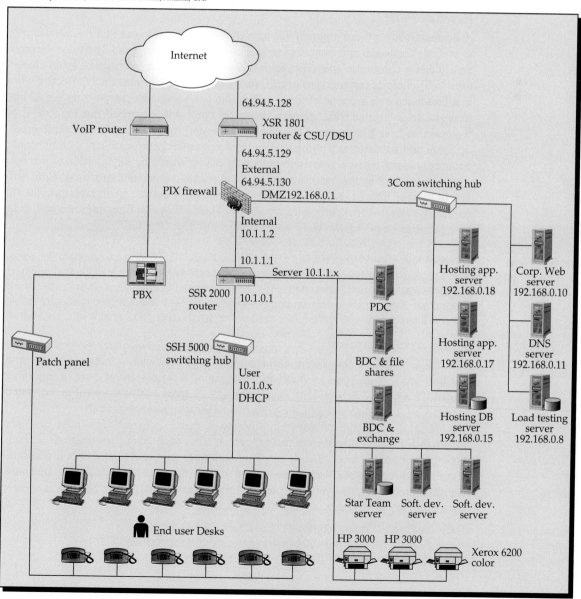

that allow one to delete e-mail based on source or words in the subject line. The following discusses various methods of detection.

Blacklist

A **blacklist**[62] is a database of known Internet addresses (or IPs) used by persons or companies sending spam, for example, *unsolicited commercial e-mail (UCE)*. Various ISPs and bandwidth providers subscribe to these blacklist databases in order to filter out spam sent across their network or to their subscribers.

Heuristics[63]

Heuristics is the application of experience-derived knowledge to a problem and is sometimes used to describe software that screens and filters out messages likely to contain a computer virus or other undesirable content. A heuristic is a "rule-of-thumb." Heuristics software looks for known sources, commonly used text phrases, and transmission or content patterns that experience has shown to be associated with e-mail containing viruses.

Bayesian Filter

A **Bayesian filter**[64] is a program that uses Bayesian logic, also called *Bayesian analysis,* to evaluate the header and content of an incoming e-mail message and determine the probability that it constitutes spam. Bayesian filters aren't perfect, but because spam characteristically contains certain types of text, such a program can be amazingly effective when it is fine-tuned over a period of time. A Bayesian filter works by categorizing e-mail into groups such as "trusted" and "suspect," based on a probability number (ranging from 0, or 0 percent, to 1, or 100 percent). The categories are defined according to user preference.

Spammers are constantly trying to invent new ways to defeat spam filters. Certain words, commonly identified as characteristic of spam, can be altered by the insertion of symbols such as periods or by the use of nonstandard but readable characters such as Â, Ç, Ë, or Í. But as the user instructs a Bayesian filter to quarantine or delete certain messages, the filter incorporates these data into its future actions. Thus, a Bayesian filter improves with time, so it becomes more likely to block spam without also blocking desired messages.

Clarification

McAfee® SpamKiller's® multilayered filtering process uses state-of-the-art Bayesian technology plus signature-based filters to eliminate spam. This extremely effective multilayered approach combines machine-based, user-based, and community-based learning capabilities to identify and block spam. Together these filtering processes analyze blocked and accepted messages to train the core filtering engine to make future decisions on what is spam and what is legitimate e-mail.

Filters versus Blacklists[65]

The real test of any technique for eliminating spam is not how much spam you can stop, but how much spam you can stop without stopping a significant amount of legitimate e-mail. That is, how do you design a defense against spam so that the error in the system is nearly all in the direction of *false negatives* rather than *false positives*?

One great advantage of Bayesian filtering is that it generates few false positives. Bayesian filters, because they're just programs, don't take spam personally. As a result, they make fewer mistakes. Simply blocking mail from any server listed on a blacklist, as some ISPs do now, is in effect a clumsy form of filtering—one that generates a large number of false positives, and yet only catches a small percentage of spam. Spammers seem to have little trouble staying a step ahead of blacklists.

---

[62] http://www.spam-blockers.com/SPAM-blacklists.htm.

[63] http://searchsecurity.techtarget.com/sDefinition/0,,sid14_gci876400,00.html.

[64] http://whatis.techtarget.com/definition/0,,sid9_gci957306,00.html.

[65] http://www.paulgraham.com/falsepositives.html.

Clarification

A whitelist is a list of e-mail addresses or **domain names** from which an e-mail blocking program will allow messages to be received.[66]

Home Network Security

We have noted above the various threats to which a network is vulnerable, whether the network is organizational or home-based. Network management must impress upon users the necessity of **home network security,** as these personal assets may well be the entry point to a corporate network. As we will note in the next chapter, the home wireless network is possibly the most dangerous entry point for corporate networks when users have access.

Intrusion Detection Systems[67]

Intrusion is when someone breaks into or misuses your system. An **intrusion detection system (IDS)**[68] is a system for detecting such intrusions. An IDS inspects all inbound and outbound network activity and identifies suspicious patterns that may indicate a network or system attack from someone attempting to break into or compromise a system. There are several ways to categorize an IDS:

- *Misuse detection versus anomaly detection.* In *misuse detection,* the IDS analyzes the information it gathers and compares it to large databases of attack signatures. Essentially, the IDS looks for a specific attack that has already been documented. Like a virus detection system, misuse detection software is only as good as the database of attack signatures that it uses to compare packets against. In *anomaly detection,* the system administrator defines the baseline, or normal, state of the network's traffic load; breakdown; protocol; and typical packet size. The anomaly detector monitors network segments to compare their state to the normal baseline and looks for anomalies.
- *Network-based versus host-based systems.* In a *network-based system, or NIDS,* the individual packets flowing through a network are analyzed. The NIDS can detect malicious packets that are designed to be overlooked by a firewall's simplistic filtering rules. In a *host-based system,* the IDS examines the activity on each individual computer or host.
- *Passive system versus reactive system.* In a *passive system,* the IDS detects a potential security breach, logs the information, and signals an alert. In a *reactive system,* the IDS responds to the suspicious activity by logging off a user or by reprogramming the firewall to block network traffic from the suspected malicious source.

Though they both relate to network security, an IDS differs from a firewall in that a firewall looks out for intrusions in order to stop them from happening. The firewall limits the access between networks in order to prevent intrusion and does not signal an attack from inside the network. An IDS evaluates a suspected intrusion once it has taken place and signals an alarm. An IDS also watches for attacks that originate from within a system.

Clarification

Intrusion detection is the art of detecting inappropriate, incorrect, or anomalous activity. The most common approaches to ID are statistical anomaly detection and pattern-matching detection. IDS are like a burglar alarm for your computer network . . . they detect unauthorized access attempts. They are the first line of defense for your computer systems.[69]

IDS can be broken down into the following categories:

- **Network intrusion detection systems (network monitor** or **NIDS)** *monitor packets* on the network wire and attempt to discover if a hacker is attempting to break into a

[66] http://searchsecurity.techtarget.com/sDefinition/0,,sid14_gci896131,00.html.
[67] http://www.robertgraham.com/pubs/network-intrusion-detection.html.
[68] http://networking.webopedia.com/TERM/I/intrusion_detection_system.html.
[69] http://www.intrusion-detection-system-group.co.uk/, http://www.sans.org/resources/idfaq/what_is_id.php.

system (or cause a denial-of-service attack). A typical example is a system that watches for large number of TCP connection requests (SYN) to many different ports on a target machine, thus discovering if someone is attempting a TCP port scan. A NIDS may run either on the target machine that watches its own traffic (usually integrated with the stack and services themselves) or on an independent machine promiscuously watching all network traffic (hub, router, probe). Note that a "network" IDS monitors many machines, whereas the others monitor only a single machine (the one they are installed on).

- **System integrity verifiers (SIV)** *monitor system files* to find when a intruder changes them (thereby leaving behind a backdoor). The most famous of such systems is "Tripwire." A SIV may watch other components as well, such as the Windows registry, in order to find well-known signatures. It also may detect when a normal user somehow acquires root/administrator-level privileges. Many existing products in this area should be considered more "tools" than complete "systems"; that is, something like "Tripwire" detects changes in critical system components but doesn't generate real-time alerts upon an intrusion.

- **Log file monitors (LFM)** *monitor log files* generated by network services. In a similar manner to NIDS, these systems look for patterns in the log files that suggest an intruder is attacking. A typical example would be a parser for HTTP server log files that looks for intruders who try well-known security holes, such as the "phf" attack.

Intrusion Prevention Systems

Intrusion detection systems are more passive in their methodology, monitoring and informing network administrators of any intrusive presence. **Intrusion prevention systems (IPSs)** try to take a proactive measure in network security so as to stop the attack before it starts. Both of these systems monitor networks, whether they are WLANs or LANS, over extended periods of time.

An *intrusion prevention system*[70] is any device (hardware or software) that has the ability to detect attacks, both known and unknown, and prevent the attack from being successful. Now that firewalls can keep track of TCP sequence numbers and have the ability to block certain types of traffic (such as Code Red or Nimda), even they can act as intrusion prevention systems. There are five different categories of IPS that focus on attack prevention at layers that most firewalls are not able to decipher, at least not yet. The five types of IPSs that we will look at are inline NIDS, application-based firewalls/IDS, layer seven switches, network-based application IDSs, and deceptive applications.

In the enterprise, intrusion detection and intrusion prevention[71] technologies are gaining traction as companies realize that the traditional signature-based defenses of antivirus software leaves a dangerous window of opportunity between the detection of a new threat and antivirus firms' responses. Instead, intrusion detection/prevention software tries to spot so-called zero-day threats—for which there is no defense—or other behaviors that may presage an attack, then block those activities.

Honeypots and Traps[72]

Honeypots[73] are closely monitored network decoys serving several purposes: they can distract adversaries from more valuable machines on a network, they can provide early warning about new attack and exploitation trends, and they allow in-depth examination of adversaries during and after exploitation of a honeypot.

[70] http://www.securityfocus.com/infocus/1670.
[71] http://www.techweb.com/wire/security/52200149.
[72] http://www.wittsend.com/mhw/1999/securing_linux/txt025.html.
[73] http://www.honeypots.net/.

The purpose of honeypots and **honey traps** is to perform one or more of several functions, with an objective of protecting a genuine network. The first is to set up fake "vulnerable" services to lure intruders into honeypots. Alternatively, they configure legitimate services with fake headers and traps to trigger alarms when attacked. Administrators may trap routine, but unused, services to provide early warning alarms. As attacks can be deflected or intruders detected, the real network workload is decreased.

Honeypots are a highly flexible security tool with different applications for security. They don't fix a single problem. Instead, they have multiple uses, such as prevention, detection, or information gathering. Honeypots all share the same concept: a security resource that should not have any production or authorized activity. In other words, deployment of honeypots in a network should not affect critical network services and applications.

Clarification   A **honeypot** is a security resource whose value lies in being probed, attacked, or compromised.

There are two general types of honeypots:

* *Production honeypots* are easy to use, capture only limited information, and are used primarily by companies or corporations.
* *Research honeypots* are complex to deploy and maintain, capture extensive information, and are used primarily by research, military, or government organizations.

An example of a honeypot is a system used to simulate one or more network services that you designate on your computer's ports. An attacker assumes you're running vulnerable services that can be used to break into the machine. This kind of honeypot can be used to log access attempts to those ports, including the attacker's keystrokes. This could give you advanced warning of a more concerted attack.

Implication   *Malicious Code Protector*® is a patent-pending technology designed to detect buffer overflow attacks.[74]

Sniffers[75]

Packet sniffers (also known as network analyzers or Ethernet sniffers) are software programs that can see the traffic passing over a network or part of a network. As data streams travel back and forth over the network, the program captures each packet and eventually decodes its content following a specification. Depending on the network structure (hub or switch), one can sniff all or only parts of the traffic from a single machine within the network; however, there are some methods to avoid traffic narrowing by switches. The special network device driver used for some packet sniffing software is said to operate in "promiscuous mode" as it listens to everything (on the wire). The versatility of packet sniffers means they can be used to

* Troubleshoot a network.
* Detect network intrusion attempts.
* Monitor the network usage and filter for suspect content.
* Spy on other network users and attempt to collect their passwords.

Clarification   *Capsa*® is a powerful yet easy to use network monitor and analyzer designed for packet decoding and network diagnosis. With the abilities of real time monitoring and data analyzing, you can capture and decode network traffic transmitted over a local host and a local network. In addition to Packet Analysis Module, Capsa also has three advanced analysis modules:

[74] http://www.checkpoint.com/.
[75] http://en.wikipedia.org/wiki/Sniffer.

Email Analysis Module, Web Analysis Module and Transaction Analysis Module, which enable you to easily analyze and view detailed information of network activities.[76]

Caveat

Sniffers[77] can be used both for legitimate network management functions and for stealing information off a network. Unauthorized sniffers can be extremely dangerous to a network's security because they are virtually impossible to detect and can be inserted almost anywhere. Wireless sniffers require no insertion point; they just listen to traffic between an access point and a client. This makes them a favorite weapon in the hacker's arsenal.

Penetration Testing

The detection methods previously listed are designed to find threats from others; **penetration testing** seeks to take overt action to determine if your own system is secure. The point is *quality assurance,* that is, am I as secure as I think I am? After the protection schemes are installed, the organization should test them to see if they are working. This can be done with in-house staff or by a third-party organization.

Implication

When a regional health care company called in network protection firm Neohapsis to find the vulnerabilities in its systems, the Chicago-based security company knew a sure place to look. Retrieving the password file from one of the health care company's servers, the consulting firm put "John the Ripper," a well-known cracking program, on the case. While well-chosen passwords could take years—if not decades—of computer time to crack, it took the program only an hour to decipher 30 percent of the passwords for the nearly 10,000 accounts listed in the file.[78]

After-the-Fact Detection

Protection has three phases: (1) install systems and assets that are intrusion-resistant; (2) implement capabilities that thwart intrusion attempts; and (3) make sure any intruder or world-be hacker leaves a trail and makes a record. This entails three methods: (a) monitoring, (b) logging, and (c) auditing. These are all passive events, although they may be coupled with active measures.

Monitoring is done by network analysis and sniffer tools, in firewall methods, and with IDS software. The purpose is to actively watch network traffic and, to the extent possible, deflect undesirable traffic.

Logging is the recording of events. This provides a searchable file of activity that can then be analyzed for patterns.

Accounting and Auditing

Accounting allows us to measure and record the consumption of network or system resources. Although this would seem to simply be a financial consideration, reviewing the accounting records may discover unusual activity. Even though review of these records is not the same as auditing the network logs, it may provide an indication of suspicious actions. These records should be protected, as they show all the financial activities of the organization, but periodic graphic reports will generally be sufficient to highlight areas of interest.

Auditing is the process of reviewing the logs created by the network resources. We are interested in doing this for two reasons: (1) it helps to determine high usage of resources, indicating problems such as constraints of bandwidth, and (2) users/intruders generally leave tracks that may alert us to problems. Periodically displaying number of accesses, data

---

[76] http://www.colasoft.com/products/capsa/index.php?id=75430g.

[77] http://www.webopedia.com/TERM/s/sniffer.html.

[78] http://news.com.com/Passwords+The+weakest+link/2009-1001_3-916719.html.

TABLE 11.9
Securing Workstations and Servers

Securing Workstations	Securing Servers
Patches and virus definitions	Log activity and review it often
Restrict installations & configuration altering	Backup data regularly & protect the backup tape
Enforce password policy	Shut down any service not needed
Kill accounts & services not needed; change Administrator account	Kill accounts not needed; change Administrator account
Don't show last person logged on; don't use default share.	Don't show last person logged on; don't use default share.

transferred, and bandwidth used will be an easy way to see discontinuities; doing a machine search or visual review of the records will be more labor intensive but may provide greater insight into suspicious activity.

Audits are formal, periodic investigations of activities to ensure that policies are being carried out. It is not enough to perform risk analysis, create a disaster recovery plan, and implement security devices that perform monitoring and logging. Audits, often performed by a third-party organization, establish that all of these have been effective. It is a good idea to include an IT auditor or the system development project team so that auditability can be built in from the beginning.

Table 11.9 shows some of the tasks that are vital to the security of personal workstations and servers. The tasks are competence-specific, with some left to a support group.

LEGAL IMPLICATIONS

Damage to enterprise data, theft, and espionage are only part of the potential problems once hackers get into a network. Government-mandated legislation provides severe penalties for breach of privacy of confidential records, or loss of mandated documentation. The major laws are discussed here:

• The **Gramm-Leach-Bliley Act (GLBA)** was enacted to assure that financial institutions protect the privacy of their customers' financial records. The act requires that such institutions take all reasonable measures to detect, prevent, and respond to attacks, intrusions, or other system failures. Failure to do so can result in fines of up to $11,000 per day or $10,000 per violation.

• The **Health Insurance Portability and Accountability Act (HIPAA)** affects virtually all health care entities in the United States. It is (in part) intended to protect the privacy of individuals' health records. Any electronic transaction that contains confidential health information of an individual is covered by HIPAA. The act requires clear control of access, policies, procedures, and technology to restrict who has access to the information, and requires establishment of security mechanisms to protect data that are electronically transmitted. Violations of HIPAA could cost $25,000 per incident.

• The **Sarbanes-Oxley Act of 2002 (SOX),**[79] on the surface, doesn't have much to do with IT security. The law was passed to restore the public's confidence in corporate governance by making chief executives of publicly traded companies personally validate financial statements and other information. Congress passed the law in quick response to accounting scandals surrounding Enron and other companies. SOX deals with many corporate governance issues, including executive compensation and the use of independent directors.

[79] Edward Hurley, "Security and Sarbanes-Oxley," *SearchSecurity.com News Writer,* September 25, 2003, http://SearchSecurity.com.

• Yet, in the law, there is a provision mandating that CEOs and CFOs attest to their companies' having proper "internal controls." It's hard to sign off on the validity of data if the systems maintaining them aren't secure. "It's the IT systems that keep the books. If systems aren't secure, then internal controls are not going to be too good." Sarbanes-Oxley doesn't mandate specific internal controls such as strong authentication or the use of encryption. "But, if someone can easily get in your system because you have a four-character password, for me, that is a no-brainer [as a sign of noncompliance]," Saidman[80] said.

Implication

With the recent federal regulation of information security policy through legislation such as the Health Insurance Portability and Accountability Act (HIPAA), Gramm-Leach-Bliley Act (GLBA), and Sarbanes-Oxley Act (SOX), enterprises are charged with protecting data residing in mail servers and on other internal systems. Security breaches violate these regulations, exposing sensitive data and opening the door to serious sanctions and costly litigation.[81]

• The **CAN-SPAM Act of 2003 (Controlling the Assault of Non-Solicited Pornography and Marketing Act)**[82] establishes requirements for those who send commercial e-mail, spells out penalties for spammers and companies whose products are advertised in spam if they violate the law, and gives consumers the right to ask e-mailers to stop spamming them.

• The law, which became effective January 1, 2004, covers e-mail whose primary purpose is advertising or promoting a commercial product or service, including content on a Web site. A "transactional or relationship message"—e-mail that facilitates an agreed-upon transaction or updates a customer in an existing business relationship—may not contain false or misleading routing information, but otherwise is exempt from most provisions of the CAN-SPAM Act.

• The Federal Trade Commission (FTC), the nation's consumer protection agency, is authorized to enforce the CAN-SPAM Act. CAN-SPAM also gives the Department of Justice (DOJ) the authority to enforce its criminal sanctions. Other federal and state agencies can enforce the law against organizations under their jurisdiction, and companies that provide Internet access may sue violators, as well.

• The law requires the following:

1. It bans false or misleading header information. Your e-mail's "From," "To," and routing information—including the originating domain name and e-mail address—must be accurate and identify the person who initiated the e-mail.

2. It prohibits deceptive subject lines. The subject line cannot mislead the recipient about the contents or subject matter of the message.

3. It requires that your e-mail give recipients an opt-out method. You must provide a return e-mail address or another Internet-based response mechanism that allows a recipient to ask you not to send future e-mail messages to that e-mail address, and you must honor the requests. You may create a "menu" of choices to allow a recipient to opt out of certain types of messages, but you must include the option to end any commercial messages from the sender.

4. Any opt-out mechanism you offer must be able to process opt-out requests for at least 30 days after you send your commercial e-mail. When you receive an opt-out request, the law gives you 10 business days to stop sending e-mail to the requestor's e-mail address. You cannot help another entity send e-mail to that address or have another entity send e-mail on your behalf to that address. Finally, it's illegal for you to sell or transfer the e-mail addresses

---

[80] Gary Saidman is an attorney specializing in information security matters with Atlanta-based law firm Kilpatrick Stockton.

[81] http://www.CipherTrust.com.

[82] http://www.ftc.gov/bcp/conline/pubs/buspubs/canspam.htm.

of people who choose not to receive your e-mail, even in the form of a mailing list, unless you transfer the addresses so another entity can comply with the law.

5. It requires that commercial e-mail be identified as an advertisement and include the sender's valid physical postal address. Your message must contain clear and conspicuous notice that the message is an advertisement or solicitation and that the recipient can opt out of receiving more commercial e-mail from you. It also must include your valid physical postal address.

• The *California Security Breach Law* (SB-1386)[83] requires organizations that maintain personal information about individuals to inform those individuals if the security of their information is compromised. The act stipulates that if there's a security breach of a database containing personal data, the responsible organization must notify each individual for whom it maintained information. The act, which went into effect July 1, 2003, was created to help stem the increasing incidence of identity theft. According to the Federal Trade Commission, the organization received 214,905 complaints of identity theft in 2003, up 40 percent from 2002.

• The new regulations may become a model for federal legislation. U.S. Senator Dianne Feinstein introduced federal legislation based on the California law. If the legislation is enacted, companies and agencies would have to provide a notice to each person whose data were compromised. Entities that fail to comply with the law could be sued in court or face FTC fines of up to $25,000 per day while the violation persists.

• The E-Government Act (Public Law No. 107-347), passed in 2002, recognized the importance of information security to the economic and national security interests of the United States. Title III of the E-Government Act, entitled the **Federal Information Security Management Act (FISMA),**[84] requires each federal agency to develop, document, and implement an agencywide program to provide information security for the information and information systems that support the operations and assets of the agency, including those provided or managed by another agency, contractor, or other source.

NASDAQ
(originally an acronym for National Association of Securities Dealers Automated Quotations) is a U.S. electronic stock exchange. It was founded by the National Association of Securities Dealers (NASD).

• The **National Association of Securities Dealers (NASD)** requires, under *Conduct Rule 3010,*[85] that member firms establish and maintain a system to "supervise" the activities of each registered representative, including transactions and correspondence with the public. In addition, NASD Conduct Rule 3110 requires that member firms implement a retention program ("Books and Records") for all correspondence involving registered representatives.

• *FDA 21 CFR part 11*[86] is a specific regulation that deals with the use of electronic records and signatures in manufacturing processes regulated by the FDA.

Case 11-1

Next-Generation Virus Defense: *An Overview of IronMail Zero-Day Virus Protection*[87]

In January 2004, the computer virus known as "MyDoom" created mass disruption to corporate resources and reputation as it spread quickly through e-mail networks worldwide.

---

[83] http://searchcio.techtarget.com/.

[84] http://csrc.nist.gov/sec-cert/ca-background.html.

[85] http://www.ziplip.com/solutions/NASD.html.

[86] http://www.matrixone.com/fda-21-cfr-part11.htm.

[87] © 2004 CipherTrust, Inc. All Rights Reserved. October 2004. This paper is used with permission from CipherTrust, http://www.ciphertrust.com/.

At its peak, MyDoom infected one in every five e-mails transmitted over the Internet.[88] The worm broke records set by previous malware, such as SoBig.F, to become the fastest-spreading virus ever. This incredible propagation speed left many networks vulnerable despite the presence of anti-virus software. Anti-virus programs rely on signatures created by the anti-virus vendor, but this process takes time. The independent testing laboratory AV-test.org found the response times to range from just under seven hours to almost 30 hours,[89] with the four leading vendors (Sophos, McAfee, Symantec, and Trend Micro) clocking in at no fewer than 12 hours.

Unfortunately, the speed of MyDoom and subsequent virus and malware attacks ensures that the damage is done by the time the anti-virus vendors have published their signature. For corporations, this means that simply deploying a signature-based solution is no longer good enough. They need to take additional steps to prevent these rapidly deploying new threats, or "Zero-Day" attacks, from causing serious disruption.

A CLEAR AND PRESENT DANGER

CipherTrust's IronMail Threat combines best-of-breed e-mail security technologies, including deep attachment filtering, anomaly detection, anti-spam protection, automatic Threat Response Updates (TRU), the CipherTrust Anti-Virus engine (powered by Authentium) and a secure platform. The end result of this multi-faceted approach is IronMail Zero-Day Virus Protection, the most effective and comprehensive protection available for enterprise e-mail environments.

VIRUS AND MALWARE ATTACK CONSEQUENCES

Attacks from viruses and other malicious code have serious consequences for businesses and other organizations providing e-mail. These include:

System Downtime

E-mail has evolved to become the primary communication method for most organizations, and the loss of e-mail due to attack can severely affect enterprise operations. Beyond the immediate financial expenses involved in restoring the network, an attack on your enterprise e-mail system can also result in lost hours and days for employees.

Resource Depletion

The costs of cleaning up an e-mail system after an attack are significant. IT teams are forced to spend considerable time and money repairing damage to servers and the network. Worse still is the prospect of checking and cleaning individual workstations.

Administration

In the past, when a new vulnerability was discovered, network administrators scrambled to apply security patches from the makers of their anti-virus software and manually reviewed quarantine lists for virus-infected messages. Software manufacturers release patches so frequently that network administrators cannot reasonably be expected to keep up with them all, particularly within a change-controlled environment. As stated by Gartner Research, "Enterprises will never be able to patch quickly enough. After all, attackers have nothing else to do."[90] The staggering damage caused by recent virus and malware attacks is clear

---

[88] http://www.nationmaster.com/encyclopedi/MyDoom

[89] http://www.esecurityplanet.com/views/article.php/3316511

[90] Pescatore, John—Gartner. "Management Update: Mount a Solid Defense Against Worms and Viruses," 9/15/04

evidence that manual intervention to institute emergency measures or review quarantined messages is rarely effective against rapidly propagating threats.

Compliance and Liability

With the recent Federal regulation of information security policy through legislation such as the Health Insurance Portability and Accountability Act (HIPAA), Gramm-Leach-Bliley Act (GLBA) and Sarbanes-Oxley Act (SoX), enterprises are charged with protecting data residing in mail servers and on other internal systems. Security breaches violate these regulations, exposing sensitive data and opening the door to serious sanctions and costly litigation.

Credibility

Falling victim to a Zero-Day attack can also result in lost trust from business partners and customers. According to Gartner, "Enterprises that spread viruses, worms, spam and denial-of-service attacks will find not only that malicious software can hinder their profitability, but also that other businesses will disconnect from them if they are considered to be risky." While an attack may not be your fault, it is most certainly your problem.

ZERO-DAY VIRUS PROTECTION

In light of the emergence of rapidly propagating threats, the most critical part of any anti-virus solution is now its ability to identify these new threats, regardless of whether they are viruses, worms, hybrids or something entirely new. IronMail utilizes a comprehensive solution to ensure detection of all malicious threats, even those that have not yet been identified.

Attachment Filtering

Attachments frequently contain the "payload" of a virus. Detecting and blocking them at the gateway, before they ever reach the mail server, is critical. IronMail provides "deep inspection" of all attachments to protect against dangerous file types. To ensure accuracy and security, IronMail examines the binary code contained in attachments to determine the actual file type, rather than relying on signature files and file extensions like other gateway appliances. IronMail also inspects the contents of .zip files, going 16 levels deep if necessary.

Anomaly Detection

IronMail utilizes a unique Anomaly Detection Engine (ADE), which dynamically identifies and responds to abnormal behavior in mail flow. By monitoring "normal" e-mail traffic rates across the Internet, the ADE allows IronMail to identify spikes in traffic that are often the first signal of a Zero-Day attack. Once these spikes are recognized, IronMail units take appropriate action to prevent infiltration of the network.

Anti-Spam

Increasingly, spam is a common carrier of viruses and other malicious code. The techniques used by spammers to evade detection are just as effective for virus writers. The anti-spam feature built into IronMail analyzes every characteristic of an e-mail message to differentiate between spam and legitimate messages. IronMail's award-winning anti-spam technology uses an array of detection and analysis techniques to "score" messages based on their likelihood of being spam. It provides deep inspection of messages and identifies many different types of deception used to hide attacks. This allows IronMail to be extremely effective at detecting and blocking viruses and worms that use spamming techniques.

Threat Response Updates (TRU)

CipherTrust provides updates to IronMail units in the field for the purpose of responding to new threats and continually optimizing the IronMail software. These updates are based on data gathered from active IronMail units in over 1000 different companies and organizations, Internet tracking and scanning, and research and analysis. Some key contributors to TRU include:

TrustedSource—By constantly observing and analyzing e-mail traffic across the Internet, CipherTrust identifies good e-mail sender behavior—those senders who consistently prevent spam, viruses, and other unwanted e-mail from being sent from their servers. TrustedSource allows IronMail to achieve the highest level of accuracy in determining good e-mail and thus reduce false positives for IronMail customers.

Statistical Lookup Service (SLS)—A service focused on identifying bulk e-mail messages, SLS leverages CipherTrust's extensive network of over 7.5 million enterprise users, including large corporations, government sites and educational institutions, to maintain a database of spam signatures and allow real-time queries.

Content Filtering—CipherTrust identifies keywords used in virus attacks and provides updated policies to allow immediate response. IronMail dictionaries are populated with words, phrases, weighted word lists and text patterns for use in content filtering.

Implication
Protect end-users from themselves.

E-mail security differs greatly from traditional network security, which uses tools such as firewalls, VPN and intrusion detection. Because e-mail is deemed "safe" traffic, every e-mail—good or bad—passes through the firewall and enters the enterprise network. And because most e-mail threats—spam, phishing, viruses etc.—are sent with a profit motive, spammers and hackers continually create new threats.

Secure Platform

An increasingly common technique, particularly for attacks on large corporations, is to make a conventional hacker attack on an organization's e-mail defenses prior to a targeted virus attack. IronMail is designed to protect the entire e-mail system from attack via an e-mail firewall and intrusion prevention capabilities. IronMail detects and blocks attacks including denial-of-service and buffer overflows.

Secure WebMail

For corporations that use them, corporate WebMail products are another vector of attack for e-mail systems. Compatible with Lotus iNotes, GroupWise WebAccess and Outlook Web Access, the Secure WebMail component of IronMail protects corporate WebMail systems from attack. It provides comprehensive protection against a range of mail threats, including cross-packet attacks, directory traversal attacks, path obfuscation, shell access attacks and database access attacks.

Policy Enforcement

IronMail's unified policy manager enforces content and policy compliance across the entire e-mail system and provides comprehensive content filtering, monitoring and reporting capabilities. IronMail improves productivity, reduces liability and saves valuable network resources.

RECYCLED THREATS

While Zero-Day protection is the most critical component of effective virus control, there is still a need for a more traditional signature-based anti-virus engine. Deployed correctly, this system provides assurance that older, known viruses will not trouble your network.

IronMail offers three options, which can be deployed singly or in combination. CipherTrust anti-virus, powered by Authentium, is IronMail's standard anti-virus engine. We also offer the Sophos and McAfee engines.

All of these engines are updated through IronMail's Threat Response Updates, as often as hourly. By deeply integrating these solutions with IronMail, CipherTrust reduces the administrative burden of managing multiple platforms and allows anti-virus to become part of a unified e-mail policy.

MANAGEMENT & USABILITY

IronMail was built for enterprise deployments. Within the IronMail solution, CipherTrust includes an internally built, high-performance Message Transfer Agent (MTA) that effectively manages and routes millions of e-mail messages a day in some of the largest enterprise messaging networks in the world. This extensible technology platform allows the IronMail solution to be continually updated with new e-mail security technologies and tools as they are developed.

SCALABILITY & RELIABILITY

With 30 percent of the Fortune 100 relying upon IronMail for messaging security, IronMail has continually demonstrated its scalability and reliability in the most demanding enterprise environments. More than seven million users depend upon IronMail for e-mail security every day.

LOGGING & REPORTING

IronMail's high-performance MTA provides administrators with automated logging and reporting, which saves time and money that would be required to perform equivalent management and monitoring functions. The IronMail dashboard provides a comprehensive view of the entire e-mail system, giving administrators full visibility of message traffic.

CONCLUSION

"IronMail paid for itself by stopping MyDoom."—Iain Liddel, information technology manager at Brunel University.

Protecting e-mail systems from viruses and malware is imperative for enterprises. With IronMail Threat Control from CipherTrust, enterprises can provide anti-virus protection at the gateway and form a partnership that will ensure threat protection into the future.

IronMail is the leading provider of robust protection to stop viruses and other malware before they reach the mail server. IronMail combines and integrates multiple security technologies to detect and block the most dangerous attacks, virus and malware outbreaks. IronMail Zero-Day Virus Protection is proven, tested every day in the most demanding environments in the world, including businesses, government agencies and educational institutions. IronMail's success as the complete messaging security solution has made CipherTrust the leader in messaging security.

Source: "Next-Generation Virus Defense: An Overview of IronMail Zero-Day Virus Protection,"
© 2004 CipherTrust, Inc. All Rights Reserved. This paper is used by permission from CipherTrust.
http://www.ciphertrust.com.

Case 11-2

IP Spoofing: *An Introduction*

Matthew Tanase

Please view this case online at http://www.securityfocus.com/infocus/1674.

Summary

The concerns for security and disaster preparedness go hand-in-hand. Security is a concern for the safety of assets; disaster preparedness begins with risk assessment and collimates in a disaster recovery, or, more appropriately, a business continuity plan. It is doubtful if the two concerns can be separated.

In 2005, IT leaders viewed network security, storage, and wireless as key technologies. Thus, there was acknowledgment of the evolution to a network-centric emphasis. The storage topic was dominated by storage-area-network (SAN) initiatives to satisfy the demand imposed by legislative mandates and e-business growth. As wireless networks were proliferating rapidly, there was a focus on employee mobility, sensors, location-tracking, and network security. IT managers were concerned about more and more devices that connected to the Internet where infections and hackers were lurking. New emphases were on stanching the spread of viruses and DoS attacks in the face of faster-moving and more-malicious threats. Networks serve as organization enablers, but they also enable the hackers and malware. As a result, the IT manager is left feeling more and more vulnerable. This has led to more holistic approaches to security. Part of this emphasis is the result of wireless networking security having become a notoriously prominent issue.

In an environment where an organization's very existence depends on connectivity and data storage and access, network security must be as important as any mission-critical application. Networks must transport data reliably from secure and reliable storage media; they both must be accessible, often 24*7; highly responsive; and, of course, secure. Network security ranges from the denial of DoS attacks to the thwarting of spam, detection and eradication of viruses, encryption of packets, physical securing of assets, deflection of adware and spyware, assurance of adequate continuous power, and denial of phishing for information. Security is the front line in the environment of information warfare; the first line of defense against pranksters, evil-doers, and competitors.

While much of the infrastructure of security is the world of technology and network administrators, a coordinated management and funding plan must be in place else it may be ineffective. Too little protection, too late is a recipe for disaster; a fragmented attack plan will likely leave holes in the defense where an enemy that has proven to be highly intelligent can flack the firewall by invading an employee's home network and entering through the front door. Thus, the concern for organizational security must extend to the actions of the members of that organization and to every mode of access, for example, their personal computers that are allowed to connect to the corporate network.

The consideration of wireless threats and defenses, for the most part, have been left to the next chapter. Wireless means a different medium and very different modes of attack. It makes the interception of packets easier because they are broadcast into the radio or IR spectrum. Wireless technologies, however, present many of the same dangers to the corporate network; they just have a different way to evoke them. Wired or wireless, security is at the very heart of risk assessment and disaster planning for business continuity.

Key Terms

24*7, *p. 308*
AAA framework, *p. 326*
accounting, *p. 326*
active threats, *p. 311*
adware, *p. 321*
architecture, *p. 324*
audit, *p. 339*
auditing, *p. 338*
authentication, *p. 325*
authorization, *p. 326*
Bayesian filter, *p. 334*
biometric
identification, *p. 327*
blacklist, *p. 334*
broadcast, *p. 309*
browser hijacking, *p. 322*
CAN-SPAM Act of 2003
(Controlling the Assault of
Non-Solicited
Pornography and
Marketing Act), *p. 340*
certificate of authority
(CA), *p. 326*
cold site, *p. 307*
compression, *p. 309*
cookie, *p. 322*
cracker, *p. 303*
crosstalk, *p. 312*
cryptography, *p. 328*
data security, *p. 309*
demilitarized zone
(DMZ), *p. 332*
denial-of-service
(DoS/DDoS)
attack, *p. 315*
digital certificate, *p. 326*
disaster, *p. 303*
disaster planning, *p. 307*
disaster recovery, *p. 306*
domain name, *p. 335*
electronic mail
(e-mail), *p. 314*
e-mail spoofing, *p. 317*
encryption, *p. 328*
environmental
hazards, *p. 311*
Federal Information
Security Management Act
(FISMA), *p. 341*
filter, *p. 332*
firewall, *p. 331*

Gramm-Leach-Bliley Act
(GLBA), *p. 339*
hacker, *p. 303*
hacking, *p. 314*
Health Insurance
Portability and
Accountability Act
(HIPAA), *p. 339*
heuristics, *p. 334*
home network
security, *p. 335*
honey trap, *p. 337*
honeypot, *p. 336*
hot site, *p. 307*
information
warfare, *p. 313*
Internet Protocol Security
(IPsec), *p. 325*
intrusion, *p. 335*
intrusion detection system
(IDS), *p. 335*
intrusion prevention
systems (IPS), *p. 336*
Kerberos, *p. 329*
Layer Two Tunneling
Protocol (L2TP), *p. 325*
log file monitor
(LFM), *p. 336*
logging, *p. 338*
logic bombs, *p. 322*
malware, *p. 321*
management, *p. 303*
modifying
factors, *p. 305*
monitor, *p. 338*
National Association
of Securities Dealers
(NASD), *p. 341*
network address, *p. 325*
Network Address
Translation
(NAT), *p. 333*
network intrusion
detection system
(NIDS), *p. 335*
network monitor, *p. 335*
outsourcing, *p. 305*
packet filter, *p. 331*
packet sniffer, *p. 337*
passive threats, *p. 311*
passwords, *p. 325*

patch, *p. 311*
penetration
testing, *p. 338*
pharming, *p. 320*
phishing, *p. 318*
physical security, *p. 308*
power outage, *p. 312*
prevention, *p. 303*
privacy, *p. 303*
proprietary, *p. 303*
proxy server, *p. 331*
public key infrastructure
(PKI), *p. 326*
radiation, *p. 311*
range, *p. 305*
regulation, *p. 312*
risk, *p. 303*
risk analysis, *p. 305*
risk assessment, *p. 305*
risk management, *p. 305*
rogue users, *p. 314*
Sarbanes-Oxley Act of
2002 (SOX), *p. 339*
script kiddies, *p. 314*
Secure Sockets Layer
(SSL), *p. 328*
security, *p. 303*
sniffer, *p. 338*
social engineering, *p. 309*
spam, *p. 309*
spoofing, *p. 317*
spyware, *p. 322*
system hijacking, *p. 323*
system integrity verifier
(SIV), *p. 336*
technology, *p. 303*
telecommute, *p. 310*
threat, *p. 303*
traffic, *p. 308*
Transport Layer Security
(TLS), *p. 328*
Trojan horse, *p. 321*
tunneling, *p. 325*
unsolicited commercial
e-mail (UCE), *p. 320*
virtual private network
(VPN), *p. 324*
virtual security, *p. 309*
virus, *p. 309*
worm, *p. 321*
zombie, *p. 323*

Recommended Readings

http://www.securitymagazine.com/

http://infosecuritymag.techtarget.com/

Mitnick, Kevin, and William L. Simon. *The Art of Intrusion.* New York: Wiley, 2005.

Patterson, Tom, and Scott Gleeson Blue. *Mapping Security: The Corporate Security Source Book for Today's Global Economy.* Cupertino, CA: Symantec Press, 2004.

Symantec Internet Security Threat Report 6 (September 2004).

Winklen, Ira. *Spies among Us: How to Stop the Spies, Terrorists, Hackers, and Criminals You Don't Even Know You Encounter Everyday.* New York: Wiley, 2005.

Management Critical Thinking

M11.1 To what extent should middle and upper management be involved in risk assessment and disaster recovery planning

M11.2 When does an organization choose to provide significant physical and virtual security for its server-based, network-accessible data or to outsource this function?

M11.3 Is the decision to support remote access to the organization's network a management or technical question?

M11.4 Should managers emphasize disaster recovery or business continuity? Why?

M11.5 What are the implications of allowing a home network to be attached to the corporate network?

Technology Critical Thinking

T11.1 How does an organization test a disaster recovery plan?

T11.2 Should the IT organization strive to support enhanced security or recommend outsourcing?

T11.3 When should penetration testing be done and who should do it?

T11.4 What are the implications of allowing a home network to be attached to the corporate network?

Discussion Questions

11.1 Network security is a field in high demand. Does this interest you?

11.2 Whose responsibility is security in a bank? At your school? In your home?

11.3 Should an animal shelter perform risk assessment? Should it have a disaster recovery plan?

11.4 Discuss risk assessment for a fast-food restaurant.

11.5 Have you ever experienced a power outage while in a retail store? What happened?

11.6 Have you ever experienced a virus on your home network or a hacker attack? What did you do?

11.7 Does a laptop computer require security protection?

11.8 Think about the last time you were in a large retail store. What networks do you think were required? What support was necessary to make them RAPS-compliant?

11.9 Have you ever seen or used a UPS, line conditioner, or backup power? Where? What happened?

11.10 What are the implications for a denial-of-service (DoS) attack on Yahoo.com? Your bank? Your school? Your home network? Could your home network nodes be used to support a distributed DoS?

11.11 List and discuss the major disasters that have occurred over the past 10 years. What have been their effects on businesses in the area? Which of them had the greatest impact on telecommunications systems?

11.12 What information do you possess that you should protect? Has anyone in the class been social engineered?

Projects

11.1 Contact a bank and determine the mission-critical applications that are supported by data communications. What security measures are in place?

11.2 Interview the security manager of a local firm. Determine how they ensure reliability.

11.3 For project 11.2, how do they ensure assessability?

11.4 For project 11.2, how do they ensure performance?

11.5 For project 11.2, how do they ensure security?

11.6 Choose a local firm and determine from an Internet search its security capability.

11.7 Develop a disaster recovery plan for *yourself,* assuming that your residence burns completely while you are in class or a meeting. List all of the things that will have to be decided and make the decisions, now.

11.8 Determine if your school or organization has performed risk assessment. Does it have a disaster recovery plan? Has it ever been tested?

11.9 Develop a disaster plan for your school or organization to be used in each of the following cases: hurricane, tornado, earthquake, flood, tsunami, or fire in the buildings. Do not look at the official one.

11.10 Go to an animal shelter and perform a risk assessment. Develop a disaster plan for each case indicated in project 11.9.

11.11 Go to your community's financial institutions and determine what sort of business continuity planning is employed. Determine any regulations or laws that require continuity planning for these institutions. Do you consider the plans adequate?

11.12 Investigate, in your community or on the Internet, the existence of organizations that provide backup data storage, hot sites, cold sites, shared installations, and so forth. What does it cost to use these services? What is the role of telecommunications in providing these services?

11.13 Investigate the Pacific Stock Exchange, located in California. Where exactly is it located, and where is the backup site? What power and telecommunications facilities does it have? How safe is it?

11.14 Read the paperback book *The Cuckoo's Egg,* by Cliff Stol. Discuss how the intrusion might differ today.

11.15 Look up *social engineering* on the Internet. Write a one-page paper on the subject.

Chapter Twelve

Wireless Network Security

In this chapter, the focus is on the special security issues and threats to wireless networks. The rapid rate of change and introduction of new threats must be addressed as wireless networks assume more of the organization's workload. The impact of breaches of wireless network security measures holds critical implications for the ultimate security of the organization's wired networks and data.

Wireless technology can provide numerous benefits in the business world. By deploying wireless networks, customers, partners, and employees are given the freedom of mobility within and outside of the organization. This can help businesses to increase productivity and effectiveness, lower costs and increase scalability, improve relationships with business partners, and attract new customers. Indeed, there are numerous reasons to deploy wireless technology, but, like other technologies, it is not without its risks and drawbacks.

In June of 2004, *WorldWide Wardrive*[1] reported that an alarming 61.6 percent of all submitted wireless **access points (APs)** were broadcasting data with no encryption enabled. That is, the data (or **packets**) being sent by their wireless hardware could be easily intercepted and understood by anyone listening in. This could include usernames, passwords, credit card numbers, or other sensitive information. The study also showed that 31.4 percent of the logged APs were using the default SSID (*service set identifier*) (which makes them easy to find and access) and that 27.5 percent were using no encryption with default SSIDs. The study found that the amount of APs using no encryption decreased by 6.04 percent from the previous year's endeavor. However, the number of wireless networks broadcasting default SSIDs while using no encryption and default SSIDs actually increased by 3.6 percent and 2.5 percent, respectively.

Clarification

Service set identifier (SSID) is a code attached to all packets on a wireless network to identify each packet as part of that network. The code consists of a maximum of 32 alphanumeric characters. All wireless devices attempting to communicate with each other must share the same SSID. Apart from identifying each packet, SSID also serves to uniquely identify a group of wireless network devices used in a given "Service Set."[2]

Wireless networks require the same considerations for security as their wired counterparts. All of the concerns of the preceding chapter apply to wireless. Additional concerns are introduced because the networks, by their very nature, broadcast data, making them available for ready detection and interception. These broadcasts can be received by any wireless client within the **range** of the broadcasting wireless AP; even the broadcasts of the wireless clients can be intercepted. The nature of wireless networks, rogue interception, and the nature of security, or lack of security for wireless networks, are the topics of this chapter.

Implication

Wireless networking has rapidly become the new way to upgrade systems and networks. It offers freedom of movement and flexibility in changing enterprise environments. Unfortunately, when the IEEE 802.11 protocol was developed, little thought went towards security. What security was applied—specifically WEP—was quickly broken. Today, it's widely recognized that WEP provides little to no security.[3]

---

[1] *WorldWide WarDrive* is an effort by security professionals and hobbyists to generate awareness of the need by individual users and companies to secure their access points. The goal of WorldWide WarDrive (or WWWD) is to provide a statistical analysis of the many access points that are currently deployed. See http://www.worldwidewardrive.org for more information.

[2] http://en.wikipedia.org/wiki/SSID.

[3] http://www.netstumbler.com/2004/03/15/wi_fi_security_review_airmagnet/.

Clarification

Wired Equivalent Privacy (WEP), part of the IEEE 802.11a standard, is a scheme used to secure WiFi wireless networks.

Wi-Fi Protected Access (WPA) is a system to secure wireless networks, created to patch the security weaknesses of WEP. As a successor, WPA implements the majority of the IEEE 802.11i standard and was intended as an intermediate measure to take the place of WEP. WPA has authentication and encryption; WEP has only encryption.[4]

RADIATION (INTERCEPTION)

All wireless devices radiate their information and are, therefore, subject to interception. Thus, due to this **radiation** of signals, they can be intercepted by any receiving station, client or access point. If the radiated signals contain default identification, the transmission in general and the packets in particular are easy prey. The untethering furnished by wireless systems provides great value and great **risk.**

Implication

In office 411 on the fourth floor of the Herbert L. Smith Plaza,[5] Diane Johnson, an accountant, was eagerly opening a package that had been delivered to her moments ago. She tore open the cardboard box and removed her new 802.11g wireless router from the packaging. The instructions inside read clearly and were easily understandable even by Diane, who was a novice at best when it came to computers and networking. Within an hour, she was connected to the company network using her personal laptop computer, which had come equipped with factory-installed 802.11g capability. Diane had recently seen a friend use wireless networking at their home and had decided to set up this wireless AP so that she could move about the office easily and still stay connected with the company network, as well as access her important accounting documents from her laptop. Diane was able to accomplish this without shelling out a lot of cash, and without having to learn a lot of networking jargon and skills. As she browsed the Internet in amazement of this intriguing technology, she convinced herself of her technological prowess and smirked at the haughtiness of the folks in her company's IT department.

Meanwhile, at the Cool Beans Coffee Shop across the street from the Herbert L. Smith Plaza sat a curious and devious individual. Taking a sip from his large double latte, he fired up his laptop computer and watched a myriad of startup command lines appear on his LCD in standard green and black monochrome. Using the antenna plugged into the PCMCIA slot of his computer and various open-source software programs, he began to scan the area for wireless networks. As the software began listing all of the detected APs in the range of his antenna, one in particular caught his attention. It was broadcasting itself with the name of "linksys" and it was determined by his software that the AP was broadcasting unencrypted data. From his experience, the man knew instantly that he was dealing with a wireless router that was using a factory configuration. With a few keystrokes, the man began to capture and examine the broadcasted packets from this transmitter, and, with much amusement, he scanned through numerous pieces of confidential accounting information that were originating from somewhere in the office building across the street.

This scenario illustrates just a few simple vulnerabilities that exist within the realm of wireless networking. We saw how confidential accounting data were compromised due to the actions of a well-intentioned employee with a simple lack of knowledge in what she was doing. It could have been much, much worse. Had our mysterious hacker been more proficient, he could have disabled critical software, initiated a denial-of-service attack, erased or destroyed data, or even wiped out a whole computer system, resulting in the complete stoppage of business functions.

---

[4] http://en.wikipedia.org/.

[5] Thanks to Joshua Burke, Brad Hartselle, Brad Knueven, and Brad Morgan of Auburn University for assistance in several parts of this chapter.

Access or Interception—Wireless

Wireless networks encompass any two nodes connected by one or more wireless channels. Broadcast radio and television were the first example. As these were for personal use, there was no consideration of security because the transmitter was cloistered and users were in receive-only mode. The next wireless network was radiotelephones, where eavesdropping was prohibited but not difficult. About 1983, cellular telephones replaced radiotelephone and also used analog technology, so eavesdropping was not difficult. About this time, portable/cordless telephones using one or two analog channels entered the market. Again, eavesdropping was not difficult, so **privacy** was still an issue.

Ever since the **radio telegraph,** wireless communications have been vital to military operations and security has always been a paramount issue. Beginning in the Second World War, Allied and Axis armies routinely used radio communications to send coded messages. In these cases, interception was very easy as all communications were in the standard radio bands. Security was possible only thought encryption of the messages; no secure voice was possible unless a coded language was used. Nowadays, cellular telephones as well as military radio use digital technology to carry the messages on an analog carrier and use encryption of the digital data to secure them. Now, the military can have an encrypted message that is carried on an encrypted channel, providing greater security. Even cellular communications that are digital are not impossible to intercept and decode.

With the advent of wireless networks in the workplace, the environment was ripe for industrial espionage for theft, hacking, and destruction via the implanting of viruses to intercepted messages. While industrial espionage is a definite concern, it is generally considered a special case of hacking.

Wiretaping (eavesdropping) is the unauthorized interception of data from transport media. It started with wired telephone lines and has extended to all wireless communications. In the wireless domain, this is as simple as interception of packets.

Clarification   Securing wireless networks requires a multitiered approach[6] that includes

- Hiding your presence.
- Encryption.
- Intrusion prevention technology.
- User education.
- Securing all laptops and mobile devices.
- Staying abreast of advances in technology that provide hackers with new opportunities.

THE FOUR MAJOR MISTAKES

The primary wireless network with which we come in contact is **IEEE 802.11, Wireless Fidelity (Wi-Fi).** The number of shipped 802.11-enabled hardware devices was estimated to exceed 40 million units by the year 2006.[7] To address the security of these devices, we now consider some of the common mistakes people make when setting up an 802.11 **wireless local area network (WLAN).**

Mistake #1: Using the Default Configurations for the AP

Many manufacturers of wireless networking devices set their devices with **default** configurations for one reason: to make it easy for users to establish the first WLAN connection.

[6] http://www.AirMagnet.com.

[7] Andrew Vladimirov, Konstantin V. Gavrilenko, and Andrei A. Mikhailovsky. *Wi-Foo: The Secrets of Wireless Hacking* (Boston: Addison-Wesley, 2004).

This means they avoid getting a lot of support calls. This allows for a minimum need for configuration of the AP; however, if you can connect so easy, then so can others who are unwanted. *Caution:* Web sites exist that list the default SSID and access **password** for most popular hardware.

Solution: Change the default settings, especially the SSID. Don't make the SSID something that identifies the user or the location. Of special note is that the default is to broadcast the SSID and encryption is disabled. Both will be discussed later.

Mistake #2: Preconfigured Passwords

The wireless hardware's configuration software requires that you log in before you can make any changes to the current, **preconfigured** settings. Most devices are shipped with a default of no password! If you don't change (establish) it after you log in, anyone from the outside will be able to connect via the WLAN and change the settings!

Solution: Enter a unique password based on the rules from Chapter 11, such as MMi60amfi65 (My Mother is60 and my father is 65).

Mistake #3: Broadcasting SSIDs

Anytime multiple access points exist in an area, there must be a way to direct the client to the correct AP that the user should be connecting to. Therefore, each wireless network uses a service set identifier (SSID). This is a unique 32-character identifier attached to the header of packets sent over a WLAN that acts as a password when a mobile device tries to connect to the network. The SSID[8] differentiates one WLAN from the next so all APs and all devices attempting to connect to a specific WLAN must use the same SSID. Because an SSID can be sniffed in plain text from a packet, it does not provide any security to the network. For easy connection of WLAN clients, APs can be configured to **broadcast** this SSID, allowing clients to scan for available WLAN networks. If the SSID is broadcast, *everyone* will be able to scan for your network, giving away its presence. If the SSID is not broadcast, the client must know the AP SSID in order to connect.

Solution: Uncheck "broadcast SSID."

Mistake #4: Require No Encryption or Authentication

Most **wireless access points** ship with security disabled, requiring no **encryption** or **authentication** to access the WLAN. Considering that there are encryption modes available to everyone such as WEP (less secure) and WPA, one should always use some type of encryption and authentication scheme for the WLAN.

Solution: Enable WEP, at least, WPA if available. This will be discussed in more detail later

Additional suggestions and options are contained in the router/AP's user manual. The following material on remote management and Web filtering is extracted from the Linksys WRT54G Wireless-G Broadband Router Users Guide. Some believe that remote management should be on the list of the top items to check/change upon setup.

Remote Management

This feature allows you to manage your Router from a remote location, via the Internet. To disable this feature, keep the default setting, Disable. To enable this feature, select Enable, and use the specified port (default is 8080) on your PC to remotely manage the Router. You must also change the Router's default password to one of your own, if you haven't already. The same rules for strong passwords apply.

---

[8] http://www.webopedia.com.

Web Filters

Using the Web Filters feature, you may enable up to four specific filtering methods.

- **Block Proxy**—Use of WAN proxy servers may compromise the Router's security. Denying Proxy will disable access to any WAN proxy servers. To enable proxy filtering, click the Block Proxy box.
- **Block Java**—Java is a programming language for Web sites. If you deny Java, you run the risk of not having access to Internet sites created using this programming language. To enable Java filtering, click the Block Java box.
- **Block ActiveX**—ActiveX is a programming language for Web sites. If you deny ActiveX, you run the risk of not having access to Internet sites created using this programming language. To enable ActiveX filtering, click the Block ActiveX box.
- **Block Cookies**—A cookie is data stored on your PC and used by Internet sites when you interact with them. To enable cookie filtering, click the Block Cookies box.

WHAT DO HACKERS WANT?[9,10]

We introduced the term **hacker** in the last chapter as one who, in a positive sense, is proficient with computers, or, in a negative sense, seeks to go where s/he is not meant to and do things s/he is not meant to do. For the wireless world, hackers are the negative kind and a significant threat to wireless networks, generally by various forms of **intrusion.**

Implication

A wired network can only be hacked through a physical connection, usually through the Internet. It requires a fairly high level of skill to break through a company's security gateway, firewall, and intrusion detection system. Wireless connectivity opened up a whole new wonderful world to hackers because of the ease of access.

HOW HACKERS ATTACK THE WIRELESS NETWORK

Being an innovative group, hackers are continually coming up with new ways to penetrate network defenses. Their attacks can generally be grouped into these categories:

- MAC address spoofing.
- Denial of service.
- Malicious association.
- Man-in-the-middle attacks.

MAC Address Spoofing[11,12]

Media access control (MAC) addresses act as personal identification numbers for verifying the identity of authorized clients on wireless networks. Hackers use **MAC address spoofing** to impersonate a legitimate network user. All network interface cards

---

[9] http://www.wardrive.net/wardriving/tools/.

[10] http://www.AirMagnet.com.

[11] *Media access control address* is a hardware address that uniquely identifies each node of a network. In **IEEE 802** networks, the data link control (DLC) layer of the OSI reference model is divided into two sublayers: the *logical link control (LLC) layer* and the *media access control (MAC) layer.* The MAC layer interfaces directly with the network medium. Consequently, each different type of network medium requires a different MAC layer. http://www.webopedia.com.

[12] *Spoofing*—To fool. In networking, the term **spoofing** is used to describe a variety of ways in which hardware and software can be fooled. *IP spoofing,* for example, involves trickery that makes a message appear as if it came from an authorized IP address. http://www.webopedia.com.

(NICs), like PCMCIA or PCI cards, provide a mechanism for changing their MAC addresses, issued by the manufacturer, that identify a specific device. In many cases, the MAC address is used as an authentication factor in granting the device access to the network, or to a level of system privilege to a user.

The hacker will change a device's MAC address within a packet to that of a legitimate user and, thus, gain access to the network. If the hacker is using the MAC address of a top executive with access privileges to sensitive material, a great deal of damage can be done. Attackers employ several different methods to obtain authorized MAC addresses from the network. A simple method is to scan packets and pick off authorized MAC addresses. A brute-force attack uses software that will try a string of random numbers until one is recognized by the network.

Denial-of-Service Attacks

DoS attacks may be launched merely as a form of vandalism to prevent legitimate users from accessing the network, or they may be carried out to provide cover for another type of attack. In layer one DoS attacks, the hacker uses a radio transmitter to jam the wireless network by emitting a frequency in the 2.4 GHz or 5 GHz **spectrum.** As 802.11 equipment operates at a certain signal-to-noise ratio, when the ratio drops below that threshold, the equipment will not be able to communicate. In a more common type of DoS attack, the hacker uses a laptop or PDA with a wireless NIC to issue floods of associate frames to take up all available client slots in the AP, severing the AP's association with legitimate users. Alternatively, the hacker issues floods of de-association frames, forcing clients to drop their association with the AP. Either way, if the attack is successful, the hacker now controls access to the network.

Malicious Association

In this environment, the hacker configures his device to behave as a functioning AP. When a user's laptop or station broadcasts a probe for an AP, it encounters the hacker's device, which responds with an association—a **malicious association.** At this point, the legitimate user's computer can be mined for any and all information, including MAC address, SSID, pass codes, and so forth.

Caveat   This technology is used in coffee shops.

Man-in-the-Middle (MITM) Attack[13]

Hailing from the early days of **cryptography, man-in-the-middle (MITM) attacks** are an old strategy applied to a new technology. In an MITM attack, one entity, with malicious intent, intercepts a message between two communicating entities. The hijacker can then send the message onto the receiver as if it had never been delayed, or alter the message's content. Used in war, this could be a valuable tool for intercepting and altering the enemy's message to suit the opposing side's purposes. In World War II, if the Axis forces needed to send information to deployed troops, they would send it with a decryption key. This key would be the primary tool for decoding the message and properly deciphering it. If the Allies could intercept this message and break the code, then they would be executing an MITM attack. Upon successful completion of the attack, they'd have three options as to how they'd like to exploit the position:

- The message could be intercepted, altered, and sent on to the recipient with fraudulent information.

[13] In cryptography, a **man-in-the-middle attack (MITM)** is an attack in which an attacker is able to read, and modify at will, messages between two parties without either party knowing that the link between them has been compromised. http://www.webopedia.com.

- The message could be blocked and prevented from proceeding any further.
- The message could simply be read and sent on its way without the recipient's knowledge.

The concept for MITM attacks on wireless networks is the same. In the OSI model, layer two deals with the **data link** portion of an application. Ethernet networks rely heavily on this layer in order to communicate with other machines. The primary method for doing this is through the MAC address. Ethernet uses the Address Resolution Protocol (ARP) to resolve an IP address to a system's MAC address. This concept has been the primary method for message communication among machines and also serves as the chief target area for many types of attacks, not just the MITM method.

The method of attack is simple in application, provided you have the right kind of software. Fortunately for a hacker (unfortunate for a director in charge of security), there are several applications available as freeware that perform the tasks necessary to execute a proper MITM attack.

The first task associated with an MITM attack comes into play after the initial tasks for hacking a wireless network have been performed. That is to say, we are assuming that a target network has been located and that the attacker is within acceptable range of a target AP. Once these tasks have been performed, the MITM process can begin.

All hosts on a network store a list of acceptable MAC addresses for their respective network. Using the ARP method, those systems can validate IP addresses that request access to their network by resolving them against a known table of valid IP/MAC addresses. MITM attacks occur by piggybacking on another attack method, called *ARP spoofing*. Many systems easily accept ARP commands and freely allow their MAC lists to be updated. Using software like *ARPoison,*[14] attackers can add their own MAC address to the list or trick the system into sending that table to the hijacker's system. ARPoison uses ARP spoofing to trick the network into sending all of their ARP requests to the hijacker's system instead of a valid host. Once the request comes in, the hijacker can reroute the information to a valid host, but only after having had extensive access to the transaction. Once attackers have established themselves as valid members of the network, they can[15]

- Execute a **denial of service (DoS)** by sending all host requests to invalid host addresses, thus causing bounce backs.
- **Monitor** all transactions between the hosts (hence "man-in-the-middle").
- Join the network by adding the attacker's MAC address to the acceptable list.

In an alternative form of MITM, the hacker sends a de-authorization to a network device, which drops its association to its AP and begins searching for a new AP. It finds the hacker's station (configured to look like an AP) and associates with it. Using the information garnered from the legitimate device, the hacker's device now associates with the legitimate network AP and the network passes through the rogue user's device, allowing him/her to change or steal data at will.

Clarification   It has been estimated that internal users create up to a third of the vulnerabilities of the enterprise WLAN. There is the occasional bad apple that steals information or money from his or her employer, but the majority of internal users are not trying to harm the company; they are just ignorant of the consequences of their actions or failure to take precautions.

---

[14] ARPoison, http://web.syr.edu/~sabuer/arpoison/.
[15] Robert Wagner, "Address Resolution Protocol Spoofing and Man-in-the-Middle Attacks," SANS Institute, http://www.sans.org/rr/papers/60/474.pdf.

ROGUE USERS[16]

Wardriving
the practice of driving around with a laptop and good antenna to locate open wireless access points. It also has been done in low-flying aircraft.

Rogue users are not just hackers and outside intruders who are wardriving through a parking lot with 802.11 antennae made out of a *Pringles*® can. Most likely, they're the organization's own employees who are unaware of company wireless usage policies. Perhaps they are experimenting with an inexpensive personal access point in the office, having grown impatient with IT's pace in deploying wireless tools. Maybe they've connected that AP to the wired network, inadvertently creating a huge security hole. In any case, the corporate information is at risk, unless you take control.

Implication

While your wired network may be a walled fortress guarded by multiple firewalls, your WLAN is much more vulnerable. A single rogue wireless user can gain entry, bypassing the firewalls and opening the floodgates for others to come in and access your corporate data.

Users love the freedom of mobility. They dislike waiting for IT's official approval to bring in WLANs. *As with PCs, wireless is a user-driven revolution.* It's simple enough for an employee to unplug his/her laptop from the Ethernet wall jack, plug in an unsanctioned wireless AP, and reconnect the laptop, wirelessly. The employee has created an ad-hoc WLAN for himself/herself or other employees in the department. They may collaborate in a conference room, link their laptops to their PDAs, or fire up a game of Quake during lunch.

When creating a rogue AP, the employee has inadvertently created a security hole through which an intruder can enter. In fact, the Gartner Group estimates that one in five companies has a WLAN that the CIO doesn't even know about. If a company has deployed WLANs, this private rogue may cause **interference,** open a new security hole, and degrade the sanctioned WLAN's performance.

An alternative threat is when an intruder has set up an AP as a "bug light," luring legitimate WLAN users to connect with the unauthorized AP. Once associated with the rogue AP, the hacker can gain access to a legitimate user's PC. Or the intruder may use an unauthorized private WLAN to mount a man-in-the-middle attack, thereby gaining full network access.

Free Access to Networks via Wi-Fi

Wardriving is detecting, and using, unsecured Wi-Fi by driving around with a laptop and antenna; **warchalking** is leaving chalk marks on the sidewalk or side of a building to indicate unsecured APs (see Figure 12.1). If one holds with the freedom of the 1960s, then warchalking and its free access to Wi-Fi networks is expected. If, however, you are the owner of these same networks, then wireless network use by others outside of your organization, at a minimum, is the theft of bandwidth and, at a maximum, provides an entry point to the total organizational network.

Implication

Walking down Yonge Street and Bay Street (heart of the Financial District in Toronto), one finds countless warchalking markings (warchalking, see Figure 12.1, refers to the "chalk marks" that people leave to indicate the proximity of open wireless networks). Wardriving, the act of looking for and using open, unsecured wireless networks, is increasing with little-to-no legal action being taken. Until laws are set to deal with this, companies will need to deal with issues themselves.[17]

Clarification
Advertisement

Wardriving Kit High Power Long Range, $82.00—WDNL2511CDPLUSEXT2
Works perfectly with *Netstumbler*. Completely tested with Windows 2000 and Windows XP. Tested by a long time wardriver. 200mw power outperforms other low powered cards. Up to

[16] http://www.trapezenetworks.com/technology/whitepapers/detectingrogue/detectingrogue.asp.
[17] http://www.netstumbler.com/2004/03/15/wi_fi_security_review_airmagnet.

FIGURE 12.1
Warchalking
Symbols

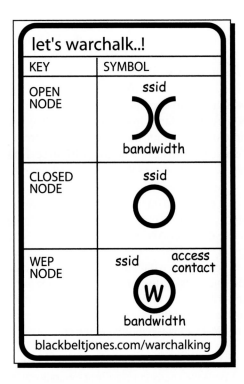

23 dBm (200mw) transmit output power. Prism 2.5 chipset. Long operating range, up to three times range of standard products. This will detect 802.11b and 802.11g networks.[18]

Identify the Rogue

The first step in protecting corporate resources from misuse is to define and identify what constitutes a rogue. While various types of threats may occur from both authorized and unauthorized users, the most common WLAN rogues include

- An employee who connects wirelessly to the wired enterprise network using his/her own unapproved, nonenterprise-grade wireless AP.
- A group of employees who set up a stand-alone AP for their workgroup, which they don't plug into the wired network.
- An unauthorized user, such as an intruder or hacker, who uses his/her own wireless tools and attempts to access the WLAN network from the parking lot, street, or other location physically nearby.

Implication   A traditional network operating system has no mechanism to detect or locate rogue users, either on the wired or wireless LAN.

The first instance—an internal, unauthorized threat—is the most common misuse of WLANs. For example, an employee who has an 802.11 WLAN at home to connect his/her laptop, printer, and PDA decides to bring his/her own AP into the office so s/he can more easily transfer data to and from the office desktop to the mobile computer. The employee buys an AP that's suited for home use at the local electronics store or off of the Internet. But this AP lacks the security built into an enterprise-grade AP, such as *Wi-Fi Protected Access* encryption. The employee may not understand that this personal

[18] http://wlanparts.com/.

WLAN is a threat to corporate network security, so s/he doesn't seek approval from the IT manager. Nor does s/he need assistance from the IT helpdesk since wireless networking at this level is plug-and-play, thereby eliminating an opportunity for the IT team to discover this rogue WLAN.

The second type of rogue is a private WLAN user group. They may be using an AP or even a "soft AP," which is software that gives AP functionality to a wireless laptop. Although this WLAN may be isolated from the corporate WLAN, the users are stealing the air time (bandwidth reduction due to collisions) from the legitimate WLAN users. They also may cause a significant amount of **RF interference (RFI),** which will cause a dramatic slowdown in the entire WLAN, or may open the door to an uninvited guest.

An uninvited guest may eavesdrop on the private LAN and can now get into the network either through the employees' LAN connections or by intercepting their user names and passwords on the official wireless LAN. A breach would have occurred and the administrator would never know it.

Many wired breaches come from within the organization's perimeter; wireless breaches come from both inside and outside the organization. Once intruders are in, they may launch a man-in-the-middle attack to gain full network access or launch a denial-of-service attack that jams the bandwidth for all users. An unauthorized use of the network or ISP connection also can create a legal liability for the enterprise.

Implication

An external attack is a real threat, especially if your WLAN security settings, such as 802.1X authentication and WEP or WPA encryption, are not operational or configured to prevent unauthorized intrusions. A knowledgeable intruder with an 802.11 device or other wireless access tool can easily pick off the SSIDs and MAC addresses and steal the identity of an authorized AP or users.

Clarification

One of the scariest attacks in wireless networking can be mounted by a man-in-the-middle rogue AP. This type of attack could grant full network access to the rogue in a way that is very hard to detect, even in networks configured to use Transport Layer Security (TLS).

An MITM rogue AP makes a PEAP Part 1 connection to a corporate AP, masquerades as a user, and trivially authenticates the corporate AP. This first step of this attack results in an encrypted TLS session between the rogue and the authenticator. Next the rogue attracts a legitimate user (called "bug lighting") and asks for TLS authentication. The rogue tunnels the TLS authentication exchange between the legitimate user and the authentication server. The authenticator thinks it's completing PEAP-TLS Part 2. Once the legitimate user is authenticated, the rogue can derive the session encryption keys, since they are based on information exchanged in the original PEAP Part 1 phase. The rogue disconnects the legitimate user and turns its bug light off.

The rogue now has complete network access. The authentication server doesn't think anything went wrong. The legitimate user retries authentication, connects to a corporate AP the next time, and gets authenticated, so the user doesn't realize anything bad has happened other than a slight delay that could have been caused by temporary RF interference.

Clarification

PEAP (Protected Extensible Authentication Protocol) is a protocol for transmitting authentication data, including passwords, over 802.11 wireless networks. PEAP authenticates wireless LAN clients using only server-side digital certificates by creating an encrypted SSL/TLS tunnel between the client and the authentication server that protects the ensuing exchange of authentication information from casual inspection.

The tunnel then protects the subsequent user authentication exchange. Note that PEAP is not an encryption protocol; as with other EAP types, it only authenticates a client into a network.[19]

[19] http://www.webopedia.com/TERM/P/PEAP.html.

Once users are mobile, they are no longer connected to a specific port on a switch. Yet, port security is predicated on that physical connection. The network operating system knows who they are but doesn't know where they are. And the switches previously knew where the user devices were located, but now the devices are moving. Legitimate wireless users move—and so do rogues. The network operating system can't detect the move. Nor can your wired intrusion detection system (IDS) or **simple network management protocol (SNMP)** application detect a rogue user or AP since they lack awareness of the air.

Using External Tools for Rogue Detection

To detect rogues, IT managers have two choices. They can do an *internal wardrive* to manually discover the rogues or they can *install a network of rogue AP sensors*. In an internal wardrive, an IT manager physically walks around the building with a wireless laptop or handheld device and wireless analysis software. Several tools are available to capture the 802.11 packets of WLAN transmissions. For example, *NetStumbler* and *AirSnort* can scan the airwaves for WLAN signals, list what is available, and reveal their descriptors and vital statistics.

Selling in the $3,000 to $4,000 range, WLAN analyzers such as *AirMagnet, WildPackets' AiroPeek,* and *Sniffer Wireless* are able to capture 802.11 packets and analyze the layer one and layer two information. They report transmission data such as signal strength, channel, and data rates. Some require expert WLAN network and security analysts to understand the data and locate the threats detected. These types of products usually come in both laptop and handheld formats. These tools usually can't pick up signals from microwaves or portable phones operating in the same spectrum that also can cause interference. So if you're looking to resolve channel conflicts between 2.4 GHz cordless phones and your WLAN, you will need a spectrum analyzer.

802.1x Authentication Is Your Best Security Offense

While expensive sniffing and monitoring tools can help with rogue user and AP detection, the best defense is a good offense. Adding 802.1x authentication combined with 802.11i encryption or even IEEE-approved standard 802.11 techniques like Wi-Fi Protected Access (WPA) will deliver a strong offense. If only authenticated users can communicate on the network and all communication is encrypted, the chances that a rogue will be able to penetrate and do damage are greatly reduced. By using 802.1x AAA[20] prevention techniques, the IT manager can severely limit if not completely eliminate rogue attacks in the enterprise.

Phreaks Love Bluetooth[21,22]

Mobile phones are more ubiquitous than PCs, and mobile workers are storing more data than ever on their handheld gadgets. And, like their corded, and weighty, desktop predecessors, *Bluetooth*-enabled phones are potential targets for widespread attacks. **Bluetooth,** the specification for wirelessly connecting with devices up to 10 meters' distance, is becoming

---

[20] Authentication, authorization, and accounting (AAA) is a term for a framework for intelligently controlling access to computer resources, enforcing policies, auditing usage, and providing the information necessary to bill for services. http://searchmobilecomputing.techtarget.com.

[21] A **phreak** is someone who breaks into the telephone network illegally, typically to make free long-distance phone calls or to tap phone lines. The term is now sometimes used to include anyone who breaks or tries to break the security of any network. Recently, the phone companies have introduced new security safeguards, making phreaking more difficult. http://searchsecurity.techtarget.com/.

Phreaking—Closely related to hacking; using a computer or other device to trick a phone system. http://www.webopedia.com.

[22] Mark Baard, contributing writer, searchSecurity@lists.techtarget.com.

the capability of choice for next-generation phone phreakers, who use the technology to swipe files from other users, make calls, or render mobile phones useless. Much of the malicious code and attacks can be stopped by users making their devices "undiscoverable." In *undiscoverable mode,* mobile device users can connect to Bluetooth headphones, for example, while remaining invisible to other devices. But fast-spreading worms and viruses may outpace the release of more secure devices—and efforts by IT to educate users. Not to mention that users of Bluetooth devices don't always want to remain invisible.

The recent news about the emergence of the SymbOS.Cabir worm, which affects devices using the Symbian OS, will only invite more trouble for mobile device users, said the IT director of one highly mobile organization. "Just as exploits of [Microsoft's software] follow the announcements of new vulnerabilities, news about *Bluetooth* viruses and exploits will get more hackers interested in its weaknesses," said Steve Conley, IT director for the Boston Red Sox. "If in a few months a lot of people get caught with their pants down, it would not surprise me one bit." Antivirus experts at Finland-based F-Secure Corp. were among those surprised by the outbreak of a virus affecting users of mobile phones running the Symbian OS.

RFID Security Issues

Implementations of **radio frequency identification (RFID)** have a range of issues, ranging from social, to financial for implementation, to security. It is this last concern we address here, as the others have been covered elsewhere. RFID tags respond to any transmitter within range; that's their function, that is, to respond to an active enquiry and reply with stored data. Consider the RFID tag on a credit card or employee access badge—without safety measures, a rogue transmitter can access the data contained in the chip without the holder being aware. To thwart this, devices such as these must be in an off state until the holder presses a part of the card/tag that allows access. Then, the tag can be interrogated only during the short time that the holder depresses the security button, greatly reducing the risk of use of RFID in this situation.

RFID tags contain specific data, ranging from a simple number such as a SKU[23] to descriptions, price, ownership, and state of security inspection.[24] If the data are changeable, that is, not read-only, a rogue transmitter could place new data onto the tag at time of interrogation. This could be as simple as changing the SKU of the item, substituting the number of a much less expensive item that goes unnoticed, to noting that a suitcase has passed through security when it has not. Thus, while RFID has significant possibilities in inventory control and cost reduction, security measures must be in place to avoid tampering. (Barcodes contain a check digit that is used to determine if the barcode lines have been altered.)

WIRELESS SECURITY FEATURES

Wired Equivalent Privacy (WEP)[25]

A good deadbolt lock will keep out the amateur crooks and slow down the professional thief; they might look elsewhere for easier pickings. The same is true for WEP.

WEP is a security protocol, specified in the IEEE Wireless Fidelity (Wi-Fi) standard, 802.11b, that is designed to provide a wireless local area network (WLAN) with a level of security and privacy comparable to what is usually expected of a wired LAN. A wired local area network (LAN) is generally protected by physical security mechanisms (controlled access to a building, for example) that are effective for a controlled physical environment but may be ineffective for WLANs because radio waves are not necessarily bound by the

[23] SKU = Stock-keeping unit, a uniform number that uniquely identified the object.

[24] The Denver airport is using RFID tags on luggage to replace barcodes. The RFID tags indicate ultimate destination and data on present state of security inspection.

[25] http://searchsecurity.techtarget.com/sDefinition/0,,sid14_gci549087,00.html.

walls containing the network. WEP seeks to establish similar protection to that offered by the wired network's physical security measures by encrypting data transmitted over the WLAN. Data encryption protects the vulnerable wireless link between clients and APs; once this measure has been taken, other typical LAN security mechanisms such as password protection, end-to-end encryption, virtual private networks (VPNs), and authentication can be put in place to ensure privacy.

WEP was integrated into wireless devices with a primary goal of preventing casual eavesdropping on a network. Much like crosstalk can occur among wireless telephones, the same effect could take place in getting packets mixed among common pathways on a wireless network. WEP performs this function rather well, but the second purpose of WEP is where the protocol falls short: to prevent unauthorized access to wireless networks. Now, don't be mistaken. WEP will prevent uninformed and unskilled crackers from accessing a wireless network, but it won't keep out the skilled hacker.

WEP encryption provides minimal security compared to newer security features for wireless communications. There is up to 128-bit encryption with WEP, but compromising the encryption algorithm can occur with off-the-shelf products since all communicating parties share the same key and no automatic key distribution occurs.

Implication

We cracked 64-bit WEP encryption within one minute after sniffing 500,000 packets using a laptop computer; people have cracked 128-bit keys in less than five minutes.

WEP2

WEP2 is an attempt to evolve WEP using new Advanced Encryption Standard (AES) instead of RC4 and Kerberos authentication. Many believe that the increase to 128-bit encryption will provide little improvement on security.

Wi-Fi Protected Access (WPA)

Wi-Fi Protected Access (WPA)[26] is an interim Wi-Fi standard that was designed to improve upon the security features of WEP. The technology is designed to work with existing Wi-Fi products that have been enabled with WEP (i.e., as a software upgrade to existing hardware), but the technology includes two improvements over WEP:

* *Improved data encryption* through the **temporal key integrity protocol (TKIP).** TKIP scrambles the keys using a hashing algorithm and, by adding an integrity-checking feature, ensures that the keys haven't been tampered with.

* *User authentication,* which is generally missing in WEP, through the extensible authentication protocol (EAP). WEP regulates access to a wireless network based on a computer's hardware-specific MAC address, which is relatively simple to be sniffed out and stolen. EAP is built on a more secure public-key encryption system to ensure that only authorized network users can access the network.

Updating to a *Wi-Fi Protected Access (WPA)* encryption is more secure and less vulnerable to decryption. TKIP, which is a part of the WPA standard, always includes user authentication, which is not always present with WEP. It also creates a new encryption key every 10,000 data packets, which is safer than WEP.

Temporal Key Integrity Protocol (TKIP)[27]

TKIP is part of the IEEE 802.11i encryption standard for wireless LANs; it is the next generation of WEP. TKIP is a software grade and uses a key scheme based on RC4, but, unlike

---

[26] http://www.webopedia.com.

[27] http://www.devx.com/wireless/Door/11455, http://en.wikipedia.org/wiki/TKIP.

**Advanced
Encryption
Standard**
(AES) is a symmetric
128-bit block data
encryption technique.

WEP, it encrypts every data packet sent with its own unique encryption key. It also hashes the initialization vector (IV) values that are sent as plaintext in the current release of WEP. This means that IVs are now encrypted and are not as easy to sniff out of the air (addressing one of WEP's largest security weaknesses). TKIP provides per-packet key mixing, a message integrity check, and a rekeying mechanism, thus fixing the flaws of WEP.

IEEE 802.11i

IEEE 802.11i[28] is a standard for wireless LANs that provides improved encryption for networks that use the popular 802.11a, 802.11b, and 802.11g standards. The 802.11i standard requires new encryption key protocols, TKIP and Advanced Encryption Standard (AES).

DEFENSIVE STRATEGIES

It is more difficult and costly to remove a problem after detection than prevent one. Correction is a very poor third choice.

We noted in the previous chapter that the key features of a secure environment are *prevention, detection,* and *correction.* The mistakes discussed previously and the defensive strategies herein are aimed at **prevention.** Because some of the threats are passive, wireless systems must make sure that if the data are intercepted, they cannot be used. As with wired networks, detection may be active or after-the-fact with **auditing** and accounting.

The first and foremost step in ensuring that a network will not be compromised is to draft policies and procedures for installing and maintaining wireless technology within your organization. These policies and procedures should be easily accessible to all network users and administrators, and should be enforced with some degree of accountability for those who choose to ignore them. By creating a well-thought-out strategy, organizations can minimize human error and lack of knowledge about the technology that is "half the battle" in wireless security.

Karygiannis and Owens[29] from the National Institute of Standards and Technology detail a list of management practices to adhere to in trying to win the "first half" of the battle:

- Identify who may use WLAN technology in an agency.
- Identify whether Internet access is required.
- Describe who can install APs and other wireless equipment.
- Provide limitations on the location of and physical security for APs.
- Describe the type of information that may be sent over wireless links.
- Describe conditions under which wireless devices are allowed.
- Define standard security settings for APs.
- Describe limitations on how the wireless device may be used, such as location.
- Describe the hardware and software configuration of all wireless devices.
- Provide guidelines on reporting losses of wireless devices and security incidents.
- Provide guidelines for the protection of wireless clients to minimize/reduce theft.
- Provide guidelines on the use of encryption and key management.
- Define the frequency and scope of security assessments to include AP discovery.

WEP

Even with its inherent weaknesses, WEP is still the minimum step for preventing attackers from capturing your network traffic. Less-experienced hackers will probably not even

[28] http://searchmobilecomputing.techtarget.com.

[29] Tom Karygiannis and Les Owens, "Wireless Network Security: 802.11, Bluetooth, and Handheld Devices," NIST Special Publication 800-48, http://csrc.nist.gov/publications/nistpubs/800-48/NIST_SP_800-48.pdf, November 2002.

attempt to capture data packets from a wireless network that is broadcasting using WEP. Even if a hacker possesses the skills and tools necessary to crack WEP, it can be an extremely time-consuming process, especially when dealing with the newer 128-bit specification, which requires in excess of 500,000[30] captured data packets to even begin the cracking process. Not only is WEP a good way to ward off many would-be attackers; it is strengthened when used with other security techniques.

MAC Address Blocking

For smaller, more static networks, you can specify which computers should be able to access your wireless APs. A MAC ACL (MAC access control list) is created and distributed to APs so that only authorized NICs can connect to the network. While MAC address spoofing is a proven means to hacking a network, **MAC address blocking** with MAC ACLs can be used in conjunction with additional security measures to increase the level of complexity of the network security, decreasing the chance of a breach.

Ditch the Defaults

Most wireless devices are sold with default configurations that are easily exploited. The three main areas to watch are the router administration passwords, SSID broadcasting, and the channel used to broadcast the signal. Upon installation, users would do well to immediately change the router's administration password. The default passwords are easy to locate; just get the user manual associated with each device. Turning off the SSID broadcast option will prevent unintentional wireless hijacking because rogue wireless devices will not be able to automatically detect the SSID without extra action. Changing the default-broadcasting channel also will make a WLAN more unique in its **architecture** and thus more difficult to detect based on default vulnerabilities.

Beacon Intervals

The **beacon** interval is a frame that is sent out to announce the presence of the AP. Client stations use this to configure parameters to join a network. This is separate from the SSID broadcast in that a beacon frame appears as a random data packet without an SSID label. These intervals should be maximized to make it more difficult to find the network.[31] The network appears quieter and any passive listening devices are not as productive at gathering and cracking encryption keys. Even when reducing the beacon and SSID intervals, valid users already have sufficient information to attach to the network.

Controlling Reset of the Access Point

Something as simple as controlling the reset function of a device such as an access point can add a great deal of security and reduce the risk of potential **hacking** to your network. After all the security measures are in place and the proper encryption settings are enforced, the factory-built "reset" button, available on nearly all wireless routers/APs, can, in an obvious way, wipe out everything. This capability, often a recessed button, should, at a minimum, be covered with tape.

Disable DHCP[32]

Disabling the use of **Dynamic Host Configuration Protocol (DHCP)** in a wireless network is, again, a simple but effective roadblock to potential hackers. In the event that a

---

[30] *Caveat:* An open-source program advertises that it can now do this in 50,000 packets.

[31] Karygiannis and Owens, "Wireless Network Security."

[32] **Dynamic Host Configuration Protocol,** a protocol for assigning dynamic IP addresses to devices on a network. With dynamic addressing, a device can have a different IP address every time it connects to the network. In some systems, the device's IP address can even change while it is still connected. DHCP also supports a mix of static and dynamic IP addresses. (www.webopedia.com)

threat breaks through your encryption, they would then have immediate access to the network if they were assigned an IP address by DHCP. This may not be feasible in a large corporate environment where thousands of IP addresses are leased throughout the day, but in a home space this is a must for all users.

Implication

"Since mobile devices are subject to all sorts of threats including both technological [viruses, worms, spam] and physical [lost or stolen], it is essential that organizations that allow the use of these instruments devise corporate policies regarding their use and further document courses of action if exposed to these kind of threats," said Dave Wreski, CEO of Guardian Digital in Allendale, New Jersey. Wreski believes that the policies and procedures should include the following:

1. Utilize advanced encryption and security standards, including Wired Equivalent Privacy (WEP) to minimize the occurrence of WLAN-related vulnerabilities.
2. Password-protect all mobile devices.
3. Encrypt sensitive documents that are stored on the device.
4. Minimize access to sensitive internal information by using firewalls.
5. Back up data regularly on all mobile devices.
6. Implement antivirus software on all mobile devices.

Why? Recently, a laptop computer of a university alumni association was misplaced with unencrypted data of 100,000 alumni on it, that's why.

Network Auditing and Intrusion Detection

In the last chapter, we discussed wired network *intrusion prevention systems (IPSs)*, which take a proactive measure in network security so as to stop the attack before it starts, and *intrusion detection systems (IDSs)*, which are more passive in their methodology, monitoring and informing network administrators of any intrusive presence. Both of these systems monitor networks, whether they are WLANs or LANS, over extended periods of time. This frees up network administrators to perform other tasks associated with network management until they are made aware of a need for action.

There are also a limited number of software tools available that allow you to deploy wireless intrusion detection systems that can automatically monitor the network and report to system administrators suspicious events that occur. These suspicious events can include unusual data packets, the presence of new wireless transmitters in the area, or traffic encrypted with unknown WEP keys.

HUMAN ERROR AND NETWORK ADMINISTRATION

It is understood that an individual with no understanding of networks can easily set up a flawed and vulnerable network. However, executives need to be aware that even their system administrators could be lacking in an understanding of wireless network implementations. With the broad number of floating, corporate **hotspots** being found everyday, it has to be assumed that some of those hotspots were put into place by less-than-knowledgeable IT staff within those corporations. Maybe a manager gave some of his development staff permission to install a wireless router while providing no oversight to the installation. Though developers may know something about software architecture and design, they may or may not know anything about network security. Perhaps the local system administrator doesn't thoroughly understand networking principles. Maybe s/he lacks the tools necessary to carefully monitor network traffic and detect anomalies that could indicate the presence of a rogue AP. Worst case, s/he might not even care. This can be extremely problematic in that the network may be compromised and no one in the organization may even know it.

TABLE 12.1
LAN Auditing Tools

Commercial	Open Source
AirDefense®—Uses devices called IDS sensors, which are placed around the network and report information to a central management server or console	*HotSpot-Defense Kit®*—Monitors the MAC address, SSID, and various other indicators that have been picked up by a rouge AP such as sudden fluctuations in signal strength. This application was released in part by the overwhelming demand for something that would prevent the attack, as hacker tools like *AirShnarf* become increasingly easier to get and use.
WiSentry®—An entirely software-based IDS that distributes a small client process to detect suspicious activity	*AirIDS®*—Offers basic IDS capabilities
AirMagnet®—Offers a suite of software tools used to diagnose security issues and other wireless network problems	*Kismet®*—Can detect suspicious hosts such as clients running *AirJack* and dictionary attacks
AiroPeek®—A wireless protocol analyzer with IDS sensor functionality	*WIDZ®*—Offers rogue AP detection and monitors the network for possible hostile traffic

The company should make sure that system administrators are well trained with a strong background in computer and network security.

In any system, the human components are the weakest link. Wireless networking is certainly no exception. Organizations should define strict policies and procedures related to wireless networking within a well-publicized company document or, preferably, on the intranet. It is especially important with regard to wireless networking that employees are made aware of these rules. Unwitting employees with good intentions can compromise company data without even knowing they have done so. Because wireless hardware is cheap and relatively easy to use, the risk of an organization's network containing rogue APs is great. Management must set standards for any wireless hardware configurations within the company network and perform routine network audits to ensure that there are no open doors. (See Table 12.1 for representative network auditing tools.)

Case 12-1

Best Practices for Wireless Network Security[33]

Susan Kennedy

Susan K. Kennedy is the information systems audit manager at the University of Pennsylvania. She has more than 13 years' experience in the IT assessment of security, computer facilities and networks; pre- and postsystem implementations; and business processes and application reviews. She holds an MBA degree and Certified Information Systems Auditor and Certified Internet Webmaster certifications.

Wireless technology is dramatically changing the world of computing, creating new business opportunities but also increasing security risks. Wireless LANs, which use radio frequencies to broadcast in the unlicensed 2.4-GHz frequency band, can be as simple as two computers equipped with wireless network interface cards or as complex as hundreds of computers outfitted with cards communicating through access points. They're relatively inexpensive and easy to install.

[33] http://www.computerworld.com/mobiletopics/mobile/story/0,10801,86951,00.html.

But, they also introduce a number of critical security risks and challenges, and it's important to implement strong security measures to mitigate these risks. What follows are potential risks and associated best practices to help you secure your network and understand WLAN characteristics:

RISK NO. 1: INSUFFICIENT POLICIES, TRAINING AND AWARENESS

Although establishing policies to govern wireless networks would appear to be a basic requirement, institutions often fail to take this step or to inform employees of the risks associated with not using a wireless network in accordance with the policies. Once policies are implemented, it's critical to communicate them to increase users' awareness and understanding.

How to mitigate:

Develop institutionwide policies with detailed procedures regarding wireless devices and usage. Maintain these policies and procedures to keep current with technology and trends. While each institution will have specific requirements, at a minimum require the registration of all WLANs as part of overall security strategy. And because a policy isn't effective if users aren't in compliance, monitor the network to ensure that users are following the policy as intended.

Conduct regular security awareness and training sessions for both systems administrators and users. It's important to keep systems administrators informed of technical advances and protocols, but it's equally important for users to understand the reasons for the protocols. An educated user will more likely be a compliant one, without as much protest. These education sessions should stress the importance of vigilance.

RISK NO. 2: ACCESS CONSTRAINTS

Wireless access points repeatedly send out signals to announce themselves so that users can find them to initiate connectivity. This signal transmission occurs when 802.11 beacon frames containing the access points' Service Set Identifier are sent unencrypted. (SSIDs are names or descriptions used to differentiate networks from one another.) This could make it easy for unauthorized users to learn the network name and attempt an attack or intrusion.

How to mitigate:

1. **Enable available security features.** Embedded security features are disabled by default.
2. **Change the default settings.** Default SSIDs are set by the manufacturer. For example, Cisco's default SSID is "tsunami," and Linksys' is "linksys." Not changing these makes it easier for an unauthorized user to gain access. Define a complex SSID naming convention. Don't change the SSID to reflect identifiable information, since this too could make it easy for an unauthorized user to gain access. Instead, use long, non-meaningful strings of characters, including letters, numbers and symbols.
3. **Disable Dynamic Host Configuration Protocol** and use static IP addresses instead. Using DHCP automatically provides an IP address to anyone, authorized or not, attempting to gain access to your wireless network, again making it just that much easier for unauthorized penetration.
4. **Move or encrypt the SSID and the Wired Equivalent Privacy (WEP) key** that are typically stored in the Windows registry file. Moving these privileged files makes it more difficult for a hacker to acquire privileged information. This step could either prevent an unauthorized intrusion or delay the intrusion until detection occurs.

5. **Use a closed network.** With a closed network, users type the SSID into the client application instead of selecting the SSID from a list. This feature makes it slightly more difficult for the user to gain access, but education on this risk-mitigation strategy can reduce potential resistance.

   To gain maximum advantage of a closed network, change the SSID regularly so that terminated employees can't gain access to the network. Develop and implement an SSID management process to change the SSID regularly and to inform authorized employees of the new SSID.

6. **Track employees who have WLANs at home** or at a remote site. Require that wireless networks are placed behind the main routed interface so the institution can shut them off if necessary. If WLANs are being used at home, require specific security configurations, including **encryption and virtual private network (VPN) tunneling.**

RISK NO. 3: ROGUE ACCESS POINTS

Rogue access points are those installed by users without coordinating with IT. Because access points are inexpensive and easy to install, rogue installations are becoming more common. Rogue access points are often poorly configured and might permit traffic that can be hard for intrusion-detection software to pinpoint.

How to mitigate:

1. **Conduct extensive site surveys regularly** to determine the location of all access points. Ensure that access points aren't near interfering appliances such as microwave ovens, electrical conduits, elevators or furniture.

2. Plan for access-point coverage to **radiate out toward windows,** but not beyond.

3. Provide **directional antennas** for wireless devices to better contain and control the radio frequency array and thus prevent unauthorized access.

4. Purchase access points that have **"flashable" firmware only,** to allow users to install security patches and upgrades in future releases.

5. **Disable Simple Network Management Protocol community passwords** on all access points. SNMP is used as an access-point management mechanism, and while it offers operational efficiencies, it increases the risk of security breaches.

6. **Set Authentication method to OPEN rather than to shared encryption key.** This seems contrary because using encryption for authentication is typically preferred. However, when using the shared encryption key feature, the challenge text is sent in clear text. This could help an unauthorized party calculate the shared secret key using the encrypted version of the same text. So ironically, using the default OPEN authentication actually reduces the possibility of an unauthorized party discovering your WEP encryption key.

7. **Use Remote Authentication Dial-In User Service (RADIUS),** which can be built into an access point or provided via a separate server. RADIUS is an additional authentication step. Interface this authentication server to a user database to ensure that the requesting user is authorized.

8. **Force 30-minute reauthentication for all users.** *The benefit is that the interloper may be denied access after a short period.*

RISK NO. 4: TRAFFIC ANALYSIS AND EAVESDROPPING

Without actually gaining access to the network, unauthorized parties can passively capture the confidential data traversing the network via airwaves and can easily read it because it's sent in clear text. So an attacker could alter a legitimate message by deleting, adding to, changing or reordering the message. Or the attacker could monitor transmissions and retransmit messages as a legitimate user.

By default, WLANs send unencrypted or poorly encrypted messages using WEP over the airwaves that can be easily intercepted and/or altered. Currently, wireless networks are beset by weak 802.11x Access Control Mechanisms, resulting in weak message authentication.

How to mitigate:

1. **Encrypt all traffic over the WLAN.** There are a variety of methods to select from:

    - **Use application encryption** such as Pretty Good Privacy, Secure Shell (SSH) or Secure Sockets Layer.

    - **Enable WEP,** an encryption method that's intended to give wireless users security equivalent to being on a wired network but that has been proved to be insecure (its RC4 stream cipher, which is used to encrypt the data, has been cracked). Both 40- and 128-bit keys have been cracked—the 128-bit encryption only prolongs the cracking process. Despite its weaknesses, the WEP security that's built into wireless LANs can delay an unauthorized user's intrusion or possibly prevent a novice hacker's attacks entirely. **(Note: The WEP factory default is OFF.)**

    - **Require the use of a VPN running at least FIPS-141** triple Data Encryption Standard and encrypting all traffic, not only the ID and password. Segment all wireless network traffic behind a firewall and configure each client with a VPN client to tunnel the data to a VPN concentrator on the wired network. Configure so users communicate only with the VPN concentration point. Evaluate the following features when purchasing VPN technologies: interoperability with existing infrastructure, support for a wireless and dial-up networking, packet-filtering or stateful-inspection firewall, automatic security updates and a centralized management console.

2. **Implement two-factor authentication scheme** using access tokens for users accessing critical infrastructure.

3. **Utilize 802.11x for key management and authentication standards.**

4. **Use Extensible Authentication Protocols.**

5. **Activate the Broadcast Key Rotation functionality.** Set a specific amount of time (usually 10 minutes or less) on the access point; each time the counter runs out, the access point broadcasts a new WEP key, encrypting it with the old, thus reducing the amount of time available to crack the key.

6. **Restrict LAN access rights by role.** *This is a means to constrain use of the LAN to those whose jobs require it.*

RISK NO. 5: INSUFFICIENT NETWORK PERFORMANCE

Wireless LANs have limited transmission capacity. Networks based on 802.11b have a bit rate of 11Mbit/sec. while networks based on 802.11a have a bit rate of 54Mbit/sec. **Media Access Control overhead alone consumes roughly half of the normal bit rate.**

Capacity is shared between all the users associated with an access point, and since load balancing doesn't exist on access points, network performance can be improved dramatically if the appropriate number of access points are available to users.

Frequently, unauthorized users' intentions are to steal bandwidth rather than view and alter the data passing along the wireless network. Therefore, these unauthorized users can significantly reduce network performance for authorized users. Finally, DoS attacks can disable or disrupt your operations. A DoS doesn't have to be intentional. For example, users can transfer large files that can cause a network outage.

Implication

As an example of intentional but nonmalicious semi-DoS attack, consider the case of a former student of one of the authors. He found that there was very limited parking at the beginning of football season due to significant changes to campus configuration, even though the athletic department's Web site indicated otherwise. After sending an e-mail to the designated person four times, he set up three servers to send that same e-mail every 30 minutes, which created a queue of 3,000 e-mails before the athletic department woke up to the request and personally called the former student to provide help.

Another unintentional DoS can occur when legitimate traffic uses the same radio channel. Conversely, a DoS can also be an intentional overflow, such as a ping flood to intentionally cause network disruptions.

How to mitigate:

1. Continually monitor network performance and investigate any anomalies immediately.
2. Segment the access point's coverage areas to reduce the number of people using each access point.
3. Apply a traffic-shaping solution to allow administrators to proactively manage traffic rather than react to irregularities.

RISK NO. 6: HACKER ATTACKS

Because wireless networks are insecure, they're prone to attacks. Such attacks can include spreading viruses, loss of confidentiality and data integrity, data extraction without detection, privacy violations and identity theft.

How to mitigate:

1. Deploy a network-based **intrusion-detection system** on the wireless network; review logs weekly.
2. Use and **maintain antivirus software.** Push out antivirus software upgrades to clients from servers.
3. **Create frequent backups** of data and perform periodic restorations.

RISK NO. 7: MAC SPOOFING/SESSION HIJACKING

Wireless 802.11 networks don't authenticate frames, which may result in frames being altered, authorized sessions being hijacked or authentication credentials being stolen by an imposter. Therefore, the data contained within their frames can't be assured to be authentic, since there's no protection against forgery of frame source addresses.

Because attackers can observe Media Access Control addresses of stations in use on the network, they can adopt those addresses for malicious transmission. Finally, station addresses, not the users themselves, are identified. That's not a strong authentication technique, and it can be compromised by an unauthorized party.

How to mitigate:

1. Limit access to specific MAC addresses that are filtered via a firewall. This technique isn't completely secure, because MAC addresses can be duped, but it does improve the overall security strategy. Another difficulty with this technique is the maintenance effort required. A MAC address is tied to a hardware device, so every time an authorized device is added to or removed from the network, the MAC address has to be registered into the database.
2. Monitor logs weekly and scan critical host logs daily.
3. Use proven data link layer cryptography such as SSH, Transport-Level Security or IPsec.

RISK NO. 8: PHYSICAL SECURITY DEFICIENCIES

Commonly used wireless and handheld devices such as PDAs, laptops and access points are easy to lose or to steal because of their small size and portability. In the event of a theft, the unauthorized party can compromise such devices to obtain proprietary information about your wireless network configuration.

How to mitigate:

1. **Implement strong physical security controls,** including barriers and guards to prevent the theft of equipment and unauthorized access.
2. Label and **maintain inventories** of all fielded wireless and handheld devices.
3. Use **device-independent authentication** so that lost or stolen devices can't gain access to the WLAN.

CONCLUSION

After examining just a few risks associated with WLANs, their high-risk nature becomes quite evident.

To moderate risks, management and systems administrators must **perform ongoing risk assessments** to ensure not just that they understand the risks that they face, but that they also **take appropriate steps to mitigate the risks.**

Overall, the greatest weakness with wireless security isn't the technical short-comings but out-of-the-box insecure installations. This risk can be overcome with attention to detail. But remember that the human factor is the weakest link and that this risk needs to be considered when appointing a network administrator and funding suitable review procedures.

In optimistic summary, risk provides opportunity that just needs to be managed. It's an inspiration for progress and should be a welcome challenge, as long as it's given the proper consideration.

Source: Printed with permission by the author. Copyright 2004, Information Systems Control Journal, Information Systems Audit and Control Association® (ISACA®), Rolling Meadows, IL, USA.

Case 12-2

Understanding Firewalls

Drew Hicks

A **firewall** controls access between two networks or between a node and the outside world. Usually, the first place a firewall is installed is between your local network and the Internet. This software or hardware prevents the rest of the world from accessing your private network and the data on it. Try to think of a firewall as a comprehensive way to achieve maximum network privacy, with the secondary goal of minimizing the inconvenience authorized users experience when accessing the network. The ultimate firewall is created by not connecting your network to the Internet.

Implementing a firewall is a difficult task that requires a balance of security with functionality. For example, your firewall must let users navigate both the local network and the Internet without too much difficulty. However, it must securely contain unrecognized users in a small area. This is called the *region of risk*. The region of risk describes what information and systems a hacker could compromise during an attack. The greatest risk would be connecting your network to the Internet without a firewall; the entire network becomes the region of risk.

Other than containing your region of risk, one must understand the numerous limitations of firewalls. Firewalls cannot defend against viruses. For example, a firewall on a large network with a high volume of incoming and outgoing packets will be asked to examine each and every packet with no chance of stopping thousands of incoming viruses. Firewalls cannot protect your data from disasters. The only way to protect data from disaster is to implement risk management, which includes backing up data to safe locations, various forms of RAID. Finally, a firewall will do nothing to guard the confidentiality of data on the internal network.

TYPES OF FIREWALLS

In order to choose the most practical firewall to meet the needs of a wide range of users, you must understand the different types of firewalls. There are three basic types: network-level, application-level, and circuit-level.

A network-level firewall is typically a screening router that examines packet addresses to determine whether to pass the packet to the local network or to take blocking action, such as denying access to the system or passing it to a protected area, called the **honeypot.** Since the packet includes the sender's and recipient's IP addresses, the firewall uses the information from each packet to manage the packet's access. This is done by instructing the screening router to block packets with a file that contains the IP address from restricted areas (blacklisting).

Depending on how the screening router file is constructed, the network-level router will recognize and perform specific actions for each request type. For example, a router could be programmed to let Internet-based users view your Web pages, yet not let those same users use FTP to transfer files to or from your server. A screen router can consider the following information before deciding to send a packet through: source address of incoming data, destination address, data's session protocol (TCP, UDP, ICMP), source and destination application port, and whether the packet is the start of a connection request. A properly installed network-level firewall will be very fast and almost transparent to users.

An application-level firewall is a host computer running proxy-server software. **Proxy servers** communicate for network users with the servers outside the network. In other words, a proxy server controls traffic between two networks. In some cases, a proxy server manages all communications of some users with a service or services on the network. This is how most of the security on Web sites is controlled today. When using an application-level firewall, your local network does not connect to the Internet. Instead, the traffic that flows on one network never interacts with traffic on the other network because the two network cables do not touch. The proxy server sends a copy of each approved packet from one network to the other. Application-level firewalls mask the origin of the initiating connection and protect your network from Internet users who may be trying to hack into your network.

As with screening routers, you can configure proxy servers to control which services you want on your network. For example, you can direct clients to perform FTP downloads but not perform FTP uploads. Unlike a router, you must set up a different proxy server for each service you provide. Application-level firewalls provide an easy means to audit type and amount of traffic on a site. They also make a physical separation between your local network and the Internet, which makes them a great choice for security. However, because a program must analyze the packets and make decisions about access control, application-level firewalls tend to reduce network performance.

A circuit-level firewall is similar to an application-level firewall in that both are proxy servers. The difference is that circuit-level firewalls do not require special proxy-client applications. As named, a circuit-level firewall creates a circuit between a client and a server without requiring that either application know anything about the service. In other words, a client and a server communicate across a circuit-level firewall without communicating with the circuit-level firewall. Circuit-level firewalls protect the transaction's commencement without interfering with the ongoing transaction. The main advantage of the circuit-level firewall is its ability to provide service for a wide variety of protocols. Unlike the application-level firewall, the circuit-level firewall can be used for HTTP, FTP, or Telnet without changing your existing application or adding new application-level proxies for each service. In addition, circuit-level firewalls only use one proxy server, making **maintenance** a snap.

SECURITY RATINGS

The Department of Defense (DoD) provides the Orange Book for further specifics of network security. The Orange Book describes four general divisions of security, with each security division having classes resulting in seven security-rating classes.

- Division D Security Rating. The Class D1 security rating is the lowest rating of the DoD's security system. Essentially, a Class D1 system provides no security protection for files or users (Windows 95).
- Division C Security Rating. Classes in the Division C security provide discretionary (need-to-know) protection and provide audit capabilities for tracking users' actions and accountability. There are two subcategories, C1 and C2.
- Division B Security Rating. Division B secure systems must have mandatory protection, instead of discretionary protection as in Division C. Mandatory means that every level of system access must have rules. In other words, every object must have a security rating attached. The system will not let a user save an object without a security rating attached. Division B security is broken down into three subcategories, B1, B2, and B3.
- Division A Security Rating. Division A security ratings are the highest ratings the Orange Book provides. Division A only contains one secure class, A1. The use of formal security verification methods characterizes Division A. Formal security methods assure that mandatory and discretionary security controls employed throughout the system effectively protect classified information the system stores or processes. Division A ratings require extensive documentation to demonstrate that the system meets all security requirements in all aspects of system design.

CONCLUSION

A firewall combines hardware and software, which you design to protect your network from unauthorized access. You can use different types of routers to provide simple security, and as a foundation for your firewall. You can use firewalls to protect your network from without and to protect internal departments from other departments.

You must use virus protection software to protect against viruses, because a firewall will not stop viruses by itself. Before you design a firewall, develop a security plan outlining the type of access your employees and users have. The three main types of firewalls are network-level, application-level, and circuit-level firewalls. The Department of Defense Orange Book defines seven levels of operating system security. You should ensure that your system has a security level of C2 or higher for maximum protection.

Case 12-3

General Computer and Network Security

Brandon T. Billingsley

There are many topics that need to be covered when discussing computer and network security. These topics are as follows: physical security, operating system security, and network security. **Physical security** covers the actual physical computer and access to the machine and networking equipment. Operating system security covers locking down the operating system. Network security covers general network security principles and security issues. All of these topics will be covered in the scope of this paper, and the best practices will be recommended

PHYSICAL SECURITY

All of the operating system and network security in the world will not help users if unauthorized individuals can walk to a computer and start changing information. Physical access to a computer is often one of the most overlooked facets of computer security. There are many different ways to physically secure a computer. One of the easiest and least expensive is to turn on the power-on password setting in the Basic Input Output System (BIOS). To do this, go into the BIOS during the computer's startup routine. This is done on most computers by hitting the delete key, but the user needs to watch the screen carefully to see if the entry key is different from computer to computer. In the BIOS, navigate to the security section, and then set a power-on password. This password will not allow intruders to start up the computer unless they have the correct password. Next, buy a computer case that has a locking front panel. If all of the drives and the power and reset buttons are inaccessible to a user, then the computer is astronomically more secure than if it were not locked.

The next step is to have locking enclosures for the computers. To make all the physical parts inaccessible, use these enclosures to lock the entire computer or any network equipment located inside another enclosed space. Only the keyboard and mouse need to be accessible to the user. The next step to consider is physical access to the room where the computer is located. For home users, this does not seem very important, but in a business environment, computers can have very sensitive information stored on them and may need to be locked away.

There are many systems, varying in complexity, that can be put in place to secure a room. The first is a simple lock. A lock will keep out anyone with a casual interest in getting into a room, but will not stop a determined intruder. The lock can be further secured by installing a card swipe or a biometric system to ensure proper identification of the person entering the room. These steps will ensure that the computer is physically inaccessible to anyone who is unauthorized to use it.

OPERATING SYSTEM SECURITY

The operating system is the heart of a computer. It is what allows the computer's resources to be used easily and efficiently; nevertheless, there are a number of flaws in the operating system that an attacker can use to infiltrate the system and cause damage. The first step to take in securing an operating system is to ensure that **all program updates (called patches)** have been downloaded and installed; the most important of these is the Windows Service Pack 2. Be sure that the computer is up-to-date with the latest patches to ensure that all of the errors in the software are corrected. Next is to manage user accounts correctly. The Windows operating system is installed with a Guest account enabled and an Administrator account enabled. The Guest account should be deleted to prevent unwanted access, and the Administrator account should be renamed and given a very secure password. To delete the Guest account and manage the administrator account, you must go to Start | Run, type in **control userpasswords2**, remove any unwanted accounts, and manage the administrator password.

Passwords are very important when users lock down an operating system. Passwords should be a combination of uppercase and lowercase letters, numbers, and symbols. A password should not be anything that is easily known about the user, such as a pet's name or one's birthday. It is also important to set the user's computer screensaver to lock after a few minutes of use and therefore require a password to log back onto the computer. To complete these tasks, right-click on the desktop, and go to Properties. Then go to the Screen Saver tab, and check the box next to "On resume, password protect." Also, set the "Wait" time to approximately three to five minutes.

Day-to-day activities should also be done using an account that does not have administrator privileges. Restricting administrator privileges can also be done from the control userpasswords2 dialog. Instead of removing an account, add an account that does not have administrator privileges and apply an appropriate password. Then, instead of logging on as the administrator, log in under the new account. This action will not allow programs from the Web or e-mail to install themselves without the user's knowledge. Only the user will have the rights to install programs.

The next thing that should be done is to turn off any services in the operating system that are not being used. If files or printers are not being shared, for example, then the file and print sharing service should be turned off. To turn off the file and print services, go to Network Connections in the Control Panel, right-click on your network connection, and click on Properties. Under "This connection uses the following items:" disable any services that are not needed. Some services allow others to connect to the computer; if these services are turned off, the computer will be more secure.

The final step to ensure that the operating system is secure is to review the security audit logs on a regular basis to ensure that unwanted access has not been attempted. To review these logs, go to the Administrative Tools in the Control Panel and then go to the Event Viewer. All of the events that are logged are shown here with warnings shown by a yellow caution sign and errors shown by a white X on a red background.

NETWORK SECURITY

Connecting a computer to a network is very useful and also very dangerous. Just by connecting one computer to other computers, the user opens the computer to other users. This danger of infiltration increases exponentially when a computer is connected to the Internet. There are two classes of networking: wired and wireless. Wireless security will be covered in another section of this paper. A wired network environment is more secure than a wireless network environment, but there are still some issues to consider. The use of a switched network over a hubbed network is generally preferable. Switches transfer packets based on the MAC address of the computer that they are meant for, and hubs broadcast packets to all of the computers on the network. In a hubbed network, someone can very easily sniff the packets going across the network because they are sent out to everyone. Sniffing can still be done in a switched network but is much harder to do and requires more technical skill.

A MAC access control list (ACL) should also be instituted throughout the network. This will allow only known MAC addresses to connect to the network and not allow someone to just plug into a network and access its resources. To set up an ACL, go to the router's Web-based control application. On Linksys routers, the user opens a Web browser by entering 192.168.1.1 in the address field. Then, a username and password are entered. (The default is admin and admin.) Next go to the Access Restrictions tab to create a list of MAC addresses of the computers that should be connected to the network. To find out a computer's MAC address, go to each individual computer, and then go to Start | Run, and type in **cmd.** In the command window, type in **ipconfig/all.** This action will give the user the MAC address of the network card in the computer.

While no network will ever be 100 percent secure, these steps will allow a user to set up a network with confidence.

Summary

Wireless means different media and very different modes of attack. It makes the interception of packets easier because they are broadcast into the radio or IR spectrum. Wireless technologies, however, present many of the same dangers to the corporate network; they just have a different way to evoke them. Wired or wireless, security is at the very heart of risk assessment and disaster planning for business continuity.

A concern for security, in general, and wireless security, in particular, extends beyond the walls of the organization. As employees work from home or on the road, customers access organizational data and networks, and wireless networks become pervasive, access to the firm's networks and data becomes even easier.

Managers and network administrators must be vigilant and establish frequent review of their wireless security processes and procedures as the technologies and threats continue to evolve. Without continuous attention, wireless networking can become a huge hole in the security blanket that's needed to protect the firm and its data from both intentional and unintentional, internal and external threats.

Clarification

IEEE 802.11x refers to a group of evolving WLAN standards that are under development as elements of the IEEE 802.11 family of specifications but that have not yet been formally approved or deployed. This incomplete standards set included the following:[34]

- Security
  802.1x.
  802.11x—Encryption standard replacing WEP; allows for dynamic shared encryption keys.
  WPA2—Standard to assure interoperability among 802.11i-based devices (WPA1 was interim standard).
- Transport
  802.11a/b/g—Standards for transmission of wireless.
  WiFi—Interoperability for 802.11a/b/g devices.
  802.11n—Seeks 100 Mbps transmission rate; completion 2007.
  802.11h—Resolves interference issues with existing 802.11 family specifications.
  802.11j—Japanese regulatory extensions to 802.11 family specifications.
  802.11k—Radio resource measurement for 802.11 specifications so that a wireless network can be used more efficiently.
  802.11m—Enhanced maintenance features, improvements, and amendments to existing 802.11 family specifications.

[34] http://searchmobilecomputing.techtarget.com.

802.16d/e—Seeks to standardize wide-area high-speed wireless zones. 802.11d is aimed at fixed-wireless and 802.11e is for mobile use.

WiMAX—Attempt to standardize 802.16 interoperability; intended to be for MAN.

- Management/Roaming/QoS

  802.11e—Defines basic levels of QoS; completion mid-2005.

  802.11f—Defines communications between APs for layer two roaming.

  802.11r—Standard for handoff for fast roaming between APs; completion 2006.

  802.11s—Standard for AP connection for backhaul and mesh networks; completion 2006.

- WME/WSM—Wireless media extensions of Wi-Fi media standards to ensure interoperability across vendor products.

- CAPWAP (Control and Provisioning of Wireless Access Point)—Standard taxonomy of mechanisms for control/program of wireless APs.

- LWAPP (Light Weight Access Point Provisioning).

Key Terms

access points (APs), *p. 351*
architecture, *p. 365*
auditing, *p. 364*
authentication, *p. 354*
beacon, *p. 365*
Bluetooth, *p. 361*
broadcast, *p. 354*
cryptography, *p. 356*
data link, *p. 357*
default, *p. 353*
denial of service (DoS), *p. 357*
Dynamic Host Configuration Protocol (DHCP), *p. 365*
encryption, *p. 354*
firewall, *p. 372*
hacker, *p. 355*
hacking, *p. 365*
honeypot, *p. 373*
hotspot, *p. 366*
IEEE 802, *p. 355*
IEEE 802.11, *p. 353*
interference, *p. 358*
intrusion, *p. 355*

MAC address blocking, *p. 365*
MAC address spoofing, *p. 355*
maintenance, *p. 374*
malicious association, *p. 356*
man-in-the-middle (MITM) attack, *p. 356*
media access control (MAC) address, *p. 355*
monitor, *p. 357*
packet, *p. 351*
passwords, *p. 354*
phreak, *p. 361*
physical security, *p. 375*
preconfigured, *p. 354*
prevention, *p. 364*
privacy, *p. 353*
proxy server, *p. 373*
radiation, *p. 352*
radio frequency identification (RFID), *p. 362*
radio telegraph, *p. 353*
range, *p. 351*

RF interference (RFI), *p. 360*
risk, *p. 352*
rogue users, *p. 358*
service set identifier (SSID), *p. 351*
simple network management protocol (SNMP), *p. 361*
spectrum, *p. 356*
spoofing, *p. 355*
temporal key integrity protocol (TKIP), *p. 363*
warchalking, *p. 358*
wardriving, *p. 358*
Wi-Fi Protected Access (WPA), *p. 352*
Wired Equivalent Privacy (WEP), *p. 352*
wireless access point, *p. 354*
Wireless Fidelity (Wi-Fi), *p. 353*
wireless local area network (WLAN), *p. 353*

Recommended Readings

Symantec AntiVirus for Handhelds, http://enterprisesecurity.symantec.com/products/products.cfm?ProductID=237&EID=0.

WLAN Management Solutions, http://airwave.com/prodserv_features.html.

Wireless Security FAQ, http://www.iss.net/wireless/WLAN_FAQ.php.

The Unofficial 802.11 Security Web Page, http://www.drizzle.com/~aboba/IEEE/.

Wireless Security Blackpaper, http://arstechnica.com/articles/paedia/security.ars.

Management Critical Thinking

M12.1 With the ever-expanding list of WLAN threats, should management appoint a *Wireless Security Officer?* If so, what should the job description entail?

M12.2 If an organization has a large requirement for wireless networks, how can risks be balanced with benefits?

Technology Critical Thinking

T12.1 How can a firm ensure that WLAN technology does not compromise company proprietary data?

T12.2 How can rogue access points, man-in-the-middle clients, and rogue clients be detected and neutralized?

Discussion Questions

12.1 Have you ever connected to a wireless access point other than your own? Was it difficult? Did the owner find out?

12.2 List defense mechanisms against DoS. Which are the minimum that should be implemented?

12.3 How can rogue access points be detected?

12.4 Is eavesdropping on wireless communications illegal? Unethical? Difficult?

12.5 Can someone eavesdrop on your telephone conversations, wired, cordless, or cellular?

12.6 How can a totally passive eavesdropping device be detected?

12.7 Compare the data content of a barcode and an RFID tag. Do barcodes still have a valid use once RFID tags reduce in cost?

12.8 What are the basic risks associated with RFID implementation?

12.9 If you have a wireless network in your home or apartment, have you secured it? What did you do? What should you now do?

12.10 Discuss the pros and cons of a wireless media center at home.

Projects

12.1 Make up a list of WLAN technologies and associated risks for each. Rank-order the technologies based on technical vulnerability.

12.2 Divide the class into teams and divide the local area/city so each team has an area. Find a Web site that shows local wireless access points (hotspots). Map the team's area using these data.

12.3 Obtain a wireless access point detector for each team. (This may be shareware on a laptop computer.) Have them drive around town and (a) locate the access points in project 12.2 and (b) find access points not listed in project 12.2. Pay particular attention to access points on the streets of class members.

12.4 Using the data in project 12.3, attempt to use the bandwidth of each access point.

12.5 In project 12.3, describe each access point, listing as much information as can be gleaned from intercepting their signals. Determine how vulnerable each access point is. Try to find its owner and tell him/her what you found and how. Record the reactions of the owners for the class.

12.6 Interview managers of local WLAN implementations and ascertain their (a) security concerns, (b) security measures, and (c) relative confidence in the security of their WLAN. (If this person is responsible for one of the access points in project 12.5, perform the interview scan in project 12.5 first; then tell him/her what you determined in project 12.5)

12.7 Determine if firms in your community have integrated wireless networking into their telecommunications infrastructure and their IT strategic plans. How has this impacted their risk assessment and disaster recovery planning?

12.8 Search on the Internet for additional articles on securing your wireless access point. Compare the articles with Case 12-1.

12.9 Map the wireless access points in your school. See what it would take to add a rogue access point for your class. Get permission before implementing it.

12.10 Determine all of the cities that have a stated policy of making their city a Wi-Fi city, such as Singapore and Philadelphia. Who would be the proponents and opponents of such a plan?

Network Management and Control

The control and administration of data communications includes management concerns as well as the technology itself. Failing to consider the areas of monitoring, performance, and security, and new projects, places the organization in jeopardy. If all goes well, the network maintenance task will be transparent to the user community; if all does not go well, the topics in this section become very important.

Chapter **Thirteen**

Monitoring and Control of Network Activity

Once we deploy and depend upon data networks, we must assume the responsibility to monitor and control activity on the networks. Here, we address the management and technical aspects of control of the networks. There must be operations, administration, management, measurement, and control over telecommunications resources. This also includes the measures needed for configuration control, capacity management, and risk management in order to assure reliable network service.

Telecommunications capabilities include voice and data systems.

Two primary objectives of *network management* are (1) to satisfy system users and (2) to provide cost-effective solutions to an organization's telecommunications requirements. These resources are installed to allow greater communications and sharing of resources, to provide strategic links to suppliers and customers, and to enable centralization of decision making in a decentralized organization. Often, these applications allow the firms to stay in business and remain competitive. When considering data communications and MIS projects, managers must consider the costs and the benefits. Benefits may be either tangible and measurable or intangible and nonquantifiable. For example, many strategic projects relate to the future nature of business and are visionary. Thus, while the cost may be readily apparent, the benefits may not be as they deal with highly intangible outcomes, such as different methods of conducting business. This may be the cost of doing business, as opposed to having a proper cost-versus-benefit consideration.

There are two sides to the management of telecommunications. One side is managing the technology; the other is managing the organization. In both cases, the objective is to have an organization and infrastructure in place that will support the voice and data communications needs of the parent organization. Given that, we now will look further at the organization and functions that support the technology.

THE ORGANIZATIONAL SIDE OF TELECOMMUNICATIONS MANAGEMENT

The **data communications** organization, like its parent MIS organization, has the same functions as its parent, that is, the same functions that any business has. The most obvious is the operations function of installation and operation of the network. Before that can happen, there must be marketing to customers to determine needs, research and development to plan for and engineer these needs, customer support once new equipment is installed, and administration to keep track of it all. As we will show, not all of these functions require the same level of technical ability. All, though, are required.

Data communications and MIS projects and continuing operations of the resultant capabilities are always undertaken to support or solve a business problem or opportunity. These projects and operations require classical functions of management—planning, organizing, staffing, directing, and controlling.

Planning requires a perspective of the timing and life of the project, and the organization being supported. Planning involves preparing to deal with events and including other facets of the organization to make the plans come to fruition. The three levels of planning are

* *Operational planning,* which deals with the day-to-day operations of the organization and its resource usage. Supervisors organize and oversee the daily activities of their people to accomplish actions in support of the data networks and resources. To do this, they plan for the day's or week's activities. The time horizon of organizational planning, as noted, is immediate, often spanning only days or weeks.

- *Tactical planning,* which deals with a longer time span and pays particular attention to the acquisition of resources for future operations. The time horizon may lengthen to a year and is concerned with personnel acquisition and training, inventory purchase and storage, and funds acquisition. It is budget-oriented and closely linked with the equivalent planning of the parent organization.

- *Strategic issues,* which are those that charter the direction of the organization. They answer questions such as What business should we be in? Where should we build a new facility? and Who are our present and future competitors? The time horizon of strategic thinking is from one to five years. Strategic planning, a prime function of upper management, prepares for the future. Because the data communications group supports the strategic planning of the parent organization, data communications management should be included in the parent's plans. The more the data communications function knows about the vision, direction, and plans of its parent, the better it is able to support these plans with architecture, reserve capability, and trained personnel.

Caveat

The idea of a strategic planning time horizon being up to five years is a historical one: that is to say, this is the time span traditionally used or used with traditional organizations. As one data communications executive noted: "How can I make a five-year plan? I cannot forecast well enough to make a one-year plan." In the communications industry, technical changes occur so rapidly that we do not have the luxury of believing we can plan five years hence or depend on our present competitive advantage lasting 12 months without change. If you are dealing with Internet-based business, change is even faster.

The next function of management is that of **organizing,** which includes several considerations, the main of which is establishing relationships among the entities of the organization. Some functions are of operational importance, while others that are used for vendor negotiations may be more tactical in nature. The creation of new networks that will change how the organization competes or supports its personnel is of strategic importance. While every project must be justified, the manner of this justification will relate to the importance and time orientation of the outcome.

The management function of **staffing** relates to the acquisition, retention, and training of qualified personnel who can plan, define, install, operate, and maintain the communications technology in support of business problems and opportunities. The level of staffing, that is, the number and quality, depends on the **complexity** of the voice and data systems resource and the extent to which their operation and **maintenance** are provided by vendors or outsourced. As the organization gains total control of the data communications resource, it requires more and better qualified personnel.

The first such member of a data communications group is the *data communications manager,* the person with an ability to grasp technical subject matters and business needs. Although s/he need not personally be a technician or former technician, s/he must be able to converse with and stand his/her ground with the technical people. This person will interpret business needs and direct other people and vendors to develop a data communications plan.

The next group of staff members is the *designers and implementers* of data communications capabilities. These engineers find solutions to business problems and opportunities and design networks and capabilities to support the solutions. Once implemented, the *network operations staff* operates them on a daily basis. This group needs to be technically competent and experienced as they must be able to define problems and resolve them quickly. The last members of the group are the *administrative support staff,* who take care of accounting, inventory, and general administrative tasks. In some cases, consultants will be hired to complement the other members of the group on a short-term basis.

The next management function, **directing,** involves the supervisory task of getting people to successfully perform the required tasks. The primary skill is dealing with people to get complex jobs done; the requirement is for leadership.

Finally, the data communications manager performs the function of **controlling** the data communications resources of the organization. This means inventory management of equipment, circuits, and networks. The concern is not just the inventory of physical parts but also the inventory of what is really important about data communications: communications circuits and channels, now referred to generically as **bandwidth.** Equipment must be acquired and funded and the costs must be charged or allocated. Some costs are obvious as are their allocation, such as when a circuit or piece of equipment is obtained specifically for a given function or project. Other costs must be allocated based on use or some other criteria. Some believe that no service should be seen as a free good, so charge-back for use, based on some criteria such as amount of data transferred or permanent bandwidth, is in order. The problem of such chargeback is that it may actually inhibit use if the method is wrong.

Another aspect of control is security, against the threat of theft or intrusion into the network. The first deals with physical locks and accountability, and the latter deals with software locks and training. In either case, because of the importance of data communications and information systems capabilities, **disaster recovery** is an important aspect of security and control. It deals with the ability to continue operations upon the catastrophic interruption of organization facilities. A secure data communications capability is one that works continuously, reliably, and without intrusion.

Design and Implementation of New Facilities and Services

If you consider what skills you would find in a data communications group, you would expect to see those who design and install the new equipment and facilities, others who make the existing telecommunications infrastructure work, and people who take care of all the nontechnical details. Thus, you will find a group with a lot of diversity, ranging from engineers to purchasing agents.

The first group to consider is the most engineering-oriented; these are the individuals who design new voice and data communications facilities. Many employees might never come in contact with these people unless they are on a project that is improving an existing capability or planning a new facility. Then they will work with this group as they strive to determine the requirements of the new system, design additions to existing capabilities, and make sure it all works together. These people do not work in a vacuum as they must know everything about the present installed plant. They must know present equipment, wiring, circuits, terminals, switches, and vendors who might interface with or maintain them.

The designers start with the total requirement for the new or improved telecommunications capability and end up engineering and planning for the detailed physical components. If, for example, the problem was a high busy rate for long-distance voice lines, the objective would be to increase the access to long-distance circuits. For this, the designers might use the installation of Voice over Internet Protocol to achieve it:

- Determine coverage of organizationally owned IP networks.
- Ascertain the accessibility of customers and suppliers if VoIP is used for long-distance service.
- Determine suitability of present **private branch exchange (PBX)** to handle handoff of calls to the IP gateway.
- Consider the risk of using VoIP compared with the use of standard long-distance lines.
- Calculate the cost of the conversion, the benefits, and any savings.

This example is a current one as companies move away from services of traditional long-distance providers. What does this group do if you want to add an IP VPN, be able to communicate via EDI to a supplier or customer, or upload large files across continents? In each case, the designers will determine the service change, define and investigate the alternatives, and present their recommendations. Once a plan of action is set, the designers will engineer or buy the required capabilities and install them.

Network Diagramming and Design Tools

In the client-server world, it has been common for new applications to place a significant load on the network, slowing both the new and existing applications. Simply adding more servers and LAN segments will not necessarily solve the problem. The older methods of prediction about the impact of new applications were mostly seat-of-the-pants estimates. Problems have arisen because, in many instances, unmonitored LAN construction has created a large web of networks with segments connected by simple bridging, growing like weeds. Network expansion without a strategy or operational plan ensures that problems will arise because of conflicts, bandwidth restrictions, and security issues.

Especially in client-server environments, corporate networks have grown far too complex for any single manager to understand the true impact of changes. Now many network managers resort to **simulation,** software that can give them a graphic view or a working model of the network, to determine the impact of changes to their networks. Several software programs are available that can show LAN managers the traffic flow on their networks and assess the impact of adding applications and new users. The software allows "what-if" scenarios, that is, network performance simulations, to be performed, such as the impact of adding nodes, routers, or servers; splitting the network into segments; and so forth. This capability is important as organizations move to scale their client-server applications up to serve the entire enterprise.

The simulation helps with capacity planning, that is, planning the network with the bandwidth to support the anticipated applications but without acquiring unneeded **capacity.** In the past, many networks began small, were built on an ad hoc basis as the LANs were relatively low-cost, were stable, and may have played a relatively minor role. As LANs have become an essential component in the organization's computing strategies, the lack of accurate measures of network performance and the piecemeal architecture of networks become significant roadblocks. Many of the networks beg for redesign or replacement. One factor that makes networks more difficult to model than mainframe computers is the *"bursty" nature of traffic,* the mode where traffic is random in occurrence and highly variable in structure. The terminals in mainframe computing usually generate predictable and easily measured traffic, but network traffic varies greatly and often follows no pattern. Because processing is occurring at both the client and the server locations, it is far harder to get a clear picture of an application's impact on overall network performance. Several vendors provide LAN simulation products. The products usually cost over US$10,000 and require both an understanding of the network and a fair amount of training.

In addition to simulation packages, there are several *network diagramming tools* available. These tools make it easy to document networks and visualize future configurations. Network diagramming software allows one to design and construct a diagram of the network quickly. Information about each device can then be input into a database that can be used as the basis for reports. Thus, these programs become very useful in network management tasks. Some popular products are *netViz®* (see Figure 13.1), *SysDraw®, Visio®,* and *ClickNet Professional®*. Network managers have an obligation to understand what these networks look like and the components that must function to enable them.

Falling in between network simulators and diagrammers are network designers such as *NetSuite Professional Design®*. Network designers help the network manager by indicating

FIGURE 13.1
**A Typical Network
Management
Architecture
Maintains Many
Relationships**

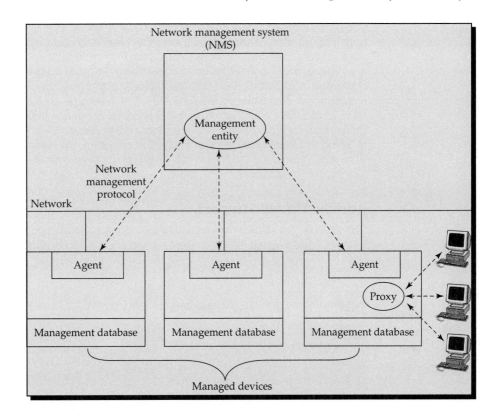

whether the design of the network has the right physical configuration to work. Altogether, simulations, diagramming tools, and **network design** tools can provide the framework for understanding the telecommunications infrastructure.

Network Operations and Technical Support

Once the designers have engineered new capabilities and have seen the installation to completion, these capabilities are turned over to the group who cares for them on a day-to-day basis. These are the people called when something does not work, and who perform routine and emergency maintenance to keep the system operational. They upgrade hardware and software, answer calls at the help desk, and work when you don't so maintenance does not interfere with normal service.

The service level described above is fine if the firm is in a business that operates for 8 to 12 hours a day. The operations and support group has people on duty to answer trouble calls and fix equipment. They have spare parts and can move nodes. Additionally, a group of the people work at night and on weekends to perform maintenance that necessitates bringing the system down. Much of this latter work is totally transparent to the users. In fact, if the maintenance is done right, there will be no downtime and no need for trouble calls during the day. However, since equipment tends to fail in spite of the best of preventative maintenance, firms do need repair personnel.

Now what happens in the situation where the firm is in a 24-hour-a-day business that must have its systems operational at all times, such as an airline reservation system, the phone company, or an emergency division of a hospital? Organizations either must have sufficient redundancy built in so a failed part has a backup part that takes over the function automatically, or they must tolerate down time. When does Total System Services, Inc., of Columbus, Georgia, perform telecommunications and computer maintenance on their network and mainframes that handle VISA® and MasterCard® credit checks all day and all night?

Caveat

One of the most important characteristics of any system is reliability. Organizations want the system to operate when they need it, and they want it to operate properly. Additionally, they want a minimum of delays. For this to occur, the firms need sufficient capacity to have the minimum delay, and they need to use a stable design of good hardware and software so that breakdown is infrequent. The measure of frequency of breakdown, or interruption of service, is called *mean time between failures (MTBF)*. The hard drive on a personal computer has an MTBF of 250,000+ hours, which, at 1,000 hours for a six-week period, means it should work continuously for 30 years before failure. (This is on an average basis, of course.) The higher the MTBF figure for any given piece of equipment or system, the higher the cost. Therefore, you either need good design and good equipment, or good repair facilities. This is where the technical support group fits in.

Administrative Support

Just in case you think telecommunications groups are all peopled with engineers and "techies," ask yourself how they buy equipment, pay their bills, keep track of all their

TABLE 13.1
Telecommunications Job Categories for a Small- to Medium-Size Organization

Position	Education	Duties
Technical		
Design engineer	BS or MSEE	Performs the analysis and engineering required to design and create the new or enhanced capability. This includes selection of media, protocol, and architecture and involves network maps that show components and capabilities that will support the organization. Works with Telco and vendors on new design and changes.
Operations and troubleshooting	Technical training plus vendor certification	Works with the system on a daily basis to keep it operational. Maintains and repairs system parts as required, generally using troubleshooting guides and component replacement. Installs hardware and software for system and end users. Works with Telco and vendors on daily operations.
Webmaster	Technical plus design	Creates, maintains, and continuously updates the organization's presence on the Internet, for example, WWW pages and interface design, database, and so forth.
Network administrator (involves management and technical duties)	BS plus network engineering certification	Concerned with the daily operation of LANs and WANs, especially the volume of traffic, speed of response, storage of data, and migration of data to avoid bottlenecks.
Management		
Chief information officer (CIO)	MBA	Overall responsibility for the total telecommunications capabilities of the organization, paying particular attention to the future needs as contained in the strategic plans. Maintains strategic linkages with vendors and partners.
Telecommunications manager	BS/BA plus MBA	Ultimate responsibility for telecommunications architecture and the organizational implications of any change or new capability. Directs the activities of all other members of the telecommunications group. Evaluates possible vendors and providers.
Administration	BS/BA	Performs the administrative functions required to support other functions, such as databases of assets, chargeback for services, and management reporting. Responsible for procurement of systems components, change configuration, and problem management.

equipment, and do other such administrative tasks. With administrative people, of course! These are the nontechnical people who have such jobs as

1. Ordering and purchasing communications products and services.
2. Receiving equipment.
3. Inventorying equipment.
4. Checking and paying communications bills.
5. Determining chargeback methods to users, that is, who pays for what?
6. **Coordinating** adds, moves, and changes of equipment, including maintaining blueprints of present installations, handling the paperwork to make changes, and so forth.
7. Preparing and publishing a phone directory, whether it be paper or electronic.
8. Registering new telecommunications users for telecommunications access and for computer applications access, that is, security maintenance.
9. Training users.
10. Maintaining telecommunications procedures.
11. Providing telephone operator services.

The administration group performs the type of work that business administration graduates are qualified to perform. While it takes an engineer or engineering-trained person to do design and heavy-duty troubleshooting, much of the other work takes a more broadly trained and educated person. Both functions are very important if the total system is to be reliable. (See Tables 13.1, 13.2, and 13.3.)

TABLE 13.2
Telecommunications Group Activities

I. System creation and upgrade
  A. Design and configuration
    1. Node equipment
    2. Media and bandwidth
    3. Software
    4. Tariffs
  B. Testing
    1. Initial
    2. Continuous
    3. Reporting
  C. Diagnosis
    1. Meeting spec
    2. For problems
    3. Fine-tuning
  D. Documentation
    1. Assets (database)
    2. Operations
    3. Repair

II. Operations
  A. Monitoring
  B. Control
  C. Diagnostics
  D. Problem-reporting system
  E. Repair
  F. Documentation

III. Administration
  A. Personnel
    1. Attract and retain qualified personnel
    2. Training
  B. Asset management
  C. Purchasing
  D. Chargeback for asset usage

TABLE 13.3
Telecommunications Job Categories for a Large-Size Organization

I.   Planning and development
   A. Director, telecommunications planning
   B. Manager, network planning
   C. Data network design technician
   D. Voice network design specialist
   E. Business applications development specialist

II. Service and support
   A. Director, network services and support
   B. Data communications service manager
   C. Help desk technician
   D. LAN service manager
   E. Office automation applications specialist

III. Operations
   A. Director of network operations
   B. Network security manager
   C. Data network operations manager
   D. LAN manager
   E. Voice systems technician
   F. Voice network operations technician

Telecommunications Management versus Telecommunications Capability

If you get a feeling that we are trying to separate the management of the telecommunications function from the telecommunications capability, you are right. The capability of telecommunications tends to be very hardware and engineering oriented, and requires people with special skills and inclinations. The management of the telecommunications function is much like the management of any specialized group of people—it takes the same organization and planning skills to direct the people and achieve the goals and objectives. As with any group of technical people, those in charge need to understand the technology at a high level, but not necessarily at a detailed level. For example, the manager of the telecommunications group needs to be able to understand what his/her designers and installers do, but not necessarily how they perform the tasks in detail.

THE TECHNICAL SIDE OF TELECOMMUNICATIONS MANAGEMENT

This part of telecommunications management is concerned with **network management** (or operations)—the set of activities required to keep the communications network operational and reliable.[1] These are the day-to-day tasks that must be done to keep the capability up and running and to keep the users satisfied. It is the part that must go right for telecommunications, as it is installed, to be successful.

The scope, or **mission,** of network operations responsibilities includes everything from the user's terminal to the mainframe, from a microcomputer to the shared network printer on the LAN, from the client to the server, and from the supplier's modem to the customer's modem. While the network operations group may not be responsible for all of

---

[1] Remember PARS or RAPS = Reliability, Accessibility, Performance, Security.

the equipment, they are the ones who are the point of help and must be able to work with such equipment as

A. User workstations

B. Cluster controllers

C. Modems

D. Communications lines

E. Line concentrators

F. Multiplexers

G. Front-end processors

H. Communications software

I. LANs/WANs

J. Network management software

K. Network operating systems

L. Servers

M. Wireless modems

N. Wireless access points

O. Hubs, switches, routers, bridges, gateways

When a call comes to the telecommunications help desk or a technician's cell phone, an alarm sounds on the trouble board, or the network operating system displays a message on the operator's terminal or wireless PDA, someone from the network operations group must respond. These are the people who take the heat, the ones who keep it all together with spit and baling wire,[2] experience, and tenacity.

Consider that computer and communications networks in many companies are growing at a rate of 15 to 50 percent per year compounded. If your organization is expanding and taking on new markets, customers, and competitors, you may double your telecommunications capability every other year. The budget for this group and its equipment tends to be increasing, with an average communications budget for companies in the United States growing 10 to 20 percent per year. With this type of physical and fiscal growth, not only will you need to pay attention to telecommunications evolution, you will have a fair amount of visibility. To be more specific, it is important to manage the telecommunications capability because

- It is an expensive asset or resource.
- Competition is often impossible without it.
- It can create, maintain, or counter a competitive advantage.
- It may define how you do business.
- It may be your primary link to suppliers and customers.

Implication   Wal-Mart, the largest retailer in the United States, has the nation's largest data center, second only to the Department of Defense, from which it manages all stores, worldwide. Wal-Mart gives ready access to its suppliers with near **real-time** data on their items, allowing these suppliers to react to market conditions at any given store. Wal-Mart noted, from mining its centralized data (all transactions are stored for two years), that there was a significant increase in

---

[2] Please forgive these metaphors, but as is often the case, historical, colloquial references serve best. With the introduction of the hay baler on U.S. farms came the wire used in the machine to hold the bales together. The wire found many uses, from the barn to the kitchen, as it fixed so many things. Alas, the wire has been replaced by twine and gone is this wonderful resource, but not the metaphor.

sales of Pop-tarts® during hurricanes. Kelloggs® uses these data to make supplemental deliveries of this product in times of these storms. To do this, Wal-Mart watches the weather closely, having *The Weather Channel*® on display in their server room. Wal-Mart and other businesses watch the news and weather to react to environmental conditions, ranging from significant lighting strikes, which may damage data communications equipment, to hurricanes, which would necessitate moving assets and people in response. *Home Depot,* for example, moves truckloads of plywood in anticipation of hurricane arrival.

Caveat

Early in the book, we described what we called *computer-based telecommunications-intensive information systems.* Because computer-based information systems in organizations are so intertwined with telecommunications, you often cannot separate them. While there is one group of people who worry about software problems and another group who worries about telecommunications problems, the total system is vital[3] for the ongoing operations of the organization. Neither part is more important than the other, and neither part works without the other. This is a symbiotic relationship.

As the organization uses and relies on telecommunications, it tends to define how it does business and competes. *SABRE*® is one of the two most extensive airline reservation systems for reservation agents and airline asset management. It was created initially to schedule maintenance and to load the aircraft. It quickly became the premier method of competition for its creator, generating more profit than the carrying of passengers. Consider American Airlines without their *SABRE* reservation system. The day they lost the pointers to their database system (of 1,100 disk drives), they operated and competed completely differently than the day before or the day afterwards. When the 800 service to Delta Airlines that ran through the central office in one part of Atlanta, Georgia, failed due to a failure of battery power, they were forced to operate and compete differently. Each of these companies depends on their telecommunications-intensive information systems to operate and compete. Without a part of the system, they have to change what they do and how they do it. As technology changes, the organization must be aware of the changes and adapt. *SABRE* is now threatened by Delta and other competitive systems such as *ORBITZ.com*®. While American Airlines made *Micro-SABRE*® available from *America Online*® and other online sources, Delta, along with most airlines, has placed its reservation capability on the **Internet.** The Internet reservation capability has fundamentally changed the way that airline reservations are made, dramatically reducing the role of travel agencies.

Network Operations

This is the management of the physical network resources. Personnel in the group work in activation of components, lines, and controllers; rerouting of traffic when circuits fail; and execution of normal and problem-related procedures. These people, in larger organizations, tend to sit in "control rooms" and watch system monitors. While larger mainframe network monitoring and operating systems, such as IBM's *NetView*®, have intelligence in the system to do much of the control, humans are still needed to direct unusual actions. For example, IBM's *NetView* was originated as a mainframe-based capability that could manage several other mainframes and the networks attached. The operator had access to all resources on the total network. When a node failed, the system would generally be designed to intercept the message to the operator and try to execute reinitialization procedures, as contained in an expert system's lookup table. If this failed, the message was passed on to the operator for action. In either case, the event was logged for future reference. This environment encompasses large networks and possibly many mainframe computers.

---

[3] The definition of vital is "necessary for life."

One view of a network operator is a person sitting at a monitor in a control room, sending out commands to components on the network through the terminal keyboard. The other view is the LAN *network administrator* who spends most of his or her time working with users and checking on equipment in closets. Both are responsible for the physical wired network, the cards in microcomputers, modems, terminals, and anything that can stop or go bad. In reality, either of these roles can be supported remotely with telecommunications equipment. Just as a programmer can dial into the mainframe or server from home for a late-night problem resolution, the network operator or administrator may be able to log into the network remotely to take action. But unlike the programmer, the network administrator may have to personally inspect equipment and control panels and, thus, be physically present.

Network Management

Scalable network management systems are available to give network managers real-time total visibility to everything on the network, no matter how sophisticated or dynamic it is. The distributed nature of modern networks requires remote monitoring of LANs across WAN connections. In order to understand and resolve network issues before they impact, it is essential to have a real-time view of activity. Some capabilities, called *sniffers,* allow the administrator to see the level of traffic by circuit and node, balking,[4] and collisions. S/he even can inspect packets. These applications are valuable in recording and reporting for **service level agreement (SLA)** compliance. In choosing network management, it is important that the overhead does not add significant traffic so that it impacts user performance on the network. Network management software becomes vital to those organizations that require 24*7 reliability since their telecommunications infrastructure undergirds their business. One source for network management software can be found at http:// www.chevin.com.

Clarification ***Network Management and Discovery Tools***	The SolarWinds Network Management Toolset (see Figure 13.2) contains applications ranging from Configuration Management, Bandwidth, and Network Performance Monitoring to Discovery and Fault Management. The tools have been bundled into five toolsets. The SolarWinds Network Management Tools were designed by Network Engineers with the emphasis placed on ease of use, speed of discovery, and accuracy of information displayed. This can best be experienced by running the IP Network Browser Discovery application. The detailed information it returns includes: Details of each Interface, Port Speed, IP Addresses, Routes, ARP Tables, Memory, sysObjectIDs, and much more.[5]
Clarification	**OpenNMS** is an open-source[6] project dedicated to the creation of an enterprise grade network management platform. OpenNMS provides three main functional areas:

- Service Polling: the system monitors services on the network and reports on their "service level."
- Performance: data is collected from the remote systems via SNMP in order to measure the performance of the network.
- Event Management and Notifications: OpenNMS includes a robust notification system, including escalations, that can be generated by network events.[7]

---

[4] *Balking* is the refusal of service due to the node being busy. Telephone switches balk on Mother's Day, refusing service and giving a busy signal.

[5] http://www.solarwinds.net/Toolsets.htm.

[6] *Open source* denotes a product whose sources or design documents are open for use, modification, and redistribution, often open-source software (OSS). Open-source software generally denotes that the source code of computer software is open source as to study, change, and improve its design through the availability of its source code under an open-source license. http://en.wikipedia.org/wiki/Open_source.

[7] http://wiki.opennms.org/tiki-index.php.

FIGURE 13.2
Network Management and Discovery Tools

Source: http://www.solarwinds.net/.

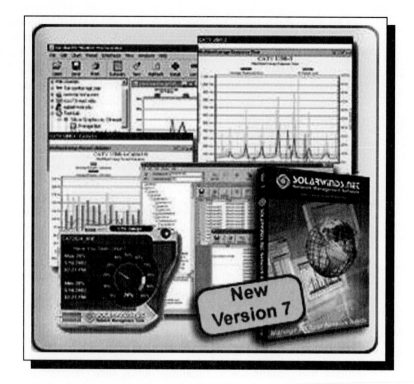

SNMP and CMIP

A problem arises when the total network contains several gatewayed networks of differing protocols. The problem is one of providing information about one network to another. To facilitate this exchange of management data among network nodes, network management protocols are essential. Two such protocols are **simple network management protocol (SNMP)** and **communications management information protocol (CMIP).**

SNMP is an application layer protocol that outlines the formal structure for communication among network devices. It is the mechanism that enables network management by defining the communications between a manager and an object (the item to be managed). SNMP is for use in an environment in which multiple management stations control the different manageable devices remotely over the network. It is composed of four components, containing details of how every piece of information regarding managed devices is represented, defines the hardware and software elements to be monitored, and contains a control console to which network monitoring and management information is reported. SNMP allows network managers to get the status of devices and to set or initialize them. It is a simple protocol with a limited command set. With limited provisions for security and lacking a strict standard base, there is some inconsistency among different vendors' implementation.

The International Standards Organization has defined the *communication management information protocol (CMIP),* which is a more complex protocol for exchanging messages among network components. Though more recently developed than SNMP for the seven-layer OSI model, it has the potential for better control and ability to overcome SNMP limitations. There is, however, no **interoperability** between SNMP and CMIP.

Network Management Information System

Network Management Information System (NMIS) is a network management system that performs multiple functions from the OSI network management functional areas. The original NMIS evolved quite rapidly to meet demands of production environments. The backend, polling engine uses SNMP to collect interface and health statistics for Cisco routers, certain Cisco catalyst switches, and generic SNMP devices every five minutes. The backend stores the statistics in round robin databases (RRDs) and ensures that devices are up, issues alerts, and so forth. The front end accesses the information stored in the RRDs and displays statistics through graphs, reports, and so on. Both the front and back ends are highly extensible and features are easy to add as the structure is learned. For example, the backend was just collecting interface statistics every poll cycle, so it was easy to add collection of health (CPU, memory, buffer, etc.) and response time availability.[8]

Service Management

Service management is a set of technologies and organization principles for monitoring the interrelationships of network components. Instead of monitoring the operation of the individual switches, routers, servers, and so forth, service management has emerged as a means to monitor interactions for ensuring that the underlying network platforms can adapt to changing requirements. The concept of service management dates from the 1980s, but at that time it was difficult to gather the configuration data and modify a broad range of network devices. Now, common databases provide a more holistic view of the network and several vendors provide various levels of service management capabilities. The impact of service management is potentially enormous. Many firms have achieved increases in productivity and user satisfaction. Most have reduced downtime, allowing network employees to be used in a more proactive, strategic role rather than just keeping the network running.

Problem (Fault) Management

This is the process of expeditiously handling a problem from its initial recognition to its satisfactory resolution. A typical scenario is when a call comes to the help desk. The person on duty listens to the problem and makes an entry in the log, noting the problem, equipment, person calling, and to whom the problem resolution was assigned. This gives a historical picture of the network and can show problem trends. If the network operator cannot clear the problem personally, the problem is further assigned to a technician, who checks with the caller for any clarification, diagnoses the equipment, makes the repair when possible, and clears the log entry.

Some would say that a major task of help desk members is to assist users in correcting configuration errors as opposed to fixing broken parts. Logging these events provides a database of institutional knowledge about the network and the user environment. Mining the data may show trends that will be present in future environments, necessitating installation changes, technical help, and a need for training for both the users and the technicians.

With many vendors supplying equipment for a telecommunications system, a problem may have to be referred to one of them for resolution. This can lead to finger pointing, as one vendor points to another's equipment as the offending apparatus. The network operations group should try to resolve the problem with the parties involved. Failing this, the problem should be escalated to a higher management level, within or outside of the organization. The first priority is to get the network operational; the second is to be sure the correct agency makes the repair or funds the resolution. Through all of this, the group must apply management to the problem environment to assure speedy resolution and **documentation** of problems and actions. Documentation is very important as it provides a record of the incident and furnishes the basis for after-action review. This review is the way for the organization to learn and the basis for refined procedures and checklists for the help desk and technicians.

[8] http://www.sins.com.au/nmis/.

Security Management

Security management[9] is the set of functions that (a) protects telecommunications networks and systems from unauthorized access by persons, acts, or influences and (b) includes many subfunctions, such as creating, deleting, and controlling security services and mechanisms; distributing security-relevant information; reporting security-relevant events; controlling the distribution of cryptographic keying material; and authorizing subscriber access, rights, and privileges.

The goal of security management[10] is to control access to network resources according to a set of policies and guidelines. The creation of an acceptably secure environment and its continuance require the design of architecture, the installation of equipment and technologies, creation of policies and procedures, and the establishment of a group of people to enforce the policies. This group will manage the firewall and IPS/IDS capabilities, ensure that backup systems work, conduct training seminars, and check to see that users are following policy in their use of the resources. Just having policies and guidelines cannot provide a secure environment—it takes a concerned effort to monitor and enforce their provisions. It is important for managers and supervisors to ensure that procedures are kept up-to-date through continuous monitoring and periodic review.

Monitoring Employees

Although the primary motive in network monitoring and measurement is to determine out-of-bounds conditions of network technology, a special case involves the actions of employees. Specifically, the level of activity and content of e-mail, instant messaging, and Web surfing are of concern. Stated simply, organizational resources belong to the company and are provided to the employees for business uses. While personal use may seem harmless, the organization has the right to **monitor employee** activities *unless* it has given the employee a reasonable expectation of privacy.[11]

Problems arise when user activities place demands on bandwidth or go where and do things that are not in the organization's best interest. For example, much e-mail is unencrypted and users may expose private information by including it in such e-mail. Visiting Web sites at lunch time may seem a simple act, but going to the wrong sites may bring back malware or subject the organization to liability or embarrassment. These visits, when done on company time, are simply wasting time, a corporate resource, and have been grounds for termination.

Clarification
Creating an Acceptable Use Policy (AUP)

It is essential that organizations create an AUP policy stating how employees are expected to use business communications. This will inform them about when (if at all) they may use business communications for personal use. If the organization intends to monitor e-mails and telephone calls (regularly or occasionally), the employees should be made aware of the fact; they should be informed that their e-mails may be reviewed. Employees should be made aware of this policy and, in turn, it should be agreed to by all employees.[12]

Performance Measurement and Tuning

Service Level

The network environment has historically been composed of one or more large mainframe computers, or servers, and their connecting circuits and equipment, supplying service to functional departments within an organization. In this environment, the question arises as

---

[9] http://en.wikipedia.org/wiki/Security_management.

[10] http://www.cisco.com/univercd/cc/td/doc/cisintwk/ito_doc/nmbasics.htm.

[11] This has been tested in U.S. federal court.

[12] http://www.bizhelp24.com/business_law/employee-monitoring.shtml.

to the service level that will be provided; that is, what will the network operations group guarantee to their customers? Suppose you are the manager of the accounting department that relies completely on mainframe-resident programs. You must have these programs, your stored data, and the connecting network. As a customer of this data services environment and in an environment of competition for your money, you would seek out a vendor[13] who would guarantee you a specified level of service at a low price. Here the term *service* means that the total system, computers and networks, is operational and reliable (PARS-compliant) during a specified percent of the time, with a stated response time. For example, you agree to a system that is available 99 percent of the time from 7:00 a.m. through 6:00 p.m., Monday through Friday, with a response time for a simple transaction of 1.5 seconds, 95 percent of the time, and for complex transactions of 4.5 seconds, 90 percent of the time. If you want a higher service level, given that was possible, you would likely pay an increasingly higher price for incremental increases. In each case, you are discussing a service level. To receive this level of service, you would negotiate a service agreement with data services.

Service Level Agreement—Quality of Service

A service comprises a set of broadly repeatable functions and processes. In the area of information technology and telecommunications, **outsourcing** has become an industry trend that has expanded geometrically over the past decade. The ubiquitous environment of the Internet and World Wide Web has fostered the connectivity of wide area networks to local area networks, with thin-client architecture evolving a pay-as-you-go business application clientele. Factors such as **total cost of ownership (TCO),** speed of deployment, focus on core competencies, scalability with flexibility, and qualified IT manpower are driving the different outsourcing model options. One constant among the IT outsourcing models, which range from application service providers (ASPs), through Internet service providers (ISPs), to virtual private networks (VPNs), is the level of services agreed upon between a client and the service provider, or *service level agreement (SLA)*. An SLA can be summarized as a series of commitments by a vendor to a customer. Unfortunately, the SLA is a very complex document to qualify, quantify, and manage.

"An SLA is, in effect, pricing. You ask for a certain commitment for a certain price," says Ellen Van Cleve, data communications director for a Fortune 500 company and vice president of the Communications Managers Association (CMA). To clarify what services are described and how those services are measured, including **quality of service (QoS),** is the challenge of an effective SLA, and SLAs cover an assortment of data services: frame relay networking, leased lines, Internet access, Web hosting and outsourcing, and specific end-user applications from analytical and vertical integration to collaborative and personal-specific.

"There is nothing special about an SLA. It's a legal contract like any other," says David Simpson, founding partner in a San Francisco law firm specializing in telecommunications contracts. As with all contracted documents, negotiations between client and service provider determine the technology agreement and verification that promises made are kept.

SLA commitments are made to convince skeptical customers to employ specific services in an atmosphere of collaboration and partnership, so what is the challenge to creating an SLA? As with any negotiated document, standards and the interpretation of standards are determined by the individual negotiators. With respect to SLAs, the concept of standardization across the IT and telecommunications industry is a desirable and attainable goal. Any

---

[13] In this case, the term *vendor* may imply an internal service organization or external organization.

negotiated SLA should address each of the following basic questions equally to form a robust contract between the service provider and client:

- What is the provider promising?
- How will the provider deliver on those promises?
- Who will measure delivery, and how?
- What happens if the provider fails to deliver as promised?
- How will the SLA change over time?

Committed Information Rate (CIR)

One should not simply lease a medium, that is, a T1 circuit, from a vendor. The object of acquiring a medium is not to simply have the medium but to have the desired functionality. Specifically, the objective is to have a given, reliable service. A **committed information rate (CIR)** is an agreement where the vendor guarantees a specific service and states penalties. For example, the vendor may run a T1 line from your server to the ISP, but you agree in the CIR to either six DS0s or a bandwidth of 384 Kbps, 22 hours per day, with the proviso that you will be allowed to exceed this bandwidth for bursty traffic on a limited basis (up to the maximum allowable rate [MAR]). Additionally, the CIR will state whether the additional packets will be discarded or whether you will pay additionally for them. As time passes, the organization may choose to rewrite the CIR to expand the bandwidth, with new guarantees.

Measurements

For the operations group to show that they or their service providers are meeting specified service agreements, or any particular environment, they should continuously or periodically measure parameters of this service. These parameters include transaction times, response time, circuit and processor utilization (busy time), queue lengths, equipment failures, circuit errors, transaction mixes, and overall service levels. With such parameters, service providers can not only demonstrate their performance but take actions to improve service by tuning present resources and by adding equipment, circuits, and computing power. For example, database administration employs the usage history of databases to migrate the most-used data closer to the disk heads and seldom-used data to offline storage.[14] The telecommunications operations group could move circuits so high-usage customers do not compete for the same resource (such as the same router or communications channel) and service requests are balanced across individual resources. As individual usage increases, resources are dedicated, such as provision of a high-speed line, distributed controller or processor, or local processor.

Configuration management (discussed below) continuously displays the present state of the network; a baseline of reports will show the present operational status of this network. This baseline, which should evolve to be kept up-to-date, will show what equipment would be required if a disaster, such as the tsunami in Micronesia or Hurricane Katrina, destroys an operations center or server room. Using data to justify the management of the cost of the data communications capability is one thing; having a baseline to recreate that capability in time of disruption or disaster is quite another.

Clarification   Many, and often expensive, tools exist to aid in network and network traffic management. For a list of such tools, refer to http://www.topology.org/comms/netmon.html.

[14] Historically, organizations have moved seldom-used data to less-expensive tape reels or cartridges. They were offline because it took intervention to load the reel or cartridge for access. Due to the decrease in the cost of hard drives, many companies are buying terabytes of hard drive storage, keeping all of their data online.

Management Reporting

Management reporting means that the operations group takes their measurements and reports to the customer and their own management the activity and success of the telecommunications capability of interest. Much of the management reporting will be by exception. That is, a report is produced only when the system or parameter goes out-of-bounds. If a circuit is expected to be operational 98 percent of the time from 6:00 a.m. through midnight and records for this month show 98.5 percent and last month, 98.6 percent, then this statistic would not need to be shown. If a historical operational statistic has been 90 percent with a 98 percent requirement, and the system makes 97 percent this month, it may be well to show this to let others know the situation is improving and closer to meeting the promised, or required, level of service. Further, if it can be shown why the former level was low and equipment has been installed or procedures instituted that have achieved the new level, it is best to let this be known.

Remember that most repetitive historical data are boring, so statistics should be reported by exception or by trend graphs. If the situation is improving, show a year's worth of data with a band of acceptability drawn in. Also, if you are reaching the limit on capability, show that along with the trend. For example, if the trend of busy time on a circuit has been increasing for 16 months and a reasonable projection shows that the circuit will be saturated or that unacceptable queues will develop within four months, use the graph of historical data to (1) alert management to this situation, (2) defend yourself if the case warrants, or (3) propose additional equipment to ward off the situation when it affects the company's competitive position or its ability to service customers.

Configuration Control

Because telecommunications capabilities change, **configuration control,** accurate knowledge of what equipment is installed and its location, is very important. The maintenance of records that track all equipment, media, and circuits is an administrative task that requires less technical skill but no less dedication. This task often falls by the wayside as more important tasks take precedence and as time passes. Take, for example, a college campus that has evolved over two centuries, such as the University of Georgia. When they began the planning for a new AT&T System 85 switch, it became more practical to lay all-new wiring than to depend on records documenting existing wiring. Given that some of the wiring on campus was over four decades old, some circuits were lost due to the age and condition of documentation. At the end of the $2 million contract for excavation, new conduits, new wiring and fiber circuits, and a complete inventory and wiring diagram database existed. This is now considered a valuable resource and is something every organization should have and keep up-to-date. A portfolio of all information technology systems, capabilities, and projects should be maintained and be available to the **chief information officer (CIO).**

Configuration management gives data to use in cost allocation and information for expansion and improvements. Documentation must include not only a database of equipment, but an inventory of circuits, channels, and bandwidth,[15] showing all equipment and points of high risk that may require special attention or redundant resources. The point of configuration management is to support design, evolution, repair, and maintenance.

Implication Network configurations are generally drawn on a two-dimensional diagram in the form of wiring diagrams. Considering the effects of flooding, perhaps a three-dimensional diagram showing physical placement of equipment is now in order. It is one thing to say a server or back-up generator is at risk for flooding; it is another to show the elevations and sources of the threat.

---

[15] *Bandwidth* should be considered a resource, asset, just as a circuit, channel, or node. The consequences are the same if any of these are not in sufficient supply.

Contingency, Expansion, and Growth

A primary reason for configuration management is to plan for normal growth and expansion and to provide contingency resources needed during unplanned events. Expansion and growth can be forecasted, based on present use, trends, and known plans. No network project should be implemented with *just enough* assets to satisfy the intent of that project; there must be a determined look into the future. While financial managers will want to hold the cost of a project to the provision of the system and the intent stated, when developing a network project plan is the time for a review of future plans so adequate growth potential can be included. It is far easier and cheaper to include growth projections in a project being installed than to retrofit it in later.

As noted in an earlier chapter, it is important to conduct continuing risk assessment, resulting in an updated continuity (or disaster) plan. Part of this planning is the inclusion of resources, as shown in the configuration database, to be used in the case of the contingency occurring. These contingency network assets may not be owned by the organization, but they must be available quickly to respond to the unforeseen event and support business continuity.

Change Management

When projects are implemented to add to or change existing configurations, it is vital to support change management. **Change management** provides monitoring of all changes to the installed network in a planned, coordinated way. When the environment is highly dynamic and users are requesting changes rapidly, lack of coordination of changes can lead to disaster. Change management includes a process for approving requests and documenting them to always know the installed environment. The person in the position of change coordinator needs to understand the nature of this technology. The end point of a change its the inclusion in the configuration file.

Capacity Management and Planning

Network administrators must know their installed base of resources and their capacity. This means documenting what is presently being used and what growth capability exists. The task involves forecasting nodes, circuits, channels, and bandwidth to accommodate present needs and future requirements. This planning is continuous and the administrator must have knowledge of operational, management, and strategic plans in order to support them.

The planners should gather data on current capacity and usage for nodes, servers, and media, and forecast the impact on computing and bandwidth in light of business growth and anticipated new applications. There is a need to model the solutions at the required performance level. The solutions next need verification at the forecasted levels.

Unfortunately, there is no tool or suite of tools that does the entire **capacity management** task. Some firms, for example, *American Airlines,* have created labs to simulate nodes, servers, and bandwidth on the network to create an environment nearly identical to the real one so modeling and volume testing can be accurate. Some network management tools, such as *Distributed Sniffer System*® from Network General Corporation, do provide net traffic statistics. Ameritech Network Services (ANS) (now SBC) used this tool to monitor LAN and WAN traffic in a five-state area. Other vendors offer tools as well. *TrendTrack*®, *LANAlert*®, and *Spectrum*® are examples.

The *Institute for Computer Capacity Management* has provided three elements needed for capacity planning: workload characterization, forecasting, and performance prediction.

Workload Characterization

- Obtain user views of the business model and an idea of the number of users and types of transactions currently and likely for the future.
- Collect data daily; do trends weekly, monthly, quarterly, and annually to reveal monthly and seasonal cycles.
- Create a baseline of the current workload for forecasting.
- Categorize nodes by type based on equipment and function to obtain information on the number of servers and their locations. Identify the feeder and major switching centers.

Forecasting (Estimating Future Workload)

- Determine the desired end result capability.
- Estimate the growth rate of the system. Don't underestimate! Be sure to reflect new users, increased use by current users, and additions, especially at the client level. FileNet always forecasts 25 to 45 percent more growth than their customers estimate.
- Determine and factor in applications that are planned or are under development.

Performance Prediction (Recommend a Configuration)

- Build a model.
- Size-test workloads to peak periods rather than average or sustained demand.
- Document all assumptions and calculations and double-check.
- Fit calculations into a long-range plan rather than being content with the immediate capability additions.
- After installation, evaluate actual versus predicted performance and adjust both the network and the capacity planning system.
- Perform capacity planning at least at annual intervals.

If all three of the above elements are incorporated into capacity planning, it is likely that network managers will avoid rude surprises. An examination of the details involved in each element reveals that the activity is nontrivial but of significant importance to the organization's operation.

Risk Assessment and Management

In the use of any technology, process, or methodology, someone should determine where things are likely to go bad. Managers must think about objectives, the system and procedures they have installed to achieve the objectives, and the weak points in the equipment, staffing, and procedures.

Risk management is the science and art of recognizing the existence of threats, determining their consequences to resources, and applying modifying factors in a cost-effective manner to keep adverse consequences within bounds. **Risk assessment** and **analysis** involve a methodological investigation of the organization's resources, personnel, procedures, and objectives to determine points of weakness. Finding such points, managers overtly manage the risk by passing it to someone else (insurance or outsourcing the task) or strengthening the weak points by making changes or building redundancies.

Risk management is the analysis and subsequent actions taken to ensure that organizations can continue to operate under any foreseeable conditions, such as illness, wars, accidents, labor strikes, hurricanes, earthquakes, fires, power outages, heavy rains, mud slides, oppressive heat, or flu epidemics. Telecommunications capabilities support all facets of the company; therefore, risk analysis and management are always in order.

Disaster Planning

The beginning of risk management is assessment, which leads to management on a continuous basis. A specific point is the creation of a disaster plan in case of a catastrophic occurrence. Because organizations such as banks, stock exchanges, and airlines depend on telecommunications-intensive computer-based information systems, what do they do in case of a total power outage, telephone line cuts, and fires? A disaster plan requires procedures that occur every day to allow recovery after a disaster and permit the organization to continue operations. For a bank, the daily operations involve placing backup copies of data and programs in an absolutely safe place. The disaster plan provides the place, procedures, and equipment to make use of those data for continued operations. Disaster plans are often referred to as "business continuity plans" for good reason. (See Case 13-2 for examples of **disaster planning.**)

We mentioned the need for a baseline of resources and capabilities under *measurements*. The disaster plan is the procedure that calls for this baseline report in order to have business continuity.

Implication

The world recoiled at the destruction by the earthquake and subsequent tsunami on December 26, 2004. There are reports of banks in the path of the flood that kept their backup copies of client data onsite. The flood destroyed both the original and backup copies of data, leaving clients and the banks alike unable to determine proof of assets.

Redundancy to Mitigate Risk

The first consideration to recover from a disaster is to put the disaster planning document procedures into action. A way to avoid this situation is to have redundancy in place. This would mean at least duplicate channels for data flow, taking different physical paths. If these are provided by suppliers, having different suppliers provide each, in addition to having them travel different paths, gives an additional level of redundancy. This situation would entail redundant routers and switches. For physical assets such as servers, it would mean having more than you need, to the point that the failure of one would not halt operations. With redundant switches and routers, an added level of security can be achieved by using redundant parts within the equipment, such as dual power supplies. This is like using RAID to have redundancy in a server, which would be strengthened by dual power supplies and disk controller cards.

Implication

"We had a situation when using RAID 5 on a NAS system that two hard drives failed within days of each other and before the first failed drive was replaced. One of the failures was not real and when we forced the drive back online with the other drive replaced, things worked, but you can believe that we worked harder to get our new multitape backup system working earlier than we had planned after that."[16]

Backup of Files

Backing up files on a regular basis is a routine way to mitigate risk. Placing them at a site remote from the central location increases the security this backup provides. Data communications provides a cost-effective way to back up mission-critical data to a site, anywhere in the world. With such a system, a new computer could be operational very quickly, at a site remote from the central location, in case of disaster. That is, an application service provider could quickly provide the processing power and the data could be downloaded to local servers to begin operations quickly.

---

[16] Unidentified reviewer of this text.

Case 13-1

Network Management Basics

Please view this case online at http://www.cisco.com/univercd/cc/td/doc/cisintwk/ito_doc/nmbasics.htm.

Source: All contents are Copyright © 1992–2002 Cisco Systems, Inc. All rights reserved.

Case 13-2

Disaster Recovery

Denise Johnson McManus

A disaster recovery plan is a series of procedures to restore normal data processing following a disaster, with maximum speed and minimal impact on operations. A comprehensive plan will include essential information and materials for necessary emergency action. Planned procedures are designed to eliminate unnecessary decision making immediately following the disaster. Disaster recovery planning begins with preventative measures and tests to detect any problem that might lead to a disaster. If this planning process is completed, the chance of experiencing a total disaster is lessened. The severity of a disaster determines the level of recovery measures. [1] Disaster classifications are helpful in organizing procedures for a disaster plan. There are nine essential and required steps for a successful implementation of disaster recovery planning, which are displayed in Table 13.4.

The key to beginning a successful disaster recovery plan is to gain a commitment from top-level management and the organization. To obtain the required support, the CEO and top managers need to understand the business risk and personal liability if a disaster recovery plan is not developed and a disaster occurs. Although many companies have excuses for not developing a plan, a corporate policy should be mandated requiring disaster recovery planning. The corporate policy would assist in defining the charter for contingency planning, while encouraging cooperation with internal and external staff. Unfortunately, very few organizations have a corporate policy on disaster recovery. Therefore, it is imperative that corporations begin developing a disaster recovery plan immediately.

FINANCIAL RISK

Furthermore, statistics indicate that if a company's computers are down for more than five working days, 90 percent will be out of business in a year; however, this can be avoided if a coherent disaster recovery plan is developed and implemented. [5, p. 65] "The disaster recovery process generally is much longer than the duration of the disaster itself." [2, p. 27]

TABLE 13.4
Disaster Recovery
Planning Process

1. Obtain top management commitment.
2. Establish a planning committee.
3. Perform risk assessment and impact analysis.
4. Prioritize recovery needs.
5. Select a recovery plan.
6. Select a vendor and develop agreements.
7. Develop and implement the plan.
8. Test the plan.
9. Continue to test and evaluate the plan.

TABLE 13.5
Budgeting a Disaster Recovery Plan

Daily	Normal No Plan	Disaster No Plan	Normal with Plan	Disaster with Plan
Revenue	$100,000	$0	$100,000	$100,000
Expenses	−80,000	−80,000	−80,164	−80,164*
Outages/expenses	0	−10,000	0	−3,000
Contingency plan	0	0	0	−5,000
Profits	$20,000	$−90,000	$19,836	$11,836

* Calculation based on $60K/365 days = $164/day.

The company experiences immediate problems from the disaster and continues to experience difficulties for several months. The inability to communicate with customers and suppliers is devastating, which can prevent the company from staying in business. Therefore, an effective disaster recovery plan directly affects the bottom line, staying in business.

Disaster recovery planning costs are feasible and can be budgeted. Not only can they be allocated across many business units, but they also can be amortized over many years. Many costs must be considered when developing the disaster recovery plan. Not only the time invested by the team members, but also implementation costs must be considered when developing the budget. Table 13.5 displays a hypothetical mid-sized business model example.

The initial cost of contingency planning would include startup costs and development of the plan. After the initial development has been completed, the yearly cost would be $60,000. [4, p. 44] Although $60,000 seems expensive, Table 13.5 reveals that one day of lost business equals $90,000. Thus, the cost of the contingency plan can be recovered quickly, if a disaster should occur.

Costs are a major concern for disaster recovery plans. Some of the costs incurred for disaster recovery plans include costs of insurance; fees for hot-site backup, stockpiled equipment, supplies, forms, redundant facilities, or cold sites; communications networks for recovery purposes; testing; training; and education. These costs are often used as the excuse, for top-level management, not to develop and implement a plan. Budgetary constraints are one of the main obstacles in disaster recovery planning. Not only will the company incur monetary losses, but also possible loss of customer confidence, should the business be affected by a disaster.

Since most organizations are very dependent on computer systems to support vital business functions, such as customer support, the need for a disaster recovery plan is critical. Financial and functional losses increase rapidly after the onset of an outage. Corrective action must be initiated quickly, and disaster recovery methods should be functioning by the end of the first week of an outage. Loss of revenues and additional costs rise rapidly and become substantial as the outage continues. The financial costs vary between industries, but all industries show progressive increase as the length of outage continues. Furthermore, if the financial impact to the business does not warrant the financial support of the corporate executives, an analysis of The Foreign Corrupt Practices Act of 1977 should get the required attention and support of the officers. The act deals with the fiduciary responsibilities, or "standard of care," of the officers, which may be judged legally. "In the legal publication 'Corpus Juris Secundum' the 'Standard of care' is defined as follows: 'A director or officer is liable for the loss of corporate assets through his negligence, fraud, or abuse of trust.'" [4, p. 43]

However, the most convincing reason for having a business disaster recovery plan is that it simply makes good business sense to have a company protected from a major disaster.

TABLE 13.6
Disaster Recovery Team

1. Top management.
2. Functional and operations managers.
3. Service providers.
4. Recovery team.
5. Disaster recovery coordinator.
6. Outside vendors.

Additional reasons to have a recovery plan include a potential for greater profits and reduced liabilities to the company and the employees. Thus, a formal or informal risk review provides a powerful argument for recovery planning. It tells you where your needs are not met and helps you determine the critical areas that must have backups. Recovery from a major disaster will be expensive. However, the inability to recover quickly and support primary business functions would be significantly more costly and destructive to the company.

PLANNING

The process of developing a recovery plan involves management and staff members, as displayed in Table 13.6. Each member of the disaster recovery team has a specific role that is defined in the plan. Disaster recovery planning is a complex process. The corporation must utilize a structured approach in determining the scope, collecting the data, performing analysis, developing assumptions, determining recovery tasks, and calculating milestones. The issues displayed in Table 13.7 must be considered during the planning process. This highly interactive process requires information from throughout the organization. The plan requires continuous revision. It is out of date whenever a change occurs in the organization, the software, or the equipment.

The process of building a plan is extremely valuable to the company. The purpose of identifying problems and developing a recovery process forces the organization to examine the impact of a disaster on the company and the business. Thus, the end result should be a plan that can be utilized for all levels of disasters. Recovery from a major disaster requires the efficient execution of numerous small plans that comprise the master plan. These subplans include acquiring hardware, reinstalling communication lines, and many other functions. Recovery managers select the plan, assign responsibility, and coordinate resources to execute the plan.

CONCLUSION

In a society where individuals are linked to each other through media technology, many individuals have experienced or witnessed the vast devastation of a disaster. Many disasters that have occurred in the United States in recent years have driven many companies to recognize the importance of disaster recovery planning. A disaster recovery plan appears to be a cost-effective, but underutilized, tool. Organizations that have prepared for an extended outage through insurance and a contingency plan reported significantly lower expected loss of revenues, additional costs, and loss of capabilities. In the last six years, a disaster has been reported somewhere in the United States and the world approximately every year.

TABLE 13.7
Recovery Planning Issues

1. Unanticipated interruption of routine operations.
2. Identification of key risks and the exposure to risk.
3. Identification of consequences if existing plan fails.
4. Identification of recovery strategy.
5. Identification of test and evaluation process.

The size of the disaster is not the determining factor of staying in business; it is the disaster recovery plan that will determine if the doors will stay open or closed. "Smart companies make it their business to have a disaster recovery plan in place." [2, p. 32] "If a disaster does strike, being prepared can make the difference between a smooth recovery and a slow terrifying struggle to survive." [2, p. 32]

BIBLIOGRAPHY

1. Collins, Mike. "Motorola ISG Disaster Recovery Plan for Huntsville Facility." *Policy and Procedures,* February 28, 1995. (Typewritten.)
2. Howley, Peter A. "Disaster Preparedness Is Key to Any Telecommunications Plan." *Disaster Recovery Journal* 7 (April/May/June 1994), pp. 26–32.
3. Lewis, Steven. "Disaster Recovery Planning: Suggestions to Top Management and Information Systems Managers." *Journal of Systems Management* 45 (May 1994), pp. 28–33.
4. Powell, Jeanne D. "Justifying Contingency Plans." *Disaster Recovery Journal* 8 (October/November/December 1995), pp. 41–44.
5. Preston, Kathryn. "Disaster Recovery Planning." *Industrial Distribution* 83 (December 1994), p. 65.

Summary

Every organization has to consider the requirement for management of its telecommunications assets. Even a small firm has to exert management and control of its network. As organizations grow, their dependence on telecommunications becomes more pronounced. Consequently, the various facets of telecommunications management are increasingly essential. Every organization needs to understand the nature of their telecommunications in terms of the infrastructure. This means the design and implementation of all components should be known and integrated to serve the organization. Once the infrastructure is designed and implemented, there are several ongoing activities that are required in order to ensure that the networks are properly operated, administered, monitored, and managed. All of this means that there must be procedures, standards, risk assessment and management, configuration control, management of change, and every other element needed for assuring that network operations are sustainable in the organization. Active management of all of the network resources is required if the telecommunications infrastructure is to support the organization's purposes.

Key Terms

bandwidth, *p. 385*
capacity, *p. 386*
capacity management, *p. 400*
change management, *p. 400*
chief information officer (CIO), *p. 399*
committed information rate (CIR), *p. 398*
communications management information protocol (CMIP), *p. 394*

complexity, *p. 384*
configuration control, *p. 399*
controlling, *p. 385*
coordinating, *p. 389*
data communications, *p. 383*
directing, *p. 385*
disaster planning, *p. 402*
disaster recovery, *p. 385*
documentation, *p. 395*
Internet, *p. 392*
interoperability, *p. 394*

maintenance, *p. 384*
mission, *p. 390*
monitoring employees, *p. 396*
network design, *p. 387*
network management, *p. 390*
organizing, *p. 384*
outsourcing, *p. 397*
planning, *p. 383*
private branch exchange (PBX), *p. 385*

quality of service
(QoS), *p. 397*
real-time, *p. 391*
risk analysis, *p. 401*
risk assessment, *p. 401*

risk management, *p. 401*
service level agreement
(SLA), *p. 393*
simple network management
protocol (SNMP), *p. 394*

simulation, *p. 386*
staffing, *p. 384*
total cost of ownership
(TCO), *p. 397*

Management Critical Thinking

M13.1 What should be the major components included in a firm's telecommunications management procedures?

M13.2 What sort of managerial/network operations issues arise from such legislation as the Sarbanes-Oxley Act and HIPAA?

M13.3 Are there priorities for network management other than the two listed at the beginning of the chapter?

Technology Critical Thinking

T13.1 How can a firm ensure that it builds redundancy into its network so that failure of multiple nodes does not cause failure of the network? What are the alternatives that should be evaluated?

T13.2 What has been learned about network security and continuity from Hurricane Katrina?

Discussion Questions

13.1 What are some significant factors for consideration in backup for critical communications capabilities?

13.2 What should the management duties of a network administrator include?

13.3 How often should SLAs be reviewed and/or renegotiated?

13.4 What are the benefits of employing network diagramming tools?

13.5 Name the different areas of network management.

13.6 What are the goals of performance management?

13.7 What are the pros and cons of outsourcing the management of your networking capabilities?

13.8 What are the goals of configuration management?

13.9 What are the goals of security management?

13.10 What is the impact of the MTBF of components of a system on the reliability of the system?

13.11 What are the goals of fault management?

Projects

13.1 Look up the topic of information technology risk analysis and management and find a model. Discuss this model in class.

13.2 Go to your community's financial institutions and determine what sort of capacity planning is employed. Determine any regulations or laws that require continuity planning for these institutions. Do you consider the plans adequate?

13.3 Interview a company and determine its network management system. What reports are generated?

13.4 Interview a company and determine its security management plan. Is it tested?

13.5 Interview a company and study its configuration and capacity plans. Are they strategic or just operational?

13.6 Divide the class into groups, each group finding an organization with a telecommunications group. What are the qualifications of its members?

13.7 Interview the manager of a data communications group and determine the most important job qualifications of the various individuals of the group. How is the organization meeting its rapidly changing staffing needs, through internal staffing, external consulting, or "growing their own"?

13.8 Draw a network diagram to show the locations/connections in an organization, such as your school, college, or local firm. Use netViz or a similar program in this project.

13.9 Interview a network manager and determine how s/he has documented the networks of the organization. Determine the degree of the existing documentation.

13.10 Interview local business managers responsible for controlling the data communications resources of their organization and determine the major issues. Why are they considered issues?

13.11 Determine if your school has documented its telecommunications assets. If so, is the documentation accurate?

13.12 Choose a topic from this chapter and research it on the Internet. Using the search engine of your choice, find additional material on the subject. See if you can find material from the major suppliers of data communications equipment, such as Cisco, 3Com, Nortel, IBM, and so on.

13.13 Investigate and classify the major hardware appliances and software that a network administrator might employ to fight spam and viruses.

Chapter Fourteen

Network and Project Management

FIGURE 14.1 **Mission, Operating Philosophy, and Vision**

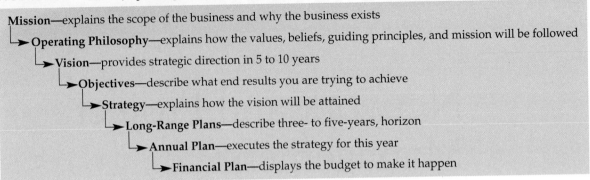

Mission—explains the scope of the business and why the business exists
→ **Operating Philosophy**—explains how the values, beliefs, guiding principles, and mission will be followed
 → **Vision**—provides strategic direction in 5 to 10 years
  → **Objectives**—describe what end results you are trying to achieve
   → **Strategy**—explains how the vision will be attained
    → **Long-Range Plans**—describe three- to five-years, horizon
     → **Annual Plan**—executes the strategy for this year
      → **Financial Plan**—displays the budget to make it happen

In order to ensure that proper data networks are built and operated, there is a great deal of effort required. These efforts start with organizational objectives, strategies, and architecture. We provide an outline of telecommunications systems analysis and design methodology to assist in developing, implementing, and operating data networks. The need for qualified personnel with the correct skill profile leads to a review of several professional certification programs.

In this chapter, we discuss the management of telecommunications projects and the certification of individuals who will administer the resultant systems. The systems development life cycle (SDLC) of a telecommunications project is concerned with the stages involved, and the activities and deliverables of each stage. Use of the SDLC approach is borrowed from the development of computer-based systems as telecommunications design and implementation follow much the same course of definition, design, and discovery.

The concern for certification of individuals in the telecommunications area is of interest because this is one way for these people to have their skills tested by recognized agencies. This may be the first step in the career of network administration and security or the demonstration of skills obtained by experience during the career. Certification can be seen as a surrogate for experience or the acknowledgment of achievement by colleagues in the field.

The management of telecommunications, like the management of an organization, starts with a vision and mission, is activated with implementation, continues with operation and maintenance, and ends with disposal. The vision must come from upper management, as it defines its business, customers, and ways of competing. From vision must come a set of objectives that bring vision to fruition. The objectives, like vision, come from the strategic and executive level of management. After objectives, a strategy is developed to make the objectives operational; then comes planning for the execution of operational actions. (See Figure 14.1.)

Vision

Vision translates the **mission** of the company and its objectives into a clear picture of simple principles. Vision is a picture of the future, an easily understood statement about a practical and desirable, if not fully predictable, goal. It should include an indication of values, by answering the question, "what's important around here?"[1]

---

[1] Peter G. W. Keen, *Shaping the Future* (Cambridge, MA: Harvard Business School Press), p. 19.

FIGURE 14.2
**TC (Telecommuni-
cations) Architecture
Must Be Congruent
with and Provide
Support to
Organizational
Strategy**

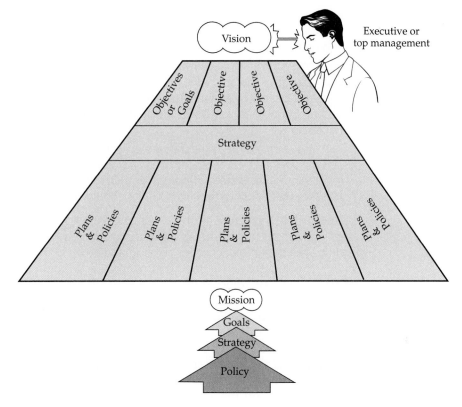

Objectives

From objectives will come an idea of the major function of the organization, who its customers are, and what activities it will be involved in. These activities will affect the functioning of the MIS and telecommunications activities.

Strategy

Strategy means large-scale, future-oriented plans for interacting in a competitive environment to optimize achievement and organizational objectives. Strategy is the organization's game plan that reflects the company's awareness of how to compete, against whom, when, where, and for what.

Architecture

The **architecture** reflects the technical strategy. It is roughly like a city plan, a mixture of fixed main routes, zoning regulations and ordinances, and procedures for extending and modifying existing buildings and expanding or adding roads. It is a model of the desired end result that is used to guide the efforts that should achieve the end result (see Figure 14.2). Architecture should be defined during the planning stages and refined throughout the life cycle of the system. According to James Martin,[2] there are three "enterprise architectures": information systems architecture, business systems architecture, and technical architecture. Each of these is described below:

    Information systems architecture is the aggregation of data, functions, and their interaction that represents business requirements. This architecture is

[2] IE-EXPERT, Version 4.0.

represented by data models such as entity-relationship diagrams and by functional decomposition diagrams accompanied by definitions, matrices, and detailed process definitions and diagrams.

Business systems architecture is the aggregation of individual business applications and descriptive characteristics of business systems. This architecture should show dependencies between business systems and define the category of each business system. Examples of categories are strategic planning, monitoring, **control,** and transactions.

Technical architecture is the defined collection of technologies and components. This includes high-level descriptions and definitions of the required technological infrastructure for systems development. Usage information, policies, and statistics that apply to the technologies should be included.

Network architecture[3] is the set of conventions or standards that ensure all the other components are interrelated and can work together. This technical architecture is the overall design blueprint for creating and evolving the network over changes in time, technology, uses, volumes, and geographic locations. The architecture has to reflect the business vision. There are a number of often-conflicting criteria for selecting the standards, scale, and type of technology and the technical policies for the selection of vendors and equipment.

Network architecture is a primary part of any computer-based system because of distributed departments, processing, and data storage. This portion of the total information systems architecture specifies structure descriptions, relationships of network senders and receivers, logical paths, protocols, and communication functions that manage and support the transfer of information. Previously, an information systems developer might consider the transfer of computer digital data, but today the consideration of networks must include voice, images, text, media, and data.

Policy

A **policy** is the set of mandates and directives from the top of the firm summarized by the ground rule, "This is how we do things around here." It addresses questions of authority and accountability. Policy is not the same as planning. It sets the criteria for planning and establishes the ground rules for it. Both are needed to define the architecture that provides the framework for using the technology.

Planning

Planning has long been cited as a primary function of top managers. In reality, **planning** can be thought of as existing on various levels: at the top level where overall company planning is done; at the middle level where intermediate plans, programs, and budgets are made for the company; and at the project level where specific projects, budgets, schedules, and technical performance targets are planned. The evolution might be viewed as moving from plans directly concerned with organizational goals, through functional planning areas, to the specific work package level. Thus, planning levels move from the general to the specific and from long to intermediate time horizons.

CEOs are challenging their staffs to create 10-year telecommunications plans. Staffs are not successful in putting together 5-year plans, much less 10-year plans, that work financially and strategically. This is extremely difficult due to the dynamic nature of the industry/market and the explosion of broadband technologies. The

[3] Keen, *Shaping the Future,* p. 19.

industry/market is accelerating as opposed to stabilizing. Video and bandwidth requirements drive the change. You have to balance short-term and long-term goals.

Planning for telecommunications may seem an impossible task because of the rate of change in technology and the uncertainty surrounding the move to Net-based business. As early as 2001,[4] it was noted that ". . . any company hoping to survive the tsunami of innovation and change that the Web threatens to unleash will need more than a digital business plan." He sees a need for "habitual and radical innovation."

For telecommunications, long-range planning involves architecture. Mid-level planning determines components that fit the architecture. At the lower levels, planning details the configuration, features, procedures, and training for specific operations.

Implementation

Implementation is the action phase of the planning process. Often implementation involves adaptation to changes that have occurred after the planning decision was made; consequently, the entire planning cycle may be considered dynamic and iterative to the extent that the environment dictates. One can easily visualize the dynamic nature of the process by considering this example. A great deal of strategic-, intermediate-, and operational-level planning had been done before implementation on D-Day when the Allied forces invaded Europe near the end of World War II. Because of the dynamic and volatile environment, military commanders at the implementation levels had to make many adaptive decisions. Similarly, telecommunications managers must be prepared to make adaptive decisions in the dynamic telecommunications environment of today.

Security

As organizations progress through the facets listed above, **security** should remain at the forefront of their thinking. Security is not added on; it is designed into any system or structure, including organizational structures. It is only when the managers and designers consider the threats, resources, and consequences that they can incorporate modifying factors as part of the overall design. (Refer back to Figure 11.1 in Chapter 11.)

TELECOMMUNICATIONS DESIGN AND IMPLEMENTATION

The one thing one can depend on in life and in business is change. You can expect telecommunications and computer vendors to change equipment capabilities, users to change their needs and expectations, and the environment to change in terms of the complexion of competition. With change comes the need to enhance present facilities and install new capabilities. A simple way of looking at this from a telecommunications standpoint is that it often means providing more bandwidth between points or to new points. Of course, new standards and advances in hardware can spur the need to enhance telecommunications facilities and capabilities. If there are no digital channels, firms will be installing or enhancing the ability to transmit digital data on analog circuits, or they must create digital channels. If they have digital data circuits, they will increase their transfer ability (bandwidth) with faster circuits or more circuits, or by multiplexing existing circuits. In some cases, wireless networks will be built. All of the changes will require network **analysis and design.**

Telecommunications networks that are put together in a piecemeal fashion are rarely adequate. The process of determining needs, analyzing **alternatives,** delineating system specifications, and designing the system that will properly serve the organization is usually referred to as systems analysis and design. Through the past 2+ decades, the principles of

[4] Gary Hamel, *Fortune,* February 5, 2001.

analysis and design have been refined in the realm of computer-based systems so that there are several standardized approaches and methodologies. Network analysis and design also have been subject to some refinement, however, not nearly to the extent that computer-based or **management information systems (MIS)** have. Because of the maturity of the systems and analysis process and **methodology** of MIS, we will use that system's development life cycle as a reference for our investigation of network analysis and design.

Network Analysis and Design

This is the process of understanding the requirements for a communications network, investigating alternative ways for implementing the network, and selecting the most appropriate alternative to provide the required capacity. Because it is so very important to understand the business problem or opportunity for which the enhanced or new network capability is being installed, much of this chapter discusses the process of analysis, design, and implementation of such capabilities. At this point, it is important to understand that the better you define the problem[5] and describe its solution, the easier your life will be in the future as you operate, maintain, and change the capability under consideration.

Telecommunications systems contain all components and capabilities needed to move information or data to serve organizational needs. The components include hardware, software, procedures, data, and (most importantly) people. The system connects the components to the telecommunications network, with each geographic location containing one or more nodes.

Network Implementation

Implementation is the process of installing and making the network operational. After you have determined what to do and how to do it, this is the "then do it" stage. In many respects, implementation is the easiest part. With proper planning, all you have to do is put the components where they go, test them, and turn them on. We will talk more about this later.

Sources of Requests for Projects

There are several reasons for projects, which is to say several sources of change. For example, one source of recommendations for change is the telecommunications group itself, specifically the network operations staff. These people, as they work with the network and handle problems daily, see (1) better and more cost-effective ways of doing the present jobs, (2) present capabilities being taxed to their limits, and (3) worn-out equipment being called upon to perform too long.

If the organization is moving data and digitized voice from site-to-site and the data must wait as the pipeline[6] is filled during normal work (business) hours, you can readily state the implications of these delays and recommend improvements. Possibly your organization requires greater bandwidth on existing circuits, new and redundant circuits, or fiber instead of microwave or satellite channels. Being the closest to the network, the network operations people have the greatest knowledge of what is needed, based on the present requirements. Their major limit often is restricted knowledge of what new requirements are desired from a business needs perspective. Thus, the telecommunications group should be kept aware that they should recommend enhancements and additions to make sure that present service is adequate, efficient, cost-effective, and reliable. Functional areas that presently use the

---

[5] Most of the time, we assume the solution to a problem. In business, taking advantage of an opportunity may be of greater value and uses the same process as solving a problem. We will use problem definition and solution also to include defining and creating a solution to take advantage of an opportunity.

[6] The term *pipeline* is often used to describe a data channel. The larger a water or oil pipeline, the more fluid can flow. In like manner, the wider the bandwidth (higher the speed) of a data circuit, the more data can flow. Thus, the digital channel is a pipeline for data. Because we can visualize physical pipelines and most data circuits look the same, the data pipeline concept is quite useful.

telecommunications capabilities, the user groups, can offer opportunities for change. As the production managers see the value of communicating with their suppliers to ensure that inventory is kept at a minimum and arrives when required, they may see how they need similar channels of communications with their customers. The finance department may want to move to electronic funds transfer (EFT), allowing more efficient handling of funds, at a lower cost, by moving monies through a network as opposed to checks through the mail. As sales and marketing departments provide laptop computers to their representatives, these users will require support for the equipment and the wired and wireless channels to support them. This may mean more **inward WATS** (area codes 800, 877, 888, etc.) lines, or it may mean a dedicated digital network. *The people who are the closest to the operation of the company know the most about their operations and processes.* As they become educated in communications and telecommunications, they are in the best position to request capabilities and enhancements to improve their business processes. However, they must understand the business needs or logic as a starting point.

Because senior managers are chartered with working at the strategic level, looking into the future, they should request information about possibilities that can address future problems and opportunities. Even if they are in the functional areas mentioned above, they will have a different view of the nature of business than their subordinates who deal with day-to-day operational problems. These managers consider long-term and far-reaching concerns. Senior managers should look at

- Who will our customers and competitors be next year, and the next?
- Who should our customers be?
- How should we operate?
- How should we compete?
- What should we be doing?

The last group of people to make demands that require new facilities or recommend new capabilities are customers, vendors, suppliers, and governmental agencies. This first cluster, customers, deserves and demands a lot of your company's attention. Firms that listen to customers and can fulfill requirements using telecommunications technology create partnerships and exclude possible competitors. Because telecommunications has the ability to define competition and customer service, telecommunications managers should expect comments and suggestions from customers that will cause change and require new or enhanced capabilities.

Technology vendors will persist in showing you their latest products. If, however, your firm has formed partnerships with them, you will be working as allies. As allies or partners, vendors will help develop the products that will assist you. Increasingly, firms forge partnerships with suppliers. This often entails sharing access to databases, ERP systems, and so forth.

Finally, governmental agencies create demands that may need to be met with technology. One of the latest cases is electronic filing of tax data. Be aware that when you work with governmental agencies, even though they may seem to be demanding expenditures from you unfairly, you may find them to be new customers. Meeting government demands may show you different and better ways to do the jobs you are doing.

A METHODOLOGY FOR DESIGNING, DEVELOPING, AND IMPLEMENTING TELECOMMUNICATIONS

Capabilities

The process of design and development of enhanced or new capabilities follows a formal development process. The process includes the members of technical services interacting with requesters and users, as well as a significant amount of work that does not involve them. One

FIGURE 14.3 **Process Flow, Sources, and Process Outputs of the TC (Telecommunications) SDLC**

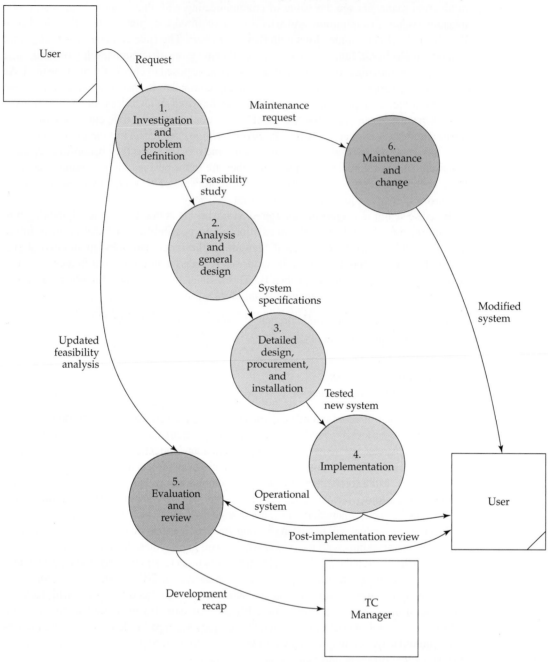

process of creating and maintaining a **telecommunications system (TC)** is based on the MIS life cycle approach, called the **systems development life cycle (SDLC).** As shown in Figures 14.3 and 14.4, the SDLC is composed of the major **phases** of problem definition: feasibility, systems analysis, design, procurement, implementation, test evaluation and review, and **maintenance** and change. This approach represents a systems view to solve problems.

In recent years, systems analysts have been concerned with the topic of network modeling. The move to **client-server architecture (C/SA)** has been a major reason for this concern. Some graphical tools have evolved to aid in the design task. We will examine network modeling briefly in the following section.

FIGURE 14.4
Time/Effort Duration
of SDLC Phases

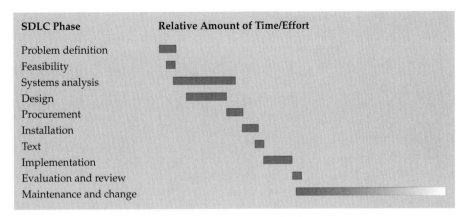

Network Modeling

Network modeling has been defined as a diagrammatic technique that is employed to document the system in terms of the users, data, and processing locations. Frequently, it is useful to focus on logical or essential network modeling, that is, the modeling of the network requirements independent of the actual implementation. In the process of design, these models will be converted to implementation models that specify the particular system.

Using capabilities such as the netViz(r) software package, several diagrams of the telecommunications system can be displayed to help the designers and managers get a picture of the system. Several presentations are employed to assist in the process.

Clarification

The following makes reference to JEI (Johnson Enterprises, Inc.), a multilocation company specializing in security and lighting capabilities. JEI is much like most multidivision and multilocation organizations that require data communications among the corporate parts.

In Figure 14.5 created with netViz, we see JEI's overall logical network between its major sites. Next is a geographic view of the network from the owner's perspective (Figure 14.6). This gives a clear picture of the overall connectivity. The system designer's view (Figure 14.7) is more detailed and provides basic information needed to solve the basic technological connections.

Another view of the system is the network topology DFD (data flow diagram) in Figure 14.8. The next example (Figure 14.9) shows netViz-created network topology. Figure 14.10 provides a logical order data flow for JEI.

Formal development is an approach to satisfying the requestor's telecommunications requirements that relies predominantly on the various groups within data services to supply the tools, resources, and expertise required. Typically, it is implemented using a life cycle approach—specific tasks occur at specific places in the total life of the system. Development is partitioned into activities with defined roles and responsibilities for users and data services personnel. Activities have specified end products, and the life cycle is subject to frequent formal review. Formality may ease political problems because roles and responsibilities are defined. Partitioning of tasks facilitates management of development. Refer to Figure 14.4; it shows the macro phases of the systems development life cycle and the relative amount of time or effort required for each. The discussion that follows breaks each of these larger phases into smaller tasks for discussion. These are the tasks during the life of a telecommunications project where specific actions take place and specific and often-visible outcomes are produced. Remember that a telecommunications project may concern only wider bandwidth or it may be part of a larger information system project and involve programmers as well as members of the telecommunications group.

FIGURE 14.5 **JEI and Its Subsidiaries**

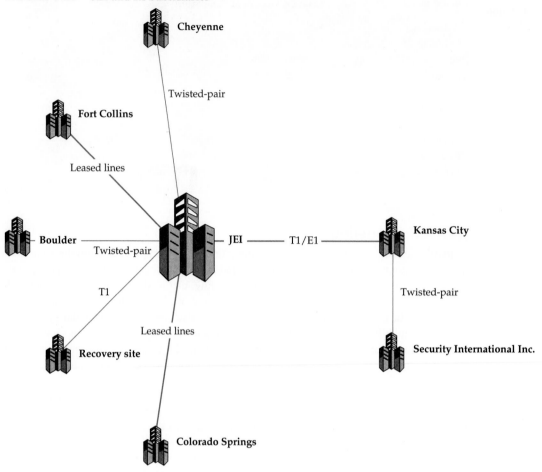

FIGURE 14.6 **System Owner's View of the Network**

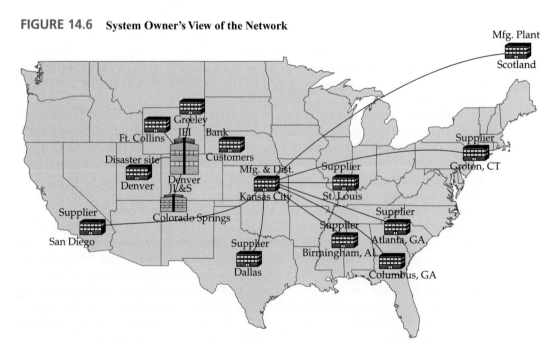

FIGURE 14.7   System Designer's View of the Network

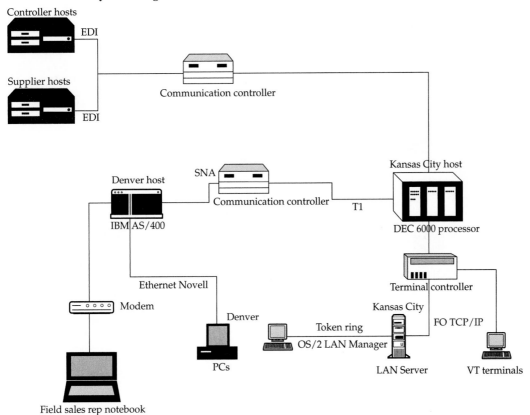

FIGURE 14.8   Network Topology

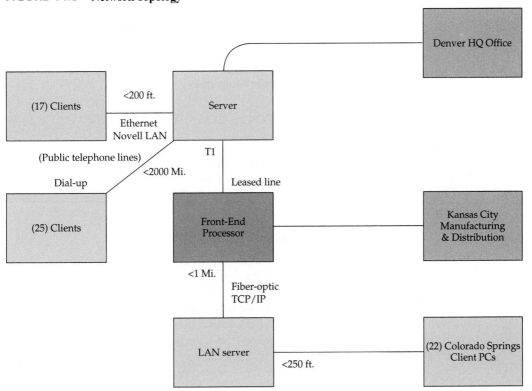

FIGURE 14.9 **JEI's Subsidiaries and Suppliers**

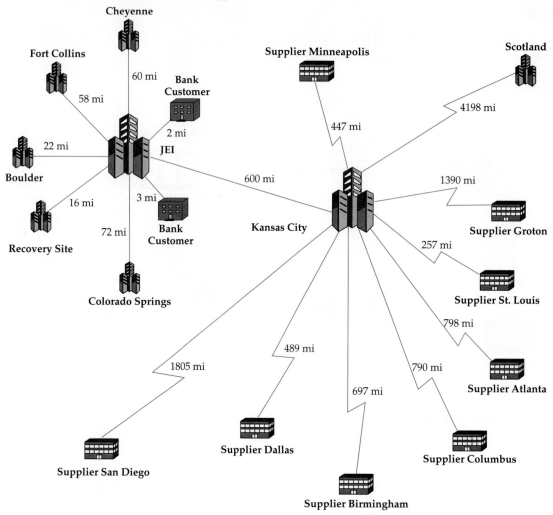

PHASES OF TELECOMMUNICATIONS ANALYSIS AND DESIGN

The Request

The beginning of a telecommunications project is the request for change, quite often called the problem definition stage. This is where one of the groups discussed previously requests an enhancement or addition that requires a change in the system or network. The request will usually be in writing to make it formal. This starts a project.

Phase I. Problem Identification, Definition, and Objective Statement

Any project to install a new telecommunications capability begins with a statement of the problem, opportunity, or threat at hand. The process should not begin with the statement, "The boss said do it, so here goes." It should begin, after a problem, opportunity, or threat has surfaced, with a technical and user-oriented discussion and description of the object of investigation. The effort can start as a request from users for quicker response from a remote site, a statement by management that it will be necessary to support a remote site with significant data transfer, or a member of a technical staff noting that a present capability has

FIGURE 14.10
Order Data Flow

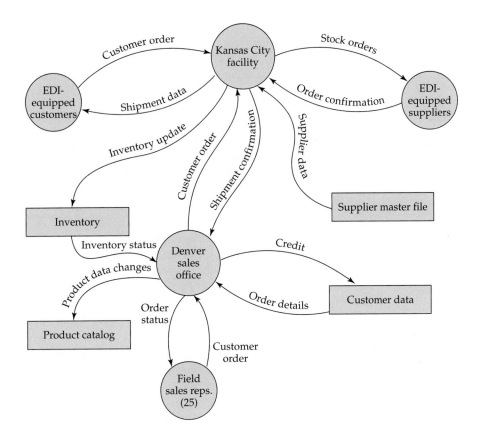

outlived its usefulness and should be replaced because of greater capabilities available, and so forth. In any case, a team of people should be assembled to identify, discuss, and describe the problem, opportunity, or threat and a proposed solution. See Table 14.1 for deliverables for this phase.

At this point, the request will be turned over to someone to review. If the request is reasonable and one that should be taken on, a team would be assembled to work on the project. The team may be gathered during problem definition or after feasibility, depending on the **complexity** of the project. In any case, the person who will be the team leader will be designated during problem definition.

Team Composition

While it would be assumed that the development team assembled will discuss and describe a proposed solution, remember that the solution will be for a user-defined business need. Therefore, the **project team** should be composed of people who not only can implement the capability technically, but can understand why the capability was requested and can use it effectively. For these reasons, the team should be composed of a leader, person(s) competent in the specific technology to be utilized, interested user(s), Telco and IXC representatives, and a consultant if the complexity of the system warrants.

Before discussing the team leader, let's discuss the concept of a champion. If we were discussing leading-edge technology or technology that would drastically change the nature of competition, structure of business, or the way jobs are performed, you should readily see the need for someone to champion the cause. A champion is one who strongly believes in the crusade at hand and spends much of his/her time educating colleagues and marketing the idea. (If the capability to be enhanced or added is considered somewhat mundane, you might question the need for such interest.) There must be at least one person who strongly believes in the capability, preferably from a business standpoint. If such is not the case, the support for the idea may soon wane or even disappear. In the case of telecommunications technology, it may be an interested user who is the champion, and not the team leader. Furthermore, the level of interest may mean that the capability is included in the overall architecture of the organization. In either case, there must be a champion for the business cause early on so that the architecture will support all capabilities.

Team Leader The end result of a telecommunications project will be a new or enhanced business-supporting capability. For this reason, the team leader should be a person who will ultimately own or utilize the capability. Because of the technical nature of telecommunications capabilities and the fact that the telecommunications capability may well be transparent to the ultimate users, the team leader will generally be someone from the telecommunications group, for example, one of the telecommunications specialists. In the rare case of the change of the total telecommunications system or the addition of several million dollars in enhancements, the leader may well be a member of upper management. The purpose of the team leader in the problem definition stage is to be sure that all participants understand and agree on the task. The leader will be responsible for **organizing** the schedule of meetings and tasks, and producing the deliverables necessary to achieve the end results. The first deliverable is usually a white paper describing the problem, opportunity, or threat to be addressed.

Technical Specialist(s) At the initial meeting, only one technical specialist may be required as it is desirable to keep the group small. The technical representative provides the expertise to discuss the technical realities of the project. He or she acts as the in-house consultant on present capabilities and advises how technology can be enhanced or added to achieve the project goals. With complex projects, several technical specialists will likely be required to ensure all technical facets of the project will be covered.

Interested User If the project resulted from a request from a user, the user or someone with a similar interest or competency will be valuable in describing the end desired result of the project. While the user may not initially understand the technical solution, involvement in the project during its life should result in an understanding of the technical features of the project and the ability of the technology to solve the business problem. As users understand the technology itself, they can apply it better and more appropriately.

Telco and IXC Representative When the project involves an enhanced or different interface with the local phone company or long-distance network, a Telco or long-distance company representative can be very valuable, even necessary, at this stage of discussion. At this phase, they participate at no cost and can provide a valuable view in addition to specific information on their capabilities and associated costs. After the implications of the U.S. Telecommunications Act of 1996 have taken full effect, the vendor may be the same for local and long-distance services.

Consultant If the project is of significant magnitude, money for a consultant may be well spent. Even though consultants entail out-of-pocket costs, they should bring valuable and specific experience without having specific vendor bias. They are hired on the basis of having worked on similar projects and created solutions to similar problems.

Risk Analysis and Management

We discussed risk analysis and management earlier, making an assumption that we were encompassing **risk** in the management of the overall telecommunications capability. When you enhance or add to the telecommunications capability, questions should be asked to determine the level of risk you are adding. With the addition of capabilities comes added complexity, which also increases the risks.

Security Risk

A critical part of the risk assessment, and one that requires continuous consideration, is that of security. As equipment is installed, it must be physically secured with proper power and cooling. As components attach to networks, especially those having Internet access, they require virtual security. Any resource that is added or upgraded that contains data must include a consideration for data security. Finally, training users is vital as they are the weakest link in this infrastructure. Physical, virtual, and data security plus user training are vital to the prevention of threats to the organization.

Preliminary Determination of Requirements

When addressing the potential telecommunications project, the team should ask and answer several questions. You might consider these questions as the first-cut requirements specification and a part of the preliminary feasibility study. You must, however, have another feasibility study (which we will call the updated feasibility analysis) after you conduct the analysis stage in which you determine what the requirements really are and the technology needed to implement the solution. The following questions, at a minimum, should be addressed in the problem definition phase.

1. *What is the purpose of this project?* What is the changed telecommunications system supposed to do for the organization? What applications will be affected? What are the strategic implications of not having the systems? In the JEI case, the telecommunications system will be used to move clients' data to the backup sites, to move information and data between the various nodes of JEI for required management control of operations, and to link to the Kansas City manufacturing site, as well as several other applications. The purpose should be clearly stated.

2. *How many nodes are to be connected?* What are the points that need connecting in order to serve organizational purposes? It is important to note that customers, suppliers, and vendors also may have nodes that must be considered. What is the distance between the nodes? Do connections cross LATA and/or national boundaries?

3. *What is the volume of traffic between nodes?* What is the traffic volume within a particular time period? For a PBX, how many lines are making how many calls? For a data network, what is the volume of data being moved, in megabytes per minute?

4. *What is the mix of voice and data usage?* The volume within a time period needs to be classified by voice, data, video image, and fax. Peak loads need to be assessed.

5. *How time-sensitive are the messages transmitted?* Can you allow queues to develop because of high demand relative to **capacity?** What is the consequence of delay? What is the consequence of not having a circuit line available?

6. *Can you create the solution with existing bandwidth, or do you need to create new bandwidth?* What excess bandwidth should you create as a result of this project? When any capacity that has a volume parameter, such as the bandwidth of a channel, is enhanced or added, there should always be some extra included, a reserve amount. First, you need some safety in your estimate of the need. Second, this is the least expensive time to create growth capability, to generate a reserve that can be used for future needs. The cost of adding such expansion capability now is slight in comparison with the cost of adding it later. A heuristic is: add 20 percent more bandwidth than you presently need and then add another 20 percent. Forecasting expected volume, users, and so on, for the presumed life of the system gives a "rough cut" of the minimum needs. By adding a safety factor, the future needs should be accounted for. In considering the amount to put in as a safety factor, look at the incremental cost of adding 25 percent or twice the number of fiber strands. If the incremental cost is acceptable, the safety should be added.

7. *What is the risk to the total telecommunications capability of adding this project and what is the risk of not adding the project?* Adding this project will increase capability and complexity. Not completing the project will leave the organization with the present level of capability. Which involves the greater risk?

8. *Geographic dispersion of nodes must be considered.* Where are the users located? How far is it between the points that must be connected? Are there transnational data flows involved? Each dimension is important for the project. As you progress through the project, you will deal with entities, events, and considerations that can impose risk on the ultimate outcome. Although the existence of vendors is of great concern and will involve significant amounts of time and analysis, the least variable component of the overall cost equation is the vendor equipment. If this is true, then what are the more variable components? They are

- System parameters
- Labor issues
- Topology
- Power
- Structures
- Rights-of-way

It is necessary to address risk initially and throughout the project. Part of the risk involves not completing the project within budget. A greater risk may be due to not completing the project by the project need date.

TABLE 14.1
SDLC Phase I
Deliverables

Phase	Deliverable
Problem definition	White paper describing the problem to be solved or the opportunity to be capitalized upon by the telecommunications capability and the objective of the project.

Phase II. Preliminary Investigation and Feasibility Study

This phase of the project will most likely occur at the first meetings of the team. Its purpose is to understand the purpose of the project and to ask very specific questions as to its viability. Specifically, it is necessary to find out if you can, and should, do the project, before you determine details about the task at hand. This is achieved by addressing specific areas of **feasibility.** Table 14.3 lists the deliverables to the phase.

Technical Feasibility

Technical feasibility is the analysis performed to determine if the planned computer and telecommunications facilities are available for the solution. Can we technically build the systems given the current environment and state-of-the-art technology? All too often, we assume that the answer is yes. It is vital to examine the technology carefully and determine if indeed the enhancement or addition is technologically possible given the present

TABLE 14.2
Cost Estimate
Variance by
SDLC Phase*

SDLC Phase	Cost Variance
Feasibility analysis	50%
Systems analysis	20%
Systems design	10%
Installation and test	5%

*These figures are not based on research but on the experience of the authors.

capabilities, resources, and skills. Although networks are readily available, the organization may not have the personnel to work with them. Network administrators are more than programmers with new titles. On the other hand, if your needs call for extreme bandwidth for data movement, such as 10 Gbps for medical or **computer-aided design (CAD)** imaging, these technologies are just entering the market.

Behavioral Feasibility and Implications

Behavioral feasibility and implications refer to the impact of the new system on operators and users. The implication of a new MIS system is that of a harbinger of change and is often viewed as a replacer of jobs. Telecommunications may be totally transparent to the ultimate users, but not to the members of the telecommunications group. If the project changes the way users do business, we must realize that change to people's jobs and human relationships is an important behavioral issue and questions such as the ones that follow must be addressed. For example, for users, the installation of telecommuting may have a negative effect on the social environment. Changing the vendors on routers or switches may have consequences for the technical staff. Thus, the organization should ask itself

* Are the users able to adapt to the changes brought on by the new system?
* How will the system change the flow of work, the nature of work, the interactions of the people, and the ability to serve the customer?
* Will personnel be able to learn to use the system easily?
* Will fear of replacement of personnel be an issue?

Economic Feasibility

Economic feasibility considers the question "Can we afford the application?" What are the costs of building, operating, and maintaining the system in view of organizational resources? Before one can determine the feasibility in all seven areas, one must understand the purposes of the telecommunications system. This implies an analysis of the problems and the tasks. The analysis will be done in great detail during the systems analysis phase (which follows feasibility analysis), but a reasonable amount of analysis must be done during feasibility to ensure that the task is indeed economically feasible. In part, this is because at the end of the feasibility analysis stage, the reviewing analyst or team will most likely be required to give preliminary cost and schedule estimates. A fixed-cost figure for the project may be demanded to attain a cost versus benefit ratio or a return-on-investment value. Again, the team will have only a basic idea of the task but will be required to give fairly accurate schedule and cost estimates. This may mean that, if you are designing a new local area network for an organization that has yet to be defined, you must be able to estimate the cost of the project prior to the analysis phase and the definition of the organization. Will there be 20 nodes or 200, one file server or five, what will traffic be like, and, thus, which network operating system is best?

As unfair as this may seem, it is a reality. Another reality is that the range of this estimate should be considered good to only plus or minus 50 percent. (See Table 14.2.) Thus, management should realize that a project estimated during the feasibility phase to cost $100,000 will cost somewhere between $50,000 and $150,000. It will not be until the completion of the systems analysis stage that the estimate can be more precise. Even in the design phase, the range of accuracy will likely remain at 10 percent.

For every investment in telecommunications technology, managers need to evaluate the benefits expected as returns for the costs. In this regard, it should be noted that initial expense is not the appropriate cost. There should be an estimate of the total cost of ownership (TCO) for the outlays. This means that there is a need to determine the costs over the life of the technologies. As noted in Chapter 10, these costs include initial equipment costs, training costs, maintenance, software, licenses, installation, power, and so forth, that the system requires. The benefits usually will be evaluated in tangible and intangible categories. Tangible benefits are those that can be objectively measured. Examples include personnel savings, time reductions, direct returns on investment, and so on. The intangible benefits are more difficult to quantify; however, they can be significant. If a system changes the way business is done, how decisions are made, and customers' perceptions, these may be vital to company success even if not easily measured.

Part of the economic consideration of installing the project is the opportunity cost; for example, the fact that the organization has limited recourse and the resources applied to the project in question cannot be applied elsewhere. If we need two new networks, and new routers and switches, and have only enough people to do one project, what is the implication of choosing a particular project relative to the value of the other projects?

Operational Feasibility

Operational feasibility addresses the question of the ability to operate the telecommunications system in the organization's environment. Do we have the requisite employees with the right skills? Do we have the people to design, install, operate, and maintain the network? You need one group to design and install the equipment but a different group to operate and maintain the systems. For the former, you can develop the group or you can hire the work done. In the latter case, you can develop and train the group or you can outsource the task to an operations and maintenance group. Many times you can operationally install a capability and not have or be able to attract and retain the people to operate and maintain the system. This factor is so important some analysts separate it into a human resources feasibility category. Do we have the appropriate facilities to properly house equipment and staff? Do we have, or can we create, procedures to make the system operable and reliable? The answers to these questions will determine the success of the project. If you allow management to edict that you have the operational ability because you outsource the design and installation, but they don't provide the resources for continuous operation, you are at great risk.

Time Feasibility

We now arrive at the question of the project schedule. We must know if the organization can build and implement the telecommunications system on time for it to meet the intended need. The analysis and design of the system must precede the implementation and operation of the system, and all of this requires time. If the system must be available by a certain date, there is a definite lead time required. If sufficient time to finish the project is not available, other considerations are of little value. If the project must be completed on time, alternatives must be found. As an example, if the organization was starting operations in a new manufacturing building in 30 days, requiring 50 telephones, but you only had 22 ports left on the PBX and it takes six months to upgrade the PBX, what should you do?

Regulatory Feasibility

The organization must determine if the proposed telecommunications system is within the provisions of the existing regulatory environment and whether the regulations will inhibit it. Does the telecommunications system comply with appropriate FCC and state PSC regulatory rules? If a global network is involved, the regulations of the various nations (PTTs) must be complied with as well. Examples of regulatory impact might be in obtaining right-of-way, obtaining FCC permits for microwave links and radio frequencies, determining that tariffs exist to support the services needed, and other licenses, permissions, and so forth.

Since the telecommunications environment has historically been heavily regulated, and constantly changing, this issue of feasibility cannot be ignored. Not only will the regulatory environment affect progress, but changes in the laws will provide additional alternatives. As

CATV adds digital channels, it may provide an excellent network around a city that has no other digital channels. New competitors will offer new options for managers and the regulatory constraints on each must be ascertained.

Ethical Feasibility

A recurring theme in the academic world is the **ethical** implications of business processes and practices. This includes the general areas of information systems and technology because of the ability to gather, store, retrieve, report, and transmit information about individuals and organizations. In the purest sense, telecommunications technology is exemplified in the cable television system supplying news and entertainment to millions of homes. The technology is amoral: it is without morality or ethics. Its use, however, is subject to considerations of both. Because of recent US laws, such as HIPAA, when a project creates or extends telecommunications capabilities, the team must consider the ethical implications of the result of the project in the form of privacy of information Team members must determine if the new capability will invade privacy, provide exposure to protected information, violate security, or cause harm to the organization, its members, its customers, suppliers, and vendors, or the environment. While we are sensitive to privacy and security of the organization's papers and property, we also must be sensitive to the effect of projects on the personal lives of its members and the environment.

Project Management

Once feasibility analysis has been completed, and the team has given or received approval to begin development of the project, some form of **project management** must be applied. The personnel of the telecommunications group may use a simple cardex file or a computer program for this. The objective is to describe the tasks involved, the personnel to be used, the resources required, the sequences of events, and the time required for each event. Significant programs with a large number of events and people may require a computer-based project management capability. PERT or critical path techniques may be valuable for complex projects.

The intent of project management is to control the progress, resources, and costs of the program. Another intent is to gain customer concurrence of major milestones of effort. For example, at different times during the systems analysis phase, at the end of design, at the completion of implementation and **testing,** and finally at cutover, the team will ask the customer organization's management to sign off on completed tasks. The intent is to have agreement between the team and the customer as to what has been done. If the work were being accomplished under a contract with an outside firm, these sign-off instances would be accompanied by progress payments. Internally, it just gives an indication of milestones completed.

Project management is required in order to complete the development on schedule and within budget. The project team leader can go far in accomplishing these objectives with adequate definition and management of the project.

TABLE 14.3
SDLC Phase II
Deliverables

Phase	Deliverable
Feasibility analysis	Report based on the technical, economic, ethical, behavioral, operational, regulatory, and time feasibility of developing and utilizing the proposed capability. Preliminary cost estimate of the project.

Phase III. Systems Analysis—Detailed Understanding and Definition

This stage has gained greater importance as technicians, users, and management alike have realized that the more we know about the task, up-front, the better the capability can be designed, created, and maintained. Management often agrees only hesitantly with this view because the more time spent in analysis, the later the "real work" (procurement, implementation, and test) begins.

Systems analysis picks up where feasibility analysis left off and has the specific objective of providing sufficient information in the form of a system specification so that a correct and complete design can be created and maintained. This involves extensive interaction with the requester to determine what is available now; what new capability, change, or enhancement is desired; and what the end result should be. Here the technical experts may converse at length with users, managers, and executives. This is neither an easy nor a trivial task because each participant uses a different point of reference and often a different form of communication. For example, users are looking for the end result, while technicians are considering the technical means of the solution. Deliverables for this phase are shown in Table 14.4.

System Specification

The **system specification** (or **target specification**) should provide information about the final system in sufficient detail for development of either **requests for proposal (RFPs)** or **requests for quotation (RFQs).** Therefore, the system specification needs to be as specific as possible so vendors can make proposals or quotes for the project. While the final details will be done later, there must be sufficient detail for the overall system blueprint at this stage. It is also advisable to provide vendors with a specified format for their proposals so the comparisons can be easily done. For example, if a specific protocol is required to be compatible with the existing telecommunications architecture, this fact must be stated clearly. If existing networks are token ring, to require a token-ring proposal, the specification must state this or else proposed systems may be very complex and not fit your needs. If the specification is too general, the vendors will interpret it differently and the resulting proposals will not be easily compared.

During systems analysis, you must determine as much about the capability as you can. Specifically, you need to know the following to create an adequate system specification:

A. Requirements specification
   1. Capacity
   2. Queuing possibility
   3. Expansion versus use of present bandwidth
   4. Future growth
   5. Geographic requirements
   6. Reliability
   7. Peak and busy hours
   8. Response
   9. Availability
B. Network map and equipment list
C. Present and anticipated applications
D. Utility Requirements (cooling, power, security)

Prototyping

A method of creating computer-based systems rapidly is called **prototyping.** This method results in a good approximation of the ultimate capability. If the telecommunications enhancement or addition is not transparent to the users, a prototype may be valuable. For example, if the project is to install a totally new data path between the central mainframe and a remote site, it would be possible to use a modem and dial-up telephone lines to demonstrate, or prototype, the end result. The demonstration would be slow but would have all of the other characteristics of the ultimate capability. At times, the prototype may actually suffice for the desired capability, as the true amount of traffic is determined. Prototyping is an interactive process and involves the interaction of users with the team as the system evolves to meet their needs. This interaction and user involvement are real strengths of the prototyping approach.

Simulation

As the telecommunications project or installed base becomes more complex, **simulation** of the present or enhanced capability is of significant value. While a prototype shows the capabilities of a new computer program, simulation shows the results of increased traffic and enhanced capabilities. This entails software into which the characteristics and parameters of the system can be stored so a simulation of that system can be exercised. Transaction generators can approximate the predicted traffic, allowing the software to gather statistics about node and channel busy levels, delay times, and overall throughput. Although network measurement and simulation may entail a sizeable outlay of capital, they give viability to the system not possible by any other means.

A very basic form of simulation, one that does not require significant resources, is a computer-based program that allows you to draw the proposed capability in order to document and visually represent the final result. Some programs also will allow for dynamic activity on the system, giving an idea if the system will support the intended outcome. Our use of netViz to show diagrams of JEI's network is an example of this type simulation.

Make-or-Buy Decision

When developing a telecommunications capability, a question that should be addressed at the end of system analysis is whether to build the system with in-house personnel or purchase the installation from an outside vendor, that is, **make-or-buy.** Telecommunications components, whether hardware, software, or media, will generally be purchased. The only question is whether to outsource the task for design and implementation and even operation and maintenance. When an organization is first entering telecommunications on a significant scale, such as for global communications, it is often a good move to buy the capability and have the vendor even operate the capability and perform maintenance. The end result incurs out-of-pocket costs but provides an operable, reliable, and maintainable capability in a minimum of time and at a reasonable cost.

Architecture

Architecture, as previously discussed, is a specification that determines how something is constructed, defining functional modularity as well as the protocols and interfaces that allow communications and cooperation (i.e., **interoperability**) among modules. Specifically, network architecture specifies structure descriptions, relationships of network sender and receivers, logical paths, and communication functions that manage and support the transfer of information. Remember that architecture is a prime feature of security. The consideration of networks should include voice, images, video, and data. Hopefully, the organization has established an overall telecommunications architecture, and this project must comply with it. If not, the total task is far greater and includes a decision on the overall environment and its standards. This is a point where a consultant could be of great value.

Planning

Once the team understands the system requirements, they and other members of the parent organization must attend to details that will come into play at the time of installation of the equipment. Specifically, plans must be made for the physical placement of the equipment, installation of the media, **documentation** for all who must use the equipment and resultant capability, and training for those who need it.

Physical placement means that the industrial engineers or the members of the telecommunications group must ensure that tables, closets, or shelves exist to hold the equipment; that electrical power is available; and that the environmental conditions are correct. Often buildings have telecommunications equipment closets that do not provide adequate cooling for the equipment installed. It sounds simple, but someone must check these items or the system will not work when it arrives, or it will fail later.

Because telecommunications projects use physical paths for voice and data flow, new circuits and channels will likely be installed. This involves allocating existing twisted-pair wires or laying additional pairs, coaxial cables, or fiber strands. This can all be done in the period of preparation for the arrival of the node equipment. The circuits and channels should be tested as they are installed even though they will be tested again upon the arrival of the node equipment.

Implication

We used a third party to upgrade the wall jacks that support our bus network. They did their job and left. The first time a computer was connected to the new jack in the dean's conference room, it was discovered that the jacks had not been tested and this one didn't work.

Documentation ranges from having a place to assemble the vendor-supplied technical material to creating new, user-oriented manuals and brochures that explain what the users need, in their language. Part of this is a library function; part is a writing function. It is easy to delay documentation, but this temptation must be avoided. Documentation provides a necessary history and should be accurate and up-to-date. Inadequate or nonexistent documentation can make system maintenance and upgrade a real challenge!

Finally, training will be required, and someone must determine who needs it, at what level, how, and when. A training plan should be developed by those who will deliver the training. The users should be told of the training they will need and a preliminary schedule developed. Training should be scheduled so users will put their new knowledge into action without significant delay or the effectiveness will be lost. Frequently vendors will include training for some of the technologies and new equipment. The organization should take advantage of vendor-provided training if available.

After the equipment is installed, testing will be performed. Thus, a test plan must be developed, including a schedule of testing. Although this plan is to be used only after the equipment arrives, someone must think through what is to be done, by whom, how, and when.

TABLE 14.4
SDLC Phase III
Deliverables

Phase	Deliverable
Systems analysis	Specification of requirements, defined strategy and architecture, and, where practical, a prototype or simulation of the system. An updated cost estimate of the project, a make-or-buy decision, and test plans.

Phase IV. Investigation of Alternatives

After the team has completed the systems analysis phase, several questions must be addressed. Deliverables for this phase are shown in Table 14.5.

- First, after all of the information has been gathered during systems analysis, is the change to be an expanded capability or a new capability?

- If the change is an added node, can the network accommodate the addition, will enhancements be required, or will a new network be required?

- Can we lease part of the enhancement, must we build it all in-house, or can we contract it out to a third party (make-or-buy)?

- Are there alternatives in media? Does the consideration of growth potential (reserve) impact this choice?

- What are the cost and schedule implications for each alternative?
- Will the cost fit the budget and required return on investment?

TABLE 14.5
SDLC Phase IV
Deliverables

Phase	Deliverable
Alternatives	Statements of the alternatives available, the cost and value of each, and a recommendation as to the best alternative.

Phase V. General Network Design

The design of the new system, based on the requirements gathered during the systems analysis stage, involve the creation, on paper, of the system parts and interactions. The system specification is the basis for **network design.** This phase of activity generates the initial configuration of the system, much like a bridge design entails the creation of engineering drawings from which the parts would later be made and the bridge constructed. Deliverables for this phase are shown in Table 14.6. We must take into account such considerations as excess capacity.

A. Physical circuits
B. Logical channels
C. Availability
D. Response
E. Capacity
    1. Primary channels
    2. Redundancy
    3. Excess capacity (growth)
F. Human resources skills required and sources
G. Training required and sources
H. Location of equipment and requirements
    1. Electrical, including grounding and UPS
    2. Cooling
    3. Lighting
    4. Support
I. Maintenance requirements

There are increasingly sophisticated design tools on the market that will assist in the design task. These should aid in documenting and visualizing the final capability.

TABLE 14.6
SDLC Phase V
Deliverables

Phase	Deliverable
Network design	Diagram showing the components of the project change as they relate to the overall telecommunications capability.

Phase VI. Selection of Vendors and Equipment

A major portion of enhancing and adding to a telecommunications network is the purchase of equipment from vendors. Whether the design and development or operations and maintenance are to be done in-house or outsourced, equipment and software must be obtained from vendors. If the design just calls for additional switches of a standard type, you only have to go to an approved catalog and place an order. However, major changes will require submission of requests for proposal (RFPs) and requests for quote (RFQs). The best sources are determined after evaluation of their proposals and quotes. Deliverables for this phase are shown in Table 14.8.

TABLE 14.7   Spreadsheet to Score Vendors

Criteria	Weight	Vendor A	Vendor B	Vendor C	Vendor D	Vendor E	Vendor F
Reliability	20%						
Cost	30%						
Features	10%						
Maintainability	10%						
Reputation	5%						
Service	10%						
Performance	10%						
Other	5%						
Total score	100%						

There is a whole field of study concerning the procurement of goods and services. It is desirable that the vendors would be rated and ranked on a list of characteristics. There are surveys published by such magazines as *Network Magazine* and *Infoworld* that show vendor ratings. Such sources can be quite valuable in the rankings of vendors. First, the technical response of each vendor should be rated based on engineering merits and the capability, **reliability,** and maintainability of the equipment. Next, the vendors should be rated as to experience, reliability, and reputation. Then, taken altogether, the equipment vendor combinations should be ranked and a recommendation based on technical and corporate merit should be recorded. For example, the project team is working on an assignment to select a new fiber-connected storage facility for a network expansion. RFPs have been sent to a number of vendors and the team has received six responses, including the design specifications of their recommendations and its cost. The team might develop a spreadsheet as in Table 14.7, allowing a final decision to be justified on the composite score of each vendor.

Clarification

Companies that wish to provide goods and services (vendors and suppliers) to an organization are often willing to perform significant tasks to win a contract. Responding to an RFP, a vendor may do a substantial amount of engineering on the project and include it in his proposal to show his capabilities. For the organization, this is a way to receive multiple ideas of how to get something done. In the case of adding a NAS to the organization's infrastructure, asking several companies expert in the capability (giving them an idea of your present infrastructure) might produce several good designs. Then, the final design might have features of them all, with the winning vendor being picked based on several parameters, such as in Table 14.7.

**TABLE 14.8
SDLC Phase VI
Deliverables**

Phase	Deliverable
Vendor and equipment selection	Rating and ranking of vendors, by equipment or service to be purchased. Recommendations based on technical and corporate merit.

Phase VII. Calculation of Costs

Estimating is the process of predicting what the system will cost when it is complete. During the feasibility phase, the team is required to make such a prediction based on minimal information. Now that the team has reached a much later point in the project, they should have as much information as they will ever have assembled and, thus, the process tends to be more of a calculation than an estimation. Rating and ranking of alternatives and vendors is based on the technical merit of the solution, technical and corporate merit of the vendors, and cost of the proposed solutions. Deliverables for this phase are shown in Table 14.9. As in the feasibility phase, in estimating costs, the following must be considered at a minimum:

A. Hardware
    1. Procured hardware
    2. Media
    3. Upgraded hardware
B. Software
C. Personnel
D. Supplies
    1. Installation
    2. Ongoing
E. Maintenance
F. Conversion (from existing systems)

TABLE 14.9
SDLC Phase VII
Deliverables

Phase	Deliverable
Cost	Calculation of the cost by alternative vendor. Recommendation of final configuration.

Phase VIII. Presentation to Technical and Management Group on Recommendation

When all of the alternatives and vendors have been rated and ranked, it is time to report to the management groups that will likely make the final decision. It would be a mistake to only present the team's finding to the technical managers because the business managers, the ultimate users, must understand the implications of the solutions from both a technology and a cost standpoint. The team should assure that senior management, not the engineers, make the final decision as to the enhanced or added capability. Senior managers need to understand the business value of what they are buying, especially the value of the overt growth potential. When senior management understands the technology, they should more readily include it in their strategic planning. Deliverables for this phase are shown in Table 14.10.

TABLE 14.10
SDLC Phase VIII
Deliverables

Phase	Deliverable
Presentation to management	Report to senior management about recommendation of final configuration.

Phase IX. Final Decisions and Design

After the presentation to management, the final decision as to vendors, capabilities, and a configuration will be made. It is time to put aside all other alternatives and finalize the design based on the decisions. New facilities are mapped onto old facilities, an inventory of the prospective new facilities and capacities (old minus deleted plus new) is made, and preparation is begun for the arrival of the new equipment. It is now time to place the orders. Once the detailed design has been generated, the organization is committed to this configuration. If you don't understand the implications of these decisions, you may live to regret them. Additionally, casting your lot with the wrong vendor or the wrong system can have severe and long-term repercussions. Deliverables for this phase are shown in Table 14.11.

Clarification

In Chapter 13, we mentioned configuration and **change management.** At the end of the project, the configuration databases will be updated to reflect this project. In the meanwhile, the project manager should ensure that change management is done, that is, the diagrams and equipment lists are organized in a way to reflect the intent of the project. It is this information that will be transferred to the configuration database.

TABLE 14.11
SDLC Phase IX
Deliverables

Phase	Deliverable
Final design decision	Updated design with pending changes.

Phase X. Procurement Order—Hardware, Software, and Services

After the new capability is designed, the hardware and software selected, and vendors chosen, it is time to create purchase orders for any products or services to be provided by vendors. This function may be provided by the parent organization, or there may be a person in the administrative portion of the telecommunications group, or the MIS group, who does this. In any case, the company usually has forms and procedures so there is no question as to the fairness of the selection or ambiguity of the order. It is important to document this process in the event a losing vendor questions the process. The deliverables for this phase are shown in Table 14.12.

TABLE 14.12
SDLC Phase X
Deliverables

Phase	Deliverable
Procurement	Purchase orders for equipment selected.

Phase XI. Preparation for Implementation

Plans for installation that were made during analysis need to be executed. The areas to house the equipment should have been provided with space, cooling, electrical power, lighting, and security. Since system documentation now exists, a training plan is approved, and installation and test plans are done, the organization is ready for the equipment to arrive. The plan should detail all of the preceding facets so that no confusion will exist when the equipment arrives. Deliverables for this phase are shown in Table 14.13.

TABLE 14.13
SDLC Phase XI
Deliverables

Phase	Deliverable
Preparation for implementation	Plans for the space, cooling, electrical power, lighting, and security needs at implementation time.

Phase XII. Installation of Equipment

The plan may call for acceptance testing of each piece of equipment as it arrives or to accept it based on final testing at the shipper's facility. Since the additional media should already be installed, the equipment that is arriving only needs to be attached to the media. It is desirable to test each interface point upon installation. In some cases, the vendor performs installation and testing as part of the contract. Deliverables for this phase are shown in Table 14.14.

TABLE 14.14
SDLC Phase XII
Deliverables

Phase	Deliverable
Equipment installation	Receive, test, and install the pieces of equipment.

Phase XIII. System Testing

During this stage, the system is tested as a total entity. Where parts were tested upon acceptance, the intent of the testing stage is to ensure the integrity of the total capability that has been changed and to demonstrate that the planned changes are, in fact, in place. Once the pieces are all together, a test plan and test data are required. Deliverables for this phase are shown in Table 14.15.

The test plan, developed at the end of the systems analysis phase, should show a schedule for exercising the pieces and the paths of the system, in progressively more rigorous form. Initially, point-to-point connectivity is checked, then end-to-end connectivity. Transaction generators can be used to create transactions at a specified rate in order to demonstrate the reliability and integrity of all parts and paths of the system. The ultimate purpose is to determine bottlenecks, level of queuing, balking, and the system's ability to handle heavy workloads. It is not until the system passes the test requirements that it can be placed into productive use. Until then, enhancements and additions should be kept separate from the in-place system where possible. When segregation is not possible, testing when the system is idle or offline is necessary.

Part of the test process is to ensure that components and circuits, when installed, meet the required FCC, Underwriters Laboratory (UL), and local electrical, structural, and electromagnetic radiation standards. While all equipment provided by vendors will have gone through a certification program, the entire system must meet the same FCC and UL requirements as to radiation, interference, and safety.

TABLE 14.15
SDLC Phase XIII
Deliverables

Phase	Deliverable
Testing	A demonstration that the system with all enhancements and additions performs as expected, end-to-end, with integrity and reliability, under prescribed loads.

Phase XIV. Training

We noted in the systems analysis phase that training would be required for all persons who must use the enhancements and additions. In many cases, these enhancements and additions will be totally transparent to the ultimate users of the system. Thus, neither training nor documentation will be required for them. However, system descriptions and documentation for the telecommunications group must be updated to show the present state of the system. Manuals from vendors will be placed in the telecommunications library, and group documents will be updated to show changes. Deliverables for this phase are shown in Table 14.16.

Members of the telecommunications group will usually receive training in-place at the vendor's site. Training should cover the new equipment and capabilities, including troubleshooting and repair methods. Training on any new test equipment and media used also should be included. At times, the vendor will provide onsite training. If the installed system touches the user, s/he must be trained in use and security considerations.

TABLE 14.16
SDLC Phase XIV
Deliverables

Phase	Deliverable
Training	Training documentation, primarily for the telecommunications group personnel. This includes equipment operations and systems procedures.

Phase XV. Implementation

Once the system has been installed and tested, it is placed in a production status and available to the users. There is much that has been done up to now to prepare for this stage, such as collection of appropriate user documentation, preparation for training, placement of new equipment, and formulation of plans for switching over to the new system after an appropriate period of use. Deliverables for this phase are shown in Table 14.17.

Cutover

Additions to the existing system can be brought into use in several ways. Depending on the extent of the enhancements and additions, the telecommunications group can use the following:

1. **Pilot.** Bring the new system up for only a small group of users and applications and let them test it in a real mode. This assumes that the system can be duplicated or that a group of users can be segregated. While this mode is common for computer-based systems, it may not be possible for telecommunications systems.

2. **Parallel.** Introduce the new system and keep the old system active as it was. Again, this is common in computer-based systems but may not be possible with telecommunications systems as the expense of two systems may be too large.

3. **Phased.** This occurs when portions of the new capability are added in sequence. This is quite possible for telecommunications systems as new media are added, new switches and routers arrive, and new servers are placed into use.

4. **Phase-in, phase-out, and modular.** This approach is similar to phased. In this case, the switch to a new capability is frequently keyed to transactions so that from a key date, all new transactions use the new systems. The modular approach is a combination of phased and pilot approaches. This mode is useful if the new capability is applicable to business units in an incremental manner.

5. **Cold turkey.** This occurs when the system is changed and brought online all at once with other considerations. For example, if the system in question was a new PBX, it could be introduced as a pilot or it could be brought online cold turkey, disabling the old system totally. This **cutover** method has the greatest risk but least expense if all goes well.

Maintenance of old and new systems is a problem if old systems are preserved. In telecommunications, this is seldom the case, as removed portions are discarded or salvaged. However, when reliability is vital, new systems may be installed to replace old ones, but the old ones are retained and maintained as backup capabilities. This is expensive but may be advisable for critical systems.

TABLE 14.17
SDLC Phase XV
Deliverables

Phase	Deliverable
Implementation	Installed system as defined in specification.

Phase XVI. After-Implementation Cleanup and Audit

When all of the enhancements and changes have been installed and brought into productive use, it is time to clean up, place everything back into inventory that is not in use, and audit your performance. The question to be answered is, "How well did we do?" The point is to learn from the experience. So review what was planned, what was actually accomplished, and how it all went. What mistakes were made and what can be learned from them? What successes were achieved and how can these be institutionalized? Deliverables for this phase are shown in Table 14.18.

After the project is reviewed, gather technical and business management and present the findings. Emphasize what the enhanced system will do, mainly from a business standpoint but a little from a technical point of view. The managers should know what the system can do for them, and they should learn more and more about the technology in place so that future system changes will be easier. It is important that the new system description be incorporated into the configuration database for inclusion in the portfolio of information technologies. This will ensure that the portfolio is up-to-date.

TABLE 14.18
SDLC Phase XVI
Deliverables

Phase	Deliverable
Cleanup and audit	Paper describing the project, process, and product. Briefing to management on the enhancements, including additional training in the technology.

Phase XVII. Turning System over to Maintenance Group

This stage follows the implementation and acceptance of the system. As the enhanced system is used, problems are uncovered, new views of the system's potential use become apparent, and new possibilities may be realized. Problems require quick correction, and changes and extensions need to be considered. While the maintenance personnel are likely to have been a part of the group that created the new system, the system must be placed in a maintenance status. The maintenance and change stage may entail numerous small system enhancements, each of which must work within the total system. This phase will last until the next project makes yet another change to an evolving telecommunications system. Frequently, system component enhancements will be made during this phase. Deliverables for this phase are shown in Table 14.19.

TABLE 14.19
SDLC Phase XVII
Deliverables

Phase	Deliverable
Maintenance and change	The system is the deliverable.

CERTIFICATION FOR NETWORK ADMINISTRATION

Certification is the passing of tests by recognized organizations or vendors to show competence in a specified technology. Although these tests can be taken at any time, they are best done when the individual has experience with the equipment and systems. To take the tests without experience presents you as a *paper-qualified* person.

From a personal standpoint of the person wishing to be a *network administrator,* certification shows a *level of achievement* and may be a *surrogate for experience*. Organizations want people managing their networks who have experience in the architecture, technology, and specific vendor's equipment they use. This ranges from understanding the care of simple hubs to deploying routers and broadband channels. When one's experience is limited, demonstrating theoretical knowledge via certification is an indicator of future abilities.

For the organization, individual certification can be an initial filter for prospective employees. That is, individuals who have achieved certification, even without experience, show motivation and would be considered better candidates than those without certification, *ceteris paribus*. Certification also shows progression of the individual once s/he works for the organization. Many organizations encourage certification and will pay for training to prepare for testing. Some companies will reward the passing of a test with immediate bonuses and/or a new computer.

The two categories of new employees are *new hires* and *experienced hires*. The source of the former is generally the college campus where candidates have a broad education but limited depth in the environment of networks. This can be enhanced when the student works outside of the classroom in **network administration,** even as a student network administrator.

Experienced hires are what all organizations wish to find and hire. While very valuable, they (1) are in short supply and (2) generally require higher salaries than new hires because they come with experience. Consider the classical bell-shaped distribution shown in Figure 14.11. It demonstrates the total pool of people available for hire for any given speciality, for example, network administrators. Those to the right end of the curve are the experienced, highly competent individuals; skills lessen as you move to the left. New hires

FIGURE 14.11
Normal (Bell-Shaped)
Distribution

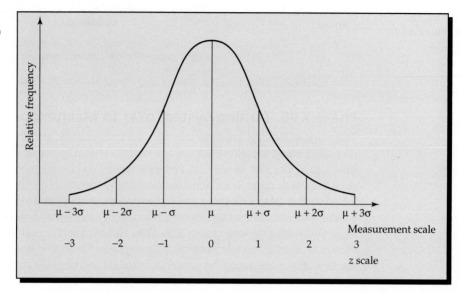

come from the middle two portions (+/–1 standard deviation from the mean), making up 68 percent of the population, those referred to as having normal competency. Experienced hires come from the total right half of the curve with the expert-knowledgeabled individuals in the rightmost portion, the top 16 percent in capability, generally the result of experience.

In a market with a shortage of experienced hires, organizations must use new hires and grow them into experienced hires, hopefully keeping them from taking this new knowledge and experience elsewhere for more money. They begin life in the company as an assistant network administrator and learn on-the-job from a senior administrator.

Certifications Available

Most fields of endeavor, especially those with professional organizations, offer certification as a way to demonstrate the individual's level of quality. Below are the ones germane to data communications. (See Table 14.20 for reference URLs.) While this list includes certifications in the field of security, see http://searchsecurity.techtarget.com/tip/0,289483,sid14_gci1044613,00.html?track=NL-358&ad=507517 for a more complete list of security certifications. For a comparison of salaries by certification, refer to http://www.certmag.com/articles/templates/cmag_feature.asp?articleid=981&zoneid=1.

Network Overview (http://www.cramsession.com, http://www.comptia.org)

CompTIA's **Network+**® certification validates the knowledge of networking professionals with at least nine months of experience in network support or administration or adequate academic training. A typical candidate would have CompTIA A+ certification or equivalent knowledge, but this is not a prerequisite. Network+ is an entry-level, vendor-neutral certification that one can achieve on the way to more advanced certifications, such as the Microsoft MCSE and the Novell CNE. In fact, it counts as an elective for the Novell CNE, and, when paired with the CompTIA A+, as an elective for the Microsoft MCSA. Earning a CompTIA Network+ certification demonstrates that a candidate can describe the features and functions of networking components, and possesses the knowledge and skills needed to install, configure, and troubleshoot basic networking hardware, protocols, and services. The exam tests technical ability in the areas of media and topologies, protocols and standards, network implementation, network support, wireless networking, and gigabit Ethernet. The certification consists of one test.

TABLE 14.20 **Data Communications Certifications**

Area	Certification	Source (URL)
Network overview	CompTIA's Network+	http://www.comptia.org/certification/network/default.aspx
Hardware	CompTIA's A+	http://www.comptia.org/certification/a/default.aspx
Microsoft	Microsoft Certified Professional (MCP)	http://www.microsoft.com/learning/mcp/benefits/default.asp
	Microsoft Certified Systems Administrator (MCSA)	http://www.microsoft.com/learning/mcp/mcsa/default.asp
	Microsoft Certified Systems Engineer (MCSE)	http://www.microsoft.com/learning/mcp/mcse/default.asp
Cisco	Certified Network Associate (CCNA)	http://www.cisco.com/web/learning/le3/le2/le0/le9/learning_certification_type_home.html
Novell	Certified Novell Administrator (CNA) 5/6	http://www.novell.com/training/certinfo/cna/index.html
	Certified Novell Engineer (CNE) 5/6	http://www.novell.com/training/certinfo/cne/index.html
Linux	CompTIA's Linux+	http://www.comptia.org/certification/linux/default.aspx
	Red Hat Certified Engineer (RHCE)	https://www.redhat.com/training/rhce/courses/
	Novell Certified Linux Professional (Novell CLP)	http://www.novell.com/training/certinfo/clp/
	Linux Professional Institute's LPIC	http://www.lpi.org/en/lpic.html
	SAIR, Linux/GNU Certified Engineer (LCE)	http://gocertify.com/certification/LCE.shtml
Security	CompTIA's Security+	http://www.comptia.org/certification/security/default.aspx
	CWSP (Certified Wireless Security Professional)	http://www.cwnp.com/cwsp/index.html
	CISSP (Certified Information Systems Security Professional)	http://www.cissps.com/
	CIW Security Analyst	http://www.ciwcertified.com/csa/default.asp?comm=CND&llm=3
Internet	CompTIA's i-Net+	http://www.comptia.org/certification/inet/default.aspx
Wireless	CWNA (Certified Wireless Network Administrator)	http://www.cwnp.com/cwna/index.html
	Certified Wireless Analysis Professional CWAP	http://www.cwnp.com/cwap/index.html

Hardware (http://www.cramsession.com, http://www.comptia.org)

CompTIA's A+® certification validates the knowledge of computer service technicians with the equivalent of 500 hours of hands-on experience. The A+ certification is a vendor-neutral certification. It has no prerequisites but is widely used as a starting point for various system administration and network administration certifications. The exams cover a broad range of hardware and software technologies but are not bound to any vendor-specific products. Earning CompTIA A+ certification proves that a candidate has a broad base of knowledge and competency in core hardware and operating system technologies including installation, configuration, diagnosing, preventive maintenance, and basic networking. The A+ certification consists of two exams, which cover installation, configuration, diagnosing, preventive maintenance, and basic networking, plus memory, bus, peripheral, operating systems, and wireless technologies.

Microsoft (http://microsoft.com)

Microsoft Certified Professional (MCP) is achieved by passing any one of the Microsoft examinations.

Microsoft Certified Systems Administrators (MCSA) administer network and systems environments based on the Microsoft Windows platforms. The MCSA exam certifies a holder's ability to implement, manage, and maintain the typically complex computing environment of medium- to large-sized companies. Specializations include MCSA: Messaging and MCSA: Security. The MCSA consists of four exams.

Microsoft Certified Systems Engineers (MCSE) design and implement an infrastructure solution based on the Windows platform and Microsoft Servers software. The MCSE exam certifies a person's ability to plan, design, and implement Microsoft Windows server solutions and architectures in medium- to large-sized companies. Specializations include MCSE: Messaging and MCSE: Security. While the MCSE certifies familiarity with Microsoft products, it is not intended to be an engineering qualification. The MCSE requires a total of seven exams, three beyond the MCSA.

Cisco (http://www.cisco.com)

Cisco Certified Network Associate (CCNA) certification indicates a foundation in an apprentice knowledge of networking. CCNA certified professionals can install, configure, and operate LAN, WAN, and dial-access services for small networks (100 nodes or fewer). A CCNA candidate must prove familiarity and expertise with Cisco's IOS, TCP/IP, LAN and WAN management, routing and routing protocols, the OSI model, and LAN troubleshooting, and be familiar with a variety of network protocols, some Cisco-specific and some not, including but not limited to IP, IGRP, Serial, frame relay, IP RIP, VLANs, RIP, Ethernet, and access lists. The CCNA is one exam. (Cisco also has the following certifications: CCIE, CCDA, CCNP, CCDP, CCIP, CCSP, and CCVP.

Novell (http://www.novell.com)

Novell's Certified Novell Administrator (CAN) 5/6 certification is meant for entry level IT professionals who have "a fundamental knowledge of networking" and want to "learn the specifics of Novell networking." It is based on Novell's NetWare networking product. The CNA is a prerequisite for Novell's mid-level certification, the Certified Novell Engineer (CNE). There is no prerequisite to the CNA certification; however, Novell "strongly recommends individuals have a CompTIA Network+ certification before pursuing the Novell CNA." In essence, candidates should have "a good understanding of network implementation, support, media, topologies, protocols, and standards." (www.cramsession.com)

Novell's Certified Novell Engineer (CNE) 5/6 certification is meant for intermediate to experienced IT professionals who "solve advanced company-wide support problems and high-level network problems." It is based on Novell's NetWare networking products. The CNE is a prerequisite for Novell's highest certification, the Master CNE. A prerequisite to the CNE certification is passing exam 50–653, "NetWare 5/6 Administration". (www.cramsession.com)

Linux (http://www.cramsession.com, http://www.comptia.org)

CompTIA's Linux+^{TM} certification is intended for "technicians with six months experience installing, operating and maintaining **Linux** operating systems." The Linux+ is an entry-level, vendor-neutral certification that makes a good starting point to higher-level certifications, such as LPI, SAIR, and Red Hat. Although there are no prerequisites for this certification, CompTIA recommends that one possess the A+ and Network+ certifications. Earning the CompTIA Linux+ designation means that the candidate can explain fundamental open-source resources/licenses, demonstrate knowledge of user administration, understand file permissions/software configurations, and manage local storage devices and network protocols.

The *Red Hat Certified Engineer (RHCE)* is a mid- to advanced-level certification for IT professionals who are able to "install and configure Red Hat Linux;

understand limitations of hardware; configure basic networking and file systems for a network; configure the X Window System; perform essential Red Hat Linux system administration; configure basic security for a network server; set up and manage common enterprise networking (IP) services for the organization, and carry out server diagnostics and troubleshooting." Job roles of those pursuing this certification typically include technician, system administrator, network administrator, system engineer, Web engineer, and developer. Red Hat Certification focuses on practical hands-on skills of the aspirants. It consists solely of practical lab exams. Therefore, just attaining the theoretical knowledge is not sufficient for clearing this exam.

Novell Certified Linux Professional (Novell CLP) certification is for people interested in being Linux administrators. Skills demonstrated by someone holding a Novell CLP certification include installing Linux servers into a network environment, managing users and groups, troubleshooting the SUSE LINUX file system, managing and compiling the Linux kernel, and troubleshooting network processes and services— just to name a few. As with all Novell certifications, course work is never required. One need only pass a Novell Practicum (050–689) in order to achieve the certification. The Novell Practicum is a scenario-based exam where students apply the knowledge they have learned to solve real-life problems—showing they not only know what to do, but they can actually do it as well.

The *Linux Professional Institute*'s *LPIC1—Junior Level Administration* is an entry-level certification for IT professionals who want to show a base level of "vendor independent and distribution neutral" knowledge of Linux. There are no specific prerequisites for the LPIC1, although candidates should be able to perform these tasks: work at the Linux command line; perform easy maintenance tasks: help out users, add users to a larger system, backup and restore, and shutdown and reboot; and install and configure a workstation (including X) and connect it to a LAN, or a **stand-alone** PC via modem to the Internet.

The *Linux Professional Institute*'s *LPIC2—Intermediate Level Administration* is a mid-level certification for IT professionals who want to show an intermediate level of "vendor independent and distribution neutral" knowledge of Linux. The specific prerequisite for the LPIC2 is the completion of LPIC Level 1. As well, candidates should be able to perform these tasks: administer a small to medium-sized site; plan, implement, maintain, keep consistent, secure, and troubleshoot a small mixed (Microsoft, Linux) network, including a LAN server (samba), Internet gateway (firewall, proxy, mail, news), and Internet server (Web server, FTP server); supervise assistants; and advise management on automation and purchases.

SAIR's *Linux/GNU Certified Engineer (LCE)* is a vendor-neutral certification offered for individuals who act as Linux system managers. This is the second (middle) level of SAIR's Linux certification program. This certification is *not* distribution specific. (http://gocertify.com/certification/LCE.shtml)

Internet (http://www.cramsession.com, http://www.comptia.org)

CompTIA's *i-Net+*® certification validates baseline technical knowledge of Internet, intranet, and e-commerce technologies, independent of specific Internet-related career roles. Those holding CompTIA i-Net+ certification can demonstrate knowledge and competency with Internet basics and clients, development, networking, security and business concepts, implementing and maintaining Internet, intranet, and extranet infrastructures and services as well as the development of related applications. It is a vendor-neutral certification. CompTIA i-Net+ certification validates the knowledge of technical and nontechnical professionals alike with at least six months' experience in Internet, intranet, extranet, and e-commerce technologies. The i-Net+ designation is achieved by passing one conventional-format exam.

Security (http://www.cramsession.com, http://www.comptia.org)

CompTIA's Security+ certification is aimed at IT professionals who have two years' on-the-job networking experience, with an emphasis on security. It is an entry-level, vendor-neutral certification that makes a great stepping stone to more advanced certifications, such as the ISC2 SSCP and CISSP, and the SANS GIAC. It also may be used in some Microsoft certification tracks. This certification is well suited to network and security administrators independent of what industry they work in. The Security+ designation is achieved by passing one conventional-format exam that covers topics such as communication security, infrastructure security, **cryptography,** access control, **authentication,** external attack, and operational and organization security.

The SANS Institute (SysAdmin, Audit, Networking, and Security) is an organization focusing on providing computer education and information security training.

SANS Institute Global Information Assurance Certification (GIAC) programs are designed to serve the people who manage and protect important information systems and networks. GIAC is an outgrowth of the SANS Global Incident Analysis Center. GIAC is designed for Individuals entering the information security industry who are tasked with **auditing** organization policy, procedure, risk, or policy conformance. Individuals who complete the GIAC IT Security and Audit Essentials will have a firm grasp of information security principles and issues and will be equipped to develop best practice audit checklists. They also will be prepared to perform limited risk assessments as well as security and conformance audits based on established best practice. (http://www.giac.org)

Certified Information Systems Security Professional (CISSP) certification is an advanced-level certification meant for IT security professionals with "a minimum of four years of PROFESSIONAL experience in the field of information security. A bachelor's degree can substitute for one of these required years. Additionally, a Master's Degree in Information Security from a National Center of Excellence can substitute for one year towards the four-year requirement." This experience requirement essentially forms the prerequisite for this vendor-neutral certification, although (ISC)2's other, lower-level certification, the SCCP (Systems Security Certified Certified Practitioner), is recommended. The CISSP certification is well suited to IT professionals who aim to be IS (information security) professionals, network security professionals, or systems security professionals. The CISSP designation is achieved by passing one exam, covering topics such as access control systems, cryptography, and security management practices. (http://www.cissps.com)

Certified Wireless Security Professional (CWSP) is a mid-level certification for IT professionals who "understand how to secure a wireless LAN from hackers" and who "know how to protect the valuable information on their network." Topics on this certification include "advanced processes and techniques for keeping enterprise wireless network data secure." Job roles of those pursuing this certification typically include network administrators, security administrators, systems security practitioners, IS (information security) professionals, and other IT professionals with wireless in their titles. To achieve the CWSP, a candidate must pass two exams.

CIW Security Analyst is a mid- to high-level, vendor-neutral certification for IT professionals who "configure, manage and deploy e-business solutions servers; and implement e-business and network security solutions." Job roles of those pursuing this certification typically include network server administrators, firewall administrators, systems administrators, application developers, and IT security officers. To achieve the rating, a candidate must pass one CIW exam as well as hold, and provide proof of, one of these certifications: CIW Master, CIW Administrator, MCSA, MCSE, CNE, CCNP, CCNA, CCIE, LPI-Level 2, or SAIR Level 2 LCE. (http://www.ciwcertified.com/csa/default.asp?comm=CND&llm=3)

Wireless[7]

Certified Wireless Network Administrator (CWNA) is an entry-level certification for those IT professionals who are new to wireless networking but who want to come "up to speed quickly," as well as for those IT professionals "already familiar with wireless LANs" and who want to fill "in any gaps in their knowledge, and prove their expertise to help their competitive edge." Job roles of those pursuing this certification typically include network administrators, system administrators, and other IT professionals with wireless in their titles.

Certified Wireless Analysis Professional (CWAP) is an advanced certification for IT professionals who are "able to confidently analyze and troubleshoot any wireless LAN system using any of the market leading software and hardware analysis tools." Topics on this certification focus "entirely on the analysis and troubleshooting of wireless LAN systems." Job roles of those pursuing this certification typically include network administrators, network architects, and other IT professionals with wireless in their titles. To achieve the CWAP, a candidate must pass two exams.

Case 14-1

Application Service Providers

Kelly Ammons

An application service provider, or **ASP,**[7] in the simplest terms, rents software. An ASP is any company that remotely hosts software applications and provides access and use of it to its clients over a network for a recurring fee. ASPs provide Web-based access to a range of business applications on a pay-as-you-go subscription basis.

With the advent of the application service provider, the cycle of computing trends has come full circle. The introduction of outsourced Web-based shared resource business applications has returned us to the days of service bureaus and time-sharing. Application service providers have emerged due to the growth of the Internet and the promise of open communications. ASP **outsourcing** differs from the traditional mainframe-based time-sharing. ASP outsourcing utilizes client-server architecture and relies on secure, cost-effective packet data communications.

The most common features of an ASP include

- An ASP owns and operates a software application.
- An ASP owns, operates, and maintains the servers that are required to run the application.

---

[7] ASP is also an abbreviation for *active server page,* but not used here.

- An ASP makes the application available via the Internet in a browser or through some type of thin client.
- An ASP charges either on a per-use basis or on a monthly/annual fee basis.

The ASP market is around two years old. It has already grown rapidly and is expected to continue to grow. Scott Heinlein of TeleChoice.com indicates that the ASP market projections range from $7.8 billion to $48 billion by 2003. There are many established ASPs, and more emerge daily. According to WebHarbor.com, there are more than 300 active ASPs. Corio, FutureLink, ServiceNet, and Usinternetworking are a few of the better-known ASPs.

ASP DRIVERS

Numerous internal factors drive the ASP model. First, an ASP allows applications to be implemented in days or weeks compared to months for the implementation of in-house or consultant-built applications. Second, it allows a company to focus on its core competencies instead of focusing on its IT development that supports the company. The ASP model also allows for scalability. A company can start small and easily expand its applications by using an ASP. The ASP also competes on price because it is considerably cheaper to rent software applications than to buy those applications.

A number of external factors drive the ASP model. One factor is the shorter application cycles of software. The application cycle of software is the amount of time between new versions of the same software. By the time a company gets their system implemented and tested, a new release is being marketed. Secondly, the Internet shows the importance of time-to-market and scalable IT infrastructures. The Internet allows users to stay informed and take advantage of the quick pace provided. Applications must be implemented quickly in order to successfully compete in the Internet's quick-paced environment.

Third, locating and retaining skilled IT personnel can be a challenge. Skilled IT personnel are highly recruited and expect extensive compensation packages in addition to exciting job assignments. It is difficult to find them and even more difficult to keep them. ASPs are responsible for implementing and maintaining the software that they lease to their customers. By using an ASP, companies can take advantage of computer applications without needing the extensive IT staff to support it. Finally, network improvements also have driven the growth of the ASP market. Increased bandwidth allows for software applications to be transferred over a network quickly. These improvements allow for a seamless blend of remotely managed shared environments and locally managed individual environments.

APPLICATION SERVICE PROVIDERS' BEST PRACTICES

In order to add value, ASPs should be based on the following "best practices":

- Availability. Some ASPs are delivering 99 percent uptime.
- Security. In order to gain customers, ASPs must be able to guarantee that a company's data and applications are secure.
- Networked storage. This should include disaster recovery.
- Management. ASPs should be able to decrease headaches associated with managing applications.

TYPES OF ARCHITECTURE

There are three types of application-hosting architecture: Web server/browser-based applications, thin client–server, and Java-based applications. Each type has its strengths and weaknesses.

The Web server/Web application can deliver an application to any browser. The common interface lowers training costs. This is best used when dealing with forms to fill out, workflow, and group scheduling. The downside is that it is difficult to manage the individual desktop experience. Security and the bandwidth requirement also limit this application.

The thin client can deliver to most client types with low bandwidth connections. It is possible to support individual users on dial-up connections or offices on dedicated data services. It provides built-in management and administration. The thin client can be used for any Windows application.

The Java architecture is platform independent and does not require installation on a desktop computer. It also can be used for any *Windows*® application. This architecture falls short on performance and also requires large amounts of bandwidth to download applications.

TYPES OF SERVICE

ASPs offer three types of service offerings. The first are core or basic services. These are services such as application updates and upgrades, continuous monitoring of the applications, support and maintenance of the network and servers on which the application runs, and, finally, customer support. ASPs also offer managed services including all core services plus additional services and guarantees around support, security, application performance, and data redundancy. These managed services include data security, technical support, and daily backup of the application and its data.

ASPs also offer extended services. These include all managed services plus professional services such as application configuration and extension, strategy and planning, and training and educational support.

TYPES OF APPLICATIONS

There are essentially six types of applications that can be offered by an ASP. First are analytical applications. These include applications to analyze business problems such as financial or risk analysis. Second are vertical applications, which are industry-specific applications such as patient billing in health care or claims processing in the insurance industry. Enterprise relationship management applications also can be offered by an ASP. These include accounting, human resources, materials management, and facilities management. Customer relationship management applications include sales force automation, customer service, and marketing applications. A fifth type of application is collaborative applications. These applications include groupware, e-mail, and conferencing applications. The last type of application is personal applications. These include office suites and other consumer applications.

SWOT ANALYSIS OF ASP

A SWOT analysis of the ASP model identifies the strengths, weaknesses, opportunities, and threats associated with the ASP model. Each category is examined below.

Strengths

An ASP provides the leasing company the latest technologies with lower risks and lower total costs. Midsize firms that could not afford high-end software solutions used by larger competitors are able to do so through ASPs. Firms can get a fully functioning, large application such as ERP, which could cost millions of dollars to implement, without paying for development, installation, hardware, or software. Firms just pay a monthly fee that amortizes the ASP's costs over time.

A basic package of a basic application starts at $30 per month per person with the fee increasing with the complexity of the application. This fee can increase drastically for the more complex applications such as a human resource application, which can cost $1,800 per month per person. *US Internetworking* has said its average customer pricing of applications ranges from $40,000 to $100,000 per month. These costs are lower than the cost to buy the software and accompanying hardware and also hire and train an IT staff on the new applications.

An ASP deploys applications quickly. ASPs can deploy basic services such as office applications throughout the company in a few hours. More complex applications such as creating an e-commerce platform can be implemented within days. Even the most complex applications such as ERP can be deployed within weeks. An ASP can decrease implementation time by over half.

The ASP also provides all the supporting technology such as networks, hardware, and supporting software and is responsible for any maintenance required within the application. ASPs provide constant updates on software. ASPs are able to connect processes across a company even when employees are in different locations without the need for expensive intranets. ASPs also can connect a company with its trading partners with lower costs and commitments. Finally, a company is able to take advantage of more complex applications without hiring an expert IT staff to support the applications.

Weaknesses

One weakness of ASP usage is that the leasing company has little control over the applications. Many are also uneasy about a company having control over a large part of their operations. A company's internal IT staff has no direct control in fixing any problems that are associated with the applications. A company must rely on a third party, the ASP, to correct problems quickly and efficiently. Sometimes an ASP's sense of urgency may not match that of the company.

Many ASPs provide little customization. The ASP offers commodity versions of flexible software that allows the ASP to install applications faster and cheaper that appeal to a greater audience. This may work for small and midsize companies that are less complex and have fewer employees to adapt to the new software. Larger companies with complex functions, many computer systems, and numerous employees need to be able to customize their applications for a smoother transition.

Connection speed is critical to the ASP application. Companies with less than high-speed connections must incur added expenses by investing in newer, faster technology in order to take advantage of an ASP application. Bottlenecks within networks can degrade real-world performance. The quality of technology for efficient network delivery of leased applications is critical to the success of the ASP and the leasing company. The presence or absence of a WAN or high-bandwidth Internet connection determines the way an ASP can provide the hosted applications.

Using an ASP provides no equity and potentially harsh consequences for opting out of a contract before it expires. A company's management must work with its IT staff to ensure that the most feasible and appropriate application is rented from the ASP.

Opportunities

Bandwidth continues to increase at a quick pace. This increased bandwidth makes it possible to lease more complex applications without the time delay caused by inadequate bandwidth.

As more applications are leased, there is a continued decrease in operational expenses due to the decrease in **information technology** personnel needed and the decrease in hardware that would be purchased if the applications were bought and run in-house.

Threats

If a continuing market shakeout does not occur within the ASP industry, it will become increasingly difficult to find a reliable ASP because it's difficult to sort through the hype created by numerous ASP companies. The market is evolving so quickly that it's difficult to understand what an ASP can offer the company. Many companies will decide to lease software in order to solve problems that cannot be solved by existing software applications.

ASPs gain significant control over important information within a business. Security remains an important issue in determining the value added by renting applications. A company that plans on leasing software should make sure that the ASP is reliable and also can guarantee that the company's information is secure.

GUIDELINES TO CHOOSING AN ASP

A company should follow a few guidelines when choosing an ASP. The first is adequate security. A company should make sure that its information is safe from hackers, employees of the ASP, and, most importantly, competitors. Secondly, with the shakeout of ASPs, a company should be concerned with the consequences of the ASP going out of business. Will the company be able to continue business, can it quickly find a replacement service, and can it gain access and possession of its data. A company also should be concerned with the frequency with which the ASP backs up data. It is also important to know how easily these data can be accessed.

A company must define the business processes that are to be outsourced and be clear on the objectives that should be accomplished by the ASP. A company also should check the track record of the ASP. Many new ASPs do not have the experience or reliability to handle the objectives that need to be accomplished.

CONCLUSION

A company should decide to use an ASP based on the bandwidth available within the company, the type of application that is to be leased, the size of the company and number of users affected by the new application, and the amount of money available for the application.

A company should make sure that the ASP conforms to "best practices" such as sufficient uptime, security, network storage, and efficient management of the application. Using an ASP that does not provide these characteristics can cost the company more than the cost of implementing the application in-house. It is important that the company research and ask the appropriate questions to determine if the ASP is reliable and if it provides the services needed by the company.

BIBLIOGRAPHY

1. Corbett, Michael F. "E-Sourcing the Corporation." *Fortune Magazine,* March 2000.
2. Gillan, Clare, and Meredith McCarty. "ASP's Are for Real . . . But What's Right for You?" An IDC White Paper. International Data Corporation. July 1999.
3. "How ASPs Deliver Value: Next Generation Portals for Business Applications." Giotto ASP White Paper. May 3, 1999.
4. Koch, Christopher. "Monster in a Box?" *CIO Magazine,* May 2000.

5. Koch, Christopher. "ASP & Ye Shall Receive." *CIO Magazine*, May 2000.

6. Ward, Lewis. "How ASPs Can Accelerate Your E-Business." *E-Business Advisor*, March 2000.

7. www.aspnews.com.

Summary

This chapter has covered a large amount of management material that applies specifically to the analysis, design, development, and operation of telecommunications capabilities and the projects that create them. Much information exists about this area for computer-based systems, but far less exists for telecommunications. There are two points to remember in the final analysis. First, telecommunications capabilities serve the parent organization in reaching its vision, goals, and objectives. They do not exist in their own right but to serve an organization or business purpose. The people who support this technology must have the same goal. Second, the design and development of a telecommunications system take time, talent, and vision. Technology, money, and people are involved. All of the factors must be managed. People management and technology management must be merged for the network to be successful.

As the organization progresses through the SDLC of a project to expand or add new capabilities or equipment (see Table 14.21), the team members must always keep in mind that the system does not exist in and of itself. The end result of this project must blend with the total existing system and fit the future needs and planned evolutionary path of the total system. This is why we keep referring to the implication of your actions. Are you following the tenets of the architecture you created? Are you choosing equipment, systems, or vendors that will fit the plan or have dire consequences? Do you indeed have a vision and plan that is compatible with the firm's? These are not idle questions; they determine the organization's ability to compete in the future. This ability may be the basis for your career. The value and use of data communications projects should only be considered within an understanding of the path that the telecommunications group is taking in supporting the organization in its vision. Among the most important phases of the SDLC is the feasibility phase because so many variables are brought to bear at one point in time. Analysis is vital to ensure that the particulars of the proposed system are adequate. The overall implications and impact of the technology and its installation must be considered. The development team must keep this point salient as they proceed throughout the project as failure to do so is almost certain to lead to suboptimal decisions and impose constraints that will prevent the telecommunications system's investment from yielding the best returns for the organization.

Finally, you must manage your career; no one else will. Certification is a surrogate for experience for the new hire employee; it is a method of demonstrating experience and skills for those who have years in the field. It is, also, a demonstration of motivation, a characteristic in great demand.

TABLE 14.21
Summary of Deliverables of Macro Phases and Effort of SDLC

SDLC Phase	Deliverables at End of Macro Phase
Problem definition	White paper-problem definition
Feasibility	TBEOTRE feasibility, preliminary cost
Systems analysis	Specification, architecture, prototype, updated cost
Design	Design showing components
Procurement	Purchase orders
Installation	Installed parts, training, documentation
Test	Tested system
Implementation	The system
Evaluation	Determine if problem is solved
Maintenance and change	Ensure system kept correct over its life

Key Terms			
	alternatives, *p. 413*	implementation, *p. 143*	planning, *p. 412*
	analysis and	information systems	policy, *p. 412*
	design, *p. 413*	architecture, *p. 411*	project management, *p. 427*
	application service	information	project team, *p. 421*
	provider (ASP), *p. 443*	technology, *p. 447*	prototyping, *p. 428*
	architecture, *p. 411*	interoperability, *p. 429*	reliability, *p. 432*
	auditing, *p. 442*	inward WATS, *p. 415*	request for proposal
	authentication, *p. 442*	Linux, *p. 440*	(RFP), *p. 428*
	capacity, *p. 423*	maintenance, *p. 416*	request for quotation
	certification, *p. 437*	make-or-buy, *p. 429*	(RFQ), *p. 428*
	change management, *p. 433*	management information	risk, *p. 422*
	client-server architecture	systems (MIS), *p. 414*	security, *p. 413*
	(C/SA), *p. 416*	methodology, *p. 414*	simulation, *p. 429*
	cold turkey, *p. 436*	mission, *p. 410*	stand-alone, *p. 441*
	complexity, *p. 421*	Network+, *p. 438*	strategy, *p. 411*
	computer-aided design	network	system specification, *p. 428*
	(CAD), *p. 425*	administration, *p. 437*	systems development life
	control, *p. 412*	network architecture, *p. 412*	cycle (SDLC), *p. 416*
	cryptography, *p. 442*	network design, *p. 431*	target specification, *p. 428*
	cutover, *p. 436*	organizing, *p. 422*	telecommunications
	documentation, *p. 429*	outsourcing, *p. 443*	system, *p. 416*
	ethical, *p. 427*	phases of SDLC, *p. 416*	testing, *p. 427*
	feasibility, *p. 424*	pilot, *p. 436*	vision, *p. 410*

Management Critical Thinking

M14.1 How does management justify the cost difference when hiring an experienced hire versus a new hire employee?

M14.2 Under what circumstances would it be acceptable/advisable to outsource the data communications function?

Technology Critical Thinking

T14.1 Can networks be kept operational and projects completed with all new-hire employees?

T14.2 Outsourcing most of the data communications function and especially the data storage capability ensures security. How can an internal organization match the outsourced capability?

Discussion Questions

14.1 Should the project team members be the same over the entire project time span? Why or why not?

14.2 What are the alternatives for developing and deploying a telecommunications system?

14.3 What are some of the considerations for the make-or-buy decision? What are the advantages and disadvantages of "make" and "buy"?

14.4 When is feasibility analysis performed, and why is it important to update the feasibility analysis after the analysis and design phases? In other words, why do feasibility analysis at least twice?

14.5 If a telecommunications vendor is willing to provide a "total systems solution," why is it necessary for the business to bother with project analysis and design?

14.6 What are the security concerns that must be addressed when planning a telecommunications system project?

14.7 What are the factors that a firm should evaluate in picking a vendor for telecommunications projects?

14.8 What are the advantages of building a small project (pilot) to discover problems that may be encountered in a large-scale telecommunications project?

14.9 Should firms that outsource telecommunications systems development have a contingency for the failure of their outsourcers? Why?

14.10 What skills sets should a telecommunications system project manager possess?

14.11 Discuss with your instructor the class substituting a data communications certification for the final exam.

Projects

14.1 Go to organizations and determine if they are increasing or decreasing their outsourcing activities. What is influencing their decisions?

14.2 Find a business that has brought up a new telecommunications system and find out what cutover methodology was used and why. What problems were encountered and would another cutover method have helped?

14.3 Put together a team to develop a cable TV system for your school or a local community. Develop a total system, using the SDLC approach.

14.4 Find an alternative systems approach to decision making and compare the steps to the phases of the TC (Telecommunications) SDLC. What are the similarities and differences?

14.5 Research the SDLC methodology on the Internet. Can you find this methodology for data communications, or just MIS? Is there another methodology for data communications projects?

14.6 Build a list of references (books, Web sites, etc.) that could help a telecommunications systems project team.

14.7 Find an organization adopting a client-server computer architecture and determine the telecommunications system's impacts.

14.8 Mark up a request for proposal (RFP) for a wireless network with 20 nodes, covering a firm's one-square-kilometer campus.

14.9 Interview the telecommunications manager of your organization and determine (1) methodology for developing new systems and (2) method of managing configuration for their systems.

14.10 Find an organization that has a significant telecommunications infrastructure and determine the extent to which their employees have network certification, the policies about such certification, and which the organization emphasizes most.

Appendix **A**

Analog Voice Capabilities

As the world moves to wireless devices, especially the wireless cellular telephone, it would appear that the wired **telephone service,** that is, the plain old telephone system (POTS), has become extinct. Such is not, however, the case, as organizations continue to rely on this tried-and-true technology. While the distinction of local service and long-distance providers have blurred, the wired **telephone** remains a valuable business resource and one that continues to provide service and incur a cost.

The wired analog telephone was the basis of organization **communications** during the 20th century. Although this technology is being supplanted by cellular and VoIP technologies, it remains a vital capability for many organizations. POTS goes everywhere in the industrial world, making it global connectivity, even though it is analog and somewhat noisy. We provide the discussion that follows as a refresher into this technology.

TELEPHONE CHANNEL CAPACITY

A circuit or medium for a telephone channel is historically UTP. The *analog bandwidth (the range of frequencies) for a telephone channel is 4,000 hertz (Hz),* or 4 kilohertz (KHz). As Figure A.1 shows, the **voice spectrum** within the telephone channel covers the frequencies from 300 to 3,000 Hz[1] and is 2,700 Hz wide, with the remainder of the channel space being allocated to guard bands that provide separation space. Note that the bandwidth is not related to the actual frequencies used for voice transmission, only with the difference between the upper and lower limits of the range. This range of 2,700 Hz is adequate for voice communications but becomes a limit when we wish to carry more analog or digital information, such as high fidelity music or digital data at a high rate of speed. While the **public switched telephone network** provides access to almost every home and office in the developed world, the size of this *information pipeline* is restricted, and, thus, so is the speed or quantity of information flow. (As with the garden hose, you can pump only so much volume through a small pipe.) In addition, much of the spectrum is not available due to the equipment and not the medium. Twisted-pair wire can handle a greater bandwidth, but the equipment limits it.

---

[1] If you read other telecommunications books, you will likely find differing views on the frequency bandwidth of the telephone voice spectrum. Most agree that the lower limit is 300 Hz, but the upper limit is stated to be 3,000, 3,100, or 3,400 Hz. Even a member of a Telco did not know the bandwidth because it was not a limiting feature for them. It is a limiting feature for companies like Hayes Corporation who make modems. The important point is that the bandwidth is very limited.

FIGURE A.1
Frequency Bandwidth for a Voice Channel

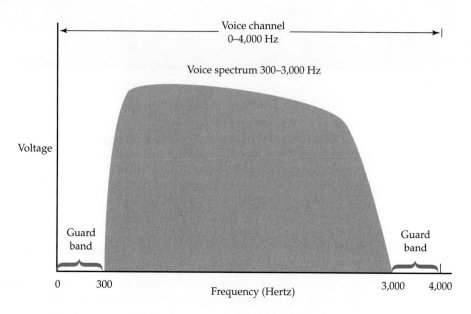

CENTRAL OFFICE, SWITCH, AND INSTRUMENT CAPABILITIES

The very first telephones were directly connected from one person's instrument to another. This meant you had to have a phone for each person to whom you talked. Then a switching office was developed so that local loops went from residences and offices to the *wiring frames*[2] in the switching office, called the **central office (CO).** A person would make the connection of one local loop to another in the central office by physically plugging a short wire from the connector attached to one local loop to the other. This required one or more human operators to make all connections. (Since the operators were often located in the central office, they were referred to as Central.) Alabama in the early 2000s had 5 million people, four major cities, 34 telephone companies, 364 central offices, 307 **exchanges,** 106,431 miles of aerial and buried copper cable and fiber, and 20,211,245 access (subscriber) lines.

When the sender is connected to one central office and the receiver to another, the call goes between the two COs via a **trunk line** or trunk cable. If the two COs are in the same calling area, there is no charge **(toll)** for this connection; otherwise the call is subject to a *long-distance toll* and the circuit is called a *toll trunk.*

With the invention of the **Strowger step-by-step switch,** human operators were replaced by switches for local connections. **Long-distance** connections continued to be made by human operators. It was this use of switch technology that allowed telephone systems to expand quickly while reducing costs. This use of switching technology continues to develop today, having gone through five distinct generations: manual switchboard, step-by-step technology, crossbar technology, solid-state technology, and digital switching.

The switching equipment in the CO is activated and operated in conjunction with the telephone instrument. The first instrument using the switch was the **rotary-dial telephone,** which created electrical pulses by opening and closing a contact in the phone when the round dial was twisted and released. The second and current equipment, called **TouchTone®** by

[2] A wiring frame is the physical structure that holds the end of the local loops. The circuit is then connected to the switch. Changing locations of telephone instruments without changing telephone numbers previously necessitated moving the connections from the switch to the proper local loop on the wiring frame. This is now done with software in the switch.

FIGURE A.2
Dual-Tone Multifrequency (DTNMF) Touchpad

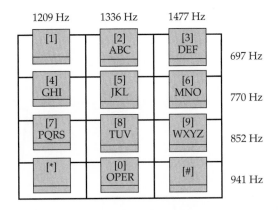

AT&T, generates tones that operate the switches. The importance of this change in technology is that this equipment is really a small 12-key terminal attached to the (now digital) central office switch. The digital switch is really a special-purpose computer; therefore, you have a small terminal on your phone connected directly to the computer. The **touchpad** telephone also allows you to command the telephone switch or other computers, using the 10 digits and the * and # keys. (See Figure A.2.) This has significant implications for telephone systems since the caller can now command the system to perform functions, such as entering an account number and responding to computer-generated questions. This increased functionality, discussed later in this appendix, enables the so-called *intelligent telephone system*. Rotary phones generally cannot perform this same function because the switch does not easily interpret the pulses as codes, rotary phones do not include the # and * keys (e.g., function keys), and they are far slower. (As shown in Figure A.2, the **dual-tone multifrequency [DTMF]** touchpad uses one tone for each column and one for each row: two per code. This is a total of 7 tones to create 12 codes.)

Whether access to the central office (wired) switch is by pulse or tone handsets, the result of the connection is a dial tone. A **dial tone** is a sound you hear on the telephone instrument when you pick up the handset (going off hook), indicating that the **CO switch** has acknowledged your request for a circuit and is ready to make the connection you request. *Dialing* is a process that signals the **network** by sending pulses or tones to the CO equipment, where they are interpreted and the proper circuit is established. This circuit connects the calling and the called telephone instruments. With a cellular telephone, there is no dial tone. This is because the cell phone sends a data packet with the desired information to the cellular switch over a data channel before a voice channel is established. On the other hand, the wired system gets the attention of the switch, which sends the dial tone.

An example of how the basic wired telephone circuit operates is depicted in Figure A.3. The calling party receives a dial tone, indicating access to the network via the local loop to her CO switch. The CO-to-CO trunk is representative of any number of connections necessary to connect one CO to another. If the call is out of the **local calling area,** this trunk will be long distance and incur a toll. If the two parties are connected to the same CO, there is no trunk used at all. The sequence of numbers she has dialed has caused the switch in her CO to pass the connection via a trunk to another CO that is connected to the called party's local loop and telephone. In the second CO, based on the numbers she has dialed, the equipment determines that the called party's telephone is available or not available and responds accordingly. If the called line is busy, a busy tone is generated and returned to the calling party. If the called party is available, the caller will hear a ring that is generated by the second CO switch and the calling party also will hear the phone ringing. The two sounds are generated by separate equipment and the called party can actually pick up before the calling party hears a ring. When the called party takes his/her instrument off

FIGURE A.3 **The Basic Telephone Voice System**

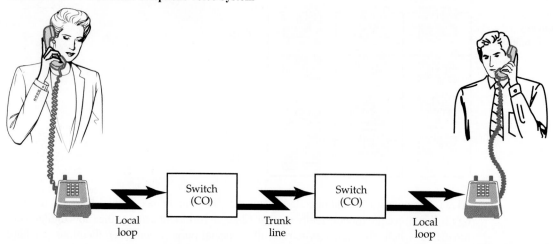

Local loop · Switch (CO) · Trunk line · Switch (CO) · Local loop

hook, the network connection is completed and the calling party may converse. Both individuals are able to hear themselves as well as the other party by way of *sidetone,* which is normal speech returned for hearing yourself as you talk. On the telephone, sidetone provides feedback on how you sound. Without sidetone, your speech will become garbled because there is no feedback.

The **plain old telephone system (POTS),** using the *Signaling System Seven* protocol, includes the telephone number of the calling party between the first and second ringing signals. If the receiving party has arranged for Caller ID®, s/he will intercept this information and the telephone number of the calling party may be displayed, along with other information. It is possible to block this information at the sending end. Some people will not accept any calls that do not include Caller ID.

The old **Bell System,** through its manufacturing division Western Electric, took great care to develop a reliable and humanly compatible telephone instrument. It evolved from the wall-mounted instrument, to the candlestick model, and then to several versions of the tabletop model. All of these were simple, or dumb, instruments as they had no switching capability within the physical telephone instrument. The exception is a *key instrument* that has the ability to switch between several lines coming to the telephone. With changes in AT&T and the Bell System, the ownership of the physical instrument became the responsibility of the user. The ability to incorporate extra features into the telephone has changed the nature of the instrument. You now have a choice in price, quality, and features, allowing customers to choose a low-priced, simple instrument or one with many added features.

Capabilities of telephone instruments now include redial of the last number called, storage of often-called numbers, display of the number called and the time, the ability of several people to hear the conversation through a speaker in the phone, placement of the caller on hold, and automatic retry of a busy number. While some of these features are also available for a fee in the switch, many telephone instruments now have a high level of capability and intelligence and provide greater flexibility; however, they are usually much more difficult to use.

As the signal propagates along the twisted-wire circuit, whether the local loop or a trunk, the circuit experiences signal *attenuation* (reduction in strength). This means that the **voltage** level of the signal is lowered as the signal moves down the wire. To keep the signal at an acceptable level, amplifiers, or repeaters, are installed about every mile of the circuit. These repeaters amplify the signal and keep it at the predetermined level. However, any induced noise is also amplified, reducing the **signal-to-noise ratio.** Thus, in an analog system, the farther the voice communication signals travel, the more noise and,

FIGURE A.4   Possible Call Circuit from Atlanta to Eugene via the Telephone Network
Dashed lines represent some alternative routes.

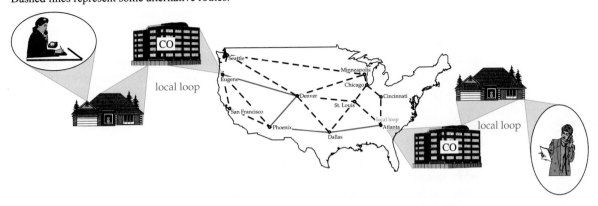

consequently, the lower the signal-to-noise ratio. The objective of a voice circuit is to minimize noise and have the desired signal strength be much greater than any noise, for example, a high signal-to-noise ratio. Therefore, it is not difficult to understand why long-distance calls were of lower quality than local calls. (When we discuss digital transfer of voice communications, we will learn that this does not have to be true).

AT&T and other companies have created a vast telephone network in the United States. A signal can be sent from any phone to any other phone in the United States and much of the rest of the world. In creating the path from one phone to another, the network connects many links. To give you an idea of how this process works, assume that you want to call from Atlanta, Georgia, to Eugene, Oregon. (See Figure A.4.) Here is what happens: The telephone is dynamically switched based on the number dialed. Even though the call could be routed different ways each time the call is made, you are not charged for incremental distances. (*Voice traffic was, historically, charged [tariffed] based on duration and distance.*)

- When you pick up the phone in Atlanta, you connect your instrument to the local loop running from your residence or office to the local telephone company's central office.
- A switch at the central office connects you with a line to the long-distance system (LDS) in much the same way you would be connected to another phone for a local call. (There may be a connection from the initial central office to another before connecting to the long-distance service. This cable is called a trunk)
- The long-distance system provides a circuit of one or more links to a central office in or near Eugene, Oregon. Because of the vast voice communications network, your circuit may not be the most direct but simply the most available at the moment of your call. This ensures a minimum delay and the greatest probability of getting a circuit. For example, Figure A.4 shows that this long-distance connection goes from Atlanta, to Dallas, to Phoenix, to Denver, and then to Eugene. However, there are literally thousands of other possible connections, including going through Chicago, Minneapolis, or Seattle. Furthermore, each point-to-point circuit may be any of the possible media.
- A line connects you from the long-distance system to the local telephone company central office wiring frame.
- Finally, a connection is made to the local loop running to the receiving instrument.

With the evolution of switching equipment (now called *the switch*) to digital-controlled computers, enhanced services not possible before can be provided. Many of the services are made possible by storing data in permanent memory in the switch. For example, the *last number calling* your number is kept in storage, and you can call this number by commanding the switch to access the memory and place the call by pressing *69. Such commands would generally be generated with codes from your touchpad phone starting with a *. Similarly, you can have a list of numbers give distinctive sounds, keep a list of numbers you do not want to receive calls from, and have your *calls forwarded* to another number. With *call waiting,* you can connect your instrument to a waiting call and place the present caller on hold. You can make conference calls and even a record of nuisance calls that will stand up in court. For a complete list of these services available from your local Telco,[3] consult your telephone directory.

911 EMERGENCY

In the United States, most communities have a simplified emergency telephone number. In order to help with speed dialing, the sequence 9-1-1 is employed. A 911 call is automatically routed to emergency response units: police, fire, and ambulance. If the call is made to a PBX (private branch exchange), using either 911 or 9-911, the PBX is smart enough to know whether to forward the call to an interior phone or outside to the standard 911 destination, even when the 900 area code is blocked.

REVERSE 911

In an extension of the emergency 911 call-in capability, there is a move to employ the system in reverse. In the case of an emergency situation such as an impending flood, storm, and so forth, emergency units can notify a large percentage of the population to take appropriate measures. This is in addition to other warning methods. This was done in smaller U.S. towns in the 1930s and 1940s when the local operator would manually connect a large number of phones and continue to repeat the message. This is how one of the authors learned of the end of World War II in Europe. Both of the above are reliable because POTS and cellular systems work even during most power outages. This is called a *life line.* Cordless phones on POTS do not.

COMMUNICATIONS CIRCUITS AVAILABLE

The most obvious choice of a communications circuit for end users is the local **telephone system (POTS[4]).** If the number you are calling is within your local calling area, you simply dial the number and make the connection at no additional cost. If the number you are calling is not in your local calling area, you have several choices. First, you may choose to dial direct and pay for the phone call at normal "long-distance" rates, which are generally lower after business hours and on weekends. An alternative for businesses that utilize the telephone heavily is to contract for **wide area telecommunications service (WATS).** The premise of WATS is that the business pays a fixed charge for a specific number of hours of phone services per month to a select group of states plus a reduced cost per minute of actual use. The total cost is lowered through this high-quantity discount method. A variation on this theme is **inward WATS** (in-WATS) service. In this case, the receiver pays all toll charges. This is the well-known 800 (and the new 888 and 877) area code and provides what appears to be a local call for a long-distance connection. It would be difficult to find a business today

---

[3] The term *Telco* is used to mean any local telephone service provider.

[4] POTS stands for the plain old telephone system, which is the nationwide voice capability often referred to as the Bell System. It is envied by less-developed nations and provides connectivity and communications between almost any home and businesses in the United States. We use the term *POTS* in this book to refer to this public switched telephone (voice) network.

that does not have in-WATS for its customers. Alternately, individuals can get in-WATS service so their friends and family can call them and pay no fee.

Another area code that has special connotations is the 900 area code. While the 800, 888, and 877 codes reverse charges, the 900 area code adds extra fees on the call. One valuable use of this feature is to allow customers whose equipment is out of warranty to call a 900 area code number for assistance, knowing they will be charged a fee for service, say $1.00/minute. As the 800, 888, and 877 numbers are all allocated, new in-WATS area codes must be created.

If a business is located in a town outside of a major city's local calling area and there is heavy telephone traffic, you can lease a telephone line (a circuit) that connects your switchboard to the city's central office switch. This procedure entails a fixed rate for the leased line but incurs no usage cost. This type of line may even be *conditioned* to provide a higher-quality, cleaner communications environment. Conditioning is important to data communications by reducing the noise environment and providing greater speed. The lessee can achieve higher quality by renting a known permanent circuit rather than utilizing a temporary, pieced-together, dial-up circuit. Replacing switched connections with permanent connections and tuning the circuit to specific electrical values achieves additional quality (a cleaner circuit). The **leased circuit** may offer the exclusive use of a twisted-pair circuit or microwave channel,[5] which gives you the total capability of that physical medium, or just a single 4 KHz analog channel or 32-to-64 Kbps circuit of any physical medium. The greater the capability and the cleaner the environment, the greater the cost. The total cost of the leased circuit, however, will generally be much lower than the equivalent dial-up circuit *if* it supports a high volume of traffic.

PRIVATE BRANCH EXCHANGE (PBX) VERSUS CENTREX

A **private branch exchange** is a multiple-line business telephone system that resides on the company premises and either supplants or supplements the Telco local services. To an extent, it is in competition with the services provided by the central office, for example, **Centrex®**. Centrex services are PBX services offered by a **common carrier,** that is, a group of lines and services owned, provided by, and located in the *CO switch* in a way that makes them appear as a separate facility for the customer. The Telco maintains the equipment and charges the customer for this service. In reality, Centrex service may be little different than standard local service except for a few added features and possibly quantity discounting. Both Telco services are regulated by state and federal agencies. In contrast, a PBX is **customer premises equipment (CPE),** is unregulated, and is owned and maintained by the customer. A company with 100 local telephone lines, standard telephone equipment, and either standard or Centrex service runs 100 lines from the CO switch to instruments on the premises. When a PBX is involved, fewer lines to the CO switch can be used. The result is lower costs for lines but a customer investment and cost in PBX equipment and maintenance. Because a large portion of an organization's communications is interior to the organization, most of the communications do not go beyond the PBX, which reduces the need for outside lines to the CO switch. There may be fewer than 30 lines running from the PBX on the premises to the CO switch, with 100 lines radiating out from the PBX to the telephone instruments. It is estimated that as of the beginning of the millennium, there are in excess of 30 million local lines terminated in PBXs. (See Figure A.5.)

---

[5] As we cover in Chapter 3, a circuit is the physical medium, such as a twisted-pair link. When you have a circuit, it is assumed that you have the total bandwidth it has to offer. A channel can be either the total circuit or a portion of it. Thus, you can divide a circuit into several channels. Any microwave channel is generally only a part of the circuit.

FIGURE A.5 **Various Forms of Phone Lines**

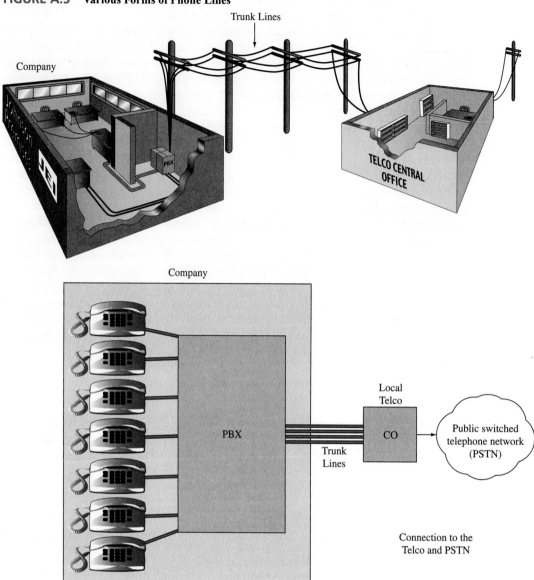

A PBX Network

A significant feature of the PBX is its ability to record call activity and create reports. For example, the equipment makes records of all calls and callers, giving an accurate account of this expense. Additionally, the reports show times of high activity and inward call blocking, noting the need for additional equipment or outside lines.

A business alternative to a PBX is to have the local telephone company provide the service. Such service can range from simple enhanced phone capabilities, such as **multiple lines,** intercommunication (*intercom*), and speaker phones, to Centrex. In Centrex service, the Telco owns and maintains the equipment, but it dedicates the capability to your business. An advantage of using the phone company and Centrex is that a third party is responsible for all service. It should be noted that some Telcos have their own registered names for Centrex; for example, BellSouth calls their service *Essex®*.

When a company progresses to a point of making a PBX-versus-Centrex decision, management must be aware of having growth built into the choice. For example, purchasing a nonexpandable PBX locks the company into a specific technology with finite limits. *Leasing Centrex means someone else (the local Telco) maintains the equipment and manages expansion and growth. Purchasing your own PBX means YOU provide this maintenance and management,* hopefully at a cost savings. This operation and maintenance, however, can be purchased from the vendor or a third-party firm. If you are risk averse, buy expertise by choosing Centrex.

Important features that could be a part of or ancillary to a PBX are **automatic call distribution (ACD)** and **voice mail.** The former adds intelligence to the receipt of incoming phone calls to channel them to a vacant instrument, often with announcements (the familiar menu you hear when calling a business). The latter is used with or without ACD to allow reception of a phone call when the recipient is busy or absent. Voice mail can be viewed as an intelligent answering machine.

Since we are studying the management of telecommunications, you should remember that the use of ACD and voice mail is first a management decision before it is a technical decision. Both features place a machine between the caller and the called party.

Have you called an airlines reservation telephone number, or the IRS, or a computer software discount house, or some other merchant lately and had the computer tell you that all agents were busy, but please stay on the line for the next available agent? You called into an automatic call distribution (ACD) computer that was controlling the instruments of a number of people. For example, when you call (800) 433–7300 for American Airlines or (800) 221–1212 for Delta Airlines, the ACD system does the following just prior to answering your call:

- Determines which reservation center around the country (Atlanta or Salt Lake City) is lightly loaded and most likely to have an available agent.
- Determines if there is an available agent at that center.
- Connects you with the agent or answers your call and gives you a message to wait on the line.

If you called into Microsoft Corporation, developer of the disk operating system (DOS) and the developer of the Windows operating system for IBM and compatible computers, the ACD always answers the phone and gives you choices as to how your call is transferred. You have to answer four questions by pressing numbers on the touchpad to finally get to the human of your choice, unless you already know the person's telephone extension. This may seem like a lot of work up front, but Microsoft is using telephone technology to ensure that its customers receive the best and quickest assistance possible. If you interact with the computer to get more information, you enter the more sophisticated realm of *interactive voice response (IVR).* As an example of IVR, let's examine a bank service of checking an account further. First the customer dials the number of the bank and the recorded voice asks that a choice be made from a menu that may include determining the account balance. If you respond by pressing the appropriate number key on your telephone, the computer asks for your account number and usually a *PIN (personal identification number).* Once the PIN is entered, the computer accesses the proper information and passes it through a voice response capability to speak the amount of the balance. The customer then may be asked to return to the menu for additional options, such as last deposit. In this case, the touchpad on the telephone is an input device, so the telephone is like a terminal and the computer uses data-to-voice translation to provide voice response in an interactive mode.

TABLE A.1
Centrex versus PBX
Considerations

Feature	Centrex	PBX
Initial cost	Low	High
Leasing cost > 500 lines	High	Attractive
Maintenance cost	None	All
Cost to move instruments	High	Low
Local control	Low	High
Selection (least-cost) routing of long-distance calls	None	Option
Software upgrades and costs	None	User responsibility
Changes to new equipment	Telco provides	Fixed
Ability to change	Telco decides	Customer decides
Move to ISDN	Telco provides	Customer funds
Insurance	Telco provides	Customer pays
Power consumption and costs	Telco provides	Customer pays
Capacity to grow	High	Low (limited)
Back-up power/processor	Telco provides	Customer responsibility
Housing space	Telco provides	Customer provides
Service mileage charge	None	Customer pays
Reliability (MTBF)	Highest	Medium-to-high

A cost consideration offered by PBX systems, and by outside vendors, is least-cost routing. With this service, the PBX connects to points-of-presence of several long-distance providers, for example, AT&T, MCI, or Sprint. The PBX will determine the lowest-cost route depending on the **destination** of the call and time of day. The person placing the call is unaware of this decision process at the PBX. Alternatively, DeltaCom®, a private corporation, provides much the same service. The customer places the call through DeltaCom, and its switch determines the lowest-cost route. DeltaCom is able to do this through negotiating rates with multiple long-distance providers, passing these rates, plus commission, to the customer.

Remember, Centrex services are not always equal. If your Telco has an AT&T **5ESS,** System 85, or equivalent *CO switch,* the switch already has the ability to offer significant flexibility and services. You may have seen advertisements for the services offered by Telcos, especially the Bell System, on television. You already may have call waiting, call forwarding, call blocking, distinctive ringing, last caller, and Caller ID, features previously offered only on PBXs. (Caller ID is available only with digital PBXs and CO switches like the digital 5ESS.) Thus, when considering Centrex versus PBX, you must first determine the capabilities or generation of the CO switch. A second consideration is the nature of regulation imposed on your Telco. If the local phone company is not authorized to offer tariffed services, you have little choice.

When choosing between Centrex and PBX, areas of concern include reliability, cost, maintenance, features, speed, vendor support, and flexibility. Table A.1 shows some of the differences between Centrex services and a PBX. Managers must carefully examine the trade-offs to make the *correct business decision.* Certainly, the actual cost figures should be used within the analysis! The value of the qualitative aspects must be assigned by the manager, based upon the specific situation.

Digital PBX

Although we have discussed digital equipment and transmission in earlier chapters, we cannot leave the topic of private branch exchanges without discussing the trend toward **digital**[6] PBX equipment.

---

[6] Briefly, digital signals are discrete values, such as 15 volts and zero volts, to represent units, for example, 1s and 0s. This is a different way to transmit information and provide control. Analog signals, on the other hand, are continuous in value.

An advantage of the PBX, in addition to increased control and lower cost, is that some systems are entirely digital. Thus, you can plug your terminal or computer into the telephone instrument and communicate with devices under the control of the PBX without using the intermediary equipment, such as modems). With this capability in-house, you could connect to a digital telecommunications network and never convert the signal. This capability would provide the cleanest and fastest switched circuit available.

Rolm, formerly a part of IBM and now part of Siemens, produces one of the digital PBX systems. When connected to a digital switch at the CO, such as AT&T's 5ESS switch, a digital path is provided through the switch, but the remainder of the local loop is analog. The Rolm system, on the other hand, provides a digital path from voice handset or computer at one end to similar equipment at the other end. This is a relatively recent improvement of equipment; the PBX has traditionally been separate from data communications. As the world moves to digital channels in order to reduce noise, increase control, and combine voice, data, image, video, and text, digital PBXs will become very important. An example of further adaptation is the connection of the PBX to local area networks (LANs), giving a digital path and control from computer to computer via the network. The extension of this is *ISDN (integrated services digital network)*.

Size matters! When switches were first installed, they occupied large rooms. You can now buy a 12-line PBX that is a circuit board and will fit in a server computer. Thus, small companies can have the value of a PBX and install it in a computer, providing additional features of computer-telephony integration, to be discussed next.

Computer-Telephony Integration (CTI)

The objective of Centrex and PBX switches is to add the decision-making ability of the computer to the switching capability of the telephone. The term *computer-telephony integration (CTI)* has been used to describe this trend. The simplest form is to have the PBX extract the Caller ID signal as it forwards the call to the appropriate person. At the same time, it sends the caller's telephone number to a computer where it is compared to a customer database. By the time the phone is answered, the computer has sent the customer's file to the computer at the same desk, giving the customer service representative knowledge of the account. Taking this further, Siemens Rolm Communications and Aspect Telecom offer products that allow monitoring of remote workers, offer advanced automated call distribution, and so forth. The Siemens Rolm software allows the using firm to track up to 250 skills and assign each a proficiency level and preference rating, so when callers indicate a problem, the system will route the call, based on information provided to the agency, to the appropriate skill level.

Leased versus Switched Lines

When determining whether to use standard, Centrex, or PBX services, consideration also must be given to which types of lines to use in high-volume situations. This is not in reference to high-volume WATS service, but to whether a leased line can reduce your cost. For example, a manager is considering installing a leased **(foreign exchange)** line to a nearby major city so his customers in that city would not have to pay long-distance charges. This means that the customer dials a local number that is connected to a distant central office. This gives you an idea of the cost trade-off of fixed-cost leased line versus switched variable cost line. An FX line gives the appearance that the organization is local, which could be a **competitive advantage**.

Depreciation and Other Concerns

Many decisions are made based on accounting standards and tax considerations as opposed to technology features. This is particularly true for the Centrex–PBX decision. If the decision is made for Centrex services, there is no concern for amortization or depreciation of equipment. The vendor, not the organization, purchases all Centrex equipment. However, with a purchased PBX, the equipment must be depreciated over its useful lifetime. While the time period for depreciation for telecommunications equipment in general formerly was 10 to 15 years, it is now from three to five years. The depreciation time or, more accurately, the useful lifespan, for PBXs is presently from four to seven years. While the equipment may easily last several times that number of years, its technological usefulness does not. The different effects of different depreciation schedules are important due to the time value of money.

Software upgrade and costs can be very significant with PBXs. In one case with which the authors are familiar, the firm had to pay over $30,000 for upgrade of a $70,000 PBX simply to accommodate new area codes.

In the modern era of rapidly evolving technology, the useful lifespan of telecommunications equipment continues to shorten. As new features are developed, they replace older ones. Staying with older technology may be cost-conscious while moving to new technology keeps you current with customers and competitors, possibly resulting in cost-effectiveness. Additionally, in a dynamic market, many vendors will fail, often leaving you with equipment that can no longer be readily supported and maintained. This may cause you to seek third-party suppliers who support such equipment at premium prices. This further presses the telecommunications manager to purchase from established suppliers, hurting the chances of new entrants in the market. Further, if you purchase new systems that do not have the ability to grow and be upgraded, you will find that you have bought into a dead-end position as the market moves. As many organizations have found, they live with their decisions for a long time.

The choice between (1) the acquisition (lease) of standard lines from the local Telco, (2) the contractual agreement with that same Telco for dedicated Centrex services, and (3) the purchase of a PBX is a management decision. It revolves around two concerns: which systems best provide the services you need now and in the near future and which systems cost least. A hidden part of this decision is the need for maintenance and upgrade. When management chooses to lease services from the local Telco, it leaves maintenance and upgrade in the hands of the lessor. When it chooses to purchase a PBX, it assumes all responsibility for, not only installation, but repair, maintenance, and upgrade. Where a PBX may indeed provide greater services at a lower cost, it does mean a greater responsibility and added cost in the form of repair and maintenance.

SUMMARY AND CONCLUSIONS

The wired, analog telephone system may seem like a historical artifact to many, but it remains valid technology. It is, at the very least, a point of reference for the way organizations progressed from written, delayed communications to electronic, immediate communications. Mobile wireless and VoIP digital telephony are supplanting wired analog telephony because they provide different and more valuable services, in addition to allowing the user to be untethered and accessible anywhere. Meanwhile, the telephone on your office desk is a mainstay of organizational communications, even though the technology supporting it may change.

Appendix B

Epilogue: Emerging Technologies, Innovation, and Risks

Forecasting the future of data communications is a dangerous temptation. Forecasting is, at best, an inexact science. After many years and countless attempts at modeling, weather predictions (temperature and precipitation) in the continental United States are accurate only about 70–80 percent of the time. This is one area that is fairly good; most others are way behind. Thus, forecasting the future of network and data communications falls in the "iffy" category. One telecommunications executive in a major TC firm who was asked for a five-year forecast by the CEO responded that he could not make even a one-year forecast with confidence. Considering the geometric acceleration of change in the field, we would make such forecasts at our peril. Instead, we address forces, in both the general area of technology and the specific field of data communications systems, that should cause change. Voice and data systems will continue to evolve and change how we conduct business, compete, and live our lives. We suggest the following as a guide, and not a specific forecast.

Telecommunications changes the home, office, and business. It can enable anyone in the world to communicate with anyone else at nearly the speed of light. The information superhighway (the Internet) connects us and businesses globally and around the corner.

Where are the customers and competitors? Geographically, they can be anywhere as long as data communications facilities are available. To access and support customers, firms use customer relationship management (CRM). To gain and maintain a competitive advantage in the global community, many firms employ enterprise resource planning (ERP) systems. In order to maintain connectivity to suppliers, organizations use supply-chain management (SCM). Most of these and other enterprise applications are not possible without broadband data communications. Data communications can reduce reaction time and increase customer service or even facilitate the creation and delivery of innovative information-based products.

In the next section, we look at factors that provide pressure to adopt technology, especially data communications systems. In addition, we note those that inhibit this movement.

THE FORCES DRIVING THE ADOPTION OF TECHNOLOGY

One way to think about future events is to look at the pressures, advances, and changes that cause technology to be adopted. The following is a list of such forces generated by a Delphi technique among faculty members in an academic department:

1. Shortening of product development cycle times.
2. Increasing global competition.

463

3. Increasing customized production versus mass production.

4. Emphasis on quality.

5. Customer orientation.

6. Proactive versus reactive adaptation.

7. Increased use of collaborative work.

8. Decentralization of functions with centralized control.

9. Work teams (cross-disciplinary versus specialized functions).

10. Constant environmental change: economic, regulatory, and technological.

11. Constant organizational change and development.

12. Just-in-time (perhaps evolving to "last minute") flexible or agile manufacturing.

13. Partnerships with suppliers and customers (EDI).

14. Deregulation, which leads to increasing competition.

15. Decrease in cost-performance ratio of information technology.

16. Increased information technology familiarization through experience, education, and training.

17. Enhanced media exposure to technological advances, for example, AOL and WWW on television, newspapers; technology issues in *The Wall Street Journal;* and so on.

18. Deployment of greater bandwidth.

19. Growing number of generic types of technology available.

20. Different technologies that can be combined to form an increasingly large number of configurations, allowing customers to find just the right combination for their needs.

21. Standardization (especially global), which works to reduce switching costs and the risk of trying a new technology.

22. Improvement in choices: after some 30 years of making information technology adoption choices, decision makers are beginning to get good at it.

23. Growing wage differential between developed and less-developed countries. Developed countries must substitute technology for labor in order to compete with low-wage countries.

24. International push for open technology standards.

25. Security for networks becomes a major effort all over the globe.

26. Greater emphasis on survivable networks and disaster recovery in the face of major calamity/catastrophe

THE CONSTRAINTS TO THE ADOPTION OF TECHNOLOGY

Not all pressures are in favor of the adoption of new technologies, especially telecommunications. The following is a list of constraints to adoption that were generated by the same Delphi experience. When you consider the inclusion of a provision for regulating decency in the U.S. Telecommunications Act of 1996, the first item on the list becomes very relevant.

1. Ethical considerations: the Internet spreads pornography and hate literature as well as terrorist manuals.

2. Security concerns: identity theft on the Internet.

3. Privacy concerns: more databases are connected; organizations and governments have more personal information.

4. Human resistance to change: the use of checks is not decreasing despite the proliferation of ATMs.

5. Regulation: the FCC continues to dictate broadcast standards.

6. Complexity of technology: telephones can be difficult to use; the telephone bills are hard to read; choices are greater.

7. Workforce literacy and skills inadequate: new technologies cannot proliferate at a fast rate.

8. Insufficient supply of qualified labor: inability to implement technology effectively.

9. Insufficient ability to increase market share: not enough to absorb higher output (productivity increase) provided by new technology, making it difficult to justify the capital costs of purchasing such technology (particularly a problem when technology also is adopted by major competitors, thus creating excess product/service supply relative to existing product/service demand).

10. Short-term management perspective: too little time for realizing expected gains from investment in new technology.

Some organizations may have a high propensity to take risks (low risk aversion) and become early adopters of technologies; others may have great risk aversion. Those who are risk averse are likely to be laggards in adopting new technologies. Managers should realize that there are risks associated with both extremes. The early adopter may bet it all on a technology that fails, while the laggard may never be able to catch up to a successful early adopter.

TECHNOLOGY AND APPLICATIONS TRENDS

Data communications technologies and the applications they support continue to change at a very rapid pace. These changes must occur to support the fundamental changes in organizations and the shifts in the way we do things in general. Leaders in information technology understand that data communications technologies enable the distributed enterprise, and they must keep pace with the increasing rate of change and the unrelenting competitive intensity of the global marketplace. Some of the centers of attention are the Internet and electronic commerce, distributed systems management, and enterprise networking, which are discussed below.

Doing Business on the Internet

In the mid-1990s, the Internet was a lightning rod attracting attention as the initial manifestation of the information highway. The World Wide Web was seeing the generation of business home pages at a phenomenal rate. Businesses were rushing to establish an Internet presence in order to stake a claim and to extend their overall communications reach. While the explosive growth of the Internet continued unabated, some critics doubted the ability of this aging technological infrastructure to keep pace with demands. Others, however, saw in the Internet the prototype of a new world of networks that was about to forever change the way business would be done. Electronic commerce appeared to have found a new and irresistible channel. The efficacy of a network of networks was just beginning to be exploited with predictions of retail sales via the Internet of more than $4 billion early in the new millennium. Electronic commerce did not mean just EDI anymore. Electronic commerce via the Internet was off and running.

The widespread popularity of the Internet led to a rapid proliferation of so-called dot.com firms with exuberant financial backers in the late 1990s. Many of these firms failed as the U.S. and world economies went into recession. The surviving e-commerce companies appeared to be those with sound business plans. Despite the contraction in the number of firms, there was continued growth in actual e-business transactions.

Rare is the business today that does not have a Web presence. If they are in retail sales, they most likely have a sales site. Those working farther up the food chain most likely have Web sites for business-to-business transactions.

Distributed Systems Management

As business systems become enterprisewide and more complex, the management of these distributed systems becomes increasingly difficult. Trends such as client-server environments presented many new challenges. The clients in this architecture often became "fat" and overloaded their servers. Managers faced very difficult tasks in performing capacity planning, disaster management, and so forth, that were essential to meeting the expectations of both top management and users. The trend toward more client-server installations will exacerbate this problem unless new technologies are implemented to handle demands. As organizations become increasingly dependent on enterprisewide distributed systems, their management will be a priority. Distributed systems require management all the way from customer contact representatives to back-office knowledge workers. Managers must keep abreast of the telecommunications requirements of distributed systems or be doomed to experience system failures.

Distributed systems can be outsourced as well as being corporate-owned assets. The dot.com experience of the 1990s showed that outsourcing is very valuable because the need for corporate servers and processing was reduced. Passing applications to ASPs and storage to server farms means greater reliability, possibly faster access, and definitely lower liability as the accountability moves from the balance sheet to the expense column.

Enterprise Networking

The distributed enterprise is absolutely dependent on a well-integrated, dependable, open communications infrastructure. This infrastructure not only must span the organization internally, but must reach out seamlessly to customers, suppliers, and partners as well. Managers must understand the choices in building the telecommunications infrastructure, the factors involved in deploying high-speed links, internetworking, evaluating and preparing for the implementation of new broadband applications, and their implications for competitive advantage. The deployment of ATM and SONET means that the managers must understand how to take advantage of the technologies or be left to play catch-up.

The trend to move data via the Internet gains momentum daily. This trend may alter the marketplace for many vendors as use continues to grow. For example, many firms now find that they can use the Internet for EDI as a replacement for VANs. Much of what we do can be delivered by the publicly available infrastructure. As firms become more comfortable with transactions via the Internet, their use of this resource will continue to grow. It appears that the networks of enterprises will be more Web-based, and tend to change the structure of the organization so it will not resemble the command-and-control hierarchies of the past. As Internet use increases, there will be pressure to modernize and improve the capacity of the Internet so that the infrastructure can support more uses and users.

It appears rather certain that nearly everything and everybody will be connected and that there will be more collaborative activity among both knowledge workers and smart devices. Geography no longer sets limits on where innovation, work, or transactions happen. Scalable, pervasive networks and new wireless technologies should eventually enable people to reach any other person or device. The infrastructure should evolve so that it will no longer be a limitation.

The virtual workplace will be more and more commonplace as "telework" gains popularity. Many employees will employ mobile computers, wireless connections, and so on, and will be "in the office" whenever and wherever they hook up to the network.

The concept of a virtual hospital is gaining acceptance. There were more than 50 medical centers in the United States offering telemedicine services at the beginning of this millennium. Improvements in the technologies of data compression, image quality, satellite, and fiber optic transmission should continue to bolster this trend. Advanced networks are providing high bandwidth and low latency, which should provide steps in the direction of scalable networks. As soon as ubiquitous, high-speed, dial-up bandwidth is widely available, scalable networks can emerge.

The addition of "smart devices" to data communications systems should increase. The installation of intelligent agents as network brokers will likely be developed to insulate users from the underlying complexity of the networks.

Technology trends such as inexpensive lasers, fiber optics, compact disc–based multimedia, and conversion to digital communications that helped shape the 1990s should evolve further. One development that is expected to have high impact is micro electrical mechanical systems (MEMS) technology. This technology is anticipated to lead to several new uses of sensors of many types combined with telecommunications links.

Telemedicine

Telemedicine (also referred to as "telehealth" or "e-health") allows health care professionals to use "connected" medical devices in the evaluation, diagnosis, and treatment of patients in other locations. These devices are enhanced through the use of telecommunications technology, network computing, video-conferencing systems, and coder-decoders (codecs). Specialized application software, data storage devices, database management software, and medical devices capable of electronic data collection, storage, and transmission are all key components of the telemedicine infrastructure.

Telemedicine customarily uses two methods to transmit images data and sound: either "live," real-time transmission, where the consulting professional participates in the examination of the patient while diagnostic information is collected and transmitted, or "store and forward" transmission, where the consulting professional reviews the data whenever time is available. Many programs employ both transmission capabilities to maximize efficient use of resources appropriate to the medical services being provided.

AMD is the worldwide leading provider of "connected" medical devices, peripherals, and software used in telemedicine. With more than 2,000 installations in over 40 countries, AMD brings a wealth of experience and expertise to your telemedicine program. AMD provides complete device and software solutions, backed up by expert integration, customer service, and training support.[1] One may find additional material at the following Web locations:

- Telemedicine Information Exchange (TIE): http://tie.telemed.org/.
- The American Telemedicine Association: http://www.atmeda.org/.
- Telemedicine Today Magazine: http://www.telemedtoday.com/.

Nanotechnology

Nano, Greek for *dwarf,* is a hybrid science: the modeling, measurement, and manipulation of matter on the nanoscale. This deals with substances that are 1 to 100 nanometers across. Nanoscale electronics, of concern to us, turn single molecules into a switch, conductor, or other circuit element.

IBM built an entire logic circuit on a carbon molecule 1.2 nanometers wide (it takes 80 thousand of these nanotubes side-by-side to be the diameter of a human hair). HP has created circuits by manipulating molecules chemically to form components and paths at this molecular size. Thus, as nanotechnology evolves, it promises very small devices that can be

[1] http://www.americanmeddev.com/about_telemedicine.cfm.

imbedded into other objects and systems to sense, manipulate, and transport data. Nanotechnology commercialization may be based on the molecular self-assembly of circuits.

Universal Description, Discovery, and Integration

Universal Description, Discovery, and Integration (UDDI) is a set of specifications that were designed to help firms publish information about themselves, their Web service offerings, and the required interfaces for linking with those services. UDDI combines an electronic "white pages" that list basic contact information, electronic "yellow pages" with details about the company and the company's electronic capabilities for trading partners, and "green pages" with standards and software interfaces to comply with an order to execute electronic functions using XML as their common language. UDDI was jointly developed by Ariba, IBM, Intel, Microsoft, and SAP.

Wireless and Mobile Technologies

Wireless technologies should continue to proliferate. Even though one fiber optic strand has more bandwidth than the entire radio spectrum, fiber will never be used to the exclusion of radio. Many believe that wireless is the driving force in technology of this decade. Personal communication systems (PCS) will continue to expand rapidly. One may find that cellular telephones with software-defined radio for automatic signal adjusting will be marketed so that they may be adaptable to any protocol or technology regardless of geographic location. The popularity of wireless is leading to so-called hot buildings that are filled with Wi-Fi hubs to attract tenants who would have free access.

Mobile positioning satellites are already employed by automobile manufacturers, boat builders, and others to provide custom services such as alerting of emergency agencies, provision of navigation assistance, and others. The technologies for such communications will be designed into the products. As an example, an award-winning product at a recent consumer electronics show was a walkie-talkie with GPS and mapping built in, allowing one user to transmit a map of his/her location to another.

Better, Faster, Cheaper

Technology should provide faster and cheaper ways to deliver data, text, graphics, audio, and video. Some refer to the idea of telecomputing as a medium of human communication. This form of networked intelligence provides vast new business opportunities. It appears that multifunction interoperable devices (in terms of codecs and file formats) are likely to proliferate in a labyrinthine marketplace.

Two-way videoconferencing promises to be widely used in medicine and education as the technologies of broadband media and compression are widely deployed. Higher-speed microcomputers enable networks to an extent unimaginable. Movement toward standardization and demands for interoperability should help in achievement of a true worldwide network. This should further underline the globalization of business.

In the telecommunications arena, networks will have increasing bandwidth capability; DSL, frame relay, and ATM may proliferate rapidly. The unrelenting demand for more bandwidth has already sparked broadband price wars. Much of the old POTS network will need to be converted so that digital transmission will become the standard; otherwise IP telephony will replace most of it. While the old POTS infrastructure is likely to remain for some time, technological enhancement should enable the system to cope while the newer technologies are implemented and eventually supplant POTS.

Business telecommunications promises to be very dynamic as managers will be faced with integration of wireless applications. Wireless has found great acceptance and continues to grow with the advent of Bluetooth, Firewire, IEEE 802.11b, IEEE 802.11a, IEEE 802.11i (for advanced wireless encryption), wireless LANs, direct satellite delivery of data, and so forth. The main business reasons for this growth are the opportunities and problem solutions found in the wireless technologies.

One of the most important forces in furthering business telecommunications is the emphasis on customer and partner relationships. We continue to see customer access to the firm increasing. Likewise, firms have noted the efficiencies of having connections to suppliers and partners. In effect, the form of the business organization has changed dramatically. Both customers and suppliers are now linked to the business for better functioning and service as well as competitive advantage. Often we see the existence of virtual LANs connecting the important players.

In the world of the consumer, data communications appears to be assuming an increasingly important role. Consumers demand such services as remote banking, shopping, and customized services that, in turn, demand more sophisticated voice and data systems. As more consumers have high-speed Internet access, this demand will likely increase.

COST TRENDS

Cost of the appliances and computer chips that enable networks will likely continue to fall as they become commodity-like. Computer chips continue to be developed that vastly outperform their predecessors and at better cost/performance ratios. These chips allow for convergence of technologies into a single unit, for example, a palmtop that contains a cell phone, an MP3 player, GPS, and a digital camera.

The recent trend toward deregulation in many nations is driving competition with an expected lowering of costs for services as vendors battle for market share. The longer-term trend is to push prices of many services to a point just above their costs. An example is long-distance telephone service. It will soon have such a competitive structure that it may develop into a fixed-cost commodity. The changing nature of the carriers' environment has led to a great number of mergers, acquisitions, and spin-offs. There has been a trend toward consolidation of the older, traditional carriers that is likely to continue in the face of decreased margins, unconventional competition, and increased customer expectations.

Equipment costs are likely to vary considerably. As new technologies protected by patents are introduced, their costs can be high. However, cost of equipment in the highly competitive marketplace should trend downward when demand and prospects of profits attract more competitors. As many of the new technologies are widely deployed, their costs should drop.

We expect the content carried by the networks to be a higher-profit-margin area. Many telecommunications providers are scrambling to add content and/or make alliances with content providers because of this. NTT's i-mode and Microsoft's Xbox point the way. The Japanese NTT's wireless architecture called *i-mode,* supporting DoCoMo, provides wide bandwidth between mobile devices. This is the platform for mobile phone communications that has revolutionized the way nearly one-fifth of the people in Japan live and work. Introduced in February 1999, this remarkably convenient, new form of mobile service has attracted over 28 million subscribers. With i-mode, cellular phone users get easy access to more than 40,000 Internet sites, as well as specialized services such as e-mail, online shopping and banking, ticket reservations, gaming, and restaurant advice. Users can access sites from anywhere in Japan, and at unusually low rates, because their charges are based on the volume of data transmitted, not the amount of time spent connected. NTT DoCoMo's i-mode network structure not only provides access to i-mode and i-mode-compatible content through the Internet, but also provides access through a dedicated leased-line circuit for added security.[2]

[2] http://www.nttdocomo.com/.

On another front, Microsoft's Xbox game machine is designed to support Internet-based, global play between competitors or colleagues. The niche in the larger game market seems to be storage (hard drive) and (more importantly) high-speed connectivity. These two personal technologies point to the future, one that is wireless, personal, and connected.

BETWEEN E-MAIL AND CHAT

Two phenomena sweeping the world are SMS and IM. SMS (short message service) is prevalent outside of the United States and moves text messages from one cell phone to another. Estimates of 17 billion SMS messages are being sent each day, at a charge of 17 cents each. SMS is often preferable to a voice call as it takes far less time for the recipient to get the message.

The alternative of SMS (although it is slowly catching on) for the United States is IM (instant messaging). This involves, for the time being, software in your desktop that determines when the members of your buddy list have an active e-mail client. The IM capability can send very short messages that pop up on the recipient's system, without the overhead of traditional e-mail. A response is short, quick, and easy. New versions, coupled with good bandwidth, allow for the transmission of video to the buddy. This system has propagated to many ISPs and is free. A major challenge for U.S. providers is to make this provide a cash stream, as is done outside of the United States with SMS.

LEGISLATIVE AND REGULATORY TRENDS

The Telecommunications Act of 1996 was the harbinger of great change in the United States. Many other nations also are making fundamental changes to their laws and regulations that will cause the environments for telecommunications to be forever altered. Many countries have now allowed the entry of domestic and foreign firms to compete with their PTTs, formerly monopolistic providers of telecommunications. The relaxation of regulation means that more firms will be going abroad. European and Asian telecommunications firms will be active in the Americas and vice versa.

The trend towards deregulation should spur more competition globally and possibly influence more common standards. The networks that exist must be adapted so they can interact as a more common infrastructure for international or global interoperability.

In order for all nations to benefit from the explosive developments in telecommunications, it will be necessary for the trend toward deregulation to continue. Legislation must be an enabler and not an inhibitor. It should facilitate networking rather than impose barriers.

CONVERGENCE

We have traced the business use of telecommunications as well as viewed its historical development to provide a better understanding of the basis of today's systems. One of the factors that is changing everything is convergence.

Technological Convergence

The convergence of technologies is rapidly taking place. The most obvious is the convergence of computing and telecommunications, which has been accelerated by business applications. Firms see the potential of combining the power of computer-based information systems and telecommunications networks. Some of the driving forces are the Internet and intranet evolution; the World Wide Web has been the greatest change stimulus in modern times.

Digital Divide

Recent trends have seen developments aimed at reducing the digital divide. Several efforts are underway to produce inexpensive laptops (<$100) that could be widely distributed to less-developed populations. Some world organizations and national governments are focused on bringing the promise of Internet connectivity to millions of people who have insufficient resources and infrastructure. As the digital divide is made narrower, there are great potential impacts on the large populations who have had no access to the "information highway."

RISK ASSESSMENT AND DISASTER RECOVERY

Network security has become a top priority in the face of multiple threats. We must guard against intrusion and internal threats. Great amounts of money have been spent to combat spyware, phishing, pharming, and spam. Firms and individuals are likely to demand better protection from virtual criminals, hackers, and crackers as they become more dependent on their data networks. They also must be prepared to cope with natural disasters such as earthquakes, fires, floods, tsunamis, hurricanes, tornados, typhoons, volcanic eruptions, blizzards, and so forth.

Hurricane Katrina—A Wake-up Call

Tables 11.1 and 11.2 (repeated here for convenience) show the responses to two surveys concerning threats to computers and networks. The first was completed in 1990 and shows *natural disasters* as the top threat by a large margin. The second survey, completed in the spring of 2004, moved *natural disasters* to sixteenth place, with viruses taking the top spot. This seems to indicate a changing comfort level with physical security concerns and an increased awareness for virtual security concerns. As computer and network systems have matured, perhaps administrators have a better grasp of physical security and its components, such as physical safety, data backup, redundancy, and so on.

It is easy and convenient to entitle this part of the last chapter as a wake-up call due to the worst natural disaster to hit the United States in recorded history. We list the major natural disasters from June 2004 (after the research reported in Table 11.2) through September 12, 2005. These data comes from http://www.infoplease.com/ipa/A0001437.html

Year 2004

June, July, and August, South Asia. The worst monsoon flooding in 15 years in India, Nepal, and Bangladesh left up to 5 million homeless, killed more than

TABLE 11.1
Ranking of Threats to Computer Systems, circa 1990

Source: Karen Lock, Houston Carr, and Merrill Warkentin, "Threats to Information Systems: Today's Reality, Yesterday's Understanding," *MIS Quarterly,* June 1992, pp 173–86.

Threats to Information Systems Weighted (circa 1990)	Rank	Votes
Natural disasters	1	324
Accidental entry of bad data by employees	2	270
Accidental destruction of data by employees	3	252
Weak or ineffective controls	4	149
Entry of computer viruses	5	128
Access to system by hackers	6	123
Inadequate control over media	7	96
Unauthorized access by employees	8	93
Poor control of I/O	9	67
Intentional destruction of data by employees	10	58
Intentional entry of bad data by employees	11	36
Access to system by competitors	12	31

TABLE 11.2
Overall Mean Importance/Ranking of Threats, Spring 2004

Source: Houston H. Carr and Jason Deegan, unpublished research, Auburn University.

Rank	Mean	Threat	1990 Rank
1	5.6	Virus	5
2	5.2	Unauthorized access by hackers	6
3	4.8	Network failure	
4	4.4	Unauthorized access by employees	8
5	4.3	Utility failure	
6	4.2	Unauthorized access by competitors	12
7	4.1	Spam	
8	4.1	Intentional data destruction by employee	10
9	4.0	Denial of service	
10	4.0	Accidental data destruction by employee	3
11	3.9	Terrorist attack (virtual)	
12	3.8	Improper system use by employee	
13	3.7	Inadequate control over hardware	4
14	3.5	Inadequate control over media	7
15	3.3	Natural disaster	1
16	3.1	Terrorist attack (physical)	

1,800 people with at least 600 in Bangladesh, and destroyed much of the infrastructure. Many people died from drinking polluted water.

August 12, Florida. *Tropical storm Bonnie* made landfall in the Florida panhandle.

August 13, Punta Gorda, Florida. *Charley, a category 4 hurricane,* struck the southwest coast of Florida with 145-mile-per-hour winds, killing 36 people in storm-related deaths in the United States. Damages were estimated at US$7.4 billion. More than 2 million people were evacuated and tens of thousands were left homeless as thousands of buildings were damaged and mobile home parks were demolished. It was the worst hurricane in Florida since Andrew in 1992. Electricity remained out for more than 350,000 customers up to five days after the storm.

August 29–31, South Carolina, North Carolina, and Virginia. *Tropical storm Gaston* made landfall in McClellanville, South Carolina, and then moved through North Carolina and Virginia. Heavy rains dumped up to 14 inches on Richmond, Virginia, and the surrounding area, devastating the historic downtown area.

September 5–8, east and west coast of Florida. *Hurricane Frances,* downgraded from a category 4 to a category 2 storm, made landfall on the Atlantic coast of Florida, near Stuart, and moved across the state dropping up to 13 inches of rain. Frances made a second Florida landfall, as a tropical storm, at St. Marks on the panhandle on September 8. More than 2.8 million people were evacuated and damages were estimated at $9 billion.

September 8, Caribbean. *Hurricane Ivan,* a category 4 hurricane, ripped through Grenada, damaging an estimated 90 percent of the homes on the island. Ivan continued north, hitting Jamaica and battering the Caymans. Both islands experienced storm surges, major flooding, and destroyed homes.

September 16, Alabama, Florida, Louisiana, and Mississippi. *Hurricane Ivan* made landfall at Gulf Shores, Alabama, as a category 3 hurricane, causing severe flooding and power outages from Louisiana to Florida, with some of the worst damage in the Florida panhandle, where the storm spawned deadly tornadoes. Damages are estimated at $12 billion. Heavy rainfall and tornadoes spread into North Carolina and Tennessee.

September 18, Gonaives, Haiti. *Tropical storm Jeanne* brought torrential rains and flooding to Haiti, killing more than 2,500 with more than 1,000 still missing.

September 25, Atlantic Coast, Florida. *Hurricane Jeanne,* the *fourth* hurricane to hit Florida in six weeks, made landfall at almost the same area as Frances. Jeanne brought strong winds and more flooding to an already saturated state and then continued north through Georgia and South Carolina with heavy rains. Total damages for the four hurricanes are estimated to be more than for Andrew in 1992.

September 28, Parkfield, California. An earthquake registering 6.0 shook central California.

October 20, Japan. *Typhoon Tokage,* the deadliest typhoon to hit Japan in more than two decades, killed at least 80 people as heavy rains flooded tens of thousands of houses and triggered numerous landslides. The typhoon produced a record (since 1970) 80-foot- (24-m) high wave, eight stories high. A record 10 typhoons have struck Japan this season, causing damages estimated at $6.7 billion.

October 23, Niigata, Japan. A 6.6-magnitude earthquake, the deadliest in more than a decade, hit Japan, killing 40, injuring more than 3,100, and destroying more than 6,000 buildings. A series of quakes triggered more than 1,000 landslides, derailed a high-speed train, disrupted power, and damaged many roads in the area. Communications to the area were cut off, leaving many without food and supplies. Tens of thousands of people were evacuated to shelters.

November 11, Kepulauan Alor, Indonesia. A 7.5 earthquake, the largest in 2004, killed 28 people.

November 20–21, El Campo, Texas. Fifteen inches of rain flooded El Campo and parts of southeast Texas.

November 26, West Papua province, Indonesia. A magnitude 7.1 earthquake destroyed 328 buildings.

November 27–28, Sierra Nevadas and Rocky Mountains. Holiday storms dumped up to three feet of snow in California, New Mexico, Colorado, and Nebraska, bringing holiday travel to a halt.

November 29, eastern coast and Quezon province, Philippines. Flash floods and landslides from *Typhoon Winnie* killed more than 500 people and hundreds are still missing. Because of the many landslides, many areas were inaccessible to rescue and cleanup crews. The landslides were caused by the deforestation from both illegal and legal logging.

December 19, Mercer County, Pennsylvania. Heavy snows and a whiteout caused a pileup of 70 cars on Interstate 80 in western Pennsylvania.

December 26, Sumatra, Indonesia. A 9.0-magnitude earthquake (the Indian Ocean earthquake, known by the scientific community as the Sumatra-Andaman earthquake)—the largest earthquake since the 1964 Prince William Sound, Alaska, earthquake—caused a powerful tsunami in the Indian Ocean that hit 12 countries: Indonesia, Sri Lanka, India, Thailand, Somalia, Myanmar, Maldives, Malaysia, Tanzania, Bangladesh, Kenya, and the Seychelles. The earthquake's epicenter was off the west coast of the Indonesian island of Sumatra. By February 2005, the death toll was more than 225,000, with thousands still missing. Millions have lost their homes. Hardest hit were Indonesia (particularly the province of Aceh), with more than 166,000 deaths; Sri Lanka, with more than 30,000 deaths; India; and Thailand.

Year 2005

January 6–13, California to Pennsylvania. In California, a low-pressure system with drenching rains and heavy snows at higher elevations dumped up to 11 inches of rain and caused a large mudslide in La Conchita. Los Angeles has had 17 inches of rain since December 27, 2004. Since the end of December, the Reno–Lake Tahoe area has up to 19 feet of snow, the most since 1916. In Arizona, 7 of 15 counties were declared states of emergency. More heavy weather from Indiana to Pittsburgh caused flooding all along the Ohio River, where state emergencies have been declared for 56 of 88 counties in Ohio.

January 22–23, eastern United States. Strong snow storms swept across the Midwest to the Atlantic coast. A blizzard blanketed parts of the Northeast with snow depths up to 38 inches north and south of Boston and the entire island of Nantucket lost power. By the end of January, Boston had the snowiest month on record, with a total of 43.1 inches of snow.

February 6–13, Pakistan. Heavy flooding from snows and rain killed more than 460 people and left thousands more missing. The Shadi Kor dam in Baluchistan province broke on February 10, leaving many homeless.

February 22, Zarand, Iran. A magnitude-6.4 earthquake in central Iran shook more than 40 villages, killing at least 612 people, injuring over 1,400, and destroying villages with many mud-brick houses.

February 17–23, California. A series of storms caused flooding. landslides, and avalanches, damaging many roads and forcing the evacuation of many homes. More than nine inches of rain fell in downtown Los Angeles.

March 28, Sumatra, Indonesia. A magnitude-8.7 earthquake, off the west coast of Sumatra, killed 1,313 in the islands of Nias and Simeulue. Many buildings were destroyed and some officials feared another tsunami would occur. The same area was at the center of a huge tsunami in December that killed over 225,000 people. Officials at the U.S. Geological Survey said that yesterday's earthquake was an aftershock of December's 9.0 quake. The 9.0-magnitude earthquake was twice the power of the 8.7 magnitude quake.

June 11, Pensacola, Florida. The first *tropic storm* of the Atlantic hurricane season, *Arlene,* made landfall in Pensacola, dropping heavy rainfall on the lower Mississippi valley.

June 19, North Dakota. Strong thunderstorms and 100 mph winds battered North Dakota.

July 10, Pensacola Florida. *Hurricane Dennis* made landfall near Pensacola, Florida, with 120 mph winds and 10-foot storm surges, leaving thousands without electricity.

August 25, Ft. Lauderdale, Florida. *Hurricane Katrina* brought heavy rains and winds to southeastern Florida.

Europe. Days of heavy rains in Austria, Bulgaria, Germany, Romania, and Switzerland have inundated rivers and lakes, flooded cities and towns, and damaged roads and railways.

August 29–30, Louisiana and Mississippi. *Hurricane Katrina,* now a category 4 hurricane and one of the most powerful to hit the United States, devastated the Louisiana and Mississippi coast, landing just east of New Orleans with 140 mph winds. The high winds and massive flooding miles inland left thousands homeless, 2.3 million without electricity, roads and bridges destroyed, and communications inoperable. Storm surges, up to 25 feet, swamped the Mississippi Gulf Coast *destroying hundreds of homes, roads, and much of the coastal infrastructure.* Loss of casinos on the Gulf Coast and damage to oil rigs increased the economic devastation to the area and the entire country. The breach of levees in New Orleans, the day after the storm, submerged 80 percent of the city with flood waters that reached up to 20 feet, greatly increasing the damage from the hurricane, shutting down the entire city, and leaving thousands of New Orleans residents trapped on rooftops and in the evacuation centers without food, water, or other services. Widespread looting occurred and troops were called in to restore order. On Monday, September 5th, the levees were patched and water began to be pumped from the city, a process which may take 24 to 80 days. The mayor ordered the city evacuated because of the health risk. In all, the Coast Guard evacuated 30,000 people who were stranded by Katrina. The final number of fatalities and the amount of damages were still unknown, but it may be the most devastating hurricane in the United States.

Caveat

The point of this listing is to caution IT managers and network administrators as to the risk from Mother Nature as well as the internal organization community and the external virtual community. With a review of what has been occurring in very recent history, risk assessment now requires different scenarios least your networks and business are washed way, snowed under, burned down, broken by earthquakes, or simply made inoperable due to loss of power or inaccessibility to qualified employees. It has become necessary not only to evaluate low probability, worse-case scenarios and their impact on data communications essential to survival but to neither *delay* nor *procrastinate* the investment in security.

Implication

Delay—To postpone until a later time; defer;

- deferment
- deferral
- moratorium
- pause
- postponement
- stall
- stay
- suspension
- abeyance

Procrastinate—To postpone or delay needlessly, to

- lollygag (U.S., colloquial)
- lallygag (U.S., colloquial)
- dawdle
- delay
- dally
- mull
- dilly-dally (colloquial)
- drag one's feet (colloquial)
- haver (British)
- poke
- put off
- stall (colloquial)
- take one's time
- drag
- lag
- linger
- trail
- stick around
- tarry

BOOK SUMMARY AND CONCLUSION

Only four decades ago (circa 1960s), the world of information technology consisted of monolithic mainframes, most without any connectivity or even terminals; this was the mainframe-centric era. The 1970s brought distributed computing through the introduction of minicomputers, although many were isolated. Although Radio Shack and Apple introduced microcomputers in the late 1970s, they were specifically stand-alone systems and not seen as a business resource. The 1980s ushered in the IBM personal computer and gave the small machine a business goal; it and its clones were, originally, stand-alone machines. This was the PC-centric era as organizations realized the cost savings possible with these small machines. Networking was introduced in the late 1980s and came into its own in the 1990s. The early years became network-centric and, due to the development of the Internet in the 1970s and the introduction of the World Wide Web in the mid-1990s, this decade is seen as Internet-centric.

The above brief history shows the dramatic change in technology over only four decades, a rate of change that has been exponential. The PC, even a connected PC, is not just a cheaper mainframe; it has become a tool for information and knowledge management never anticipated nor available.

It was not until the age of connectivity, especially Internet connectivity, that external harm to computers became a problem. Table 11.1 indicates that the top four threats were internal or natural. With the advent of external connectivity through networking, the threat environment changed.

During the latter three decades, organizational life has changed to incorporate, assimilate, and depend on technology, for continuity. During this time, new threats have confronted the new technology and systems, accompanied by a significantly greater concern for security. What was once only a concern for internal physical and data security must now encompass external and virtual threats.

Despite the risks and difficulties in dealing with a myriad of data communications technologies and threats to them, the outlook for the future is bright. Managers face an ever-expanding set of choices, problems, and opportunities in dealing with the evolving multidimensional data networks. We expect that there will be many technological innovations that can help overcome present limitations and that can be exploited for greater competitive advantage. Although the road ahead may be uncertain, the traveler should anticipate a journey of discovery that promises to be both exciting and rewarding.

Glossary[1]

10BaseT The nomenclature of the medium and protocols that use unshielded twisted-pair wiring, in a baseband mode, operating at 10 megabits per second. This is the most-installed medium and protocol of the mid-1990s as most buildings have wiring installed that will support it. *See also* IEEE 802.3.

100BaseT 100 Mbps Ethernet, using category 5 cabling.

1000BaseT/FX Gigabit service (Ethernet); 1,000 Mbps using category 6 or FX cable.

100BaseT4 100 Mbps Ethernet, using four-line category 3 cable.

100BaseTX 100 Mbps Ethernet, using optic strands.

24*7 Operations 24 hours per day, 7 days per week; that is, all the time.

5ESS (electronic switching system) AT&T digital switch at the CO; an electronic switching computer for central office functions.

800 area code (in-WATS) An area code where the recipient pays all charges. *See* inward WATS.

802.11 *See* IEEE 802.11.

A

AAA framework Authentication, authority, and accounting, the primary focus areas of security.

access charge The charge LEC customers pay to access an IXC connection. After divestiture, the FCC mandated "access charges" so that every telephone customer would pay a tariff for having access to the public network.

access points (APs) A wireless transmitter-receiver that connects to a wired network.

accounting Allows us to measure and record the consumption of network or system resources.

acknowledgment (ACK) A character sent from the receiving unit to the source, confirming that a packet of data was good.

acoustic Refers to sound waves in an air medium. The maximum human acoustic (hearing) ability is in the range of 20 cycles per second (hertz) to 20,000 hertz.

active threats Some action must be taken that can be detected.

ad hoc Impromptu.

adaptive differential pulse code modulation (ADPCM) The standard that measures only the difference in adjacent values and, thus, requires less bandwidth. It measures the

signal just like the PCM method but transmits the difference between successive measured values instead of the value itself.

address Part of the header of a block of synchronous data that notes the destination of that block.

Advanced Mobile Phone Service (AMPS) Analog cellular system using the 800 MHz frequency band.

Advanced Research Projects Agency Network (ARPANET) The original model for the Internet.

adware Any software application in which advertising banners are displayed while the program is running.

alternatives The choices available, including the cost and value of each.

American National Standards Institute (ANSI) Standards group composed of industries and agencies.

American Standard Code for Information Interchange (ASCII) A binary coding convention. Seven-bit codes define 128 characters; eight-bit codes define 256 characters. The exact codes in a code set depend on the use of that code. Most microcomputers use ASCII codes. IBM mainframes and some PC software use EBCDIC.

Americans with Disabilities Act of 1990 (ADA) Stipulates that individuals with disabilities that can be accommodated will be considered without prejudice for jobs, advancement, and facilities.

amperage The strength of an electric current expressed in amperes.

amplifier For analog signals, repeaters are used to increase the signal strength. See repeater.

amplitude The distance from the trough to the top of the crest of the signal; the greater the amplitude, the "louder" the signal.

amplitude modulation (AM radio) Using the variation in amplitude in the carrier wave to carry information.

analog (to) One thing is an analog of another if it represents the former, as electrical waves represent acoustic waves.

analog (wave) A signal in the form of a continuous wave.

analogous Similar to.

analysis and design The process of determining needs, analyzing alternatives, delineating system specifications, and designing the system that will properly serve the organization, usually referred to as systems analysis and design.

antennae The interface point between a transmitter and space.

---

[1] For an extension of these and other definitions, refer to www.webopedia.com and www.wikipedia.com.

anycast   Communication over a network between a single sender and the nearest of a group of receivers.

appliance   A generic capability computer or communications device that provides limited functionality at the user level.

application layer   The highest layer in the OSI hierarchy. It provides user-oriented services such as determining the data to be transmitted, the message or record format for the data, and the transaction codes that identify the data to the receiver.

application service provider (ASP)   A company that delivers and manages applications and computer services from remote data centers to multiple users across a wide area network.

architecture   A specification that defines how something is constructed, defining functional modularity as well as the protocols and interfaces, which allow communications and cooperation among modules. It is a concept or plan that is implemented in a set of hardware, software, and communications products. It reflects the technical strategy.

area code   The three digits used to direct the call to the geographic area in which the exchange of the destination resides.

asymmetric digital subscriber line (ADSL)   Digital technology on analog POTS lines, providing high bandwidth. The circuit is digitized and divided into a telephone channel and one or more data channels.

asynchronous   Protocols and systems that have no synchronism between cells or characters. In character-based communications, one character is sent at a time with start and stop bits to signal the receiver where a character begins and ends. With cell-based systems, the cell is of fixed length and format.

asynchronous communications   The communications of one character or packet at a time with no relation or synchronism to the previous or succeeding character or packet.

asynchronous transfer mode (ATM)   A switching technology used by BISDN. It is a variation of packet-switching technology that transmits fixed-length units of data (called *cells*) at very high speeds. The speeds presently specified range from 155.52 Mbps to 2.488 Gbps, with future predictions of 100 Gbps or faster. *See also* automatic teller machine.

attenuation   A characteristic of electrical and photonic signals is that they diminish or weaken as they travel away from their source. For example, the light from a flashlight is blinding at the bulb but is barely visible a mile away.

audio   Relating to sound frequencies in the range of 20 to 20,000 Hz.

audio conferencing   Use of audio devices to aid in meetings where participants are not all at the same location.

audit   Formal, periodic investigation of activities to ensure that policies are being carried out.

auditing   Review of records to determine activity; a way to catch unauthorized or excessive activity.

authentication   The process of determining whether someone or something is, in fact, who or what it is declared to be.

authorization   The process of giving someone permission to do or have something.

automated call distribution (ACD)   A computer-based system of managing telephone calls.

automatic repeat request (ARQ)   An error-correcting technique wherein a NAK is sent if the block was determined by the receiving unit to have errors, resulting is a retransmission of that block.

automatic teller machine (ATM)   Provides 24-hour banking and puts the banking industry into a real-time mode. It is also called a bank-in-a-box.

AVAIL   A large bank transaction network.

B

backbone network   Connectivity at a "higher" speed between LANs and major nodes, such as mainframe computers. A backbone network is to LANs as a high-speed loop around a city is to the major streets that connect to the loop. The backbone network is analogous to the backbone of a human in that it acts as a central structure.

backup generators   A gas- or diesel-powered AC generator that can take over during a power outage.

bandwidth   The information-carrying capability of a channel. In digital circuits, it is measured in bits per second, denoting the speed of transmission. For analog channels, it is the difference between the highest and lowest analog frequencies, measured in hertz (Hz), of a transmission channel.

bandwidth-on-demand   Dynamic allocation of bandwidth.

barcode   An optical tag that uses vertical bars to represent numbers. Bar code technology eliminates the necessity to key data by either data entry personnel or the end users because codes can be easily read by scanners. *See* wands.

baseband   Narrow band circuit, generally not subdivided into channels. Often describes a digital circuit.

basic rate interface (BRI)   The residential/office configuration of ISDN generally configured with two 64-Kbps B channels and a 16-Kbps D channel, plus 16 Kbps for administration, for a total bandwidth of 160 Kbps.

battery life   The duration usable power from a battery; its effective lifetime.

baud   A unit of signaling speed equal to the number of discrete conditions or signal events per second. In modems, it refers to the number of changes of the analog signal per second and is generally not the same as bits per second.

Baudot code   A code of five digits using two shift characters to represent 58 characters.

Bayesian filter   A program that uses Bayesian logic, also called *Bayesian analysis,* to evaluate the header and content of an incoming e-mail message and determine the probability that it constitutes spam.

beacon   A frame that is sent out to announce the presence of the AP.

bearer (B) channel   DS-0 channel on ISDN.

Bell System   Referred originally to AT&T, which was comprised of the Bell Operating Companies, Western Electric, Bell Labs, and AT&T Long Lines. It is now only the Bell Operating Companies.

binary   The basic numbering system used by digital computers. That is, every piece of data that a computer processes or stores is encoded into a series of 1s and 0s in accordance with some standard convention, called a code.

biometric identification   The use of part of a human for identification, such as the eye's iris or retina, a fingerprint, or a hand shape.

bit   A binary digit, the basic building block for digital representation of data.

bit rate   The information transfer rate usually expressed in bits per second (not to be confused with baud). Bit rate specifies bandwidth.

bits per second   Digital bandwidth.

blacklist   A database of known Internet addresses (or IPs) used by persons or companies sending spam.

block check character (BCC)   The data field in the message trailer containing error detection characters. The larger the BCC field, the greater the probability of error detection.

Bluetooth (802.15.1)   An RF protocol for connecting devices within 10 meters' distance.

bridge   Intelligent devices that connect networks using the same protocol.

broadband   Transmission equipment and media that can support a wide bandwidth. An analog path that may be frequency division multiplexed to create several channels; a high-speed digital path that may be time division multiplexed to create multiple channels.

broadband integrated services digital network (BISDN)   A service requiring transmission channels capable of supporting rates greater than the primary rate. It also can be defined as an all-purpose, all-digital network that will meet the diverse needs of users by providing a wide range of services: high-speed data services, video phone, videoconferencing, high-resolution graphics transmission, and CATV services, including ISDN services such as telephone, data, telemetry, and

facsimile. BISDN employs SONET as the transmission medium and ATM as the switching technology.

broadcast   Simultaneous transmission to a number of stations; sending data to all IP addresses on a network or set of networks.

Brouter   A device that functions as a bridge and a router.

browser   A computer program that provides a graphical user interface for the World Wide Web.

browser hijacking   External code that changes a user's Internet Explorer settings.

bus   A transmission path or channel where all attached devices receive all transmissions at the same time.

bus network   A multiple-point network with no master-slave, as all nodes are considered equal although one node may have the network operating system software resident.

bus topology   A single communications line, or channel, to which many nodes are connected. Each node on the network is of equal status; there is no master and no slave; the failure of one node has no effect on other nodes; faults may be difficult to locate; additions are easy.

business-to-business (B2B)   Internet communications and business between business partners in support of retail trade.

business-to-consumer (B2C)   The transactions of retail business carried on over the Internet.

bypass   When an organization circumvents the standard telecommunications provider, such as the LEC, to get a telecommunications service.

byte   The combination of bits to hold one character; usually a group of eight bits.

C

cable   A group of wires or other media bundled or packaged as a single line. A coaxial circuit is referred to as a cable, although it generally has only a single circuit.

cable, or community, television (CATV)   A system for distributing TV from a central point (the head end) that typically is based on coax cables and is able to carry 50–100 television channels.

cable modem   A terminating device for CATV that uses the coaxial cable as a high-speed conduit for data, generally for access to the Internet.

campus area network (CAN)   Small local area network.

CAN-SPAM Act of 2003 (Controlling the Assault of Non-Solicited Pornography and Marketing Act)   Establishes requirements for those who send commercial e-mail.

capacity   The bandwidth of the channels. The greater the bandwidth, the greater the capacity.

capacity management   The process of documenting and managing the bandwidth and resource capacity of an organization.

carrier frequency   The basic high-frequency signal onto which the true information signal is modulated.

carrier sense   The protocol's ability to listen to a multiple-user channel and determine if it is presently unused.

carrier sense multiple access/collision detect (CSMA/CD) protocol   A scheme used on the Ethernet bus networks to ensure that transmissions do not interfere with each other, that is, collisions are handled gracefully.

carrier wave   A high-frequency signal onto which the information is modulated.

category 3 wiring/cable   Unshielded twisted-pair 24–26-gauge wiring, with a mild twist; provides 10 Mbps bandwidth.

category 5 wiring/cable   Unshielded twisted-pair 24–26-gauge wiring with a mild to severe twist to reduce interference. Used as network cabling in many buildings; provides 100 Mbps bandwidth.

category 6 wiring/cable   Upgrade of CAT 5 cabling, providing 1000 Mbps (gigabit) service. May be UPT or fiber.

cathode ray tube (CRT)   A specialized type of monitor developed for radar that displays via dots.

cell   A fixed-length packet in an asynchronous transfer mode (ATM) system.

cellular   Equipment used for cellular telephone systems.

cellular (radio) telephone   A means of dividing wireless radio telephone coverage areas into cells for reuse of frequencies.

central office (CO)   The terminal point of all local loops and the location of the switch that connects the users to the network.

Centrex   A group of services provided by the Telco as a specially priced service, giving the view that you have your own PBX. The LEC maintains the equipment and charges the customer for this service. Both are regulated by state and federal bodies.

certificate of authority (CA)   An authority in a network that issues and manages security credentials and public keys for message encryption enables users of a basically unsecure public network such as the Internet to securely and privately exchange data and money through the use of a public and a private cryptographic key pair that is obtained and shared through a trusted authority.

certification   Credentials that demonstrate proof of passing a recognized test.

change management   Organizes the process of making changes to an installed capability. Because of the complexity of telecommunications systems, a lack of change management will lead to the inability to have configuration management.

channel   A communications path between nodes.

checksum   A count of the number of bytes in a transmission unit that are included with the unit so that the receiver can check to see whether the same number of bytes arrived.

chief information officer (CIO)   The top information executive. A position at the same level as the vice president of finance, sales, and operations is being created in many organizations to control the information resource from a total organization perspective. In some organizations, the position has greater scope, and the title is chief technology officer.

circuit   A (physical) path over which a signal can travel. Some media include paths in both directions, either simultaneously or one direction at a time.

circuit media   Vary in expense, bandwidth, and immunity to noise. Circuit media are the "physical" paths of circuits, which may be subdivided into channels. The same media discussed previously (twisted-pair wire, coaxial cable, microwave, omnidirectional radio, and satellite channels) can be used for transmitting either analog or digital signals.

circuit switching   The temporary establishment of a connection between two pieces of equipment that permits the exclusive use until the connection is released, for example, a POTS call.

cladding   A layer of material surrounding the glass core of a fiber-optic strand.

clarity of the interface   The user not only understands the layout of the computer screen (the interface) but is comfortable with it and the interface is nonambiguous.

classes of IP networks   Categorization of networks based on size of networks in nodes.

client   The user or using computer that takes advantage of facilities or services of the server computer in a client-server architecture (C/SA) system.

client-server architecture (C/SA)   An extension of cooperative computing and distributed computing, where two or more computers cooperate to perform a task.

CO switch   Central office equipment that makes telephone connections without human operators. This sets up a path from the sender to the receiver. *See* central office.

coaxial cable   Called "coax"; a shielded wire that performs the same function as twisted-pair but provides broader bandwidth and more protection from interference. A coaxial cable is similar to a pair of copper wires except that one wire is a braided or solid sheath that encompasses (shields) the other wire.

code   The representation of characters by numbers or bit patterns.

code division multiple access (CDMA)   A digital cellular technology that uses spread-spectrum techniques.

codec   A device that converts analog signals to digital signals or vice versa.

cold site   Sometimes called a shell site, it consists of prepared facilities without any equipment.

cold turkey   A switch-over process when the system is changed and brought online all at once without other considerations. This involves the greatest risk but least expense if all goes well.

collision   The interference of one transmission with another, when both are in contention for the use of the channel; when two nodes transmit simultaneously.

collision avoidance   A protocol that makes the occurrence of collisions of packets on a channel impossible: CSMA/CA, token passing, polling.

collision detection   The ability to detect and react to a collision of packets on a channel, for example, CSMA/CD.

committed information rate (CIR)   The specified guaranteed bandwidth that will be delivered, regardless of the bandwidth of the channel or circuit.

common carrier   Any company that provides a service to the public. AT&T is a common carrier of long-distance services.

common management information protocol (CMIP)   An ISO standard protocol for exchange of network management commands and information between devices attached to a network. *See also* simple network management protocol (SNMP).

communications   The process of transferring information from a sender to one or more receivers via some medium.

competitive advantage   The ability to gain a disproportionately larger share of a market because of cost leadership or product or service differentiation.

complexity   The characteristics that make equipment or systems difficult to use.

compression   A process (hardware- or software-based) to reduce the amount of data communicated or stored by coding redundant data; it makes a data stream smaller by encoding schemes, by having single characters represent larger groups of redundant characters. Compression reduces the bandwidth required and may be lossy or lossless.

computer-aided design (CAD)   The use of a computer for the creation of automated industrial, statistical, biological design, and so forth.

computer-integrated manufacturing (CIM)   The application of information and manufacturing technology, plans, and resources to improve the efficiency and effectiveness of a manufacturing enterprise through vertical, horizontal, and external integration.

concentration   The process of combining multiple messages into a single message for transmission.

concentrator   Any device that combines incoming messages into a single message or places them onto a single line.

conditioned line   A communications line on which the specifications for amplitude and distortion have been tightened by adjusting the electronic parameters.

conducted   *See* wired.

conducted wire   A wire that is suitable for carrying an electrical current.

configuration control   The inventory of all resources. It is vital to ensure that you know what resources are presently available and the relationships that exist among them. Configuration management gives you data to use in cost allocation and information for expansion and improvements.

connect time   The amount of time that elapses while the user of a remote terminal is connected to a timeshare system.

connection   A voice or data path from one point to another.

connection-oriented   Communications where a true physical or a permanent or temporary virtual path is established for that data transfer. The POTS is a prime example of such a service.

connectionless   A type of communications in which no fixed or permanent path exists or is created between sending and receiving stations, and each data unit is sent on its own to traverse the network.

connectivity   The communications path between two or more nodes.

consumer-to-consumer (C2C)   Business conducted via the Internet directly between consumers without retail or wholesale intervention, for example, eBay.

contention   A situation in which two or more nodes compete for use of the same channel.

contingency planning   Preparing to deal with unexpected events; a plan to continue business in the event of unforeseen adverse occurrences.

control   (1) A project involving feedback to determine if the goals and objectives of a project are being met. (2) To exercise authority and management over an entity, e.g., to control voltage, to control the activities of a group.

controlling   Concern with the data communications resources of the organization, including inventory management of equipment, circuits, and networks.

convergence   The use of two or more media (multimedia).

cookie   A mechanism for storing information about an Internet user on his/her own computer.

cooperative processing   The sharing of computer processing tasks between two or more processors, either within or between computers.

coordinating   To perform the necessary steps in the proper sequence to keep the business operating as normal.

critical success factors (CSF)   Those (few) things that must go properly for the organization to be successful.

crosstalk   A radiation of signal from one circuit to another. Signals inadvertently transferred (induced) between elements (wires) of a cable. The result appears as noise.

cryptography   The field concerned with linguistic and mathematical techniques for *securing information*.

customer premises equipment (CPE)   The instruments and equipment on the customer's site.

customer service area (CSA)   The distance from the central office that can be serviced without engineering, for example, the 12,000-foot length for DSL.

cutover   The different methods to bring additions or modifications to the system online. The methods: pilot, parallel, phased, phase-in, phase-out, modular, and cold turkey.

cyclic redundancy check (CRC)   A standard used for error detection in synchronous communications.

D

data   Facts about the activities of the organization.

data circuit-termination equipment (DCE)   Noncommunications-oriented components of a data communications environment

data communications   That part of telecommunications that relates to movement of data between machines, usually computers.

data link   The physical means of connecting one location to another for communications.

data link layer   The OSI layer with the actual transmissions of characters and the sequence in which they are transmitted.

data security   The protection of the organization's data from physical or virtual access.

Data Services One (DS1)   A standard; any medium that carries 1.544 Mbps speed with 24 DS0 (64 Kbps) channels plus 8 Kbps for control. *See* T1.

Data Services Zero (DS0)   A standard for a data channel having a bandwidth of 64 Kbps.

data terminating equipment (DTE)   The end nodes that use data communications, often with DCE as intermediaries. DTEs talk to DTEs.

data transfer rate   The average number of bits, characters, or blocks per unit of time.

datagram   One mode of data transfer for the X.25 packet network, analogous to the U.S. Postal Service in that datagrams find their way around the network.

de facto   Existing in fact whether by lawful authority or not.

de jure   By right, according to law.

decibel (db)   A measurement of change in signal strength, based on a logarithmic scale.

decode   The process of reconversion of the message so the receiver can understand it. Changing a digital signal into its analog form or another type of digital signal.

default   Setting used "from the factory."

delta (D) channel   16 Kbps control and communications channels for ISDN.

demilitarized zone (DMZ)   A location on a network accessible to the public network by use of a firewall; other portions will be shielded from public access.

demodulation   The process of retrieving intelligence (data) from a modulated carrier wave; for example, the reverse of modulation.

denial-of-service (DoS/DDoS) attack   An explicit attempt by attackers to prevent legitimate users of a service from using that service.

destination   The final receiver.

dial tone   The sound on the handset indicating that the switch is ready to accept information for connection of a call.

differential phase shift keying (DPSK)   A modulation technique in which the relative changes of the carrier signal phase are coded according to the data to be transmitted.

digital   Binary (bi-state) codes to represent data, which means that the data are encoded into a series of 1s and 0s in accordance with some standard convention, called a *code*. Noise on a circuit appears much like digital data.

digital bandwidth   Bits per second; sometimes called digital speed.

digital cash   The use of authorization to pay for purchases on the Internet, for example, PayPal.

digital certificate   An electronic "credit card" that establishes your credentials when doing business or other transactions on the Web.

digital circuit   A circuit expressly designed to carry the pulses of digital signals. It may, however, utilize analog and modem technologies, for example, ISDN.

digital divide   The separation of society due to possession of technology, primarily computers and broadband access to the Internet.

Digital Phase Shift Keying (DPSK)   A digital modulation scheme that conveys data by changing, or modulating, the phase of a reference signal (the carrier wave).

digital signal processor (DSP)   Microprocessor especially designed to analyze, enhance, or otherwise manipulate sounds, images, or other signals.

digital subscriber line (DSL)   Digital channels over the analog POTS circuits.

digital-to-analog   The conversion or translation of a digital signal into its original analog form.

digital-to-analog (D/A) converter   A device that converts a digital value to a proportional analog signal.

direct broadcast satellite (DBS)   The transmission of commercial television from satellite directly to the home.

direct-sequence spread spectrum (DSSS)   One of two types of spread-spectrum radio, it is a transmission technology where a data signal at the sending station is combined with a higher data rate bit sequence, or chipping code, that divides the user data according to a spreading ratio.

directing   Involves the supervisory task of getting people to successfully perform the required tasks.

dirty power   Electrical power that has too many variations, such as spikes, high voltages, low voltages, and brownouts.

disaster   The results of severe adverse conditions, such as the damage from a flood.

disaster planning   A contingency plan for continuance of operations in the event of unforeseen adverse occurrences, often called *business continuity planning.*

disaster recovery   Provides alternatives for continued operations and recovery from damage. Recovery and planning are vital to modern organizations as they rely on computers and telecommunications for continued operations. It is an important aspect of security and control.

dispersion   The tendency of photons in a light pulse to spread out as they travel in any medium.

distortion   The tendency for signals to change from their originated form appears naturally as a result of the environment through which the signal must pass. Noise and distortion must be countered in order to pass data correctly.

distributed computing   Placing the computer processing power and data storage at the site requiring it; that is, localized processing with centralized control.

distributed processing   Data processing in which some or all of the processing, storage, and control functions are situated in different places and depends on telecommunications.

DoCoMo   (Means "anywhere" in Japanese.) A NTT subsidiary and Japan's biggest mobile service provider, with over 31 million subscribers as of June 2000.

documentation   To log and compile the required information, rationale, and steps to perform a task.

Dolby Laboratories   Developer of audio signal processing systems and manufacturer of professional equipment to implement these technologies in the motion picture, broadcasting, and other industries. Developer of Surround Sound and Dolby Digital technologies.

domain name   Name that identifies one or more IP addresses, such as Microsoft.com.

domain name server (DNS)   Public and private servers that translate character-based domain names to specific IP addresses.

domain name system (DNS)   *See* domain name server (DNS).

downloading   The transmission of a file of data from a mainframe or other host computer to a microcomputer.

dual-tone-multifrequency (DTMF)   A method of signaling a desired telephone number by sending tones on the telephone line. Each code is composed of two (dual) tones.

dumb terminal   A terminal that has little or no memory or processing power and is not programmable.

duplex   Bidirectional circuit capability.

Dynamic Host Configuration Protocol (DHCP)   A protocol for assigning dynamic IP addresses to devices on a network.

dynamic routing   A technique used in data networks by which each node can determine the best way for a message to be sent to its destination.

E

E-911   A location technology that enables mobile, or cellular, phones to process 911 emergency calls.

echo checking   The process of echoing each transmitted character back from the receiving unit to the sending operator, generally via a full-duplex capability.

effective bandwidth   The resultant bandwidth available after adjusting the native bandwidth for the effects of noise and compression.

electrical   Direct or alternating current used to carry a signal on a conductive medium.

electrical topology   How the network operates as a result of protocol and central node connectivity.

electromagnetic   Rapidly varying (high frequency) current used to carry another signal.

electronic business (e-business)   Conducting business electronically, involving everything from sending e-mail to advertising on the Web or creating an intranet so that your HR department can post online policies and procedures manuals.

electronic cash (e-cash)   The use of authorization to pay for purchases on the Internet, for example, PayPal.

electronic commerce (e-commerce)   A dynamic set of technologies, applications, and business processes that link enterprises, consumers, and communities through electronic transactions and the electronic and physical exchange of goods, services, information, and capital.

electronic data interchange (EDI)   A set of standards for computer-to-computer communications of standardized business documents.

electronic funds transfer (EFT)   Allows more efficient handling of funds, at a lower cost, by moving monies through a network as opposed to checks in the mail.

electronic government (e-government)   Technology that allows citizens to find information and conduct government business, such as license renewal, over the Internet.

electronic mail (e-mail)   Allows users to communicate electronically with other users as if two typewriters were connected by a channel. E-mail adds a new dimension to the office environment, replacing paper copies and reducing time of transmittal.

electronic performance support systems (EPSS)   Special systems designed to capture knowledge of experts so that the relevant knowledge may be called on by a person who needs just-in-time expert advice. Some knowledge management (KM) systems employ EPSS.

Electronic Product Code (EPC) A coding used with RFID that replaces UPC, used with barcode.

e-mail spoofing Sending e-mail with the source field changed to hide the true sender.

encapsulation The placing of (digital) data in an (analog) envelop to reduce the effects of noise.

encode The process of preparing a message for efficient transmission. Encoding means to change the idea or information into symbols during transmission; also, the transforming of a digital signal into an analog signal.

encryption Transformation of data from the meaningful code that is normally transmitted to a meaningless sequence of digits and letters that must be decrypted before it becomes meaningful again.

enterprise information systems The aggregation of data processing, functions, and their interaction across the entire organization.

environmental hazards Things and effects that obstruct data communications, such as noise, buildings, hills, and so forth.

equipment life span The useful time during which equipment will meet its intended purpose. In the past, the expected life of IT equipment was 10–20 years. Presently it is 3–5 years. While the equipment may easily last several times that number of years, its technological usefulness does not.

error A discrepancy between a computed, observed, or measured value or condition and the true, specified, or theoretically correct value or condition.

error correction Compensating for the effect of errors, generally performed by resending synchronous data.

error detection The techniques used to ensure that transmission and other errors are identified.

Ethernet The most-used network operating system protocol using CSMA/CD (carrier sense multiple access/collision detect) management (IEEE 802.3 standard) at 10 Mbps over unshielded twisted-pair wire (10BaseT). Ethernet was developed in 1980 by Xerox, DEC, and Intel.

ethical In accordance with the accepted principles of right and wrong; conforming to standards of what is right, moral, just in behavior. In data communications, the concern is for the ethical use of the technology because the technology itself is amoral.

exchange The first three digits of the seven-digit phone number is the exchange destination. One or more exchanges reside in the switch in a central office.

Extended Binary Coded Decimal Interchange Code (EBCDIC) An eight-bit mainframe-originated code used on IBM PCs.

Extensible Markup Language (XML) A pared-down version of SGML, designed especially for Web documents.

extranet An invitation-only group of trading partners conducting business via the Internet.

F

facsimile The transfer of any document one could put on a copier. It senses and transmits spots of light and dark.

fat client A client computer with large storage and processing power; a mainframe terminal is an example.

feasibility A phase in the SDLC analysis to determine the feasibility of a project, including technical, behavioral, economic, operational, time, regulatory, and ethical feasibility categories.

Federal Communications Commission (FCC) The appointed federal body governing most interstate communications activities. It has a board of governors, appointed by the president of the United States, with control of the radio and television broadcasting industries. The FCC regulates all types of telecommunications service and rates through tariffs.

Federal Information Security Management Act (FISMA) U.S. legislation to bolster computer and network security within the federal government and affiliated parties.

fiber-distributed data interface (FDDI) A network based on optical fiber media, ring topology, and token-passing protocol, operating at 100 Mbps; a major form of MAN.

fiber optics Very small one-way glass strands that have the largest bandwidth of any media used.

fiber-optic circuits Use of a laser or light-emitting diode (LED) at the source end and a light detector at the receiving end. There must be a light-to-electrical signal conversion at each end.

fibre channel Optical fiber access for new mass storage devices and other peripheral devices that require very high bandwidth (e.g., 100 Mbps).

file server A dedicated computer on a network that shares its disk space with other nodes.

file transfer protocol (FTP) The protocol for exchanging files over the Internet.

file upload Transmitting a file to the host site, or from another device that is connected via a network.

filter A hardware or software device that allows only specified information (packets) to pass.

firewall A set of related programs, located at a network gateway server, that protect the resources of a private network from users from other networks.

Firewire (IEEE 1394) A very fast external bus standard that supports data transfer rates of up to 400 Mbps (in 1394a) and 800 Mbps (in 1394b).

fixed node A node that does not move, for example, desktop computer or router.

fixed wireless   Wireless, such as microwave, where the end points are stationary.

flooding   To place a significantly large number of packets on a network for the purpose of overloading that network.

foreign exchange   A leased line from a local Telco or long-distance carrier, or both, from the firm's location to the switch of a distant central office.

format   The message format of the protocol defines the location and amount of true data contained in the message and the overhead necessary to ensure that the destination receives the data as they were sent.

forward channel   In full-duplex, when the primary movement of data is in one direction, the circuit can either be switched half-duplex or allocate the greater bandwidth to the source-to-destination (forward) direction and a minimum bandwidth to the reverse (response) channel. This makes the forward channel fast and the reverse channel slow.

fractional T1   Use of one or more DS-0 channels of a T1 channel.

frame format   The specification of data field within a frame, or packet, as determined by the standard.

frame relay   An ISDN frame-mode service based upon fast packet switching. In simplistic terms, frame relay can be thought of as "relaying" variable-length units of data, called *frames,* through the network.

free access to wireless networks   Detecting an insecure wireless access point and making unauthorized use of the bandwidth; a.k.a. bandwidth hijacking.

frequency   The number of oscillations of a signal per second; the number of repetitions or cycles that occur within a second of time is termed *hertz.* It is the pitch of the signal.

frequency division multiplexing (FDM)   The division of a transmission circuit into two or more channels by splitting the frequency band carried by the circuit into narrow bands to become distinctive channels.

frequency-hopping spread spectrum (FHSS)   One of two types of spread-spectrum radio, where the data signal is modulated with a narrowband carrier signal that "hops" in a random but predictable sequence from frequency to frequency as a function of time over a wide band of frequencies.

frequency modulation (FM)   Uses variation in the frequency of the carrier wave to carry information. In FM radio, the analog music and voice signal varies the frequency, not the strength (amplitude), of the carrier wave.

frequency shift keying (FSK)   Frequency modulation of a carrier by a signal that varies between a fixed number of discrete values.

front-end processor (FEP)   A small computer that sits between remote devices and the host mainframe, relieving the host of tasks, such as line control, message handling, and code conversion.

full-duplex   A circuit that allows data to be sent and received at the same time—a simultaneous two-way circuit or channel. This method requires two paths, that is, two twisted-pair circuits, or division of the communications channel into two parts.

G

galactic area network (GAN)   Connectivity to earth-orbiting and other nonterrestrial objects.

gateway   Intelligent devices that connect networks that use differing protocols, thus requiring protocol conversion.

General Packet Radio Service (GPRS)   A standard for wireless communications that runs at speeds up to 150 kilobits per second.

geosynchronous orbit   A satellite orbit that exactly matches the rotation speed of the earth, generally placing it at an altitude of 22,500 miles above the surface.

giga   One billion. A gigahertz is one billion Hz.

gigabit   1,000 Mbps, that is, Gigabit Ethernet service.

Gigabit Ethernet   1000baseT or 1000baseFX service.

global area network (GAN)   The extreme of the WAN.

global positioning system (GPS)   System of 22 medium-orbit satellites that provide precise location anywhere on the surface of the earth.

global system for mobile communications (GSM)   One of the leading digital cellular systems; uses narrowband TDMA, which allows eight simultaneous calls on the same radio frequency.

Gopher   A software system to make navigation around the Internet easier.

Gramm-Leach-Bliley Act (GLBA)   A law enacted to assure that financial institutions protect the privacy of their customers' financial records.

grid computing (P2P)   Technology that harnesses unused processing cycles of all computers in a network.

groupware   Software that allows groups of people to communicate and work together simultaneously.

guard channel   The space between the primary signal and the edge of the analog channel.

H

hacker   (1) Person very competent in the use of computers and networks; (2) individuals who enjoy exploring the connected world, especially the unauthorized invading of the computers of others, for fun or mischievous intent.

hacking   Activities designed to break into another's computer or network; *see* intrusion.

half-duplex   Transmission in either direction on a specific channel, one direction at a time.

hamming code   A data code that is capable of being corrected automatically by including sufficient redundancy to allow the true data to be detected through errors.

handshake   The initial process of a communications session whereby two units determine the speed of transmission and other parameters and, therefore, the receiving node knows when to sample the line to detect a bit.

hardware port   Physical interface points on a node, for example, USB port. Not the same as TCP/IP ports.

hashing   A function for summarizing or probabilistically identifying data. Hash functions (a type of one-way function) are fundamental for much of cryptography.

hate group   Organization (with a Web presence) that espouses socially repugnant beliefs, for example, Neo-Nazi.

Hayes-compatibility   Equipment that has the same protocol and configuration of the standard established by Hayes.

header   The part of a block of synchronous data that contains information about the message, such as its destination, source, and block number.

health hazard   The potential for a data communications device to harm users by close proximity to their retina or brain.

Health Insurance Portability and Accountability Act (HIPAA)   Intended to protect the privacy of individuals' health records.

hertz (Hz)   The technical term for analog frequency, that is, cycles per second. The greater the hertz, the higher the pitch of an audible signal.

heuristics   The application of experience-derived knowledge to a problem.

high-data-rate digital subscriber line (HDSL)   An early form of DSL that is bidirectional, running at about T1 speed. It is often used to create T1 circuits.

high-definition (HD) radio   Digital modulation of AM and FM radio carriers to provide enhanced sound quality.

high-definition television (HDTV)   Digital TV (in the United States) with superior video and sound, using a bandwidth of 19.2 Mbps.

high-level data link control (HDSL)   A bit-oriented data link protocol. A protocol standardized by ISO.

home network   Connectivity among multiple computers in a home environment, generally accessing the Internet via a cable or DSL modem and router.

home network security   Placing the same security on the home network as in an office environment because many home systems interact with organization networks.

honey trap   Set up fake "vulnerable" services to lure intruders. *See also* honeypot.

honeypot   Closely monitored network decoys.

hot site   Sites in disaster recovery that are set up with equipment already installed.

hotspot   *See* Wi-Fi hotspot.

hub   Network equipment that connects multiple nodes.

hybrid fiber/coax (HFC)   The combination of coax and optic fiber circuits, allowing for high bandwidth to the neighborhood and coax to the residence, giving a good balance between cost and existing media.

hybrid network   Some combination of any or all of the types of networks.

hype   Emphasize the good; downplay the bad; repeat the message.

hypermedia   An extension to hypertext that supports linking graphics, sound, and video elements in addition to text elements.

hypertext   Technology in which objects (text, pictures, music, programs, and so on) can be creatively linked to each other.

HyperText Markup Language (HTML)   Computer application/language for creating pages for the World Wide Web. A browser is required to display the page.

HyperText Transfer Protocol (HTTP)   The software capability that provides the graphical user interface for the WWW.

I

I-link   *See* firewire (IEEE 1394).

IEEE 802   Protocol for telecommunications.

IEEE 802.3   Protocol for CSMA/CD (Ethernet) and token passing.

IEEE 802.11   Protocol for wireless, packet radio networks.

IEEE 802.15.1   Bluetooth standard.

IEEE 802.15.3   Ultra wideband standard.

IEEE 802.15.4   ZigBee standard.

IEEE 802.16   WiMAX standard.

IEEE 1394   *See* firewire.

implementation   The action phase of the SDLC where the system is made operational and turned over for use.

impulse noise   A sudden spike on the communications circuit, often originating from inductive sources such as a refrigerator or air conditioner. The most-encountered noise in PC telecommunications.

in-house   The creation of goods or services within the physical organization as opposed to contracting this to be done by an outside organization.

information   Processed data that enhance the recipient's knowledge. Subjective versus objective because of processing for a user.

information systems architecture   The aggregation of data, equipment, applications, functions, and their interaction that represents the business requirements.

information technology (IT)   All equipment, processes, procedures, and systems used to provide and support information systems within an organization.

information warfare   The process of protecting your information and network resources while, potentially, denying the adversary access to his/hers.

infrared (IR)   Light below the visible range used in some wireless communications.

Institute of Electrical and Electronics Engineers (IEEE)   Organization that authorizes standards and protocols.

integrated services digital network (ISDN)   A hierarchy of digital switching and transmission systems with high-data-rate transmission channels, high speed, and high quality using digital capabilities. Digital telephone, providing a digital circuit, composed of multiple channels.

intelligent terminal   A terminal that has onboard memory, processing, and/or can be programmed.

inter-LATA   Long-distance telephone calls between LATAs, which must be handled by IXCs.

interexchange carrier (IXC)   Provider of telecommunications services across LATA boundaries; that is, a long-distance carrier.

interface   A shared boundary; the interface of the computer is the display on the monitor.

interference   The intrusion of an unwanted signal. One signal's interruption, mixing with, or replacing another.

intermediate layer   Communications beyond the organization but not including a WAN.

internal layer   Connectivity for communications within the organization.

International Telecommunications Union (ITU)   The successor to the CCITT.

Internet   A network of networks. The ultimate WAN, connecting people over most of the earth. Global connectivity. Not the World Wide Web.

Internet2   A higher-bandwidth, Internetlike network for academic and research projects.

Internet Explorer® **(IE)**   Microsoft's browser.

Internet protocol (IP)   The second layer of TCP/IP that provides addressing between nodes.

Internet Protocol Security (IPsec)   A framework for a set of protocols for security at the network or packet processing layer of network communications.

Internet Protocol Version 6 (IPv6)   *See* IPv6 addresses.

Internet service provider (ISP)   Retail point for access to the Internet backbone. Company who provides Internet access to users.

interoperability   The capability of two or more devices to transmit and receive data or carry out processes regardless of whether they are from the same or different manufacturers.

interoperate   The ability of a node to work with nodes of similar design by other manufacturers.

intraframe (I-frame)   A video compression method used by the MPEG standard.

intranet   Implementation of Internet technologies within a corporate organization. An internal, corporate Internet.

intrusion   The unauthorized, and undesirable, access to a network or stored data.

intrusion detection system (IDS)   Passive in their methodology, they monitor and inform network administrators of any intrusive presence.

intrusion prevention system   Tries to take a proactive measure in network security so as to stop the attack before it starts.

inward WATS (800, 877, or 888 service)   Capability that allows callers to call a long-distance number toll-free. Long distance, for example, 800 and 888 area codes, where the receiver pays the charges.

IP addressing   The 12-octet (IPv4) or 128-bit (IPv6) naming convention that distinguishes a node. Generally relates to the MAC address of the node unless a proxy server is being used.

IP network   Any network, public or private, that uses the Internet protocol.

IP television   Converting TV to digital packets and conveying these packets over an IP network, providing two-way communications and greater control.

IPv6 addresses   Composed of two logical parts: a 64-bit network prefix and a 64-bit host-addressing part, which is often automatically generated from the interface MAC address.

ISO-OSI model   A model for the creation of telecommunications equipment. Its primary objective is to provide a basis for interconnecting dissimilar systems for the purpose of information exchange.

isochronous communications   The accommodation of asynchronous transmission over a synchronous link, where filler bits are inserted to keep the circuit synchronized.

J

Java   A high-level object-oriented language similar to C++ developed by Sun Microsystems.

JavaScript   An object-based scripting programming language based on the concept of prototypes

just-in-time (JIT)   In a broad sense, an approach to achieving excellence in a manufacturing company based on the continuing elimination of waste. In a narrow sense, just-in-time refers to the movement of material at the necessary place at the necessary time, negating large storage quantities.

K

Kerberos   A network authentication protocol designed to enable two parties to exchange private information across an otherwise open network.

kilo   One thousand; 1 kilohertz is 1,000 hertz.

L

laser   A device that produces a pure intense light. For telecommunications purposes, the light is the carrier frequency that can be modulated or pulsed to carry information.

last mile   The distance from a neighbor or street point to the resident or office wall; considered the most expensive portion of a circuit.

Layer Two Tunneling Protocol (L2TP)   A tunneling protocol used to support virtual private networks (VPNs).

layers   Allow the network to be changed without affecting the user; layer independence is like data independence in DBMS.

leased circuit   A circuit that is owned by a common carrier but leased to another organization.

legal implications   The concern for the lawful use of technology, even when the technology is legal.

light-emitting diode (LED)   A semiconductor device that converts electrical energy to light; used for very low power numeric displays and status lights.

line access   The method that the sender node uses to gain access to send a message. The simplest method of access, though the most expensive in overhead, is polling, involving one master node and several slave nodes.

line conditioning   A means for providing higher-quality communications on a channel or circuit by adjusting parameters, thus reducing noise. Electronics are used to take the incoming power and change it to conform to more rigid specifications by balancing the capacitance, impedance, and resistance on a circuit to limit noise.

line termination equipment   The equipment that terminates local loops (wiring frame) and then connects them to the switch that makes the loop-to-loop connections.

line turnaround   A protocol for half-duplex transmission in which one modem stops transmitting and becomes the receiver, and vice versa. Although there is time required for turnaround, this allows the full bandwidth of the channel to be used in each direction.

link layer   Layer two of the TCP/IP layered model that specifies how packets are transported over the physical layer, including the framing.

Linux   A freely distributable open-source operating system that runs on a number of hardware platforms.

liquid crystal display (LCD)   Called a "flat screen display," this device is thin and has far lower voltage, power, cooling, and space requirements than a CRT display.

local access and transport area (LATA)   A geographically defined area in which LECs operate and across which IXCs operate.

local area network (LAN)   Connectivity among two or more nodes closely located, often thought of as covering an area of less than one kilometer in radius and connecting people within an organization such as a campus, building, department, floor, or work group. LANs are privately owned; thus, like the PBX (customer premise equipment), they are nonregulated.

local calling area   Those telephone exchange prefixes to which a caller on flat-rate service can call without a toll.

local exchange carrier (LEC)   The local telephone company.

local loop   The connection from the telephone instrument to the switching equipment, located at the Telco's CO. It is a pair of wires that are in reality a single (conducted) wire that forms an electrical path, a loop from the telephone to the connect point at the switching equipment.

Local Multipoint Distribution Service (LMDS)   *See* Multipoint Microwave Distribution System (MMDS).

log file monitors (LFM)   Monitor log files generated by network services looking for patterns in the log files that suggest an intruder is attacking.

logging   Recording activity of a node or channel.

logic bombs   Applications that lie dormant until one or more logical conditions are met to trigger it.

long-distance calls   Calls outside of the local service area.

long-distance provider   *See* interexchange carrier (IXC).

longitudinal redundancy check (LRC)   Method of error checking for synchronous data packets; CRC is more reliable.

lossless compression   Compression that does not discard, or lose, any of the stored or transmitted information, thus allowing an exact replication of the original signal; for example, ADPCM.

lossy compression   Achieving (greater) compression by discarding part of the original information; for example, MP3 music.

M

MAC address blocking   Restricting access to a node based on the MAC address.

MAC address spoofing   Changing the packets to make the recipient believe they came from another (legitimate) node.

machine-to-machine   The communications between nodes where there is no human intervention.

mainframe computer   A large computer that can process large programs or a very large number of transactions quickly. Users can access via terminal or via a network.

maintenance   The "care and feeding" of installed systems to keep them operational within parameters.

make-or-buy   The decision as to whether the capability or system will be created with organization personnel (make) or purchased (buy). Data communications components are always bought, but their installation may be a make-or-buy decision.

malicious association   A node joining a network for malevolent intent.

malware   Short for "malicious software." It is any program or file that is harmful to a computer user.

man-in-the-middle (MITM) attack   One entity, with malicious intent, intercepts a message between two communicating entities.

management   The control of the work of others.

management information systems (MIS)   All systems and capabilities necessary to manage, process, and use information as a resource to the organization.

material requirements planning (MRP)   A software system that provides the capability to schedule the materials required to build the end product.

media access control (MAC)   The lower sublayer of the *OSI data link layer,* the interface between a *node*'s *Logical Link Control* and the network's *physical layer.*

media access control (MAC) address   Hardware address that uniquely identifies each node of a network. These addresses act as personal identification numbers for verifying the identity of authorized clients on wireless networks.

medium   The means of movement of signals from node to node; the physical evidence of the path.

mega   One million; 1 megahertz is 1,000,000 hertz.

mesh-enabled architecture (MEA)   Establishment and extension of a wireless network by use of wireless routers in each user device that receive and rebroadcast packets.

mesh network   A network configuration in which there is one or more paths between any two nodes.

message   The content of what is being sent across the line.

methodology   The process by which something is designed, created, secured, or maintained.

Metropolitan Area Ethernet (MAE)   Major access point to the Internet backbone.

metropolitan area network (MAN)   (1) Connectivity within a metropolitan-sized area, often providing services to many companies. (2) An optical fiber network serving all or part of a city.

metropolitan optical networks (MON)   A MAN using optical fiber.

microcomputer   A small-scale, or personal, computer.

microwave   Terrestrial and satellite technology that provides communications via line-of-sight high-frequency radio waves. They use micro or short wavelength radio waves and parabolic antennas to constrain the direction and radiation of the signal, thereby giving greater distance and lower susceptibility to noise.

milli   One-thousandth; a millisecond is 1/1,000 of a second.

minicomputer   A midrange computer that is between the mainframe and the microcomputer in size and processing power.

mission   A statement of the scope of the business and why the business exists.

mobile commerce   The ability to conduct commerce while on the go, using a mobile device such as a cell phone, PDA, and so forth.

model   Abstractions or representations; a person's cognitive representation of an idea or thought process.

modem (MOdulator-DEModulator)   A piece of equipment that transforms digital codes to analog form (and vice versa) so that they can be carried by the voice telephone system.

modifying factors   Those technologies or policies that help mitigate the effects of a threat.

modulation   The process by which a signal's characteristic is varied according to the characteristics of another signal. See amplitude modulation and frequency modulation.

monaural   Single track of sound.

monitor   In telecommunications, it is software or hardware that observes, supervises, controls, or verifies the operations of a system.

monitoring employees   A valid organization security measure unless the employees have been given a reasonable expectation of privacy.

Morse code   The standard code developed by Samuel Morse using dots and dashes (long and short signals) to represent characters.

motor-generator (MG) sets   The combination of a motor driven by AC power and a generator that creates AC power to act as an isolation point in a system.

moveable node   One end of a wireless system can be moved from place to place and then connectivity is reestablished. It is not free to roam and maintain connectivity.

MSNDirect®   A Microsoft capability that piggybacks on commercial FM stations to delivery news, and so on, to a mobile device such as a wrist watch.

MTBF   The mean or average time before or between failure of equipment or components.

multicasting   Sending out data to multiple destinations.

multimedia messaging service (MMS)   Enables subscribers to compose and send messages with one or more multimedia parts from one MMS-enabled mobile phone to another.

multimode fiber   Use of several light wavelengths on a fiber-optic strand to increase bandwidth. Multimode generally has a limited transmission distance.

multiple lines   Having more than one phone line.

multiplexer (MUX)   A device that interleaves two or more data streams to be carried on a single channel.

multiplexing   A function that allows two or more data sources to share a common transmission medium so that each data source has its own "channel."

multipoint circuit   A circuit with several nodes connected to it.

Multipoint Microwave Distribution System (MMDS/LMDS)   Operates in the 28 GHz band and offers line-of-sight coverage over distances up to three to five kilometers. It can deliver data and telephony services to 80,000 customers from a single node.

multistation access unit (MAU or MSAU)   A wiring hub in a token ring LAN.

Muzak®   A system for the transmission and distribution of background music, creating a more pleasant environment for customers; workers were believed to be more productive when music was played in the workplace.

N

narrowcast   Whereas broadcast (radio) stations send a single message simultaneously to all listening receivers, narrowcast servers send an individual message to each receiver.

National Association of Securities Dealers (NASD)   The regulatory body primarily responsible for the regulation of persons involved in the securities industry in the United States.

negative acknowledge character (NAK)   A transmission control character transmitted by a station indicating that the block of data received contains errors.

Netscape Navigator®   Netscape's browser.

Netware   Network operating system (NOS) software for a LAN produced by Novell.

network   One or more channels that provide connectivity between two or more nodes. An interconnected group of systems or devices that are remote from one another.

Network+   Certification by *CompTIA*® that shows entry-level understanding of networks.

network access layer   Layer one of TCP/IP four-layer model.

network access point (NAP)   A public network exchange facility where Internet service providers can connect with one another.

network address   *See* IP addressing, media access control (MAC) address.

Network Address Translation (NAT)   An Internet standard that enables a local area network to use one set of IP addresses for internal traffic and a second set of addresses for external traffic.

network administration   (1) The act of designing, installing, and maintaining the hardware and software that comprise the network. (2) The application of policies and procedures to equipment and channels to achieve a reliable network.

network appliance   Generic hardware for user interface of a network.

network architecture   The set of conventions or standards that ensure all the other components are interrelated and can work together. This architecture is the overall design blueprint for creating and evolving the network over changes in time, technology, uses, volumes, and geographic locations.

network-attached storage (NAS)   Servers dedicated to nothing more than file sharing.

network computer (NC)   A generic computer for attaching to a network but without storage or processing power. A very thin client that depends on the server in C/SA to perform all processing and storage.

network control program (NCP)   Software that controls the operation of a front-end processor or communications controller.

network design   The process of understanding the requirements for a communications network, investigating alternative ways for implementing the network, and selecting the most appropriate alternative to provide the required capacity.

network-directed broadcast addressing   Single packets sent to all hosts on a specified network.

network interface card (NIC)   A circuit card in a microcomputer that provides the electrical interface to a network.

network interface layer   Level one of the TCP/IP four-layer model.

network intrusion detection system (NIDS)   Watches live network packets on multiple computers and looks for signs of computer crime, network attacks, network misuse, and anomalies.

network layer   The third layer of the OSI model that defines message addressing and routing methods. This layer does end-to-end routing of packets or blocks of information; collects billing, accounting, and statistical information; and routes messages.

network management   All activities required to design, create, and maintain a network that will be reliable, be accessible, perform as desired, and be secure, that is, PARS.

network monitor (NIDS)   *See* network intrusion detection system (NIDS).

network of networks   The Internet.

network operating system (NOS)   The software that manages a network.

network port   *See* TCP/IP port.

network security   Protection of a network from intrusion or eavesdropping.

network server   A computer that is dedicated to the function of providing service to all nodes (microcomputers and devices) on the network.

node   An end point or switching point in a group of devices that can communicate with each other.

noise Any unwanted signal that interferes with the desired signals.

nomadic (moving) nodes Nodes that are free to move and continue to maintain connectivity.

O

omnidirectional Transmits and receives in all directions.

online The state of being connected, usually to a computer, not requiring any intervention for access.

online services America Online, CompuServe, MSN, and others—online communications systems that can be utilized to communicate around the world via personal computer. While providing Internet access, these services offer specific computer-based services from their own mainframes.

open architecture Equipment and software design using open standards; nonproprietary that lets the hosts interoperate, basing that interoperation on standard protocols.

open standards Standards that are published for all to see and, often, make comment on or additions to.

open systems interconnection (OSI) model A model that consists of seven layers that contain specific protocols for each control level.

operating system The central control program that governs computer hardware's operation.

Optical Carrier 1 (OC-1) STS-1 on optical fiber.

Optical Carrier 3 (OC-3) An optical standard with a speed of three times STS-1 = 155.52 Mbps.

Optical Carrier 12 (OC-12) An optical standard with a speed of 622 Mbps.

optical fiber A communications medium made of very pure, very thin glass or plastic fiber that conducts light waves.

optical signals Use of light to carry information.

organizing The task of defining relationships.

orthogonal frequency division multiplexing (OFDM) More efficient use of spectrum.

out-of-band signals Signals outside of the frequency range allowed for a voice signal.

outsourcing Buying as opposed to making systems or services.

P

packet A unit of digital data with a set number of bytes, including some that act as an address code.

packet assembler/disassembler (PAD) A device that receives the total data block; breaks it into predetermined-sized packets; adds the appropriate addressing, administrative, and error-checking data to the packet; and places the packet onto the network. The process is reversed at the destination.

packet data network (PDN) Provides connectivity to many points geographically. The network appears like a cloud with entry/exit points in many locations. Internal working is obscure to users as it provides connectivity in two ways: (1) Almost exclusive use is by providing virtual circuits upon request. Commands from the user cause the network to establish a physical channel for the duration of the communications. (2) The second method of data transfer is by datagrams, where the data are packetized and sent over the network, much as a letter through the U.S. Postal Service.

packet filter Software that inspects the content of packets to determine appropriateness for entry into a network.

packet-radio network A wireless capability that sends data as packets on an RF signal.

packet sniffers Also known as network analyzers or Ethernet sniffers, they are software programs that can see the traffic passing over a network or part of a network.

packet-switched networks Packet data networks in which packets are individually routed between nodes over data links that might be shared by many other nodes.

packet switching A transmission technique that is designed to cut costs and maximize use of digital transmission facilities by interspersing packets (blocks) of digital data from many customers on a single communications channel.

pager A radio-based alerting device, generally capable of receiving textual information.

parallel circuit Circuits that use concurrent movement of all bits in a byte. In a parallel circuit, the number of wires from sender to receiver is equal to or greater than the number of bits for a character.

parallel port Multiwire port (originally designed for connecting to printers).

parameters of a valuable network *See* PARS.

parity bit The binary digit appended to a group of binary digits to make the sum of all the digits either always odd (odd parity) or even (even parity).

parity checking Bit-level error detection in an asynchronous protocol. *See* vertical redundancy checking (VRC).

PARS Network characteristics of reliability, accessibility, performance, and security.

passive threats Unlike active threats, which are overt acts of humans, passive threats may be acts of nature or inadvertent actions of well-meaning people.

passwords A "word" known only to the user to gain authentication to a system or account; may be composed of letters, numbers, and special characters.

patch The process of replacing or adding a portion of a program.

path The route between any two nodes of the network.

PayPal® A service owned by eBay for the transfer of monies.

peer-to-peer The ability of two computers to communicate directly without passing through or using the capabilities of another computer.

peer-2-peer (P2P) computing The distribution of computers over a wide area, often by the Internet, to use them for a common problem, such as is done with SETI and United Devices.

penetration testing A method of evaluating the security of a computer system or network by simulating an attack or intrusion.

performance measurement and tuning Measuring the way a system is operating and making adjustments (tuning) to make it more optimal.

peripheral equipment Equipment that works in conjunction with a communications or computer system but is not integral to them.

permanent virtual channel (PVC) A full-time connection between two nodes in a packet-switching network, created by static entries in a table.

personal area network (PAN) Wireless connectivity of equipment on the person and to nearby equipment for synchronization.

personal computer (PC) A microcomputer.

personal digital assistant (PDA) A portable (pocket-sized) device that offers the functionality of a notebook, clock, and, when connected with a modem, access to paging and the Internet.

pharming Seeking to obtain information through domain spoofing.

phase An attribute or parameter of an analog signal that describes its relative position measured in degrees.

phase modulation Modulation in which the phase angle of the carrier is the characteristic varied.

phase shift keying (PSK) A modulation technique in which the phase of an analog signal is varied.

phase shift modulation The offset of an analog signal from its previous location. This is a way to achieve phase modulation.

phases of SDLC Problem definition, feasibility, system analysis, design, procurement, implementation, test, maintenance, and change.

phishing The luring of sensitive information, such as passwords and other personal information, from a victim by masquerading as someone trustworthy with a real need for such information.

photonic Using light as the data carrier on a transparent medium. Infrared systems use space as the medium and light as the carrier. Fiber optic cable uses glass or plastic as the medium and light waves as the carrier.

Phreak Person who hacks into a telephone system for personal cost without paying fees.

physical layer The lowest layer of the OSI model, dealing with electrical properties.

physical security That part of security that deals with access by touching and protection of physical assets.

physical topology The way the cabling of a network is actually laid out.

piconet A network of devices connected in an ad hoc fashion using Bluetooth technology.

piggyback One signal "riding" another; that is, use of an unused portion of a channel to carry added information.

pilot A means to bring a new system up in the cutover phase for only a small group of users and applications and let them test it in a real mode.

plain old telephone service (POTS) The simple telephone service provided by the local exchange carrier. The local loops that connect the CO switch and residence and office telephones in a star network.

planning The preparation for the future, thought of as existing on various levels from the top level, where overall company planning is done; to the middle level, where intermediate plans, programs, budgets, schedules, and technical performance targets are planned. Planning levels move from the general to the specific and from long to intermediate time horizons.

point of failure A specific piece of equipment or procedure where loss of functionality will cause the system to fail.

point of presence (POP) The location within a LATA at which customers are connected to an IXC.

point-of-service (POS) Systems that automate the sales transaction data to a large extent. The universal product code (UPC) and bar codes are technologies used by POS.

point-to-multipoint link The attachment of many nodes to a common communications channel.

point-to-point link Single communications devices at both ends of a single communications link.

point-to-point protocol (PPP) The Internet standard for serial communications. *See* SLIP.

point-to-point protocol over Ethernet (PPPoE) A network protocol for encapsulating PPP frames in Ethernet frames. It is used mainly with cable modem and DSL services. It offers standard PPP features such as authentication, encryption, and compression.

point-to-point topology A simple network in which both nodes are computers.

policy The small set of mandates and directives from the top of the firm summarized by the ground rule "This is how we do things around here."

polling An access method that involves the master node asking, in turn or based on a priority listing, each slave if it has messages to transmit or if it is ready to receive a message the master is holding for it. With this method of line access, collisions are not possible because the slave must have the attention of the master to transmit and two slaves cannot have access at the same time.

pornography   The display of adult content material.

portable   Any device that is small and easy to move from one location to another, such as a laptop computer.

portal   A Web site or service that offers a broad array of resources and services from a single point of entry.

portlet   A Web-based component or portion of a Web page that will process requests and generate dynamic content.

ports   *See* hardware ports or TCP/IP ports.

postal, telephone, and telegraph (PTT)   An office or administration that governs telecommunications services and pricing within the total country. Non-U.S. equivalent to U.S. FCC.

power outage   The loss of nonbattery electrical power.

Power over Ethernet   Providing electrical power over two wires of an Ethernet cable in order to power remote nodes.

preconfigured   The appropriate settings provided on equipment upon arrival; these are generally different than default settings, which should be changed.

presentation layer   The layer of the OSI model that is responsible for formatting and displaying the data to/from the application. The presentation layer provides transmission syntax, message transformations and formatting, data encryption, code conversion, and data compression.

prevention   A primary focus of network security, which, when successful, negates detection and correction.

prioritization   The process of ranking the importance or precedence of a list.

privacy   The added security provided for assets, especially information, of a personal nature.

private branch exchange (PBX)   A multiple-line business telephone system that resides on the company premises and either supplants or supplements the LEC local services.

problem (fault) management   The managing of activities and capture of information required to clear problems. This should create a log file that will aid others with similar problems.

process layer   *See* application layer.

project management   The administration, organization, and direction of the parts and resources of a project.

project team   A group of people organized for the purpose of completing a project.

proprietary   Information of particular importance to an organization such that disclosure could harm competitive advantage or divulge trade secrets.

proprietary standard   A standard known only to the originator. Installing equipment built to proprietary standards often means the equipment will not work with other companies' equipment.

protocol   Rules of communication. A protocol is a standard or set of rules or guidelines that govern the interaction between people, between people and machines, or between machines.

protocol analyzer   Test equipment that examines the bits on a communications circuit to determine whether the rules of a particular protocol are being followed.

protocol converter   Hardware or software that converts a data transmission from one protocol to another. *See* gateway.

prototyping   A quick and easy method used to create an approximation of the ultimate capability.

proxy server   A server that sits between a client application, such as a Web browser, and a real server. The proxy provides the resource either by connecting to the specified server or by serving it from a cache.

public key infrastructure (PKI)   A system of digital certificates, certificate of authority, and other registration authorities that verify and authenticate the validity of each party involved in an Internet transaction.

public service commission (PSC)   *See* public utilities commission (PUC).

public-switched telephone (voice) network (PSTN)   A network that provides circuits switched to many customers. *See* POTS.

public utilities commission (PUC)   The state agencies in the United States that establish tariffs within a state. Some states use the title public service commission (PSC). In some states, the commission members are elected; in other states, they are appointed to office.

pulse code modulation (PCM)   The standard for transforming voice analog signals to digital representation using 8,000 samples per second and eight-bit codes to represent signal amplitude. This requires a bandwidth of 64,000 bps.

push-to-talk   An implementation of VoIP on walkie-talkie devices or cell phones.

Q

quadrature amplitude modulation (QAM)   A technique using a combination of phase and amplitude modulation to achieve high data rates while maintaining relatively low signaling rates, that is, high bps on a low or moderate baud rate.

quality of service (QoS)   Refers to the probability of the telecommunications network meeting a given traffic contract; guaranteed throughput level.

R

radiated   *See* wireless.

radiation   The transmission of energy into space, such as radio frequency (RF) energy for AM/FM radio.

radio   A device that employs electromagnetic radiation for wireless transmission and reception of communications.

radio frequency identification (RFID)   A barcode replacement system that uses antenna-transmitters to activate a passive tag containing data.

radio telegraph   A mode of communications that uses a radio channel for telegraph purposes.

range   The distance a wireless device will work with a partner.

RAPS   *See* PARS.

real-time   The actual time during which a physical process transpires. Information received in time to affect the decision at hand.

receiver   The destination of communications.

reception area   The physical area (boundary) in which a device's wireless signal will be received.

reciprocal backup agreement   An agreement between companies with very similar equipment and operating systems to back each other up in case of disaster.

regeneration   The remaking of a perfectly shaped signal from a weak and misformed signal. Where analog channels simply amplify the attenuated signals, digital channels utilize signal regeneration or recreation to overcome attenuation and noise.

regenerator   A device that recreates (only) the desired signal, in perfect and amplified form.

regional Bell operating company (RBOC)   One of 22 LEC corporations carrying the Bell system name and logo resulting from divestiture.

registration of a domain name   Internet domain names must be registered with an authority; they are unique and have a guaranteed lifetime of use.

regulation   (a) Maintaining voltage levels between specified limits. (b) The oversight of an organization's activity by a governmental or professional agency.

reliability   The characteristic and probability that a device or channel will operate without error.

remote terminal   A terminal attached to a computer via a telecommunications line.

repeater   Also called an amplifier. By convention, devices that amplify analog signals are called repeaters, whereas devices that regenerate and amplify digital signals are called regenerators, though they could be referred to as regenerative repeaters.

request for proposal (RFP)   Formal request to vendors for a formal response describing the proposed makeup and possible costs for a specific set of goods and services.

request for quotation (RFQ)   Formal request to vendors for best prices for existing equipment or services.

reverse channel   The secondary channel in a full-duplex circuit, generally of low bandwidth.

RF interference (RFI)   A signal from one RF source that acts as noise, or interferes with another.

rhetoric   Simplify the message to fit time and complexity constraints.

right-of-way   The legal requirements to obtain the permission of the property holder before access or construction on his/her property.

ring network   A network in which each node is connected to two adjacent nodes. The protocol most often used with the ring topology is the token-passing protocol (IEEE 802.5).

risk   The possibility and probability of an undesirable event causing loss or injury and the level of consequence.

risk analysis   A methodological investigation of the organization, its resources, personnel, procedures, and objectives to determine points of weakness.

risk assessment   The process of analyzing present operations and proposed additional capabilities to determine the risk to continued operations. *See* risk analysis.

risk management   The way risk is avoided, minimized, or transferred. The analysis and subsequent actions taken to ensure that you can continue to operate under any foreseeable conditions, such as illness, labor strikes, hurricanes, earthquakes, power outages, heavy rains, oppressive heat, or epidemics of the flu.

roaming   Receiving service outside of one's normal supplied service area.

rogue user   Person making unauthorized use of a network.

rotary-dial telephone   Third-generation telephone instrument where the switch is signaled via pulses created by the rotating dial.

router   A piece of hardware or software that directs messages toward their destination, often from one network to another.

routing   The process of moving a packet of data from source to destination.

RS-232-C   A specification for the physical, mechanical, and electrical interface between data terminal equipment (DTE) and circuit-terminating equipment (DCE). A specification for a port and cable for serial communications.

S

Sarbanes-Oxley Act of 2002 (SOX)   Passed to restore the public's confidence in corporate governance by making chief executives of publicly traded companies personally validate financial statements and other information.

satellite   An earth-orbiting transponder that receives a signal from one earth station and sends the message back to another station. Satellites also may be data gatherers that send their information to an earth station.

scatternet   A group of independent and nonsynchronized piconets that share at least one common Bluetooth device.

script kiddies   Inexperienced crackers who use scripts and programs developed by others for the purpose of compromising computer accounts and files, and for launching attacks on whole computer systems.

search engines Web sites that gather information from all the Web pages they can find (using spiders or robots), organize the information, and reply to queries.

Secure Sockets Layer (SSL) A commonly used protocol for managing the security of a message transmission on the Internet.

security The capability to defend against intrusion and, ultimately, to protect assets (from access and disclosure, change, or destruction).

sender The originator or source of communications.

serial circuit Bits of the character follow each other sequentially down a single channel.

serial line Internet protocol (SLIP) A standard for connecting to the Internet with a modem over a phone line.

serial port *See* RS-232C.

server A device on a LAN that provides shared access to file space or print capability.

service level agreement (SLA) A set of performance objectives reached by consensus between the user and the provider of a service.

service set identifier (SSID) A 32-character unique identifier attached to the header of packets sent over a WLAN that acts as a password.

session layer The OSI layer that deals with the organization of a logical session. This layer provides access procedures, rules of half-duplex or full-duplex dialogs, rules for recovering if the session is interrupted, and rules for logically ending the session.

shielded twisted-pair (STP) UTP with an extra shielding layer for noise reduction.

shielding A protective enclosure that surrounds a transmission medium. It is designed to protect and minimize electromagnetic leakage and interference.

short message service (SMS) Sending text messages on cell phones.

signal system no. 7 (SS7) A signaling system used among telephone company central offices to set up calls, indicate their status, and tear down the calls when they are completed.

signal-to-noise (S/N) ratio The relationship between the level of the desired signal and the level of the undesirable noise.

Simple Mail Transfer Protocol (SMTP) A protocol for sending e-mail messages between servers.

simple network management protocol (SNMP) A protocol for exchanging network management commands and information between devices on a network.

simplex Transmission on one preassigned path, one-way circuit, or channel, like AM or FM radio, air-broadcast TV, and cable television.

simulation A program that shows the results of increased traffic and enhanced capabilities of a new computer program, or any new project, by use of mathematical models to represent external systems or processes.

slave station A node that operates under the control of a master or control node.

small office home office (SOHO) An office at home or one having limited facilities.

smart terminal A terminal that is not programmable but that has memory capable of being loaded with information.

SMDS interface protocol Uses distributed queued dual bus (DQDB) as an access protocol between the subscriber and the network.

sniffer *See* packet sniffer.

social engineering A set of tactics and psychological tricks used by hackers on computer users in order to gain entry to computer systems.

software-defined network (SDN) A bulk pricing offered by telephone companies designed for businesses or others who make a large number of calls. Standard-switched telephone lines are used to carry the calls.

solid wire A single thread of wire of the size, or gauge, stated. It is the least expensive to make, but its stiffness increases as the size increases.

source The originator of communications, that is, the sender.

spam Any unwanted electronic mail; unsolicited commercial e-mail (UCE).

spectrum The frequency bandwidth of an analog channel. The spectrum of a telephone channel is 4 KHz; the shared spectrum of television goes from 50–900 MHz.

spoofing The creation of TCP/IP packets using somebody else's IP address.

spread spectrum A radio transmission technique in which the frequency of the transmission is changed periodically, increasing quality and security.

spyware Any technology that aids in gathering information about a person or organization without their knowledge.

staffing The attracting, retaining, and training of the human resources needed for continued operations.

stand-alone A device that is not connected to another device, such as a personal computer that is not on a network.

standard An agreed-to specification by which components and software are built, allowing for interoperability.

Standard Generalized Markup Language (SGML) A system for organizing and tagging elements of a document.

star network The topology used to connect a central node to each outlying node of a network in a star network. In this topology, all circuits radiate from a central node, point to point. It is used in low-cost, slow-speed data networks.

start of header (SOH) The first character of the header of a block of synchronous data.

statistical multiplexer (stat-MUX) A device that multiplexes based on volume of traffic. It allocates channel space or time based on historical need, providing bandwidth on demand.

statistical time division multiplexing (STDM) A technique that combines signals from several nodes based on need for bandwidth.

step-by-step (strowger) switch Use of a rotary dial on the instrument to send out electrical pulses that stepped a series of switches according to the telephone number, setting up a path between sender and receiver.

stereo Music reproduction with two channels from two origination points; it attempts to emulate to the listener" ears, the actual performance.

stop bit The bit that indicates the end of a character in asynchronous communications.

storage area network (SAN) A specialized network that deals with blocks of data. NAS is a specialized server, dedicated to serving files.

store-and-forward An application in which input is transmitted, usually to a computer; stored; and then later delivered to the recipient.

stranded wire A single wire that is composed of a group of smaller solid wires. The objective of stranding is to make the wire flexible and easy to handle, while having the same electrical properties as a single strand of the same equivalent size.

strategy Large-scale, future-oriented plans for interacting with a competitive environment to optimize achievement and organizational objectives. It is the organization's game plan that reflects the company's awareness of how to compete, against whom, when, where, and for what.

stress testing Placing a heavy load on a system to see if it performs properly.

subnet mask A bitmask used to tell how much of an IP address identifies the subnetwork the host is on and how much identifies the host.

subnetting (an IP network) Allows you to break down what appears (logically) to be a single large network into smaller ones. In order to subnet, every machine must be told its subnet mask, which defines what part of its IP address is allocated for the subnetwork ID and what part for the host ID on that subnetwork.

Subscriber Identity Module (SIM) card A smart card inside of a GSM cellular phone that encrypts voice and data transmissions and stores data about the specific user so that the user can be identified and authenticated to the network supplying the phone service.

subsidiary communications authorization (SCA) A subchannel within the assigned FM frequency (channel) of a radio station.

surge protector A device that shorts (eliminates) short-term voltage spikes from input power.

Surround Sound The addition of a rear channel to stereo sound to give greater presence.

switch Any mechanical, electromechanical, digital, or photonic device that opens or closes circuits, changes parameters, or selects paths or circuits. *See also* CO switch.

switched multimegabit data service (SMDS) A high-speed, connectionless (datagram), cell-oriented, public, packet-switched data service developed to meet the demands for broadband services.

switched virtual circuit A temporary connection between two nodes established only for the duration of a session in a packet-switching network.

synchronizing To occur at the same time; be simultaneous.

synchronous Communications that are block-of-data oriented and must have the transmitter and receiver synchronized.

synchronous data link control (SDLC) A bit-oriented data link protocol developed by IBM.

synchronous digital hierarchy (SDH) BISDN is based on transmission speeds and capacities at the 155.52 Mbps, 622.08 Mbps, and 2.488 Gbps levels. These bit rates derive from CCITT G series recommendations for synchronous digital hierarchy, synchronous optical network (SONET) standard of the T1 Committee, and the CCITT I series recommendations, which support the concept of asynchronous transfer mode (ATM)–based BISDN. The transmission medium over which BISDN will operate is described by SDH/SONET with ATM employed as the switching mode.

synchronous optical network (SONET) Uses the same basic structure as SDH and is designed to interoperate at 155.52 Mbps and higher transmission rates. SONET's main purpose is high-speed, high-reliability, serial digital transmission over optical fiber cable.

Synchronous Transport Signal (STS-1) A standard with a speed of 51.840 Mbps.

system A group of interrelated and interdependent parts working together to achieve a common goal.

system hijacking Turns it into a zombie.

system integrity verifier (SIV) Monitors system files to find when a intruder changes them (thereby leaving behind a backdoor).

system specification Document created during the analysis phase of SDLC that documents what the new capability will be able to do and how.

systems development life cycle (SDLC) A formal development process composed of logical phases that guide systems development from concept through implementation, operations, and maintenance. Its purpose is to ensure a proper and complete solution

T

T1   A standard; any medium that carries 1.544 Mbps speed with 24 64 Kbps (DS0) channels plus 8 Kbps for control; same as DS1.

T3   A standard; any medium that has approximately 45 Mbps bandwidth; for example, 44.736 Mbps, equal to 28 T1 channels.

T-carrier system   A family of high-speed, digital transmission systems, designated according to their transmission capacity.

tags   Code in a document that changes the appearance of text during display, for example, **bolding.**

target specification   The documentation from the systems analysis phase, detailing the new system in sufficient detail so as to be designable and buildable by others. *See also* system specification.

tariff   A regulated telecommunications service and rates to be charged.

TCP/IP port   An endpoint to a logical connection. The port number identifies what type of port it is. For example, port 80 is used for HTTP traffic.

technology   Devices and systems (including programming and methodology) that perform new tasks. From the Greek *tekhnikos*, meaning method.

Telco   An abbreviation meaning the local telephone service provider; the telephone company.

telecommunications   Communications via electronic, electromagnetic, or photonic means over a distance. From the Greek *tele*, meaning at a distance.

telecommunications systems   All components and capabilities needed to move information, or data, to serve organizational needs. Components include hardware, software, procedures, data, and (most important) people.

telecommute   Use telecommunications to work from home or other location instead of on the business's premises. It allows you to work at the office without going to the office.

teleconferencing   A meeting at a distance. The oldest form is audio conferencing—telephone conference call.

telegraph   A device consisting of a long-distance loop of wire that has an electric storage battery for energy, a key to open and close the circuit, and a sounding unit that responds to the electrical current with a noise. This is the earliest electrical transmission of data.

telephone   The device that converts voice (acoustic) signals into (analogous) electrical signals for transmission, and reconverts them into acoustic (voice) signals at the receiving end.

telephone number prefix   The first three digits of the seven-digit telephone number.

telephone service   Service provided by the LEC and the IXC for use to call anywhere.

telephone system   The public switched voice network; goes almost everywhere.

telephony   Transmission of speech or other sounds. A general term used to denote all voice telecommunications.

teleprocessing   The processing of data at a distant or remote location by using data communications circuits.

teletypewriter   A typewriter-like device for creating the codes used in the telegraph system. It replaced the code-skilled operator with machine encoding and decoding as well as providing the first form of message storage (tape) for the telegraph.

telnet   A program that enables you to communicate with other computers and is typically used for remote access to host computers on which they have an account or to publicly accessible catalogs and databases.

temporal key integrity protocol (TKIP)   Part of the IEEE 802.11i encryption standard for wireless LANs.

terminal   A point at which information can enter or leave a communications network.

Terminal Emulation (Telnet)   Client software that provides command-line access to mainframes.

terrestrial   Of or on the earth; as opposed to orbiting satellite.

Terrestrial One   *See* T1.

testing   The process to determine bottlenecks, level of queuing, balking, and the system's ability to handle heavy workloads.

thin client   End node (computer or terminal) with very limited storage and processing, depending on the server for these functions.

third party   A participant other than the maker or the buyer of a product.

threat   Any unwanted action to a computer or network system.

throughput   The effective (useful) bandwidth of a channel.

time division multiple access (TDMA)   A technology for radio networks that allows several users to share the same frequency by dividing it into different time slots.

time division multiplexing (TDM)   A technique for the division of a transmission channel into two or more subchannels by allotting the time slots of a common channel to several information sources. Unlike FDM, which is simultaneous (parallel), TDM is sequential (serial) transmission.

time-to-deploy   How long it takes to get something up and running.

timeshare   A computer system operating technique that provides the interleaving of two or more programs in the processor, for example, multitasking. Only one program runs at a time, but they take turns quickly.

timeshare service   Services that are shared by a number of clients. Generally, computer-oriented services such as almost unlimited computer processing and storage, output report print or printing at their premises, fee-based programming services, and the use of a wide variety of royalty-based programs.

token   A small data packet used to control conflicts and congestion with the token-passing protocol.

token-passing protocol   The protocol most often used with the ring topology (IEEE 802.5). In this operating system, a small data packet, called a *token,* continuously passes around the ring from node to node. The use of the token ensures that there will be no collisions on the network.

toll   The charge to use a telecommunications channel or service. A tariff.

topology   The physical layout and connectivity of a network. It refers to the way the wires, or, more specifically, the channels, connect the nodes; protocol refers to the rules by which data communications takes place over these channels.

total cost of ownership (TCO)   Takes into account the cost of equipment, installation, and testing plus all costs incurred over the life of the system.

Touch Tone®   Signals sent in "dialing" that involves 7 (DTMF) tones to create 12 distinct codes.

touchpad   A small 12-key terminal attached to the CO switch. The DTMF touchpad uses a tone for each column and one for each row, two per code. This is a total of 7 tones to create 12 codes.

traffic   The movement, volume, and speed over a communications link. As the level of traffic increases, the probability of a circuit being busy increases.

translating bridges   Connect two LANs that use different data link protocols.

transmission control protocol (TCP)   The upper-level protocol for TCP/IP that guarantees that data sent by one endpoint will be received in the same order by the other, and without any pieces missing.

transmission control protocol/Internet protocol (TCP/IP)   A set of transmission protocols for interconnecting communications networks such as the Internet.

transnational data flow   The movement of data across national boundaries. Because some nations treat data as a resource, much like materials, the movement of data across their national boundaries is restricted, for example, tariffed.

transparent bridges   Connect two LANs that use the same data link protocol.

transparent mode   A mode of binary synchronous text transmission in which data, including normally restricted data link control characters, are transmitted only as specific bit patterns.

transport layer   The OSI layer responsible for maintaining a reliable and cost-effective communications channel. The transport layer provides addressing to a specific user process at the destination, message reliability, sequential delivery of the data, and flow control of data between user processes.

Transport Layer Security (TLS)   A protocol that ensures privacy between communicating applications and their users on the Internet.

tree network   A form that creates a network of networks. It connects the individual networks so that the total system works together while giving the individual network independence. There is no single point of failure.

trellis code modulation (TCM)   A specialized form of quadrature amplitude modulation (QAM) that codes the data so that many bit combinations are invalid. It is used for high-speed data communications.

Trojan horse   A destructive program that masquerades as a benign application.

trunk line   When the sender is connected to one central office (CO) and the receiver to another, the call goes between the two COs via a trunk line or trunk cable.

tunneling   The encryption format that creates a secure tunnel through an insecure environment; *see* virtual private network (VPN).

twisted-pair copper wire (UTP)   A circuit of copper wires as a loop with a single wire going from the sender to the receiver and back.

U

ultra high frequency (UHF)   A band of frequencies above VHF in the range of 440–900 megahertz.

ultra wideband (UWB) (802.15.3)   Transmitter works by sending billions of pulses across a very wide spectrum of frequencies several GHz in bandwidth.

unicast   A one-to-one communications relationship between a single source and a single destination.

Unicode   A standardized coding system that has 2^{16} points that can be used to represent the characters of all languages.

unidirectional   Broadcasting or receiving in one direction only; an example is a parabolic microwave antenna.

uniform resource locator (URL)   The global address of documents and other resources on the World Wide Web.

uninterruptible power supply (UPS)   A battery-operated device for protection against unstable voltage, power outages, and power surges.

universal product code (UPC)   A printed bar code on retail goods that is used with POS.

universal serial bus (USB)   Bus, port, and cable providing high-speed access to computers; USB 1.0 has a 12 Mbps bandwidth; USB 2.0 has a bandwidth of 480 Mbps.

universal service   The provision of high-quality voice communications to home and offices in all of the United States. The vice president of AT&T, Kinsbury, stated belief in and commitment to universal service.

unlicensed spectrum Radio frequencies that are available for public use without license. There are, however, standards for use that must be followed.

unshielded twisted pair (UTP) The standard wiring used for telephony and much of data communications. *See* twisted-pair copper wire.

unsolicited commercial e-mail (UCE) Junk e-mail, generally not of a dangerous nature, similar to third class postal mail.

V

V.32bis A CCITT standard for transmitting data at 14.4 Kbps, full-duplex on a switched circuit.

validation A nonambiguous, nonsharable, noncopyable, unique identifier for an individual.

value area networks (VAN) A special form of WAN that generally provides wide area coverage and offers services in addition to just connectivity in the form of added intelligence such as speed translation, store-and-forward messaging, protocol conversion, data handling, and packet assembly and disassembly. VANs are nonregulated because of the added services they provide.

vertical redundancy checking (VRC) Used for parity bit error detection with asynchronous communications.

very high frequency (VHF) A band of frequencies in the 200–400 MHz range.

very small aperture terminal (VSAT) Small satellite earth stations, ranging in size from 18 to 36 inches in diameter.

video graphics array (VGA) A graphics display system for PCs.

videoconferencing Use of television to connect conference participants visually and audibly. This is simply the use of two-way video and audio channels between two or more sites.

virtual circuit (VC) Exists on protocols where there are multiple paths from sender to destination. The network operating system or router selects a temporary path for the duration of the communications. This is one mode for data transfer in an X.25 packet network.

virtual local area network (VLAN) A group of nodes that act as if they were connected to the same physical LAN although they are on separate LANs. VLANs provide control and security.

virtual private network (VPN) Works by using the shared public infrastructure while maintaining privacy through security procedures and tunneling protocols.

virtual security Protection of assets from intrusion via networks.

virtual telecommunications access method (VTAM) IBM's primary telecommunications access method.

virtual terminal A concept that allows an application program to send or receive data to or from a generic terminal definition. Other software transforms the input and output to correspond to the actual characteristics of the real terminal being used.

virus A program or piece of code that is loaded onto your computer without your knowledge and runs against your wishes.

vision Determining the mission of the company, including broad statements about philosophy and goals.

Voice-enabled DSL (VeDSL) A digital subscriber line that is channelized to carry multiple voice channels, providing voice and data to the SOHO.

voice mail An intelligent answering machine that stores voice messages for the receiver.

Voice over Frame Relay Transmitting digitized voice over a frame relay WAN by controlling the inherent delays.

Voice over Internet Protocol (VoIP) Transporting digitized voice over a network that incorporates Internet Protocol.

voice spectrum The 300 to 3,000 Hz band used on telephone equipment for the transmission of voice and data.

voltage The electrical potential in computers and telecommunications equipment, equivalent to water pressure.

W

wand A specialized laser reader that can read the data from the bar code and send it directly to the computer for processing.

warchalking Symbols left on walls and sidewalks to indicate free access to Wi-Fi networks.

wardriving Detecting and using unsecured Wi-Fi by driving around with a laptop and antenna.

Web address A character-based notation for a Web site. Browsers translate the Web address at a domain name server to the IP address to find the site.

Web seal Logo on a Web site, indicating that the site is secure, for example, *Verisign*®.

wide area networks (WANs) At the opposite end of the network continuum from LANs. They are networks that cover wide geographical areas. They go beyond the boundaries of cities and extend globally. The extreme of the WAN is the global area network (GAN).

wide area telecommunications service (WATS) Bulk discount pricing of long-distance services.

Wi-Fi hotspot An access point for an 802.11 client; some retail establishments have such hotspots to attract customers.

Wi-Fi Protected Access (WPA) A wireless protocol that is designed to be more secure than WEP.

WiMAX (Worldwide Interoperability for Microwave Access, IEEE 802.16) A standards-based wireless technology that provides high-throughput broadband connections over long distances to fixed, portable, and nomadic users.

wired   Using UTP, coax, or optic fiber.

Wired Equivalent Privacy (WEP)   A primary method of security for wireless devices, particularly 802.11 devices.

wireless   Using radio frequency or infrared carrier to transport data.

wireless access point (WAP)   *See* access point.

Wireless Application Protocol (WAP)   A secure protocol that allows users to access information via handheld wireless devices, that is, an Internet microbrowser on a cell phone.

wireless broadband (WiBro)   A broadband wireless Internet technology being developed by the Korean telecommunications industry.

wireless cable   The wireless equivalent to coax-based CATV.

wireless communications   Communications in which the medium is not wire or cable but is broadcast by radio or infrared waves.

Wireless Fidelity (Wi-Fi, 802-11a/b/g)   Wireless unlicensed communications in the 2.400 and 5.300 GHz range.

wireless local area network (WLAN)   A LAN that works with wireless devices instead of wired.

wireless metropolitan area network (WMAN)   The wireless equivalent of a MAN, probably using WiMAX technology.

wireless network   Networks established by either continuous radio signals or packet radio.

wireless routers   Routing in a wireless environment; *see* router.

wireless security   The application of encryption, MAC lists, other technologies, and policies to secure wireless communications.

wireless T1   The wireless equivalent of a wired DS1 circuit, providing the equivalent to 24 DS0 channels.

work-at-home (WAH)   By use of telecommunications, the user is able to work at a home office instead of traveling to a formal, urban office.

workload generator   Computer software designed to generate transactions or other work for a computer or network for testing purposes.

World Wide Web (WWW)   The graphical interface portion of the Internet. The Web uses HTTP and HTML protocols.

worm   A program or algorithm that replicates itself over a computer network and usually performs malicious actions.

X

X.12   A set of standards for electronic data interchange.

X.25   A standard for packet-switched networks.

X.400   A standard for the transmission of electronic mail.

X-ON/X-OFF   A low-level protocol for managing data flow in a point-to-point network, for example, CPU to printer.

Xmodem   An often-used PC-to-PC asynchronous communications protocol.

Z

ZigBee (802.15.4)   A proprietary set of high-level communication protocols designed to use small, low-power digital radios.

zombie   A computer that has been implanted with a daemon (a process that runs in the background and performs a specified operation at predefined times or in response to certain events).

Abbreviations

24*7 24 hours a day, seven days a week

AAA Authentication, authorization, auditing

ACD Automated call distribution

ACK Acknowledge

ADA Americans with Disabilities Act of 1990

ADPCM Adaptive differential pulse code modulation

ADSL Asymmetric digital subscriber line

AM Amplitude modulation

AMPS Advanced Mobile Phone Service

ANSI American National Standards Institute

ARPANET Advanced Research Projects Agency Network

ARQ Automatic repeat request

ASCII American Standard Code for Information Interchange

ASP Applications service provider

ATM Asynchronous transfer mode

ATM Automatic teller machine

B2B Business-to-business

B2C Business-to-consumer

BBS Electronic bulletin board system

BCC Block check character

BISDN Broadband integrated services digital network

BOC Bell Operating Company

BPS Bits per second

BRI Basic rate interface of ISDN

BTW By the way

C2C Consumer-to-consumer

CAD Computer-aided design

CAN Campus area network

CAN-SPAM Controlling the Assault of Non-Solicited Pornography and Marketing Act

CATV Cable, or community, television

CCNA Cisco Certified Network Associate

CIM Computer-integrated manufacturing

CIO Chief information officer

CIR Committed information rate

CISSP Certified Information Systems Security Professional

CLP (Novell) Certified Linux Professional

CMIP Common management information protocol

CNA Certified Novell Administrator

CNE Certified Novell Engineer

CO Central office

CPE Customer premises equipment

CRC Cyclic redundancy check

CRT Cathode ray tube

C/SA Client-server architecture

CSA Customer service area

CSF Critical success factors

CSMA/CD protocol Carrier sense multiple access/collision detect

CWAP Certified Wireless Analysis Professional

CWSP Certified Wireless Security Professional

D-A Digital to analog

db Decibels

DBS Direct broadcast satellite

DCE Data Circuit-Termination Equipment

DHCP Dynamic Host Configuration Protocol

DMZ Demilitarized zone

DoS/DDoS (Distributed) denial-of-service attacks

DP Data processing

DPSK Differential phase shift keying

DS0 Digital Signal Zero

DS1 Digital Signal One (*see* T1)

DSL Digital subscriber line

DSP Digital signal processor

DSSS Direct-sequence spread spectrum

DTE Data terminal equipment

DTMF Dual tone multifrequency

E-mail Electronic mail

EBCDIC Extended Binary Coded Decimal Interchange Code

EDI Electronic data interchange

EFT Electronic funds transfer

EPC Electronic product code

EPPS Electronic performance support systems

FCC Federal Communications Commission

FDDI Fiber-distributed data interface

FDM Frequency division multiplexing

FHSS Frequency-hopping spread spectrum

FISMA Federal Information Security Management Act

FM Frequency modulation

FSK Frequency shift keying

FTTC Fiber to the curb

FTTH   Fiber to the home

FTTP   Fiber to the premise

G2C   Government to citizen/consumer

GAN   Global area network

GIAC   Global Information Assurance Certification [SANS Institute]

GLBA   Gramm-Leach-Bliley Act

GSM   Global system for mobile communications

HDSL   High-level data link control

HFC   Hybrid fiber/coax

HIPAA   Health Insurance Portability and Accountability Act

HTML   HyperText Markup Language

HTTP   HyperText Transfer Protocol

Hz   Hertz

IDS   Intrusion detection system

IEEE   Institute of Electrical and Electronic Engineers

IP   Internet protocol

IPS   Intrusion protection system

IPsec   Internet Protocol Security

IRC   Internet Relay Chat

ISDN   Integrated services digital network

IT   Information technology

ITU   International Telecommunications Union

IXC   Interexchange carrier

JIT   Just-in-time

Kbps   Kilobits per second

LAN   Local area network

LATA   Local access and transport area

LCD   Liquid crystal display

LCE   Linux/GNU Certified Engineer [SAIR]

LEC   Local exchange carrier

LED   Light-emitting diode

LMDS   Local Multipoint Distribution Service

LPI   Linux Professional Institute

LPIC   Linux Professional Institute

MAC   Media access control

MAE   Metropolitan Area Ethernet

MAN   Metropolitan area network

MAU or MSAU   Multistation access unit

Mbps   Megabits per second

MCP   Microsoft Certified Professional

MCSA   Microsoft Certified Systems Administrator

MCSE   Microsoft Certified Systems Engineer

MEA   Mesh-enabled architecture

MHz   Megahertz

MIS   Management information systems

MITM   Man-in-the-middle

MMDS   Multipoint Microwave Distribution System

Modem   MOdulator-DEModulator

MON   Metropolitan Optical Network

MRP   Material requirements planning

MUX   Multiplexer

NAK   Negative acknowledge character

NAP   Network access point

NAS   Network attached storage

NASD   National Association of Securities Dealers

NAT   Network Address Translation

NCP   Network control program

NIC   Network interface card

NOS   Network operating system

OC-1   Optical Carrier 1 (51.84 Mbps)

OC-3   Optical Carrier 3 (155.51 Mbps)

OC-12   Optical Carrier 12 (622 Mbps)

OFDM   Orthogonal frequency division multiplexing

OSI   Open systems interconnection

P2P   Peer-2-peer

PAD   Packet assembler/disassembler

PAN   Personal area network

PBX   Private branch exchange

PC   Personal computer

PCM   Pulse code modulation

PDA   Personal digital assistant

PDN   Packet data network

PKI   Public key infrastructure

POP   Point of presence

POS   Point-of-service

POTS   Plain old telephone service

PPP   Point-to-point protocol

PSC   Public services commission

PSK   Phase shift keying

PSN   Public switched network

PSTN   Public-switched telephone (voice) network

PTT   Postal, telephone, and telegraph

PUC   Public utilities commission (*see* PSC)

PVC   Permanent virtual circuit

QAM   Quadrature amplitude modulation

QoS   Quality of service

RFI   Radio frequency interference

RFID   Radio frequency identification

RFP   Request for proposal

RFQ   Request for quotation

RHCE   Red Hat Certified Engineer

SAN   Storage area network

SCA Subsidiary communications authorization
SDH Synchronous digital hierarchy
SDLC Systems development life cycle
SDN Software-defined network
SGML Standard Generalized Markup Language
SIM Subscriber Identity Module
SKU Stock-keeping unit
SLIP Serial line Internet protocol
SMDS Switched multimegabit data service
SMS Short message service
SMTP Simple Mail Transfer Protocol
S/N Signal-to-noise
SNA Systems network architecture
SNMP Simple network management protocol
SOH Start of header
SOHO Small office, home office
SONET Synchronous optical network
SOX Sarbanes-Oxley Act
SSID Service set identifier
SSL Secure Sockets Layer
STDM Statistical time division multiplexing
STP Shielded twisted-pair
STS-1 Synchronous Transport Signal 1
T1 Terrestrial data standard 1, 1.544 Mbps
T3 Terrestrial data standard 3, 45 Mbps
TCO Total cost of ownership
TCP Transmission control protocol
TCP/IP Transmission control protocol/Internet protocol
TDM Time division multiplexing
TDMA Time division multiple access
TKIP Temporal key integrity protocol
TLS Transport Layer Security
TPS Transaction processing system

UCE Unsolicited commercial e-mail
UHF Ultra high frequency
UPC Universal product code
UPS Uninterruptible power supply
URL Universal resource locator
UTP Unshielded twisted-pair
UWB Ultra wideband
VAN Value added network
VeDSL Voice-enabled DSL
VGA Video graphics array
VHF Very high frequency
VIVID VIdeo, Voice, Image, and Data
VLAN Virtual local area network
VOD Video on Demand
VoIP Voice over Internet Protocol
VPN Virtual private network
VRC Vertical redundancy check
VSAT Very small aperture terminal
VTAM Virtual telecommunications access method
WAH Work-at-home
WAN Wide area networks
WAP Wireless Application Protocol
WATS Wide area telecommunications service
WEP Wired Equivalent Privacy
WiBro Wireless broadband
Wi-Fi Wireless Fidelity
WiMAX Worldwide Interoperability for Microwave Access; a Wireless MAN
WLAN Wireless local area network
WMAN Wireless metropolitan area network
WPA Wi-Fi Protected Access
WWW World Wide Web
XML Extensible Markup Language

Index

cellular towers, 62
cellular voice systems, 280–281
central office (CO), 47, 452, 453
Centrex
    multiple lines to, 458
    registered names for, 459
    services, 457
Centrino technology. *See also* Intel
        Centrino
    commercial flights and, 289
CEOs, attesting to internal controls, 340
CERN (European Particle Physics
        Laboratory), 196
certificate of authority (CA), 326
certification programs
    for wired networks, 291
    for wireless networks, 291–292
certifications, 437
    available, 438–443
    of individuals in telecommunications,
        410
    in the IT industry, 291
Certified Information Systems Security
        Professional (CISSP), 442
Certified Linux Professional, 441
Certified Novell Administrator
        (CAN), 440
Certified Novell Engineer (CNE), 440
Certified Wireless Analysis Professional
        (CWAP), 292, 443
Certified Wireless Network Administrator
        (CWNA), 292, 443
Certified Wireless Networking Expert
        (CWNE), 292
Certified Wireless Network Professional
        program, 292
Certified Wireless Security Professional
        (CWSP), 292, 442
CFOs, attesting to internal controls, 340
champion, 421–422
change management, 400, 433–434
change, sources of, 414
channel conflicts, between 2.4 GHz
        cordless phones and WLANs, 361
channel label, as the medium, 56
channelization capability in ADSL2, 169
Channelized Voice over DSL
        (CVoDSL), 169
channels, xxi, 3, 6, 8, 45. *See also* analog
        channels
    connection to a common, 113
    delivered by coaxial cable, 31
    divided, 27–29, 31
    forward and reverse, 8
    multiple using multiplexing, 49
    reverse, 8, 78
    UHF, 17, 58
    VHF, 17, 58
    virtual, 182
character-interactive traffic, supported by
        frame relay, 176
Charter Cable, offering
        video-on-demand, 78
checksum, 153

Chicago NAP, 206
Chick-Fil-A (CFA) portal, 235
children, Web portal for, 235
chipping, 252
chips, used by RFID, 256–257
CIFS (Common Internet File System), 158
CIM (computer-integrated manufacturing)
        movement, xxviii
CIO (chief information officer), 388, 399
CipherTrust, content filtering, 344
CIR (committed information rate),
        150, 398
circuit(s), xxi, 6, 45, 457
    available for end users, 456–457
    composition of, 10
    information-carrying capacity of, 8
    multiplexed, 30
circuit media
    bandwidth of, 49
    characteristics of, 49, 50
circuit sharing (multiplexing), 29–33
circuit switching, 142–143, 170, 171, 172
circuit-level firewall, 374
circuit-level gateway, 331
Cisco certifications, 440
Cisco Certified Network Associate
        (CCNA), 440
Cisco Systems
    success using B2B, 226
    use of portals, 235
CISSP (Certified Information Systems
        Security Professional), 442
cities, wireless connectivity in, 255
CIW Security Analyst, 443
clarity of the interface, 237
Class D1 security rating, 374
classes of networks, 99
classroom performance systems, 67
cleanup, after-implementation, 436–437
ClickNet Professional, 386
client-loadable NOS, 118
clients, 115, 117
client-server architecture (C/SA), 114–116
    clients in, 117
    network modeling and, 416
    WAP as, 295
client-server network operating systems,
        118
client-server operations, advantages of,
        116
closed network, 369
cloud, viewing a PDN as, 171
CMIP (communications management
        information protocol), 394
CNE (Certified Novell Engineer), 440
CO (central office), 47, 452, 453
coaxial cable(s), 11, 45, 49, 147
    channels delivered by, 31
    characteristics of, 49, 50
    transport of cable TV, 64
    types of, 51
code division multiple access. *See* CDMA
code length, 21
code points in the Baudot code, 20

coded messages, circuit forms for
        sending, 8
codes, 15, 19
COFDM (coded orthogonal frequency
        division multiplexing), 253
cold site, 307
cold turkey method, 436
collaboration, xxv
collaborative activity, among knowledge
        workers and smart devices, 466
collaborative applications, offered by
        ASPs, 445
college financial aid, WWW sites for, 197
collision(s), 122
    avoidance, 122–123
    detection, 123–124
collision domain (hearing range), 249
collisionless connectivity, 123
collocation facilities at a MAE, 205
colon hexadecimal in IPv6, 216
Columbus Day virus, 322
.com domain suffix, 207
COM1, 41
COM2, 41
committed information rate (CIR),
        150, 398
common carrier, 109, 457
Common Internet File System (CIFS), 158
communications, xx
    basis of organization, 451
    history of, xxii–xxiii
    rules for, 19
    as a significant cost, xxiv–xxv
    simple, 3–4
communications appliances, 80
communications channels. *See* channels
communications circuits. *See* circuit(s)
communications management information
        protocol (CMIP), 394
communications media. *See* media
communications model, 3–5
compact discs (CDs), bandwidth of, 10
Compass Bank of Alabama, 234
competitive advantage
    from an FX line, 461
    provided by technology, xxiii–xxiv
    strategies for gaining and sustaining,
        xxiv
complexity
    of a device or system, 286
    partitioning, xxvi
    of a project, 421
    of voice and data systems, 384
    of a wireless network, 280
compression, 10, 11–13, 24–25. *See also*
        data compression
    in HD radio, 67
    on the Internet, 239
CompTIA
    A+ certification, 439
    i-Net+ certification, 441–442
    Linux+ certification, 440
    Network+ certification, 438
    Security+ certification, 442